A First Course in
NUMERICAL METHODS

Computational Science & Engineering

The SIAM series on Computational Science and Engineering publishes research monographs, advanced undergraduate- or graduate-level textbooks, and other volumes of interest to an interdisciplinary CS&E community of computational mathematicians, computer scientists, scientists, and engineers. The series includes introductory volumes aimed at a broad audience of mathematically motivated readers interested in understanding methods and applications within computational science and engineering, monographs reporting on the most recent developments in the field, and volumes addressed to specific groups of professionals whose work relies extensively on computational science and engineering.

SIAM created the CS&E series to support access to the rapid and far-ranging advances in computer modeling and simulation of complex problems in science and engineering, to promote the interdisciplinary culture required to meet these large-scale challenges, and to provide the means to the next generation of computational scientists and engineers.

Editor-in-Chief
Donald Estep
Colorado State University

Editorial Board

Ben Adcock
Simon Fraser University

Daniela Calvetti
Case Western Reserve University

Paul Constantine
Colorado School of Mines

Omar Ghattas
University of Texas at Austin

Chen Greif
University of British Columbia

Jan S. Hesthaven
Ecole Polytechnique Fédérale de Lausanne

Johan Hoffman
KTH Royal Institute of Technology

David Keyes
Columbia University

Ralph C. Smith
North Carolina State University

Karen Willcox
Massachusetts Institute of Technology

Series Volumes

Vidyasagar, M., *An Introduction to Compressed Sensing*

Sipahi, Rifat, *Mastering Frequency Domain Techniques for the Stability Analysis of LTI Time Delay Systems*

Bardsley, Johnathan M., *Computational Uncertainty Quantification for Inverse Problems*

Hesthaven, Jan S., *Numerical Methods for Conservation Laws: From Analysis to Algorithms*

Sidi, Avram, *Vector Extrapolation Methods with Applications*

Borzì, A., Ciaramella, G., and Sprengel, M., *Formulation and Numerical Solution of Quantum Control Problems*

Benner, Peter, Cohen, Albert, Ohlberger, Mario, and Willcox, Karen, editors, *Model Reduction and Approximation: Theory and Algorithms*

Kuzmin, Dmitri and Hämäläinen, Jari, *Finite Element Methods for Computational Fluid Dynamics: A Practical Guide*

Rostamian, Rouben, *Programming Projects in C for Students of Engineering, Science, and Mathematics*

Smith, Ralph C., *Uncertainty Quantification: Theory, Implementation, and Applications*

Dankowicz, Harry and Schilder, Frank, *Recipes for Continuation*

Mueller, Jennifer L. and Siltanen, Samuli, *Linear and Nonlinear Inverse Problems with Practical Applications*

Shapira, Yair, *Solving PDEs in C++: Numerical Methods in a Unified Object-Oriented Approach, Second Edition*

Borzì, Alfio and Schulz, Volker, *Computational Optimization of Systems Governed by Partial Differential Equations*

Ascher, Uri M. and Greif, Chen, *A First Course in Numerical Methods*

Layton, William, *Introduction to the Numerical Analysis of Incompressible Viscous Flows*

Ascher, Uri M., *Numerical Methods for Evolutionary Differential Equations*

Zohdi, T. I., *An Introduction to Modeling and Simulation of Particulate Flows*

Biegler, Lorenz T., Ghattas, Omar, Heinkenschloss, Matthias, Keyes, David, and van Bloemen Waanders, Bart, editors, *Real-Time PDE-Constrained Optimization*

Chen, Zhangxin, Huan, Guanren, and Ma, Yuanle, *Computational Methods for Multiphase Flows in Porous Media*

Shapira, Yair, *Solving PDEs in C++: Numerical Methods in a Unified Object-Oriented Approach*

A First Course in NUMERICAL METHODS

Uri M. Ascher
Chen Greif
The University of British Columbia
Vancouver, British Columbia, Canada

Society for Industrial and Applied Mathematics
Philadelphia

Copyright © 2011 by the Society for Industrial and Applied Mathematics.

10 9 8 7 6 5 4

All rights reserved. Printed in the United States of America. No part of this book may be reproduced, stored, or transmitted in any manner without the written permission of the publisher. For information, write to the Society for Industrial and Applied Mathematics, 3600 Market Street, 6th Floor, Philadelphia, PA 19104-2688 USA.

Trademarked names may be used in this book without the inclusion of a trademark symbol. These names are used in an editorial context only; no infringement of trademark is intended.

MATLAB is a registered trademark of The MathWorks, Inc. For MATLAB product information, please contact The MathWorks, Inc., 3 Apple Hill Drive, Natick, MA 01760-2098 USA, 508-647-7000, Fax: 508-647-7001, *info@mathworks.com*, *www.mathworks.com*.

Figures 4.9a and 11.16 reprinted with permission from ACM.
Figure 7.1 reproduced with permission from World Scientific Publishing Co. Pte. Ltd.
Figure 13.7 reprinted with permission from MOS and Springer.
Figures 16.12 and 16.13 reprinted with permission from Elsevier.

Library of Congress Cataloging-in-Publication Data

Ascher, U. M. (Uri M.), 1946-
  A first course in numerical methods / Uri M. Ascher, Chen Greif.
    p. cm. – (Computational science and engineering series)
  Includes bibliographical references and index.
  ISBN 978-0-898719-97-0
  1. Numerical calculations–Data processing. 2. Numerical analysis. 3. Algorithms.
I. Greif, Chen, 1965- II. Title.
  QA297.A748 2011
  518'.4–dc22
                                        2011007041

 is a registered trademark.

*We dedicate this book
to our tolerant families.*

Contents

List of Figures		xi
List of Tables		xix
Preface		xxi

1 Numerical Algorithms — 1
- 1.1 Scientific computing 1
- 1.2 Numerical algorithms and errors 3
- 1.3 Algorithm properties 9
- 1.4 Exercises 14
- 1.5 Additional notes 15

2 Roundoff Errors — 17
- 2.1 The essentials 17
- 2.2 Floating point systems 21
- 2.3 Roundoff error accumulation 26
- 2.4 The IEEE standard 29
- 2.5 Exercises 32
- 2.6 Additional notes 36

3 Nonlinear Equations in One Variable — 39
- 3.1 Solving nonlinear equations 39
- 3.2 Bisection method 43
- 3.3 Fixed point iteration 45
- 3.4 Newton's method and variants 50
- 3.5 Minimizing a function in one variable 55
- 3.6 Exercises 58
- 3.7 Additional notes 64

4 Linear Algebra Background — 65
- 4.1 Review of basic concepts 65
- 4.2 Vector and matrix norms 73
- 4.3 Special classes of matrices 78
- 4.4 Singular values 80
- 4.5 Examples 83
- 4.6 Exercises 89
- 4.7 Additional notes 92

5	**Linear Systems: Direct Methods**	**93**
	5.1 Gaussian elimination and backward substitution	94
	5.2 LU decomposition	100
	5.3 Pivoting strategies	105
	5.4 Efficient implementation	110
	5.5 The Cholesky decomposition	114
	5.6 Sparse matrices	117
	5.7 Permutations and ordering strategies	122
	5.8 Estimating errors and the condition number	127
	5.9 Exercises	133
	5.10 Additional notes	139
6	**Linear Least Squares Problems**	**141**
	6.1 Least squares and the normal equations	141
	6.2 Orthogonal transformations and QR	151
	6.3 Householder transformations and Gram–Schmidt orthogonalization	157
	6.4 Exercises	163
	6.5 Additional notes	166
7	**Linear Systems: Iterative Methods**	**167**
	7.1 The need for iterative methods	167
	7.2 Stationary iteration and relaxation methods	173
	7.3 Convergence of stationary methods	179
	7.4 Conjugate gradient method	182
	7.5 *Krylov subspace methods	191
	7.6 *Multigrid methods	204
	7.7 Exercises	210
	7.8 Additional notes	218
8	**Eigenvalues and Singular Values**	**219**
	8.1 The power method and variants	219
	8.2 Singular value decomposition	229
	8.3 General methods for computing eigenvalues and singular values	236
	8.4 Exercises	245
	8.5 Additional notes	249
9	**Nonlinear Systems and Optimization**	**251**
	9.1 Newton's method for nonlinear systems	251
	9.2 Unconstrained optimization	258
	9.3 *Constrained optimization	271
	9.4 Exercises	286
	9.5 Additional notes	293
10	**Polynomial Interpolation**	**295**
	10.1 General approximation and interpolation	295
	10.2 Monomial interpolation	298
	10.3 Lagrange interpolation	302
	10.4 Divided differences and Newton's form	306
	10.5 The error in polynomial interpolation	313
	10.6 Chebyshev interpolation	316
	10.7 Interpolating also derivative values	319

	10.8	Exercises	323
	10.9	Additional notes	330

11 Piecewise Polynomial Interpolation — 331
- 11.1 The case for piecewise polynomial interpolation — 331
- 11.2 Broken line and piecewise Hermite interpolation — 333
- 11.3 Cubic spline interpolation — 337
- 11.4 Hat functions and B-splines — 344
- 11.5 Parametric curves — 349
- 11.6 *Multidimensional interpolation — 353
- 11.7 Exercises — 359
- 11.8 Additional notes — 363

12 Best Approximation — 365
- 12.1 Continuous least squares approximation — 366
- 12.2 Orthogonal basis functions — 370
- 12.3 Weighted least squares — 373
- 12.4 Chebyshev polynomials — 377
- 12.5 Exercises — 379
- 12.6 Additional notes — 382

13 Fourier Transform — 383
- 13.1 The Fourier transform — 383
- 13.2 Discrete Fourier transform and trigonometric interpolation — 388
- 13.3 Fast Fourier transform — 396
- 13.4 Exercises — 405
- 13.5 Additional notes — 406

14 Numerical Differentiation — 409
- 14.1 Deriving formulas using Taylor series — 409
- 14.2 Richardson extrapolation — 413
- 14.3 Deriving formulas using Lagrange polynomial interpolation — 415
- 14.4 Roundoff and data errors in numerical differentiation — 420
- 14.5 *Differentiation matrices and global derivative approximation — 426
- 14.6 Exercises — 434
- 14.7 Additional notes — 438

15 Numerical Integration — 441
- 15.1 Basic quadrature algorithms — 442
- 15.2 Composite numerical integration — 446
- 15.3 Gaussian quadrature — 454
- 15.4 Adaptive quadrature — 462
- 15.5 Romberg integration — 469
- 15.6 *Multidimensional integration — 472
- 15.7 Exercises — 475
- 15.8 Additional notes — 479

16 Differential Equations — 481
- 16.1 Initial value ordinary differential equations — 481
- 16.2 Euler's method — 485
- 16.3 Runge–Kutta methods — 493

	16.4	Multistep methods	500
	16.5	Absolute stability and stiffness	507
	16.6	Error control and estimation	515
	16.7	*Boundary value ODEs	520
	16.8	*Partial differential equations	524
	16.9	Exercises	531
	16.10	Additional notes	537

Bibliography 539

Index 543

List of Figures

1.1	Scientific computing.	2
1.2	A simple instance of numerical differentiation: the tangent $f'(x_0)$ is approximated by the chord $(f(x_0+h)-f(x_0))/h$.	6
1.3	The combined effect of discretization and roundoff errors. The solid curve interpolates the computed values of $\lvert f'(x_0) - \frac{f(x_0+h)-f(x_0)}{h}\rvert$ for $f(x)=\sin(x)$, $x_0 = 1.2$. Also shown in dash-dot style is a straight line depicting the discretization error without roundoff error.	9
1.4	An ill-conditioned problem of computing output values y given in terms of input values x by $y=g(x)$: when the input x is slightly perturbed to $\bar{x}$, the result $\bar{y}=g(\bar{x})$ is far from y. If the problem were well-conditioned, we would be expecting the distance between y and $\bar{y}$ to be more comparable in magnitude to the distance between x and $\bar{x}$.	12
1.5	An instance of a stable algorithm for computing $y=g(x)$: the output $\bar{y}$ is the exact result, $\bar{y}=g(\bar{x})$, for a slightly perturbed input, i.e., $\bar{x}$ which is close to the input x. Thus, if the algorithm is stable and the problem is well-conditioned, then the computed result $\bar{y}$ is close to the exact y.	12
2.1	A double word (64 bits) in the standard floating point system. The blue bit is for sign, the magenta bits store the exponent, and the green bits are for the fraction.	19
2.2	The "almost random" nature of roundoff errors.	21
2.3	Picture of the floating point system described in Example 2.8.	26
3.1	Graphs of three functions and their real roots (if there are any): (i) $f(x)=\sin(x)$ on $[0,4\pi]$, (ii) $f(x)=x^3-30x^2+2552$ on $[0,20]$, and (iii) $f(x)=10\cosh(x/4)-x$ on $[-10,10]$.	40
3.2	Fixed point iteration for $x=e^{-x}$, starting from $x_0=1$. This yields $x_1=e^{-1}$, $x_2=e^{-e^{-1}}$,.... Convergence is apparent.	47
3.3	The functions x and $2\cosh(x/4)$ meet at two locations.	48
3.4	Newton's method: the next iterate is the x-intercept of the tangent line to f at the current iterate.	51
3.5	Graph of an anonymous function; see Exercise 18.	63
4.1	Intersection of two straight lines: $a_{11}x_1+a_{12}x_2=b_1$ and $a_{21}x_1+a_{22}x_2=b_2$.	66
4.2	Eigenvalues in the complex plane, Example 4.5. Note that the complex ones arrive in pairs: if λ is an eigenvalue, then so is $\bar{\lambda}$. Also, here all eigenvalues have nonpositive real parts.	72
4.3	The "unit circle" according to the three norms, ℓ_1, ℓ_2, and ℓ_∞. Note that the diamond is contained in the circle, which in turn is contained in the square.	75

4.4	The Cartesian coordinate system as a set of orthonormal vectors.	80
4.5	Singular value decomposition: $A_{m \times n} = U_{m \times m} \Sigma_{m \times n} V_{n \times n}^T$.	81
4.6	Data fitting by a cubic polynomial using the solution of a linear system for the polynomial's coefficients.	84
4.7	A discrete mesh for approximating a function and its second derivative in Example 4.17.	86
4.8	Recovering a function v satisfying $v(0) = v(1) = 0$ from its second derivative. Here $N = 507$.	87
4.9	Example 4.18: a point cloud representing (a) a surface in three-dimensional space, and (b) together with its unsigned normals.	89
5.1	Gaussian elimination for the case $n = 4$. Only areas of potentially nonzero entries are shown.	99
5.2	LU decomposition for the case $n = 10$. Only areas of potentially nonzero entries in the square matrices L and U are shown.	102
5.3	Compressed row form of the matrix in Example 5.14. Shown are the vectors $\mathbf{i}, \mathbf{j}, \mathbf{v}$. The arrows and the "eof" node are intended merely for illustration of where the elements of the vector $\mathbf{i}$ point to in $\mathbf{j}$ and correspondingly in $\mathbf{v}$.	118
5.4	A banded matrix for the case $n = 10$, $p = 2$, $q = 3$. Only areas of potentially nonzero entries are shaded in green.	120
5.5	Graphs of the matrices A (left) and B (right) of Example 5.15.	125
5.6	Sparsity pattern of a certain symmetric positive definite matrix A (left), its RCM ordering (middle), and approximate minimum degree ordering (right).	126
5.7	Sparsity pattern of the Cholesky factors of A (left), its RCM ordering (middle), and approximate minimum degree ordering (right).	127
5.8	Stretching the unit circle by a symmetric positive definite transformation.	133
6.1	Matrices and vectors and their dimensions in ℓ_2 data fitting.	143
6.2	Discrete least squares approximation.	145
6.3	Linear regression curve (in blue) through green data. Here, $m = 25$, $n = 2$.	148
6.4	The first 5 best polynomial approximations to $f(t) = \cos(2\pi t)$ sampled at $0 : .05 : 1$. The data values appear as red circles. Clearly, p_4 fits the data better than p_2, which in turn is a better approximation than p_0. Note $p_{2j+1} = p_{2j}$.	150
6.5	Ratio of execution times using QR vs. normal equations. The number of rows for each n is $3n + 1$ for the upper curve and $n + 1$ for the lower one.	157
6.6	Householder reflection. Depicted is a 20×5 matrix A after 3 reflections.	160
7.1	A two-dimensional cross section of a three-dimensional domain with a square grid added.	169
7.2	A two-dimensional grid with grid function values at its nodes, discretizing the unit square. The length of each edge of the small squares is the grid width h, while their union is a square with edge length 1. The locations of $u_{i,j}$ and those of its neighbors that are participating in the difference formula (7.1) are marked in red.	170
7.3	The sparsity pattern of A for Example 7.1 with $N = 10$. Note that there are $nz = 460$ nonzeros out of 10,000 matrix locations.	172
7.4	Red-black ordering. First sweep simultaneously over the black points, then over the red.	177
7.5	Example 7.10, with $N = 31$: convergence behavior of various iterative schemes for the discretized Poisson equation.	185

7.6	Sparsity patterns for the matrix of Example 7.1 with $N = 8$: top left, the IC factor F with no fill-in (IC(0)); top right, the product FF^T; bottom left, the full Cholesky factor G; bottom right, the IC factor with drop tolerance .001. See Figure 7.3 for the sparsity pattern of the original matrix.	189
7.7	Iteration progress for CG, PCG with the IC(0) preconditioner and PCG with the IC preconditioner using drop tolerance `tol`= 0.01.	190
7.8	Convergence behavior of restarted GMRES with $m = 20$, for a 10,000 × 10,000 matrix that corresponds to the convection-diffusion equation on a 100 × 100 uniform mesh.	201
7.9	The polynomials that are constructed in the course of three CG iterations for the small linear system of Examples 7.9 and 7.14. The values of the polynomials at the eigenvalues of the matrix are marked on the linear, quadratic, and cubic curves.	202
7.10	An illustration of the smoothing effect, using damped Jacobi with $\omega = 0.8$ applied to the Poisson equation in one dimension.	205
7.11	Convergence behavior of various iterative schemes for the Poisson equation (see Example 7.17) with $n = 255^2$.	209
7.12	Example 7.17: number of iterations (top panel) and CPU times (bottom panel) required to achieve convergence to `tol`= 1.e-6 for the Poisson problem of Example 7.1 (page 168) with $N = 2^l - 1, l = 5, 6, 7, 8, 9$.	209
8.1	Convergence behavior of the power method for two diagonal matrices, Example 8.2.	223
8.2	A toy network for the PageRank Example 8.3.	224
8.3	Things that can go wrong with the basic model: depicted are a dangling node (left) and a terminal strong component featuring a cyclic path (right).	225
8.4	Convergence behavior of the inverse iteration in Example 8.4.	229
8.5	Original 200 × 320 pixel image of a clown.	232
8.6	A rank-20 SVD approximation of the image of a clown.	232
8.7	The result of the first stage of a typical eigensolver: a general upper Hessenberg matrix for nonsymmetric matrices (left) and a tridiagonal form for symmetric matrices (right).	237
8.8	The result of the first stage of the computation of the SVD is a bi-diagonal matrix C (left). The corresponding tridiagonal matrix $C^T C$ is given on the right.	243
8.9	Mandrill image and a drawing by Albrecht Dürer; see Exercise 11.	248
9.1	A parabola meets a circle.	252
9.2	The point **x**, the direction **p**, and the point **x**+**p**.	254
9.3	Two solutions for a boundary value ODE.	257
9.4	The function $x_1^2 + x_2^4 + 1$ has a unique minimum at the origin $(0,0)$ and no maximum. Upon flipping it, the resulting function $-(x_1^2 + x_2^4 + 1)$ would have a unique maximum at the origin $(0,0)$ and no minimum.	258
9.5	The function of Example 9.5 has a unique minimum at $\mathbf{x}^* = (3,.5)^T$ as well as a saddle point at $\hat{\mathbf{x}} = (0,1)^T$.	262
9.6	Convergence behavior of gradient descent and conjugate gradient iterative schemes for the Poisson equation of Example 7.1.	266
9.7	Nonlinear data fitting; see Example 9.8.	270
9.8	Equality and inequality constraints. The feasible set Ω consists of the points on the thick red line.	271
9.9	A feasible set Ω with a nonempty interior, and level sets of ϕ; larger ellipses signify larger values of ϕ.	272

9.10	One equality constraint and the level sets of ϕ. At $\mathbf{x}^*$ the gradient is orthogonal to the tangent of the constraint.	273
9.11	Center path in the LP *primal* feasibility region.	278
9.12	Two minimum norm solutions for an underdetermined linear system of equations. The ℓ_1 solution is nicely sparse.	286
9.13	A depiction of the noisy function (in solid blue) and the function to be recovered (in red) for Exercise 17.	291
10.1	Different interpolating curves through the same set of points.	296
10.2	Quadratic and linear polynomial interpolation.	299
10.3	The quadratic Lagrange polynomials $L_0(x)$, $L_1(x)$, and $L_2(x)$ based on points $x_0 = 1$, $x_1 = 2$, $x_2 = 4$, used in Example 10.2.	303
10.4	The Lagrange polynomial $L_2(x)$ for $n = 5$. Guess what the data abscissae x_i are.	304
10.5	The interpolating polynomials p_2 and p_3 for Example 10.4.	310
10.6	Global polynomial interpolation at uniformly spaced abscissae can be bad. Here the blue curve with one maximum point is the Runge function of Example 10.6, and the other curves are polynomial interpolants of degree n.	317
10.7	Polynomial interpolation at Chebyshev points, Example 10.6. Results are much improved as compared to Figure 10.6, especially for larger n.	317
10.8	Top panel: the function of Example 10.7 is indistinguishable from its polynomial interpolant at 201 Chebyshev points. Bottom panel: the maximum polynomial interpolation error as a function of the polynomial degree. When doubling the degree from $n = 100$ to $n = 200$ the error decreases from unacceptable (> 1) to almost rounding unit level.	318
10.9	A quadratic interpolant $p_2(x)$ satisfying $p_2(0) = 1.5$, $p_2'(0) = 1$, and $p_2(5) = 0$.	320
10.10	The osculating Hermite cubic for $\ln(x)$ at the points 1 and 2.	322
10.11	Quadratic and linear polynomial interpolation.	324
11.1	A piecewise polynomial function with break points $t_i = i$, $i = 0, 1, \ldots, 6$.	333
11.2	Data and their broken line interpolation.	334
11.3	Matching $s_i, s_i', $ and s_i'' at $x = x_i$ with values of s_{i-1} and its derivatives at the same point. In this example, $x_i = 0$ and $y_i = 1$.	338
11.4	Not-a-knot cubic spline interpolation for the Runge Example 10.6 at 20 equidistant points. The interval has been rescaled to be $[0, 1]$.	340
11.5	The interpolant of Example 11.6.	341
11.6	Hat functions: a compact basis for piecewise linear approximation.	346
11.7	Basis functions for piecewise Hermite polynomials.	348
11.8	B-spline basis for the C^2 cubic spline.	349
11.9	Parametric broken-line interpolation.	350
11.10	Parametric polynomial interpolation.	351
11.11	A simple curve design using 11 Bézier polynomials.	352
11.12	Bilinear interpolation. Left green square: data are given at the unit square's corners (red points), and bilinear polynomial values are desired at mid-edges and at the square's middle (blue points). Right blue grid: data are given at the coarser grid nodes (blue points) and a bilinear interpolation is performed to obtain values at the finer grid nodes, yielding values $f_{i,j}$ for all $i = 0, 1, 2, \ldots, N_x$, $j = 0, 1, 2, \ldots, N_y$.	355
11.13	Triangle mesh in the plane. This one is MATLAB's data set `trimesh2d`.	356
11.14	Linear interpolation over a triangle mesh. Satisfying the interpolation conditions at triangle vertices implies continuity across neighboring triangles.	357

11.15	Triangle surface mesh.	358
11.16	RBF interpolation of an upsampling of a consolidated point cloud.	359
12.1	The first five best polynomial approximations to $f(x) = \cos(2\pi x)$. The approximated function is in solid green. Note the similarity to Figure 6.4 of Example 6.3.	369
12.2	The first five Legendre polynomials. You should be able to figure out which curve corresponds to which polynomial.	372
12.3	Chebyshev polynomials of degrees 4, 5, and 6.	378
13.1	A linear combination of sines and cosines with various k-values up to 110 (top panel), filtered by taking its best least squares trigonometric polynomial approximation with $l = 10$ (bottom panel). This best approximation simply consists of that part of the given function that involves the first 20 basis functions ϕ_j; thus the higher frequency contributions disappear.	385
13.2	For the discrete Fourier transform, imagine the red mesh points located on a blue circle with $x_0 = x_{n+1}$ closing the ring. Thus, x_n is the left (clockwise) neighbor of x_0, x_1 is the right (counterclockwise) neighbor of x_{n+1}, and so on.	390
13.3	Trigonometric polynomial interpolation for the hat function with $p_3(x)$.	391
13.4	Trigonometric polynomial interpolation for a smooth function with $p_3(t)$ (top left), $p_7(t)$ (top right), $p_{15}(t)$ (bottom left), and $p_{31}(t)$ (bottom right). The approximated function is plotted in dashed green.	393
13.5	Trigonometric polynomial interpolation for the square wave function on $[0, 2\pi]$ with $n = 127$.	396
13.6	The unit circle in the complex plane is where the values of $e^{\iota\theta}$ reside. The $m = 8$ roots of the polynomial equation $\theta^8 = 1$, given by $e^{-\iota 2j\pi/8}$, $j = 0, 1, \ldots, 7$, are displayed as red diamonds.	398
13.7	Example 13.9: an observed blurred image b, and the result of a deblurring algorithm applied to this data. (Images courtesy of H. Huang [43].)	402
13.8	Cosine basis interpolation for the function $\ln(x+1)$ on $[0, 2\pi]$ with $n = 31$. . .	404
14.1	Actual error using the three methods of Example 14.1. Note the log-log scale of the plot. The order of the methods is therefore indicated by the slope of the straight line (note that h is decreased from right to left).	413
14.2	The measured error roughly equals truncation error plus roundoff error. The former decreases but the latter grows as h decreases. The "ideal roundoff error" is proportional to η/h. Note the log-log scale of the plot. A red circle marks the "optimal h" value for Example 14.1.	422
14.3	Numerical differentiation of noisy data. On the left panel, $\sin(x)$ is perturbed by 1% noise. On the right panel, the resulting numerical differentiation is a disaster.	425
14.4	An image with noise and its smoothed version. Numerical differentiation must not be applied to the original image.	426
14.5	Maximum absolute errors for the first and second derivatives of $f(x) = e^x \sin(10x)$ on the interval $[0, \pi]$ at the Chebyshev extremum points, i.e., using the Chebyshev differentiation matrix.	429
14.6	Maximum absolute errors for the first and second derivatives of $f(x) = e^{-5x^2}$ on the interval $[-2\pi, 2\pi]$. The two left subplots use FFT, and the right subplots use the Chebyshev differentiation matrix. The errors are evaluated at the corresponding data locations (also called collocation points).	431
14.7	This "waterfall" plot depicts the progress of two initial pulses in time; see Example 14.14. The two solitons merge and then split with their form intact. Rest assured that they will meet (and split) again many times in the future.	434

15.1	Area under the curve. Top left (cyan): for $f(x)$ that stays nonnegative, I_f equals the area under the function's curve. Bottom left (green): approximation by the trapezoidal rule. Top right (pink): approximation by the Simpson rule. Bottom right (yellow): approximation by the midpoint rule.	444
15.2	Composite trapezoidal quadrature with $h = 0.2$ for the integral of Figure 15.1.	448
15.3	Numerical integration errors for the composite trapezoidal and Simpson methods; see Example 15.3.	450
15.4	Numerical integration errors for the composite trapezoidal and Simpson methods; see Example 15.4.	451
15.5	The discontinuous integrand of Example 15.10.	459
15.6	The integrand $f(x) = \frac{200}{2x^3-x^2}(5\sin(\frac{20}{x}))^2$ with quadrature mesh points along the x-axis for $\texttt{tol} = 5 \times 10^{-3}$.	467
15.7	Approximating an iterated integral in two dimensions. For each of the quadrature points in x we integrate in y to approximate $g(x)$.	474
15.8	Gaussian points; see Exercise 10.	477
16.1	A simple pendulum.	483
16.2	Two steps of the forward Euler method. The exact solution is the curved solid line. The numerical values obtained by the Euler method are circled and lie at the nodes of a broken line that interpolates them. The broken line is tangential at the beginning of each step to the ODE trajectory passing through the corresponding node (dashed lines).	486
16.3	The sixth solution component of the HIRES model.	490
16.4	RK method: repeated evaluations of the function f in the current mesh subinterval $[t_i, t_{i+1}]$ are combined to yield a higher order approximation y_{i+1} at t_{i+1}.	493
16.5	Predator-prey model: $y_1(t)$ is number of prey, $y_2(t)$ is number of predator.	497
16.6	Predator-prey solution in phase plane: the curve of $y_1(t)$ vs. $y_2(t)$ yields a limit cycle.	498
16.7	Multistep method: known solution values of y or $f(t,y)$ at the current location t_i and previous locations $t_{i-1}, \ldots, t_{i+1-s}$ are used to form an approximation for the next unknown y_{i+1} at t_{i+1}.	500
16.8	Stability regions for q-stage explicit RK methods of order q, $q = 1,2,3,4$. The inner circle corresponds to forward Euler, $q = 1$. The larger q is, the larger the stability region. Note the "ear lobes" of the fourth-order method protruding into the right half plane.	508
16.9	Solution of the heat problem of Example 16.17 at $t = 0.1$.	512
16.10	The exact solution $y(t_i)$, which lies on the lowest of the three curves, is approximated by y_i, which lies on the middle curve. If we integrate the next step exactly, starting from (t_i, y_i), then we obtain $(t_{i+1}, \bar{y}(t_{i+1}))$, which also lies on the middle curve. But we don't: rather, we integrate the next step approximately as well, obtaining (t_{i+1}, y_{i+1}), which lies on the top curve. The difference between the two curve values at the argument t_{i+1} is the local error.	516
16.11	Astronomical orbit using $\texttt{ode45}$; see Example 16.20.	518
16.12	Oscillatory energies for the FPU problem; see Example 16.21.	520
16.13	Oscillatory energies for the FPU problem obtained using MATLAB's $\texttt{ode45}$ with default tolerances. The deviation from Figure 16.12 depicts a significant, nonrandom error.	520

16.14	The two solutions of Examples 16.22 and 9.3 obtained using simple shooting starting from two different initial guesses c_0 for $v'(0)$. Plotted are the trajectories for the initial guesses (dashed magenta), as well as the final BVP solutions (solid blue). The latter are qualitatively the same as in Figure 9.3. Note the different vertical scales in the two subfigures.	524
16.15	Hyperbolic and parabolic solution profiles starting from a discontinuous initial value function. For the advection equation the exact solution is displayed. For the heat equation we have integrated numerically, using homogeneous BC at $x = \pm\pi$ and a rather fine discretization mesh.	526
16.16	"Waterfall" solutions of the classical wave equation using the leapfrog scheme; see Example 16.25. For $\alpha = 1$ the solution profile is resolved well by the discretization.	530

List of Tables

7.1	Errors for basic relaxation methods applied to the model problem of Example 7.1 with $n = N^2 = 225$.	178
9.1	Convergence of Newton's method to the two roots of Example 9.1.	256
9.2	Example 9.5. The first three columns to the right of the iteration counter track convergence starting from $\mathbf{x}_0 = (8,.2)^T$: Newton's method yields the minimum quickly. The following (rightmost) three columns track convergence starting from $\mathbf{x}_0 = (8,.8)^T$: Newton's method finds a critical point, but it's not the minimizer.	262
9.3	Example 9.14. Tracking progress of the primal-dual LP algorithm.	284
14.1	Errors in the numerical differentiation of $f(x) = e^x$ at $x = 0$ using methods of order 1, 2, and 4.	413
14.2	Errors in the numerical differentiation of $f(x) = e^x$ at $x = 0$ using three-point methods on a nonuniform mesh, Example 14.6. The error e_g in the more elaborate method is second order, whereas the simpler method yields first order accuracy e_s.	419
16.1	Absolute errors using the forward Euler method for the ODE $y' = y$. The values $e_i = y(t_i) - y_i$ are listed under Error.	487
16.2	Errors and calculated observed orders (rates) for the forward Euler, the explicit midpoint (RK2), and the classical Runge–Kutta (RK4) methods.	496
16.3	Example 16.12: errors and calculated rates for Adams–Bashforth methods; (s,q) denotes the s-step method of order q.	503
16.4	Example 16.12: errors and calculated rates for Adams–Moulton methods; (s,q) denotes the s-step method of order q.	503
16.5	Coefficients of BDF methods up to order 6.	505
16.6	Example 16.14: errors and calculated rates for BDF methods; (s,q) denotes the s-step method of order q.	505

Preface

This book is designed for students and researchers who seek practical knowledge of modern techniques in scientific computing. The text aims to provide an in-depth treatment of fundamental issues and methods, and the reasons behind success and failure of numerical software. On one hand, we avoid an extensive, encyclopedic, heavily theoretical exposition, and try to get to current methods, issues, and software fairly quickly. On the other hand, this is by no means a quick recipe book, since we feel that the introduction of algorithms requires adequate theoretical foundation: having a solid basis enables those who need to apply the techniques to successfully design their own solution approach for any nonstandard problems they may encounter in their work.

There are many books on scientific computing and numerical analysis, and a natural question here would be why we think that yet another text is necessary. It is mainly because we feel that in an age where yesterday's concepts are not necessarily today's truths, scientific computing is constantly redefining itself and is now positioned as a discipline truly at the junction of mathematics, computer science, and engineering. Books that rely heavily on theory, or on algorithm recipes, do not quite capture the current state of this broad and dynamic area of research and application. We thus take an algorithmic approach and focus on techniques of a high level of applicability to engineering, computer science, and industrial mathematics practitioners. At the same time, we provide mathematical justification throughout for the methods introduced. While we refrain from a theorem–proof type of exposition, we construct the theory behind the methods in a systematic and fairly complete fashion. We make a strong effort to emphasize computational aspects of the algorithms discussed by way of discussing their efficiency and their computational limitations.

This book has been developed from notes for two courses taught by the Department of Computer Science at the University of British Columbia (UBC) for over 25 years. The first author developed a set of notes back in the late 1970s. These were based on the books by Conte and de Boor [13], Dahlquist and Bjorck [15], and others, as well as on notes by Jim Varah. Improvements to these notes were subsequently made by Ian Cavers, by the second author, and by Dhavide Aruliah. A substantial addition of material was made by the two authors in the last few years. A few of our colleagues (Jack Snoeyink, Oliver Dorn, Ian Mitchell, Kees van den Doel and Tyrone Rees at UBC, Jim Lambers at Stanford University, Marsha Berger and Michael Overton at NYU, Edmond Chow at Columbia University, André Weidemann at Stellenbosch, and others) have used the notes and provided us with invaluable feedback.

Most of the material contained in this book can be covered in two reasonably paced semesterial courses. A possible model here is the one that has been adopted at UBC: one course covers Chapters 3 to 9, while another concentrates on Chapters 10 to 16. The two parts are designed so that they do not heavily rely on each other, although we have found it useful to include the material of Chapter 1 and a subset of Chapter 2 in both semesterial courses. More advanced material, contained in sections denoted by an asterisk, is included in the text as a natural extension. However, with one or two exceptions, we have not taught the advanced material in the two basic courses.

Another use of this text is for a breadth-type introductory course on numerical methods at the graduate level. The target audience would be beginning graduate students in a variety of disciplines including computer science, applied mathematics, and engineering, who start graduate school without having taken undergraduate courses in this area. Such a course would also use our more advanced sections and possibly additional material.

The division of the material into chapters can be justified by the type of computational errors encountered. These concepts are made clear in the text. We start off with *roundoff error*, which appears in all the numerical processes considered in this book. Chapters 1 and 2 systematically cover it, and later chapters, such as 5 and 6, contain assessments of its effect. Chapter 3 and Chapters 7 to 9 are concerned mainly with iterative processes, and as such, the focus is on the iteration, or *convergence error*. Finally, control and assessment of *discretization error* is the main focus in Chapters 10 to 16, although the other error types do arise there as well.

We illustrate our algorithms using the programming environment of MATLAB and expect the reader to become gradually proficient in this environment while (if not before) learning the material covered in this book. But at the same time, we use the MATLAB language and programming environment merely as a tool, and we refrain from turning the book into a language tutorial.

Each chapter contains exercises, ordered by section. In addition, there is an exercise numbered 0 which consists of several review questions intended for self-testing and for punctuating the process of reading and absorbing the presented material. Instructors should resist the temptation to answer these review questions for the students.

We have made an effort to make the chapters fairly independent of one another. As a result, we believe that readers who are interested in specific topics may often be able to settle for reading almost exclusively only the chapters that are relevant to the material they wish to pursue.

We are maintaining a project webpage for this book which can be found at

http://www.siam.org/books/cs07

It contains links to our programs, solutions to selected exercises, errata, and more.

Last but not at all least, many generations of students at UBC have helped us over the years to shape, define, and debug this text. The list of individuals who have provided crucial input is simply too long to mention. This book is dedicated to you, our students.

Chapter 1
Numerical Algorithms

This opening chapter introduces the basic concepts of numerical algorithms and scientific computing.

We begin with a general, brief introduction to the field in Section 1.1. This is followed by the more substantial Sections 1.2 and 1.3. Section 1.2 discusses the basic errors that may be encountered when applying numerical algorithms. Section 1.3 is concerned with essential properties of such algorithms and the appraisal of the results they produce.

We get to the "meat" of the material in later chapters.

1.1 Scientific computing

Scientific computing is a discipline concerned with the development and study of **numerical algorithms** for solving mathematical problems that arise in various disciplines in science and engineering.

Typically, the starting point is a given **mathematical model** which has been formulated in an attempt to explain and understand an observed phenomenon in biology, chemistry, physics, economics, or any other scientific or engineering discipline. We will concentrate on those mathematical models which are *continuous* (or *piecewise continuous*) and are difficult or impossible to solve analytically; this is usually the case in practice. Relevant application areas within computer science and related engineering fields include graphics, vision and motion analysis, image and signal processing, search engines and data mining, machine learning, and hybrid and embedded systems.

In order to solve such a model approximately on a computer, the continuous or piecewise continuous problem is approximated by a discrete one. Functions are approximated by finite arrays of values. Algorithms are then sought which approximately solve the mathematical problem efficiently, accurately, and reliably. This is the heart of scientific computing. **Numerical analysis** may be viewed as the theory behind such algorithms.

The next step after devising suitable algorithms is their implementation. This leads to questions involving programming languages, data structures, computing architectures, etc. The big picture is depicted in Figure 1.1.

The set of requirements that good scientific computing algorithms must satisfy, which seems elementary and obvious, may actually pose rather difficult and complex practical challenges. The main purpose of this book is to equip you with basic methods and analysis tools for handling such challenges as they arise in future endeavors.

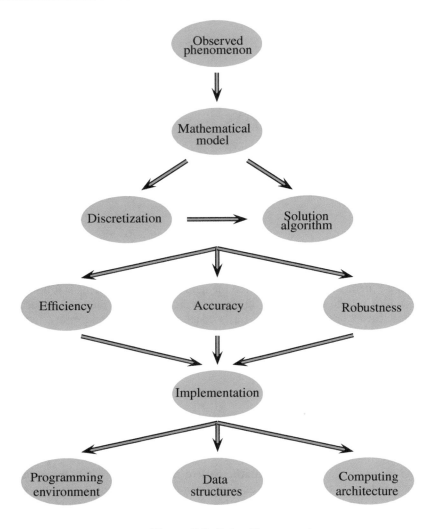

Figure 1.1. *Scientific computing.*

Problem solving environment

As a computing tool, we will be using MATLAB: this is an interactive computer language, which for our purposes may best be viewed as a convenient *problem solving environment*.

MATLAB is much more than a language based on simple data arrays; it is truly a complete environment. Its interactivity and graphics capabilities make it more suitable and convenient in our context than general-purpose languages such as C++, Java, Scheme, or Fortran 90. In fact, many of the algorithms that we will learn are already implemented in MATLAB... *So why learn them at all?* Because they provide the basis for much more complex tasks, not quite available (that is to say, not already solved) in MATLAB or anywhere else, which you may encounter in the future.

Rather than producing yet another MATLAB tutorial or introduction in this text (there are several very good ones available in other texts as well as on the Internet) we will demonstrate the use of this language on examples as we go along.

1.2 Numerical algorithms and errors

The most fundamental feature of numerical computing is the inevitable presence of error. The result of any interesting computation (and of many uninteresting ones) is typically only approximate, and our goal is to ensure that the resulting error is tolerably small.

Relative and absolute errors

There are in general two basic types of measured error. Given a scalar quantity u and its approximation v:

- The *absolute error* in v is
$$|u - v|.$$

- The *relative error* (assuming $u \neq 0$) is
$$\frac{|u - v|}{|u|}.$$

The relative error is usually a more meaningful measure. This is especially true for errors in floating point representation, a point to which we return in Chapter 2. For example, we record absolute and relative errors for various hypothetical calculations in the following table:

u	v	Absolute error	Relative error
1	0.99	0.01	0.01
1	1.01	0.01	0.01
-1.5	-1.2	0.3	0.2
100	99.99	0.01	0.0001
100	99	1	0.01

Evidently, when $|u| \approx 1$ there is not much difference between absolute and relative error measures. But when $|u| \gg 1$, the relative error is more meaningful. In particular, we expect the approximation in the last row of the above table to be similar in quality to the one in the first row. This expectation is borne out by the value of the relative error but is not reflected by the value of the absolute error.

When the approximated value is small in magnitude, things are a little more delicate, and here is where relative errors may not be so meaningful. But let us not worry about this at this early point.

Example 1.1. The Stirling approximation

$$v = S_n = \sqrt{2\pi n} \cdot \left(\frac{n}{e}\right)^n$$

is used to approximate $u = n! = 1 \cdot 2 \cdots n$ for large n. The formula involves the constant $e = \exp(1) = 2.7182818\ldots$. The following MATLAB script computes and displays $n!$ and S_n, as well as their absolute and relative differences, for $1 \leq n \leq 10$:

```
e=exp(1);
n=1:10;                          % array
Sn=sqrt(2*pi*n).*((n/e).^n);     % the Stirling approximation.
```

```
fact_n=factorial(n);
abs_err=abs(Sn-fact_n);           % absolute error
rel_err=abs_err./fact_n;          % relative error
format short g
[n; fact_n; Sn; abs_err; rel_err]'  % print out values
```

Given that this is our first MATLAB script, let us provide a few additional details, though we hasten to add that we will not make a habit out of this. The commands `exp`, `factorial`, and `abs` use built-in functions. The command `n=1:10` (along with a semicolon, which simply suppresses screen output) defines an array of length 10 containing the integers $1, 2, \ldots, 10$. This illustrates a fundamental concept in MATLAB of working with arrays whenever possible. Along with it come *array operations*: for example, in the third line ".*" corresponds to elementwise multiplication of vectors or matrices. Finally, our printing instructions (the last two in the script) are a bit primitive here, a sacrifice made for the sake of simplicity in this, our first program.

The resulting output is

1	1	0.92214	0.077863	0.077863
2	2	1.919	0.080996	0.040498
3	6	5.8362	0.16379	0.027298
4	24	23.506	0.49382	0.020576
5	120	118.02	1.9808	0.016507
6	720	710.08	9.9218	0.01378
7	5040	4980.4	59.604	0.011826
8	40320	39902	417.6	0.010357
9	3.6288e+005	3.5954e+005	3343.1	0.0092128
10	3.6288e+006	3.5987e+006	30104	0.008296

The values of $n!$ become very large very quickly, and so are the values of the approximation S_n. The absolute errors grow as n grows, but the relative errors stay well behaved and indicate that in fact the larger n is, the *better* the quality of the approximation is. Clearly, the relative errors are much more meaningful as a measure of the quality of this approximation. ∎

Error types

Knowing how errors are typically measured, we now move to discuss their source. There are several types of error that may limit the accuracy of a numerical calculation.

1. **Errors in the problem to be solved**.

   These may be approximation **errors in the mathematical model**. For instance:

   - Heavenly bodies are often approximated by spheres when calculating their properties; an example here is the approximate calculation of their motion trajectory, attempting to answer the question (say) whether a particular asteroid will collide with Planet Earth before 11.12.2016.

   - Relatively unimportant chemical reactions are often discarded in complex chemical modeling in order to obtain a mathematical problem of a manageable size.

   It is important to realize, then, that often approximation errors of the type stated above are deliberately made: the assumption is that simplification of the problem is worthwhile even if it generates an error in the model. Note, however, that we are still talking about the mathematical model itself; approximation errors related to the numerical solution of the problem are discussed below.

1.2. Numerical algorithms and errors

Another typical source of error in the problem is **error in the input data**. This may arise, for instance, from physical measurements, which are never infinitely accurate.

Thus, it may be that after a careful numerical simulation of a given mathematical problem, the resulting solution would not quite match observations on the phenomenon being examined.

At the level of numerical algorithms, which is the focus of our interest here, there is really nothing we can do about the above-described errors. Nevertheless, they should be taken into consideration, for instance, when determining the accuracy (tolerance with respect to the next two types of error mentioned below) to which the numerical problem should be solved.

2. **Approximation errors**

   Such errors arise when an approximate formula is used in place of the actual function to be evaluated.

   We will often encounter two types of approximation errors:

   - **Discretization errors** arise from discretizations of continuous processes, such as interpolation, differentiation, and integration.
   - **Convergence errors** arise in iterative methods. For instance, nonlinear problems must generally be solved approximately by an iterative process. Such a process would converge to the exact solution in infinitely many iterations, but we cut it off after a finite (hopefully small!) number of such iterations. Iterative methods in fact often arise in linear algebra.

3. **Roundoff errors**

   Any computation with real numbers involves roundoff error. Even when no approximation error is produced (as in the direct evaluation of a straight line, or the solution by Gaussian elimination of a linear system of equations), roundoff errors are present. These arise because of the finite precision representation of real numbers on any computer, which affects both data representation and computer arithmetic.

Discretization and convergence errors may be assessed by an analysis of the method used, and we will see a lot of that in this text. Unlike roundoff errors, they have a relatively smooth structure which may occasionally be exploited. Our basic assumption will be that approximation errors dominate roundoff errors in magnitude in actual, successful calculations.

Theorem: Taylor Series.
Assume that $f(x)$ has $k+1$ derivatives in an interval containing the points x_0 and $x_0 + h$. Then

$$f(x_0 + h) = f(x_0) + hf'(x_0) + \frac{h^2}{2}f''(x_0) + \cdots + \frac{h^k}{k!}f^{(k)}(x_0)$$
$$+ \frac{h^{k+1}}{(k+1)!}f^{(k+1)}(\xi),$$

where ξ is some point between x_0 and $x_0 + h$.

Discretization errors in action

Let us show an example that illustrates the behavior of discretization errors.

Example 1.2. Consider the problem of approximating the derivative $f'(x_0)$ of a given smooth function $f(x)$ at the point $x = x_0$. For instance, let $f(x) = \sin(x)$ be defined on the real line $-\infty < x < \infty$, and set $x_0 = 1.2$. Thus, $f(x_0) = \sin(1.2) \approx 0.932\ldots$.

Further, consider a situation where $f(x)$ may be evaluated at any point x near x_0, but $f'(x_0)$ may not be directly available or is computationally expensive to evaluate. Thus, we seek ways to approximate $f'(x_0)$ by evaluating f at x near x_0.

A simple algorithm may be constructed using **Taylor's series**. This fundamental theorem is given on the preceding page. For some small, positive value h that we will choose in a moment, write

$$f(x_0+h) = f(x_0) + hf'(x_0) + \frac{h^2}{2}f''(x_0) + \frac{h^3}{6}f'''(x_0) + \frac{h^4}{24}f''''(x_0) + \cdots.$$

Then

$$f'(x_0) = \frac{f(x_0+h) - f(x_0)}{h} - \left(\frac{h}{2}f''(x_0) + \frac{h^2}{6}f'''(x_0) + \frac{h^3}{24}f''''(x_0) + \cdots\right).$$

Our *algorithm* for approximating $f'(x_0)$ is to calculate

$$\frac{f(x_0+h) - f(x_0)}{h}.$$

The obtained approximation has the *discretization error*

$$\left| f'(x_0) - \frac{f(x_0+h) - f(x_0)}{h} \right| = \left| \frac{h}{2}f''(x_0) + \frac{h^2}{6}f'''(x_0) + \frac{h^3}{24}f''''(x_0) + \cdots \right|.$$

Geometrically, we approximate the slope of the tangent at the point x_0 by the slope of the chord through neighboring points of f. In Figure 1.2, the tangent is in blue and the chord is in red.

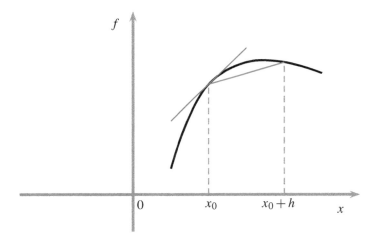

Figure 1.2. *A simple instance of numerical differentiation: the tangent $f'(x_0)$ is approximated by the chord $(f(x_0+h) - f(x_0))/h$.*

If we know $f''(x_0)$, and it is nonzero, then for h small we can estimate the discretization error by

$$\left| f'(x_0) - \frac{f(x_0+h) - f(x_0)}{h} \right| \approx \frac{h}{2}|f''(x_0)|.$$

Using the notation defined in the box on the next page we notice that the error is $\mathcal{O}(h)$ so long as

1.2. Numerical algorithms and errors

$f''(x_0)$ is bounded, and it is $\Theta(h)$ if also $f''(x_0) \neq 0$. In any case, even without knowing $f''(x)$ we expect the discretization error to decrease at least as fast as h when h is decreased.

> **Big-O and Θ Notation**
> Throughout this text we consider various computational errors depending on a discretization step size $h > 0$ and ask how they decrease as h decreases. In other instances, such as when estimating the efficiency of a particular algorithm, we are interested in a bound on the work estimate as a parameter n increases unboundedly (e.g., $n = 1/h$).
> For an error e depending on h we denote
> $$e = \mathcal{O}(h^q)$$
> if there are two positive constants q and C such that
> $$|e| \leq Ch^q$$
> for all $h > 0$ small enough.
> Similarly, for $w = w(n)$ the expression
> $$w = \mathcal{O}(n \log n)$$
> means that there is a constant $C > 0$ such that
> $$w \leq Cn \log n$$
> as $n \to \infty$. It will be easy to figure out from the context which of these two meanings is the relevant one.
> The Θ notation signifies a stronger relation than the $\mathcal{O}$ notation: a function $\phi(h)$ for small h (resp., $\phi(n)$ for large n) is $\Theta(\psi(h))$ (resp., $\Theta(\psi(n))$) if ϕ is asymptotically bounded *both above and below* by ψ.

For our particular instance, $f(x) = \sin(x)$, we have the exact value $f'(x_0) = \cos(1.2) = 0.362357754476674\ldots$. Carrying out our short algorithm we obtain for $h = 0.1$ the approximation $f'(x_0) \approx (\sin(1.3) - \sin(1.2))/0.1 = 0.315\ldots$. The absolute error thus equals approximately 0.047. The relative error is not qualitatively different here.

This approximation of $f'(x_0)$ using $h = 0.1$ is not very accurate. We therefore apply the same algorithm using several increasingly smaller values of h. The resulting errors are as follows:

h	Absolute error
0.1	4.716676e-2
0.01	4.666196e-3
0.001	4.660799e-4
1.e-4	4.660256e-5
1.e-7	4.619326e-8

Indeed, the error appears to decrease like h. More specifically (and less importantly), using our explicit knowledge of $f''(x) = -f(x) = -\sin(x)$, in this case we have that $\frac{1}{2}f''(x_0) \approx -0.466$. The quantity $0.466h$ is seen to provide a rather accurate estimate for the above-tabulated absolute error values. ■

The damaging effect of roundoff errors

The calculations in Example 1.2, and the ones reported below, were carried out using MATLAB's standard arithmetic. Let us stay for a few more moments with the approximation algorithm featured in that example and try to push the envelope a little further.

Example 1.3. The numbers in Example 1.2 might suggest that an arbitrary accuracy can be achieved by the algorithm, provided only that we take h small enough. Indeed, suppose we want

$$\left| \cos(1.2) - \frac{\sin(1.2+h) - \sin(1.2)}{h} \right| < 10^{-10}.$$

Can't we just set $h \leq 10^{-10}/0.466$ in our algorithm?

Not quite! Let us record results for very small, positive values of h:

h	Absolute error
1.e-8	4.361050e-10
1.e-9	5.594726e-8
1.e-10	1.669696e-7
1.e-11	7.938531e-6
1.e-13	4.250484e-4
1.e-15	8.173146e-2
1.e-16	3.623578e-1

A log-log plot[1] of the error versus h is provided in Figure 1.3. We can clearly see that as h is decreased, at first (from right to left in the figure) the error decreases along a straight line, but this trend is altered and eventually reversed. The MATLAB script that generates the plot in Figure 1.3 is given next.

```
x0 = 1.2;
f0 = sin(x0);
fp = cos(x0);
i = -20:0.5:0;
h = 10.^i;
err = abs (fp - (sin(x0+h) - f0)./h );
d_err = f0/2*h;
loglog (h,err,'-*');
hold on
loglog (h,d_err,'r-.');
xlabel('h')
ylabel('Absolute error')
```

Perhaps the most mysterious line in this script is that defining `d_err`: it calculates $\frac{1}{2}|f''(x_0)|h$. ∎

---

[1]Graphing error values using a logarithmic scale is rather common in scientific computing, because a logarithmic scale makes it easier to trace values that are close to zero. As you will use such plotting often, let us mention at this early stage the MATLAB commands `plot`, `semilogy`, and `loglog`.

1.3. Algorithm properties

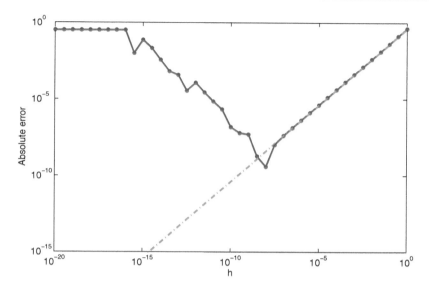

Figure 1.3. *The combined effect of discretization and roundoff errors. The solid curve interpolates the computed values of $|f'(x_0) - \frac{f(x_0+h)-f(x_0)}{h}|$ for $f(x) = \sin(x)$, $x_0 = 1.2$. Also shown in dash-dot style is a straight line depicting the discretization error without roundoff error.*

The reason the error "bottoms out" at about $h = 10^{-8}$ in the combined Examples 1.2–1.3 is that the total, measured error consists of contributions of both discretization and roundoff errors. The discretization error decreases in an orderly fashion as h decreases, and it dominates the roundoff error when h is relatively large. But when h gets below approximately 10^{-8} the discretization error becomes very small and roundoff error starts to dominate (i.e., it becomes larger in magnitude).

Roundoff error has a somewhat erratic behavior, as is evident from the small oscillations that are present in the graph in a few places.

Moreover, for the algorithm featured in the last two examples, overall the roundoff error *increases* as h decreases. This is one reason why we want it always dominated by the discretization error when solving problems involving numerical differentiation such as differential equations.

Popular theorems from calculus

The Taylor Series Theorem, given on page 5, is by far the most cited theorem in numerical analysis. Other popular calculus theorems that we will use in this text are gathered on the following page. They are all elementary and not difficult to prove: indeed, most are special cases of Taylor's Theorem.

Specific exercises for this section: Exercises 1–3.

1.3 Algorithm properties

In this section we briefly discuss performance features that may or may not be expected from a good numerical algorithm, and we define some basic properties, or characteristics, that such an algorithm should have.

Theorem: Useful Calculus Results.

- **Intermediate Value**

  If $f \in C[a,b]$ and s is a value such that $f(\hat{a}) \le s \le f(\hat{b})$ for two numbers $\hat{a}, \hat{b} \in [a,b]$, then there exists a real number $c \in [a,b]$ for which $f(c) = s$.

- **Mean Value**

  If $f \in C[a,b]$ and f is differentiable on the open interval (a,b), then there exists a real number $c \in (a,b)$ for which $f'(c) = \frac{f(b)-f(a)}{b-a}$.

- **Rolle's**

  If $f \in C[a,b]$ and f is differentiable on (a,b), and in addition $f(a) = f(b)$, then there is a real number $c \in (a,b)$ for which $f'(c) = 0$.

Criteria for assessing an algorithm

An assessment of the quality and usefulness of an algorithm may be based on a number of criteria:

- **Accuracy**

  This issue is intertwined with the issue of error types and was discussed at the start of Section 1.2 and in Example 1.2. (See also Exercise 3.) The important point is that the accuracy of a numerical algorithm is a crucial parameter in its assessment, and when designing numerical algorithms it is necessary to be able to point out what magnitude of error is to be expected when the computation is carried out.

- **Efficiency**

  A good computation is one that terminates before we lose our patience. A numerical algorithm that features great theoretical properties is useless if carrying it out takes an unreasonable amount of computational time. Efficiency depends on both CPU time and storage space requirements. Details of an algorithm implementation within a given computer language and an underlying hardware configuration may play an important role in yielding code efficiency. Other theoretical properties yield indicators of efficiency, for instance, the **rate of convergence**. We return to this in later chapters.

  Often a machine-independent estimate of the number of elementary operations required, namely, additions, subtractions, multiplications, and divisions, gives an idea of the algorithm's efficiency. Normally, a floating point representation is used for real numbers and then the costs of these different elementary floating point operations, called **flops**, may be assumed to be roughly equal to one another.

  Example 1.4. A polynomial of degree n, given as

  $$p_n(x) = c_0 + c_1 x + \cdots + c_n x^n,$$

  requires $\mathcal{O}(n^2)$ operations[2] to evaluate at a fixed point x, if done in a brute force way without intermediate storing of powers of x. But using the *nested form*, also known as Horner's rule and given by

  $$p_n(x) = (\cdots((c_n x + c_{n-1})x + c_{n-2})x \cdots)x + c_0,$$

---

[2] See page 7 for the $\mathcal{O}$ notation.

1.3. Algorithm properties

suggests an evaluation algorithm which requires only $\mathcal{O}(n)$ elementary operations, i.e., requiring linear (in n) rather than quadratic computation time. A MATLAB script for nested evaluation follows:

```
% Assume the polynomial coefficients are already stored
% in array c such that for any real x,
% p(x) = c(1) + c(2)x + c(3)x^2 + ... + c(n+1)x^n
p = c(n+1);
for j = n:-1:1
  p = p*x + c(j);
end
```

The "onion shell" evaluation formula thus unravels quite simply. Note also the manner of introducing comments into the script. ∎

It is important to note that while operation counts as in Example 1.4 often give a rough idea of algorithm efficiency, they do not give the complete picture regarding execution speed, since they do not take into account the price (speed) of memory access which may vary considerably. Furthermore, any setting of parallel computing is ignored in a simple operation count as well. Curiously, this is part of the reason the MATLAB command flops, which had been an integral part of this language for many years, was removed from further releases several years ago. Indeed, in modern computers, cache access, blocking and vectorization features, and other parameters are crucial in the determination of execution time. The computer language used for implementation can also affect the comparative timing of algorithm implementations. Those, unfortunately, are much more difficult to assess compared to an operation count. In this text we will not get into the gory details of these issues, despite their relevance and importance.

- **Robustness**

  Often, the major effort in writing **numerical software**, such as the routines available in MATLAB for solving linear systems of algebraic equations or for function approximation and integration, is spent not on implementing the essence of an algorithm but on ensuring that it would work under all weather conditions. Thus, the routine should either yield the correct result to within an acceptable error tolerance level, or it should fail gracefully (i.e., terminate with a warning) if it does not succeed to guarantee a "correct result."

  There are intrinsic numerical properties that account for the robustness and reliability of an algorithm. Chief among these is the rate of accumulation of errors. In particular, *the algorithm must be stable*; see Example 1.6.

Problem conditioning and algorithm stability

In view of the fact that the problem and the numerical algorithm both yield errors, a natural question arises regarding the appraisal of a given computed solution. Here notions such as *problem sensitivity* and *algorithm stability* play an important role. If the problem is too sensitive, or **ill-conditioned**, meaning that even a small perturbation in the data produces a large difference in the result,[3] then no algorithm may be found for that problem which would meet our requirement of solution robustness; see Figure 1.4 for an illustration. Some modification in the problem definition may be called for in such cases.

---
[3] Here we refer to intuitive notions of "large" vs. "small" quantities and of values being "close to" vs. "far from" one another. While these notions can be quantified and thus be made more precise, such a move would typically make definitions cumbersome and harder to understand at this preliminary stage of the discussion.

12   Chapter 1. Numerical Algorithms

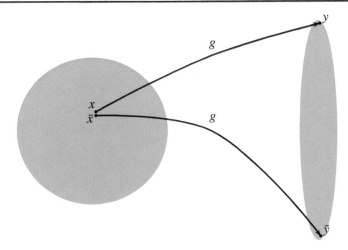

Figure 1.4. *An ill-conditioned problem of computing output values y given in terms of input values x by $y = g(x)$: when the input x is slightly perturbed to $\bar{x}$, the result $\bar{y} = g(\bar{x})$ is far from y. If the problem were well-conditioned, we would be expecting the distance between y and $\bar{y}$ to be more comparable in magnitude to the distance between x and $\bar{x}$.*

For instance, the problem of numerical differentiation depicted in Examples 1.2 and 1.3 turns out to be ill-conditioned when extreme accuracy (translating to very small values of h) is required.

The job of a **stable** algorithm for a given problem is to yield a numerical solution which is the exact solution of an only slightly perturbed problem; see the illustration in Figure 1.5. Thus, if the algorithm is stable and the problem is **well-conditioned** (i.e., not ill-conditioned), then the computed result $\bar{y}$ is close to the exact y.

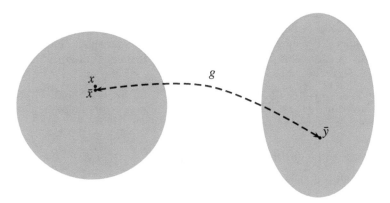

Figure 1.5. *An instance of a stable algorithm for computing $y = g(x)$: the output $\bar{y}$ is the exact result, $\bar{y} = g(\bar{x})$, for a slightly perturbed input, i.e., $\bar{x}$ which is close to the input x. Thus, if the algorithm is stable and the problem is well-conditioned, then the computed result $\bar{y}$ is close to the exact y.*

Example 1.5. The problem of evaluating the square root function for an argument near the value 1 is well-conditioned, as we show below.

1.3. Algorithm properties

Thus, let $g(x) = \sqrt{1+x}$ and note that $g'(x) = \frac{1}{2\sqrt{1+x}}$. Suppose we fix x so that $|x| \ll 1$, and consider $\bar{x} = 0$ as a small perturbation of x. Then $\bar{y} = g(\bar{x}) = 1$, and $y - \bar{y} = \sqrt{1+x} - 1$.

If we approximate $\sqrt{1+x}$ by the first two terms of its Taylor series expansion (see page 5) about the origin, namely, $g(x) \approx 1 + \frac{x}{2}$, then

$$y - \bar{y} \approx \left(1 + \frac{x}{2}\right) - 1 = \frac{x}{2} = \frac{1}{2}(x - \bar{x}).$$

Qualitatively, the situation is fortunately *not* as in Figure 1.4. For instance, if $x = .001$, then $y - \bar{y} \approx .0005$.

We can say that the conditioning of this problem is determined by $g'(0) = \frac{1}{2}$, because $g'(0) \approx \frac{g(x) - g(0)}{x - 0}$ for x small. The problem is well-conditioned because this number is not large.

On the other hand, a function whose derivative wildly changes may not be easily evaluated accurately. A classical example here is the function $g(x) = \tan(x)$. Evaluating it for x near zero does not cause difficulty (the problem being well-conditioned there), but the problem of evaluating the same function for x near $\frac{\pi}{2}$ is ill-conditioned. For instance, setting $x = \frac{\pi}{2} - .001$ and $\bar{x} = \frac{\pi}{2} - .002$ we obtain that $|x - \bar{x}| = .001$ but $|\tan(x) - \tan(\bar{x})| \approx 500$. A look at the derivative, $g'(x) = \frac{1}{\cos^2(x)}$, for x near $\frac{\pi}{2}$ explains why. ∎

Error accumulation

We will have much to say in Chapter 2 about the floating point representation of real numbers and the accumulation of roundoff errors during a calculation. Here let us emphasize that in general it is impossible to prevent *linear* accumulation, meaning the roundoff error may be proportional to n after n elementary operations such as addition or multiplication of two real numbers. However, such an error accumulation is usually acceptable if the linear rate is moderate (i.e., the constant c_0 below is not very large). In contrast, *exponential* growth cannot be tolerated.

Explicitly, if E_n measures the relative error at the nth operation of an algorithm, then

$$E_n \simeq c_0 n E_0 \text{ for some constant } c_0 \text{ represents linear growth, and}$$
$$E_n \simeq c_1^n E_0 \text{ for some constant } c_1 > 1 \text{ represents exponential growth.}$$

An algorithm exhibiting relative exponential error growth is *unstable*. Such algorithms must be avoided!

Example 1.6. Consider evaluating the integrals

$$y_n = \int_0^1 \frac{x^n}{x+10} dx$$

for $n = 1, 2, \ldots, 30$.

Observe at first that analytically

$$y_n + 10 y_{n-1} = \int_0^1 \frac{x^n + 10 x^{n-1}}{x+10} dx = \int_0^1 x^{n-1} dx = \frac{1}{n}.$$

Also

$$y_0 = \int_0^1 \frac{1}{x+10} dx = \ln(11) - \ln(10).$$

An algorithm which may come to mind is therefore as follows:

1. Evaluate $y_0 = \ln(11) - \ln(10)$.

2. For $n = 1, \ldots, 30$, evaluate
$$y_n = \frac{1}{n} - 10\, y_{n-1}.$$

Note that applying the above recursion formula would give *exact* values if roundoff errors were not present.

However, this algorithm is in fact unstable, as the magnitude of roundoff errors gets multiplied by 10 each time the recursion is applied. Thus, there is exponential error growth with $c_1 = 10$. In MATLAB (which automatically employs the IEEE double precision floating point arithmetic; see Section 2.4) we obtain $y_0 = 9.5310e - 02$, $y_{18} = -9.1694e + 01$, $y_{19} = 9.1694e + 02, \ldots, y_{30} = -9.1694e + 13$. It is not difficult to see that the exact values all satisfy $0 < y_n < 1$, and hence the computed solution, at least for $n \geq 18$, is meaningless! ∎

Thankfully, such extreme instances of instability as illustrated in Example 1.6 will not occur in any of the algorithms developed in this text from here on.

Specific exercises for this section: Exercises 4–5.

1.4 Exercises

0. **Review questions**

   (a) What is the difference, according to Section 1.1, between scientific computing and numerical analysis?

   (b) Give a simple example where relative error is a more suitable measure than absolute error, and another example where the absolute error measure is more suitable.

   (c) State a major difference between the nature of roundoff errors and discretization errors.

   (d) Explain briefly why accumulation of roundoff errors is inevitable when arithmetic operations are performed in a floating point system. Under which circumstances is it tolerable in numerical computations?

   (e) Explain the differences between accuracy, efficiency, and robustness as criteria for evaluating an algorithm.

   (f) Show that nested evaluation of a polynomial of degree n requires only $2n$ elementary operations and hence has $\mathcal{O}(n)$ complexity.

   (g) Distinguish between problem conditioning and algorithm stability.

1. Carry out calculations similar to those of Example 1.3 for approximating the derivative of the function $f(x) = e^{-2x}$ evaluated at $x_0 = 0.5$. Observe similarities and differences by comparing your graph against that in Figure 1.3.

2. Carry out derivation and calculations analogous to those in Example 1.2, using the expression
$$\frac{f(x_0 + h) - f(x_0 - h)}{2h}$$
for approximating the first derivative $f'(x_0)$. Show that the error is $\mathcal{O}(h^2)$. More precisely, the leading term of the error is $-\frac{h^2}{6} f'''(x_0)$ when $f'''(x_0) \neq 0$.

3. Carry out similar calculations to those of Example 1.3 using the approximation from Exercise 2. Observe similarities and differences by comparing your graph against that in Figure 1.3.

4. Following Example 1.5, assess the conditioning of the problem of evaluating

$$g(x) = \tanh(cx) = \frac{\exp(cx) - \exp(-cx)}{\exp(cx) + \exp(-cx)}$$

near $x = 0$ as the positive parameter c grows.

5. Consider the problem presented in Example 1.6. There we saw a numerically unstable procedure for carrying out the task.

   (a) Derive a formula for approximately computing these integrals based on evaluating y_{n-1} given y_n.

   (b) Show that for any given value $\varepsilon > 0$ and positive integer n_0, there exists an integer $n_1 \geq n_0$ such that taking $y_{n_1} = 0$ as a starting value will produce integral evaluations y_n with an absolute error smaller than ε for all $0 < n \leq n_0$.

   (c) Explain why your algorithm is stable.

   (d) Write a MATLAB function that computes the value of y_{20} within an absolute error of at most 10^{-5}. Explain how you choose n_1 in this case.

1.5 Additional notes

The proliferation in the early years of the present century of academic centers, institutes, and special programs for scientific computing reflects the coming of age of the discipline in terms of experiments, theory, and simulation. Fast, available computing hardware, powerful programming environments, and the availability of numerical algorithms and software libraries all make scientific computing an indispensable tool in modern science and engineering. This tool allows for an interplay with experiment and theory. On the one hand, improvements in computing power allow for experimentation and computations in a scale that could not have been imagined just a few years ago. On the other hand, the great progress in the theoretical understanding of numerical methods, and the availability of convenient computational testing environments, have given scientists and practitioners the ability to make predictions of and conclusions about huge-scale phenomena that today's computers are still not (and may never be) powerful enough to handle.

A potentially surprising amount of attention has been given throughout the years to the definitions of scientific computing and numerical analysis. An interesting account of the evolution of this seemingly esoteric but nevertheless important issue can be found in Trefethen and Bau [70].

The concept of problem conditioning is both fundamental and tricky to discuss so early in the game. If you feel a bit lost somewhere around Figures 1.4 and 1.5, then rest assured that these concepts eventually will become clearer as we gain experience, particularly in the more specific contexts of Sections 5.8, 8.2, and 14.4.

Many computer science theory books deal extensively with $\mathcal{O}$ and Θ notations and complexity issues. One widely used such book is Graham, Knuth, and Patashnik [31].

There are many printed books and Internet introductions to MATLAB. Check out wikipedia and what's in Mathworks. One helpful survey of some of those can be found at http://www.cs.ubc.ca/~mitchell/matlabResources.html .

Chapter 2
Roundoff Errors

As observed in Chapter 1, various errors may arise in the process of calculating an approximate solution for a mathematical model. In this chapter we introduce and discuss in detail the most fundamental source of imperfection in numerical computing: roundoff errors. Such errors arise due to the intrinsic limitation of the finite precision representation of numbers (except for a restricted set of integers) in computers.

Different audiences may well require different levels of depth and detail in the present topic. We therefore start our discussion in Section 2.1 with the bare bones: a collection of essential facts related to floating point systems and roundoff errors that may be particularly useful for those wishing to concentrate on the last seven chapters of this text.

> **Note:** If you do not require detailed knowledge of roundoff errors and their propagation during a computation, not even the essentials of Section 2.1, then you may skip this chapter (not recommended), at least upon first reading. What you must accept, then, is the notion that each number representation and each elementary operation (such as $+$ or $*$) in standard floating point arithmetic introduces a small, random relative error: up to about 10^{-16} in today's standard floating point systems.

In Section 2.2 we get technical and dive into the gory details of floating point systems and floating point arithmetic. Several issues only mentioned in Section 2.1 are explained here.

The small representation errors as well as errors that arise upon carrying out elementary operations such as addition and multiplication are typically harmless unless they accumulate or get magnified in an unfortunate way during the course of a possibly long calculation. We discuss roundoff error accumulation as well as ways to avoid or reduce such effects in Section 2.3.

Finally, in Section 2.4, the IEEE standard for floating point arithmetic, which is implemented in any nonspecialized hardware, is briefly described.

2.1 The essentials

This section summarizes what we believe all our students should know as a minimum about floating point arithmetic. Let us start with a motivating example.

Example 2.1. Scientists and engineers often wish to believe that the numerical results of a computer calculation, especially those obtained as output of a software package, contain no error—at least not a significant or intolerable one. But careless numerical computing does occasionally lead to trouble.

> **Note:** The word "essential" is not synonymous with "easy." If you find some part of the description below too terse for comfort, then please refer to the relevant section in this chapter for more motivation, detail, and explanation.

One of the more spectacular disasters was the Patriot missile failure in Dhahran, Saudi Arabia, on February 25, 1991, which resulted in 28 deaths. This failure was ultimately traced to poor handling of roundoff errors in the missile's software. The webpage http://www.ima.umn.edu/~arnold/disasters/patriot.html contains the details of this story. For a large collection of software bugs, see http://wwwzenger.informatik.tu-muenchen.de/persons/huckle/bugse.html. ∎

Computer representation of real numbers

Computer memory has a finite capacity. This obvious fact has far-reaching implications on the representation of real numbers, which in general *do not* have a finite uniform representation. How should we then represent a real number on the computer in a way that can be sensibly implemented in hardware?

Any real number $x \in \mathcal{R}$ is accurately representable by an infinite sequence of digits.[4] Thus, we can write
$$x = \pm(1.d_1 d_2 d_3 \cdots d_{t-1} d_t d_{t+1} \cdots) \times 2^e,$$
where e is an integer *exponent*. The (possibly infinite) set of binary digits $\{d_i\}$ each have a value 0 or 1. The decimal value of this *mantissa* is
$$1.d_1 d_2 d_3 \cdots = 1 + \frac{d_1}{2} + \frac{d_2}{2^2} + \frac{d_3}{2^3} + \cdots.$$

For instance, the binary representation $x = -(1.10100 \cdots) \times 2^1$ has in decimal the value
$$x = -(1 + 1/2 + 1/8) \times 2 = -3.25.$$

Of course, it should be clear that the choice of a binary representation is just one of many possibilities, indeed a convenient choice when it comes to computers. To represent any real number on the computer, we associate to x a *floating point representation* fl(x) of a form similar to that of x but with only t digits, so
$$\mathrm{fl}(x) = \mathrm{sign}(x) \times (1.\tilde{d}_1 \tilde{d}_2 \tilde{d}_3 \cdots \tilde{d}_{t-1} \tilde{d}_t) \times 2^e$$
for some t that is fixed in advance and binary digits $\tilde{d}_i$ that relate to d_i in a manner that will soon be specified. Storing this fraction in memory thus requires t bits. The exponent e must also be stored in a fixed number of bits and therefore must be bounded, say, between a lower bound L and an upper bound U. Further details are given in Section 2.2, which we hope will provide you with much of what you'd like to know.

Rounding unit and standard floating point system

Without any further details it should already be clear that representing x by fl(x) necessarily causes an error. A central question is how accurate the floating point representation of the real numbers is.

---
[4]It is known from calculus that the set of all rational numbers in a given real interval is dense in that interval. This means that any number in the interval, rational or not, can be approached to arbitrary accuracy by a sequence of rational numbers.

2.1. The essentials

Specifically, we ask how large the relative error in such a representation, defined by

$$\frac{|\text{fl}(x) - x|}{|x|},$$

can be.

Modern floating point systems, such as the celebrated IEEE standard described in detail in Section 2.4, guarantee that this relative error is bounded by

$$\eta = \frac{1}{2} \times 2^{-t}.$$

The important quantity η, which will be more generally defined in Section 2.2, has been called **rounding unit**, **machine precision**, **machine epsilon**, and more. We will use the term rounding unit throughout.

The usual floating point word in the standard floating point system, which is what MATLAB uses by default, has 64 bits. Of these, 52 bits are devoted to the mantissa (or fraction) while the rest store the sign and the exponent. Hence the rounding unit is

$$\eta = 2^{-53} \approx 1.1 \times 10^{-16}.$$

This is called *double precision*; see the schematics in Figure 2.1.

Figure 2.1. *A double word (64 bits) in the standard floating point system. The blue bit is for sign, the magenta bits store the exponent, and the green bits are for the fraction.*

If the latter name makes you feel a bit like starting house construction from the second floor, then rest assured that there is also a *single precision* word, occupying 32 bits. This one obviously has a much smaller number of digits t, hence a significantly larger rounding unit η. We will not use single precision for calculations anywhere in this book except for Examples 2.2 and 14.6, and neither should you.

Roundoff error accumulation

Even if number representations were exact in our floating point system, *arithmetic operations* involving such numbers introduce roundoff errors. These errors can be quite large in the relative sense, unless *guard digits* are used. These are extra digits that are used in interim calculations. The IEEE standard requires **exact rounding**, which yields that the relative error in each arithmetic operation is also bounded by η.

Given the above soothing words about errors remaining small after representing a number and performing an arithmetic operation, can we really put our minds at ease and count on a long and intense calculation to be as accurate as we want it to be?

Not quite! We have already seen in Example 1.3 that unpleasant surprises may arise. Let us mention a few potential traps. The fuller version is in Section 2.3.

Careless computations can sometimes result in division by 0 or another form of undefined numerical result. The corresponding variable name is then assigned the infamous designation NaN. This is a combination of letters that stands for "not a number," which naturally one dreads to see in one's own calculations, but it allows software to detect problematic situations, such as an attempt to divide 0 by 0, and do something graceful instead of just halting. We have a hunch that you will inadvertently encounter a few NaN's before you finish implementing all the algorithms in this book.

There is also a potential for an *overflow*, which occurs when a number is larger than the largest that can be represented in the floating point system. Occasionally this can be avoided, as in Example 2.9 and other examples in Section 2.3.

We have already mentioned in Section 1.3 that a roundoff error accumulation that grows *linearly* with the number of operations in a calculation is inevitable. Our calculations will never be so long that this sort of error would surface as a practical issue. Still, there are more pitfalls to watch out for. One painful type of error magnification is a *cancellation error*, which occurs when two nearly equal numbers are subtracted from one another. There are several examples of this in Section 2.3. Furthermore, the discussion in this chapter suggests that such errors may consistently arise in practice.

The rough appearance of roundoff error

Consider a smooth function g, sampled at some t and at $t+h$ for a small (i.e., near 0) value h. Continuity then implies that the values $g(t)$ and $g(t+h)$ are close to each other. But the rounding errors in the machine representation of these two numbers are unrelated to each other: they are random for all we know! These rounding errors are both small (that's what η is for), but even their signs may in fact differ. So when subtracting $g(t+h) - g(t)$, for instance, on our way to estimate the derivative of g at t, the relative roundoff error becomes significantly larger as cancellation error naturally arises. This is apparent already in Figure 1.3.

Let us take a further look at the seemingly unstructured behavior of roundoff error as a smooth function is being sampled.

Example 2.2. We evaluate $g(t) = e^{-t}(\sin(2\pi t) + 2)$ at 501 equidistant points between 0 and 1, using the usual double precision as well as *single precision* and plotting the differences. This essentially gives the rounding error in the less accurate single precision evaluations. The following MATLAB instructions do the job:

```
t = 0:.002:1;
tt = exp(-t) .* (sin(2*pi*t)+2);
rt = single(tt);
round_err = (tt - rt) ./tt ;
plot (t,round_err,'b-');
title ('error in sampling exp(-t)(sin(2\pi t)+2) single precision')
xlabel('t')
ylabel('roundoff error')

% relative error should be about eta = eps(single)/2
rel_round_err = max(abs(round_err)) / (eps('single')/2)
```

Thus, the definition of the array values `tt` is automatically implemented in double precision, while the instruction `single` when defining the array `rt` records the corresponding values in single precision.

The resulting plot is depicted in Figure 2.2. Note the disorderly, "high frequency" oscillation of the roundoff error. This is in marked contrast to discretization errors, which are usually "smooth," as we have seen in Example 1.2. (Recall the straight line drop of the error in Figure 1.3 for relatively large h, which is where the discretization error dominates.)

The output of this program indicates that, as expected, the relative error is at about the rounding unit level. The latter is obtained by the (admittedly unappealing) function call

2.2. Floating point systems

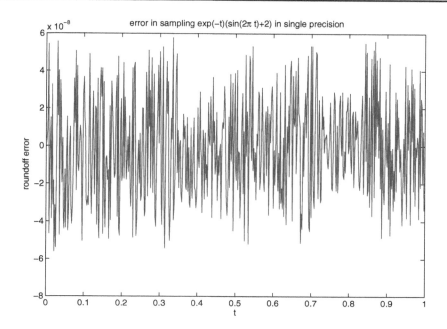

Figure 2.2. *The "almost random" nature of roundoff errors.*

eps('single')/2, which yields roughly the value $\eta_{\text{single}} = 6\text{e-}8$. The last line in the script produces approximately the value 0.97, which is indeed close to 1. ■

Specific exercises for this section: Exercises 1–2.

2.2 Floating point systems

> **Note:** Here is where we get more detailed and technical in the present chapter. Note that we are not assuming the IEEE standard before discussing it in Section 2.4.

A floating point system can be characterized by four values (β, t, L, U), where

$\beta =$ base of the number system;
$t =$ precision (# of digits);
$L =$ lower bound on exponent e;
$U =$ upper bound on exponent e.

Generalizing the binary representation in Section 2.1, we write

$$\text{fl}(x) = \pm \left(\frac{\tilde{d}_0}{\beta^0} + \frac{\tilde{d}_1}{\beta^1} + \cdots + \frac{\tilde{d}_{t-1}}{\beta^{t-1}} \right) \times \beta^e,$$

where β is an integer larger than 1 referred to as the *base* and $\tilde{d}_i$ are integer digits in the range $0 \leq \tilde{d}_i \leq \beta - 1$. The number fl($x$) is an approximation of x. To ensure uniqueness of this representation, it is normalized to satisfy $\tilde{d}_0 \neq 0$ by adjusting the exponent e so that leading zeros are dropped. Note that unless $\beta = 2$ this does not fix the value of $\tilde{d}_0$, which therefore must be stored, too. This is why the last stored digit has index $t-1$ and not t as in Section 2.1. In addition, e must be in the range $L \le e \le U$.

Chopping and rounding

To store $x = \pm(d_0.d_1d_2d_3 \cdots d_{t-1}d_t d_{t+1} \cdots) \times \beta^e$ using only t digits, it is possible to use one of a number of strategies. The two basic ones are

- *chopping:* ignore digits $d_t, d_{t+1}, d_{t+2}, d_{t+3} \ldots$, yielding $\tilde{d}_i = d_i$ and
$$\text{fl}(x) = \pm d_0.d_1d_2d_3 \cdots d_{t-1} \times \beta^e;$$

- *rounding:* consult d_t to determine the approximation
$$\text{fl}(x) = \begin{cases} \pm d_0.d_1d_2d_3 \cdots d_{t-1} \times \beta^e, & d_t < \beta/2, \\ \pm \left(d_0.d_1d_2d_3 \cdots d_{t-1} + \beta^{1-t}\right) \times \beta^e, & d_t > \beta/2. \end{cases}$$

In case of a tie ($d_t = \beta/2$), round to the nearest even number.

Example 2.3. Here are results of chopping and rounding with $\beta = 10$, $t = 3$:

x	Chopped to 3 digits	Rounded to 3 digits
5.672	5.67	5.67
−5.672	−5.67	−5.67
5.677	5.67	5.68
−5.677	−5.67	−5.68
5.692	5.69	5.69
5.695	5.69	5.70

∎

Example 2.4. Since humans are used to decimal arithmetic, let us set $\beta = 10$. Consider
$$\frac{8}{3} = 2.6666\ldots = \left(\frac{2}{10^0} + \frac{6}{10^1} + \frac{6}{10^2} + \frac{6}{10^3} + \frac{6}{10^4} + \cdots\right) \times 10^0.$$

This number gives an infinite series in base 10, although the digit series has finite length in base 3.

To represent it using $t = 4$, chopping gives
$$\frac{8}{3} \simeq \left(\frac{2}{10^0} + \frac{6}{10^1} + \frac{6}{10^2} + \frac{6}{10^3}\right) \times 10^0 = 2.666 \times 10^0.$$

On the other hand, with rounding note that $d_t = 6 > 5$, so $\beta^{1-t} = 10^{-3}$ is added to the number before chopping, which gives 2.667×10^0.

This floating point representation is not unique; for instance, we have
$$2.666 \times 10^0 = 0.2666 \times 10^1.$$

Therefore, we **normalize** the representation by insisting that $d_0 \ne 0$, so
$$1 \le d_0 \le 9, \quad 0 \le d_i \le 9, \quad i = 1, \ldots, t-1,$$
which eliminates any ambiguity. ∎

2.2. Floating point systems

The number 0 cannot be represented in a normalized fashion. It and the limits $\pm\infty$ are represented as special combinations of bits, according to an agreed upon convention for the given floating point system.

Example 2.5. Consider a (toy) decimal floating point system with $t = 4$, $U = 1$, and $L = -2$. Thus, the decimal number 2.666 is precisely representable because it has four digits in its mantissa and $L < e < U$.

The largest number here is $99.99 \lessapprox 10^2 = 100$, the smallest is $-99.99 \gtrapprox -100$, and the smallest positive number is $10^{-2} = 0.01$.

How many different numbers do we have? The first digit can take on 9 different values, the other three digits 10 values each (because they may be zero, too). Thus, there are $9 \times 10 \times 10 \times 10 = 9{,}000$ different normalized fractions possible. The exponent can be one of $U - L + 1 = 4$ values, so in total there are $4 \times 9{,}000 = 36{,}000$ different positive numbers possible. There is the same total of negative numbers, and then there is the number 0. So, there are 72,001 different numbers in this floating point system. ■

What about the choice of a base? Computer storage is in an integer multiple of bits, hence all computer representations we know of have used bases that are powers of 2. In the 1970s there were architectures with bases 16 and 8. Today, as we have observed in Section 2.1, the standard floating point system uses base $\beta = 2$; see Figure 2.1 and Section 2.4.

The error in floating point representation

The relative error is generally a more meaningful measure than absolute error in floating point representation, because it is independent of a change of exponent. Thus, if in a decimal system (i.e., with $\beta = 10$) we have that $u = 1{,}000{,}000 = 1 \times 10^6$ and $v = \text{fl}(u) = 990{,}000 = 9.9 \times 10^5$, we expect to be able to work with such an approximation as accurately as with $u = 10$ and $v = \text{fl}(u) = 9.9$. This is borne out by the value of the relative error.

Let us denote the floating point representation mapping by $x \mapsto \text{fl}(x)$, and suppose rounding is used. Then the quantity η in the formula on the current page is fundamental as it expresses a bound on the relative error when we represent a number in our prototype floating point system. We have already mentioned in Section 2.1 the *rounding unit* η. Furthermore, the negative of its exponent, $t - 1$ (for the rounding case), is often referred to as the **number of significant digits**.

> **Rounding unit.**
> For a general floating point system (β, t, L, U) the rounding unit is
> $$\eta = \frac{1}{2}\beta^{1-t}.$$

Recall that for the double precision word in the standard floating point system, a bound on the relative error is given by $|x - \text{fl}(x)|/|x| \leq \eta \approx 1.1\text{e-}16$.

Floating point arithmetic

As briefly mentioned in Section 2.1, the IEEE standard requires **exact rounding**, which means that the result of a basic arithmetic operation must be identical to the result obtained if the arithmetic operation was computed exactly and then the result rounded to the nearest floating point

Theorem: Floating Point Representation Error.
Let $x \mapsto \mathrm{fl}(x) = g \times \beta^e$, where $x \neq 0$ and g is the normalized, signed mantissa.
  Then the absolute error committed in using the floating point representation of x is bounded by

$$|x - \mathrm{fl}(x)| \leq \begin{cases} \beta^{1-t} \cdot \beta^e & \text{for chopping,} \\ \frac{1}{2}\beta^{1-t} \cdot \beta^e & \text{for rounding,} \end{cases}$$

whereas the relative error satisfies

$$\frac{|x - \mathrm{fl}(x)|}{|x|} \leq \begin{cases} \beta^{1-t} & \text{for chopping,} \\ \frac{1}{2}\beta^{1-t} & \text{for rounding.} \end{cases}$$

number. With exact rounding, if $\mathrm{fl}(x)$ and $\mathrm{fl}(y)$ are machine numbers, then

$$\mathrm{fl}(\mathrm{fl}(x) \pm \mathrm{fl}(y)) = (\mathrm{fl}(x) \pm \mathrm{fl}(y))(1 + \epsilon_1),$$
$$\mathrm{fl}(\mathrm{fl}(x) \times \mathrm{fl}(y)) = (\mathrm{fl}(x) \times \mathrm{fl}(y))(1 + \epsilon_2),$$
$$\mathrm{fl}(\mathrm{fl}(x) \div \mathrm{fl}(y)) = (\mathrm{fl}(x) \div \mathrm{fl}(y))(1 + \epsilon_3),$$

where $|\epsilon_i| \leq \eta$. Thus, the *relative errors* remain small after each such operation.

Example 2.6. Consider a floating point system with decimal base $\beta = 10$ and four digits, i.e., $t = 4$. Thus, the rounding unit is $\eta = \frac{1}{2} \times 10^{-3}$. Let

$$x = .1103 = 1.103 \times 10^{-1}, \quad y = 9.963 \times 10^{-3}.$$

Subtracting these two numbers, the exact value is $x - y = .100337$. Hence, exact rounding yields .1003. Obviously, the relative error is

$$\frac{|.100337 - .1003|}{.100337} \approx .37 \times 10^{-3} < \eta.$$

However, if we were to subtract these two numbers without guard digits we would obtain $.1103 - .0099 = .1004$. Now the obtained relative error is not below η, because

$$\frac{|.100337 - .1004|}{.100337} \approx .63 \times 10^{-3} > \eta.$$

Thus, guard digits must be used to produce exact rounding. ∎

Example 2.7. Generally, proper rounding yields $\mathrm{fl}(1 + \alpha) = 1$ for any number α that satisfies $|\alpha| \leq \eta$. In particular, the MATLAB commands

```
eta = .5*2^(-52),  beta = (1+eta)-1
```

produce the output eta = 1.1102e-16, beta = 0. Returning to Example 1.3 and to Figure 1.3, we can now explain why the curve of the error is flat for the very, very small values of h. For such values, $\mathrm{fl}(f(x_0 + h)) = \mathrm{fl}(f(x_0))$, so the approximation is precisely zero and the recorded values are those of $\mathrm{fl}(f'(x_0))$, which is independent of h. ∎

2.2. Floating point systems

Spacing of floating point numbers

If you think of how a given floating point system represents the real line you'll find that it has a somewhat uneven nature. Indeed, very large numbers are not represented at all, and the distance between two neighboring, positive floating point values is constant in the relative but not in the absolute sense.

Example 2.8. Let us plot the decimal representation of the numbers in the system specified by $(\beta, t, L, U) = (2, 3, -2, 3)$. The smallest possible number in the mantissa $d_0.d_1d_2$ is 1.00 and the largest possible number is 1.11 in binary, which is equal to $1 + 1/2 + 1/4 = 1.75$. So, we will run in a loop from 1 to 1.75 in increments of $\beta^{1-t} = 2^{-2} = 0.25$. This is the basis for one loop. The exponent runs from -2 to 3, which is what we loop over in the second (nested) loop. The resulting double loop builds up an array with all the positive elements of the system.

Here is a MATLAB script that plots the numbers of the system. Notice the format in the `plot` command.

```
x = [];

% Generate all positive numbers of the system (2,3,-2,3)
for i = 1:0.25:1.75
  for j = -2:3
      x = [x i*2^j];
  end
end

x=[x -x 0];      % Add all negative numbers and 0
x = sort(x);     % Sort
y = zeros(1,length(x));
plot(x,y,'+')
```

The resulting plot is in Figure 2.3. Note that the points are not distributed uniformly along the real line.

The rounding unit for this system is

$$\eta = \frac{1}{2}\beta^{1-t} = \frac{1}{2} \times 2^{1-3} = \frac{1}{8},$$

which is half the spacing between $1 = \beta^0$ and the next number in the system. (Can you see why?) ∎

Most annoying in Figure 2.3 is the fact that the distance between the value 0 and the smallest positive number is significantly *larger* than the distance between that same smallest positive number and the next smallest positive number! A commonly used way to remedy this in modern floating point systems is by introducing next to 0 additional, *subnormal numbers* which are not normalized. We will leave it at that.

MATLAB has a parameter called `eps` that signifies the spacing of floating point numbers. In particular, `eps(1)` gives twice the rounding unit η. Enter `help eps` to learn more about this parameter.

Specific exercises for this section: Exercises 3–8.

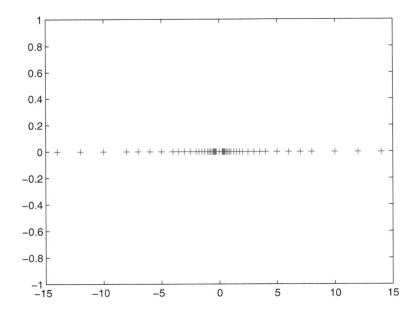

Figure 2.3. *Picture of the floating point system described in Example* 2.8.

2.3 Roundoff error accumulation

Because many operations are being performed in the course of carrying out a numerical algorithm, many small errors unavoidably result. We know that each elementary floating point operation may introduce a small relative error, but how do these errors accumulate?

In general, as we have already mentioned in Chapter 1, error growth that is linear in the number of operations is unavoidable. Yet, there are a few things to watch out for.

- Error magnification:

    1. If x and y differ widely in magnitude, then $x + y$ has a large absolute error.
    2. If $|y| \ll 1$, then x/y may have large relative and absolute errors. Likewise for xy if $|y| \gg 1$.
    3. If $x \simeq y$, then $x - y$ has a large relative error (*cancellation error*).

- Overflow and underflow:

    An **overflow** is obtained when a number is too large to fit into the floating point system in use, i.e., when $e > U$. An **underflow** is obtained when $e < L$. When overflow occurs in the course of a calculation, this is generally fatal. But underflow is nonfatal: the system usually sets the number to 0 and continues. (MATLAB does this, quietly.)

It is also worth mentioning again the unfortunate possibility of running into NaN, described in Section 2.1.

Design of library functions

You may wonder what the fuss is all about, if the representation of a number and basic arithmetic operations cause an error as small as about 10^{-16} in a typical floating point system. But taking this

2.3. Roundoff error accumulation

issue lightly may occasionally cause surprisingly serious damage. In some cases, even if the computation is long and intensive it may be based on or repeatedly use a short algorithm, and a simple change in a formula may greatly improve performance with respect to roundoff error accumulation. For those operations that are carried out repeatedly in a typical calculation, it may well be worth obtaining a result as accurate as possible, even at the cost of a small computational overhead.

Example 2.9. The most common way to measure the size of a vector $\mathbf{x} = (x_1, x_2, \ldots, x_n)^T$ is the Euclidean norm (also called ℓ_2-norm) defined by

$$\|\mathbf{x}\|_2 = \sqrt{x_1^2 + x_2^2 + \cdots + x_n^2}.$$

We discuss vector norms and their properties in Section 4.2. Any standard numerical linear algebra software package has a library function that computes the norm, and often the input vector is very large; see, for instance, Chapter 7 for relevant situations. One of the dangers to avoid in computing the norm is the possibility of overflow. To illustrate this point, let us look at a small vector with just two but widely differing components.

Consider computing $c = \sqrt{a^2 + b^2}$ in a floating point system with four decimal digits and two exponent digits, for $a = 10^{60}$ and $b = 1$. Thus, c is the Euclidean norm of the vector $\mathbf{x} = (10^{60}, 1)^T$. The correct result here is $c = 10^{60}$. But overflow will result during the course of calculation, because squaring a gives 10^{120}, which cannot be represented in this system. Yet, this overflow can easily be avoided if we rescale ahead of time, noting that $c = s\sqrt{(a/s)^2 + (b/s)^2}$ for any $s \neq 0$. Thus, using $s = a = 10^{60}$ gives an underflow when b/s is squared, which is set to zero. This yields the correct answer. (It is important to stress that "correct answer" here refers not necessarily to the "true," precise answer on the real line, but rather to the most accurate answer we can obtain given this particular floating point system.)

The above principle can be generalized to the computation of the norm of a vector of more than two components, simply scaling by the largest one.

Another issue is the order in which the values x_i^2 are summed. Doing this in increasing order of magnitude reduces error accumulation; see, for instance, Exercise 17.

In Exercise 18 you are asked to design a library function for computing the ℓ_2-norm of a vector. ∎

Avoiding cancellation error

Another trouble spot to watch out for is cancellation error, mentioned in Section 2.1. Some insight into the damaging effect of subtracting two nearly equal numbers in a floating point system can be observed by deriving a bound on the error. Suppose $z = x - y$, where $x \approx y$. Then

$$|z - \mathrm{fl}(z)| \leq |x - \mathrm{fl}(x)| + |y - \mathrm{fl}(y)|,$$

from which it follows that the relative error satisfies

$$\frac{|z - \mathrm{fl}(z)|}{|z|} \leq \frac{|x - \mathrm{fl}(x)| + |y - \mathrm{fl}(y)|}{|x - y|}.$$

The numerator of the bound could be tight since it merely depends on how well the floating point system can represent the numbers x and y. But the denominator is very close to zero if $x \approx y$ regardless of how well the floating point system can work for these numbers, and so the relative error in z could become large.

Example 2.10. The roots of the quadratic equation
$$x^2 - 2bx + c = 0$$
with $b^2 > c$ are given by
$$x_{1,2} = b \pm \sqrt{b^2 - c}.$$
The following simple script gives us the roots according to the formula:

```
x1 = b + sqrt(b^2-c);
x2 = b - sqrt(b^2-c);
```

Unfortunately, if b^2 is significantly larger than c, to the point that $\sqrt{b^2-c} \approx |b|$, then one of the two calculated roots is bound to be approximately zero and inaccurate. (There is no problem with the other root, which would be approximately equal to $2b$ in this case.)

To our aid comes the formula $x_1 x_2 = c$. We can then avoid the cancellation error by using a modified algorithm, given by the script

```
if b > 0
  x1 = b + sqrt(b^2-c);
  x2 = c / x1;
else
  x2 = b - sqrt(b^2-c);
  x1 = c / x2;
end
```

The algorithm can be further improved by also taking into consideration the possibility of an overflow. Thus, if $b^2 \gg c$, we replace $\sqrt{b^2-c}$ by $|b|\sqrt{1 - \frac{c}{b^2}}$, in a way similar to the technique described in Example 2.9. In Exercise 16 you are asked to design and implement a robust quadratic equation solver. ■

Cancellation error deserves special attention because it often appears in practice in an identifiable way. For instance, recall Example 1.2. If we approximate a derivative of a smooth (differentiable) function $f(x)$ at a point $x = x_0$ by the difference of two neighboring values of f divided by the difference of the arguments with a small step h, e.g., by
$$f'(x_0) \approx \frac{f(x_0 + h) - f(x_0)}{h},$$
then there is a cancellation error in the numerator, which is then magnified by the denominator. Recall also the discussion in Section 2.1, just before Example 2.2.

When cancellation errors appear in an identifiable way, they can often be avoided by a simple modification in the algorithm. An instance is illustrated in Exercise 2. Here is another one.

Example 2.11. Suppose we wish to compute $y = \sqrt{x+1} - \sqrt{x}$ for $x = 100{,}000$ in a five-digit decimal arithmetic. Clearly, the number 100,001 cannot be represented in this floating point system exactly, and its representation in the system (when either chopping or rounding is used) is 100,000. In other words, for this value of x in this floating point system, we have $x + 1 = x$. Thus, naively computing $\sqrt{x+1} - \sqrt{x}$ results in the value 0.

We can do much better if we use the identity
$$\frac{(\sqrt{x+1} - \sqrt{x})(\sqrt{x+1} + \sqrt{x})}{(\sqrt{x+1} + \sqrt{x})} = \frac{1}{\sqrt{x+1} + \sqrt{x}}.$$

Applying this formula (i.e., computing the right-hand expression in 5-digit decimal arithmetic) yields 1.5811×10^{-3}, which happens to be the correct value to 5 decimal digits. ∎

There are also other ways to avoid cancellation error, and one popular technique is the use of a Taylor expansion (page 5).

Example 2.12. Suppose we wish to design a library function for computing

$$y = \sinh(x) = \frac{1}{2}(e^x - e^{-x}).$$

Think, for example, of having a scientific calculator: we want to be sure that it gives an accurate result for any argument x.

While the above formula can be straightforwardly applied (assuming we have a decent library function for computing exponents), the going may get a little tougher for values of x near 0. Directly using the formula for computing y may be prone to severe cancellation errors, due to the subtraction of two quantities that are approximately equal to 1. On the other hand, using the Taylor expansion

$$\sinh(x) = x + \frac{x^3}{6} + \frac{\xi^5}{120}$$

for some ξ satisfying $|\xi| \leq |x|$ may prove useful. The simple cubic expression $x + \frac{x^3}{6}$ should give an effective approximation if $|x|$ is sufficiently small, as the discretization error can be bounded by $\frac{|x|^5}{120}$ and the roundoff error is no longer an issue.

Employing 5-digit decimal arithmetic for our cubic approximation, we compute $\sinh(0.1) = 0.10017$ and $\sinh(0.01) = 0.01$. (Believe us! Or otherwise please feel free to verify.) These are the "exact" values in this floating point system. On the other hand, using the formula that involves the exponential functions, which is the *exact* formula and would produce the exact result had we not had roundoff errors, we obtain 0.10018 and 0.010025 for these two values of x, respectively, so there are pronounced errors.

For $x = 0.01$ it is in fact sufficient to use only one term of the Taylor expansion, i.e., approximate $\sinh(x)$ by x in this toy floating point system. ∎

Specific exercises for this section: Exercises 9–19.

2.4 The IEEE standard

During the dawn of the modern computing era there was chaos. Then, in 1985, came the IEEE standard, slightly updated in 2008. Today, except for special circumstances such as a calculator or a dedicated computer on a special-purpose device, every software designer and hardware manufacturer follows this standard. Thus, the output of one's program no longer varies upon switching laptops or computing environments using the same language. Here we describe several of the nuts and bolts (from among the less greasy ones) of this standard floating point system.

Recall from Section 2.1 that the base is $\beta = 2$. In decimal value we have

$$\text{fl}(x) = \pm \left(1 + \frac{\tilde{d}_1}{2} + \frac{\tilde{d}_2}{4} + \cdots + \frac{\tilde{d}_t}{2^t}\right) \times 2^e,$$

where $\tilde{d}_i$ are binary digits each requiring one bit for storing.

The standard and the single precision floating point words

The standard floating point word used exclusively for all calculations in this book except Examples 2.2 and 14.7 requires 64 bits of storage and is called *double precision* or *long word*. This is the MATLAB default. Of these 64 bits, one is allocated for sign s (the number is negative if and only if $s = 1$), 11 for the exponent, and $t = 52$ for the fraction:

Double precision (64-bit word)

$s = \pm$	$b = $ 11-bit exponent	$f = $ 52-bit fraction

$\beta = 2, t = 52, L = -1022, U = 1023$

Since the fraction f contains t digits the precision is $52 + 1 = 53$. Thus, a given bit pattern in the last 52 bits of a long word is interpreted as a sequence of binary digits $d_1 d_2 \cdots d_{52}$. A given bit pattern in the exponent part of the long word is interpreted as a positive integer b in binary representation that yields the exponent e upon shifting by U, i.e., $e = b - 1023$.

Example 2.13. The sequence of 64 binary digits

0100000000111111010000000000000000000000000000000000000000000000

considered as a double precision word can be interpreted as follows:

- Based on the first digit 0 the sign is positive by convention.

- The next 11 digits,

  10000000111,

  form the exponent: in decimal, $b = 1 \times 2^{10} + 1 \times 2^2 + 1 \times 2^1 + 1 \times 2^0 = 1031$. So in decimal, $e = b - 1023 = 1031 - 1023 = 8$.

- The next 52 bits,

  1110100000000000000000000000000000000000000000000000,

  form the decimal fraction: $f = \frac{1}{2} + \frac{1}{4} + \frac{1}{8} + \frac{1}{32} = 0.90625$.

- The number in decimal is therefore

  $$s \cdot (1 + f) \times 2^e = 1.90625 \times 2^8 = 488. \quad \blacksquare$$

There is also a *single precision* arrangement of 32 bits, as follows:

Single precision (32-bit word)

$s = \pm$	$b = $ 8-bit exponent	$f = $ 23-bit fraction

$\beta = 2, t = 23, L = -126, U = 127$

Storing special values

Upon carefully examining the possible range of exponents you will notice that it is not fully utilized. For double precision $2^{11} = 2048$ different exponents could be distinguished, and for single precision $2^8 = 256$ are available. But only 2046 and 254, respectively, are used in practice. This is because

2.4. The IEEE standard

the IEEE standard reserves the endpoints of the exponent range for special purposes. The largest number precisely representable in a standard long word is therefore

$$\left[1+\sum_{i=1}^{52}\left(\frac{1}{2}\right)^i\right] \times 2^{1023} = (2-2^{-52}) \times 2^{1023} \approx 2^{1024} \approx 10^{308},$$

and the smallest positive number is

$$[1+0] \times 2^{-1022} = 2^{-1022} \approx 2.2 \times 10^{-308}.$$

How are 0 and ∞ stored? Here is where the endpoints of the exponent are used, and the conventions are simple and straightforward:

- For 0, set $b=0$, $f=0$, with s arbitrary; i.e., the minimal positive value representable in the system is considered 0.
- For $\pm\infty$, set $b = 1\cdots 1$, $f = 0$.
- The pattern $b = 1\cdots 1$, $f \neq 0$ is by convention NaN.

In single precision, or short word, the largest number precisely representable is

$$\left[1+\sum_{i=1}^{23}\left(\frac{1}{2}\right)^i\right] \times 2^{127} = (2-2^{-23}) \times 2^{127} \approx 2^{128} \approx 3.4 \times 10^{38},$$

and the smallest positive number is

$$[1+0] \times 2^{-126} = 2^{-126} \approx 1.2 \times 10^{-38}.$$

Rounding unit

The formulas for the machine precision or rounding unit η were introduced on page 19 in Section 2.1. Note again the shift from $t-1$ to t in the power of the base as compared to the general formula given on page 23.

Using the word arrangement for single and double precision, η is therefore calculated as follows:

- For single precision, $\eta = \frac{1}{2} \cdot \beta^{-t} = \frac{1}{2} \cdot 2^{-23} \approx 6.0 \times 10^{-8}$ (so, there are 23 significant binary digits, or about 7 decimal digits).
- For double precision, $\eta = \frac{1}{2} \cdot \beta^{-t} = \frac{1}{2} \cdot 2^{-52} \approx 1.1 \times 10^{-16}$ (so, there are 52 significant binary digits, or about 16 decimal digits).

Typically, single and double precision floating point systems as described above are implemented in hardware. There is also quadruple precision (128 bits), often implemented in software and thus considerably slower, for applications that require very high precision (e.g., in semiconductor simulation, numerical relativity and astronomical calculations).

The fundamentally important *exact rounding*, mentioned in both Sections 2.1 and 2.2, has a rather lengthy definition for its implementation, which stands in contrast to the cleanly stated requirement of its result. We will not dive deeper into this.

Specific exercises for this section: Exercises 20–21.

2.5 Exercises

0. **Review questions**

    (a) What is a normalized floating point number and what is the purpose of normalization?

    (b) A general floating point system is characterized by four values (β, t, L, U). Explain in a few brief sentences the meaning and importance of each of these parameters.

    (c) Write down the floating point representation of a given real number x in a decimal system with $t = 4$, using (i) chopping and (ii) rounding.

    (d) Define rounding unit (or machine precision) and explain its importance.

    (e) Define overflow and underflow. Why is the former considered more damaging than the latter?

    (f) What is a cancellation error? Give an example of an application where it arises in a natural way.

    (g) What is the rounding unit for base $\beta = 2$ and $t = 52$ digits?

    (h) Under what circumstances could nonnormalized floating point numbers be desirable?

    (i) Explain the storage scheme for single precision and double precision numbers in the IEEE standard.

1. The fraction in a single precision word has 23 bits (alas, *less* than half the length of the double precision word).

    Show that the corresponding rounding unit is approximately 6×10^{-8}.

2. The function $f_1(x_0, h) = \sin(x_0 + h) - \sin(x_0)$ can be transformed into another form, $f_2(x_0, h)$, using the trigonometric formula

    $$\sin(\phi) - \sin(\psi) = 2\cos\left(\frac{\phi + \psi}{2}\right)\sin\left(\frac{\phi - \psi}{2}\right).$$

    Thus, f_1 and f_2 have the same values, in exact arithmetic, for any given argument values x_0 and h.

    (a) Derive $f_2(x_0, h)$.

    (b) Suggest a formula that avoids cancellation errors for computing the approximation $(f(x_0 + h) - f(x_0))/h$ to the derivative of $f(x) = \sin(x)$ at $x = x_0$. Write a MATLAB program that implements your formula and computes an approximation of $f'(1.2)$, for $h = $ 1e-20, 1e-19, ..., 1.

    (c) Explain the difference in accuracy between your results and the results reported in Example 1.3.

3. (a) How many distinct positive numbers can be represented in a floating point system using base $\beta = 10$, precision $t = 2$ and exponent range $L = -9$, $U = 10$?

    (Assume normalized fractions and don't worry about underflow.)

    (b) How many normalized numbers are represented by the floating point system (β, t, L, U)? Provide a formula in terms of these parameters.

4. Suppose a computer company is developing a new floating point system for use with their machines. They need your help in answering a few questions regarding their system. Following the terminology of Section 2.2, the company's floating point system is specified by (β, t, L, U). Assume the following:

2.5. Exercises

- All floating point values are normalized (except the floating point representation of zero).
- All digits in the mantissa (i.e., fraction) of a floating point value are explicitly stored.
- The number 0 is represented by a float with a mantissa and an exponent of zeros. (Don't worry about special bit patterns for $\pm\infty$ and NaN.)

Here is your part:

(a) How many different nonnegative floating point values can be represented by this floating point system?

(b) Same question for the actual choice $(\beta, t, L, U) = (8, 5, -100, 100)$ (in decimal) which the company is contemplating in particular.

(c) What is the approximate value (in decimal) of the largest and smallest positive numbers that can be represented by this floating point system?

(d) What is the rounding unit?

5. (a) The number $\frac{8}{7} = 1.14285714285714\ldots$ obviously has no exact representation in any decimal floating point system ($\beta = 10$) with finite precision t. Is there a finite floating point system (i.e., some finite integer base β and precision t) in which this number does have an exact representation? If yes, then describe such a system.

(b) Answer the same question for the irrational number π.

6. Write a MATLAB program that receives as input a number x and a parameter n and returns x rounded to n decimal digits. Write your program so that it can handle an array as input, returning an array of the same size in this case.

Use your program to generate numbers for Example 2.2, demonstrating the phenomenon depicted there without use of single precision.

7. Prove the Floating Point Representation Error Theorem on page 24.

8. Rewrite the script of Example 2.8 without any use of loops, using vectorized operations instead.

9. Suggest a way to determine approximately the rounding unit of your calculator. State the type of calculator you have and the rounding unit you have come up with. If you do not have a calculator, write a short MATLAB script to show that your algorithm works well on the standard IEEE floating point system.

10. The function $f_1(x, \delta) = \cos(x + \delta) - \cos(x)$ can be transformed into another form, $f_2(x, \delta)$, using the trigonometric formula

$$\cos(\phi) - \cos(\psi) = -2\sin\left(\frac{\phi + \psi}{2}\right)\sin\left(\frac{\phi - \psi}{2}\right).$$

Thus, f_1 and f_2 have the same values, in exact arithmetic, for any given argument values x and δ.

(a) Show that, analytically, $f_1(x, \delta)/\delta$ or $f_2(x, \delta)/\delta$ are effective approximations of the function $-\sin(x)$ for δ sufficiently small.

(b) Derive $f_2(x, \delta)$.

(c) Write a MATLAB script which will calculate $g_1(x, \delta) = f_1(x, \delta)/\delta + \sin(x)$ and $g_2(x, \delta) = f_2(x, \delta)/\delta + \sin(x)$ for $x = 3$ and $\delta = 1.\text{e-}11$.

(d) Explain the difference in the results of the two calculations.

11. (a) Show that
$$\ln\left(x - \sqrt{x^2 - 1}\right) = -\ln\left(x + \sqrt{x^2 - 1}\right).$$

   (b) Which of the two formulas is more suitable for numerical computation? Explain why, and provide a numerical example in which the difference in accuracy is evident.

12. For the following expressions, state the numerical difficulties that may occur, and rewrite the formulas in a way that is more suitable for numerical computation:

   (a) $\sqrt{x + \frac{1}{x}} - \sqrt{x - \frac{1}{x}}$, where $x \gg 1$.

   (b) $\sqrt{\frac{1}{a^2} + \frac{1}{b^2}}$, where $a \approx 0$ and $b \approx 1$.

13. Consider the linear system
$$\begin{pmatrix} a & b \\ b & a \end{pmatrix} \begin{pmatrix} x \\ y \end{pmatrix} = \begin{pmatrix} 1 \\ 0 \end{pmatrix}$$

   with $a, b > 0$; $a \neq b$.

   (a) If $a \approx b$, what is the numerical difficulty in solving this linear system?

   (b) Suggest a numerically stable formula for computing $z = x + y$ given a and b.

   (c) Determine whether the following statement is true or false, and explain why:

   "When $a \approx b$, the problem of solving the linear system is ill-conditioned but the problem of computing $x + y$ is not ill-conditioned."

14. Consider the approximation to the first derivative
$$f'(x) \approx \frac{f(x+h) - f(x)}{h}.$$

   The *truncation* (or *discretization*) error for this formula is $\mathcal{O}(h)$. Suppose that the absolute error in evaluating the function f is bounded by ε and let us ignore the errors generated in basic arithmetic operations.

   (a) Show that the total computational error (truncation and rounding combined) is bounded by
$$\frac{Mh}{2} + \frac{2\varepsilon}{h},$$
   where M is a bound on $|f''(x)|$.

   (b) What is the value of h for which the above bound is minimized?

   (c) The rounding unit we employ is approximately equal to 10^{-16}. Use this to explain the behavior of the graph in Example 1.3. Make sure to explain the shape of the graph as well as the value where the apparent minimum is attained.

2.5. Exercises

(d) It is not difficult to show, using Taylor expansions, that $f'(x)$ can be approximated more accurately (in terms of truncation error) by

$$f'(x) \approx \frac{f(x+h) - f(x-h)}{2h}.$$

For this approximation, the truncation error is $\mathcal{O}(h^2)$. Generate a graph similar to Figure 1.3 (please generate only the solid line) for the same function and the same value of x, namely, for sin(1.2), and compare the two graphs. Explain the meaning of your results.

15. Suppose a machine with a floating point system $(\beta, t, L, U) = (10, 8, -50, 50)$ is used to calculate the roots of the quadratic equation

$$ax^2 + bx + c = 0,$$

where a, b, and c are given, real coefficients.

For each of the following, state the numerical difficulties that arise if one uses the standard formula for computing the roots. Explain how to overcome these difficulties (when possible).

(a) $a = 1$; $b = -10^5$; $c = 1$.

(b) $a = 6 \cdot 10^{30}$; $b = 5 \cdot 10^{30}$; $c = -4 \cdot 10^{30}$.

(c) $a = 10^{-30}$; $b = -10^{30}$; $c = 10^{30}$.

16. Write a quadratic equation solver. Your MATLAB script should get a, b, c as input, and accurately compute the roots of the corresponding quadratic equation. Make sure to check end cases such as $a = 0$, and consider ways to avoid an overflow and cancellation errors. Implement your algorithm and demonstrate its performance on a few cases (for example, the cases mentioned in Exercise 15). Show that your algorithm produces better results than the standard formula for computing roots of a quadratic equation.

17. Write a MATLAB program that

    (a) sums up $1/n$ for $n = 1, 2, \ldots, 10{,}000$;

    (b) rounds each number $1/n$ to 5 decimal digits and then sums them up *in 5-digit decimal arithmetic* for $n = 1, 2, \ldots, 10{,}000$;

    (c) sums up the same rounded numbers (in 5-digit decimal arithmetic) in reverse order, i.e., for $n = 10{,}000, \ldots, 2, 1$.

    Compare the three results and explain your observations. For generating numbers with the requested precision, you may want to do Exercise 6 first.

18. (a) Explain in detail how to avoid overflow when computing the ℓ_2-norm of a (possibly large in size) vector.

    (b) Write a MATLAB script for computing the norm of a vector in a numerically stable fashion. Demonstrate the performance of your code on a few examples.

19. In the statistical treatment of data one often needs to compute the quantities

$$\bar{x} = \frac{1}{n}\sum_{i=1}^{n} x_i, \quad s^2 = \frac{1}{n}\sum_{i=1}^{n}(x_i - \bar{x})^2,$$

where $x_1, x_2, \ldots, x_n$ are the given data. Assume that n is large, say, $n = 10{,}000$. It is easy to see that s^2 can also be written as

$$s^2 = \frac{1}{n} \sum_{i=1}^{n} x_i^2 - \bar{x}^2.$$

(a) Which of the two methods to calculate s^2 is cheaper in terms of overall computational cost? Assume $\bar{x}$ has already been calculated and give the operation counts for these two options.

(b) Which of the two methods is expected to give more accurate results for s^2 in general?

(c) Give a small example, using a decimal system with precision $t = 2$ and numbers of your choice, to validate your claims.

20. With exact rounding, we know that each *elementary* operation has a relative error which is bounded in terms of the rounding unit η; e.g., for two floating point numbers x and y, fl$(x + y) = (x+y)(1+\epsilon)$, $|\epsilon| \leq \eta$. But is this true also for elementary functions such as sin, ln, and exponentiation?

Consider exponentiation, which is performed according to the formula

$$x^y = e^{y \ln x} \quad \text{(assuming } x > 0\text{)}.$$

Estimate the relative error in calculating x^y in floating point, assuming fl$(\ln z) = (\ln z)(1 + \epsilon)$, $|\epsilon| \leq \eta$, and that everything else is exact. Show that the sort of bound we have for elementary operations and for ln does not hold for exponentiation when x^y is very large.

21. The IEEE 754 (known as the floating point standard) specifies the 128-bit word as having 15 bits for the exponent.

What is the length of the fraction? What is the rounding unit? How many significant decimal digits does this word have?

Why is quadruple precision more than twice as accurate as double precision, which is in turn more than twice as accurate as single precision?

2.6 Additional notes

A lot of thinking in the early days of modern computing went into the design of floating point systems for computers and scientific calculators. Such systems should be economical (fast execution in hardware) on one hand, yet they should also be reliable, accurate enough, and free of unusual exception-handling conventions on the other hand. W. Kahan was particularly instrumental in such efforts (and received a Turing award for his contributions), especially in setting up the IEEE standard. The almost universal adoption of this standard has significantly increased both reliability and portability of numerical codes. See Kahan's webpage for various interesting related documents: http://www.cs.berkeley.edu/~wkahan/.

A short, accessible textbook that discusses floating point systems in great detail is Overton [58]. A comprehensive and thorough treatment of roundoff errors and many aspects of numerical stability can be found in Higham [40].

The practical way of working with floating point arithmetic, which is to attempt to keep errors "small enough" so as not be a bother, is hardly satisfactory from a theoretical point of view. Indeed, what if we want to use a floating point calculation for the purpose of producing a mathematical proof?! The nature of the latter is that a stated result should always—not just usually—hold true. A

2.6. Additional notes

more careful approach uses **interval arithmetic**. With each number are associated a lower and an upper bound, and these are propagated with the algorithm calculations on a "worst case scenario" basis. The calculated results are then guaranteed to be within the limits of the calculated bounds. See Moore, Kearfott, and Cloud [54] and also [58] for an introduction to this veteran subject. Naturally, such an approach is expensive and can be of limited utility, as the range between the bounds typically grows way faster than the accumulated error itself. But at times this approach does work for obtaining truly guaranteed results.

A move to put (more generally) **complexity theory** for numerical algorithms on firmer foundations was initiated by S. Smale and others in the 1980s; see [9]. These efforts have mushroomed into an entire organization called Foundations of Computational Mathematics (FOCM) that is involved with various interesting activities which concentrate on the more mathematically rich aspects of numerical computation.

Chapter 3
Nonlinear Equations in One Variable

This chapter is concerned with finding solutions to the scalar, nonlinear equation

$$f(x) = 0,$$

where the variable x runs in an interval $[a,b]$. The topic provides us with an opportunity to discuss various issues and concepts that arise in more general circumstances.

There are many canned routines that do the job of finding a solution to a nonlinear scalar equation; the one in MATLAB is called `fzero`; type `help fzero` to see how this function is used and what input parameters it requires.

Following an extended introduction to our topic in Section 3.1 are three sections, 3.2–3.4, each devoted to a different method or a class of methods for solving our nonlinear equation. A closely related problem is that of minimizing a function in one variable, and this is discussed in Section 3.5.

3.1 Solving nonlinear equations

Referring to our prototype problem introduce above, $f(x) = 0$, let us further assume that the function f is *continuous* on the interval, denoted $f \in C[a,b]$. Throughout our discussion we denote a solution of the equation (called *root*, or *zero*) by x^*.

> **Note:** Why introduce nonlinear equations before introducing their easier comrades, the linear ones?! Because one linear equation in one unknown is just too easy and not particularly illuminating, and systems of equations bring a wave of complications with them. We have a keen interest in solving scalar nonlinear equations not only because such problems arise frequently in many applications, but also because it is an opportunity to discuss various issues and concepts that arise in more general circumstances and highlight ideas that are extensible well beyond just one equation in one unknown.

Example 3.1. Here are a few simple functions and their roots.

1. $f(x) = x - 1$, on the interval $[a,b] = [0,2]$.

   Obviously there is one root for this linear equation: $x^* = 1$.

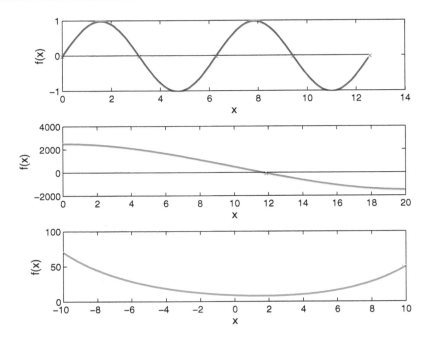

Figure 3.1. *Graphs of three functions and their real roots (if there are any):* (i) $f(x) = \sin(x)$ *on* $[0, 4\pi]$, (ii) $f(x) = x^3 - 30x^2 + 2552$ *on* $[0, 20]$, *and* (iii) $f(x) = 10\cosh(x/4) - x$ *on* $[-10, 10]$.

2. $f(x) = \sin(x)$.

   Since $\sin(n\pi) = 0$ for any integer n, we have

   (a) On the interval $[a, b] = [\frac{\pi}{2}, \frac{3\pi}{2}]$ there is one root, $x^* = \pi$.

   (b) On the interval $[a, b] = [0, 4\pi]$ there are five roots; see Figure 3.1.

3. $f(x) = x^3 - 30x^2 + 2552$, $\quad 0 \leq x \leq 20$.

   In general, a cubic equation with complex coefficients has three complex roots. But if the polynomial coefficients are real and x is restricted to be real and lie in a specific interval, then there is no general a priori rule as to how many (real) roots to expect. A rough plot of this function on the interval $[0, 20]$ is given in Figure 3.1. Based on the plot we suspect there is precisely one root x^* in this interval.

4. $f(x) = 10\cosh(x/4) - x$, $\quad -\infty < x < \infty$,

   where the function cosh is defined by $\cosh(t) = \frac{e^t + e^{-t}}{2}$.

   Not every equation has a solution. This one has no real roots.

   Figure 3.1 indicates that the function $\phi(x) = 10\cosh(x/4) - x$ has a minimum. To find the minimum we differentiate and equate to 0. Let $f(x) = \phi'(x) = 2.5\sinh(x/4) - 1$ (where, analogously to cosh, we define $\sinh(t) = \frac{e^t - e^{-t}}{2}$). This function should have a zero in the interval $[0, 4]$. For more on finding a function's minimum, see Section 3.5. ∎

3.1. Solving nonlinear equations

Example 3.2. Here is the MATLAB script that generates Figure 3.1.

```
t = 0:.1:4*pi;
tt = sin(t);
ax = zeros(1,length(t));
xrt = 0:pi:4*pi;
yrt = zeros(1,5);
subplot(3,1,1)
plot(t,tt,'b',t,ax,'k',xrt,yrt,'rx');
xlabel('x')
ylabel('f(x)')

t = 0:.1:20;
tt = t.^3 - 30*t.^2 + 2552;
ax = zeros(1,length(t));
subplot(3,1,2)
plot(t,tt,'b',t,ax,'k',11.8615,0,'rx');
xlabel('x')
ylabel('f(x)')

t = -10:.1:10;
tt = 10 * cosh(t ./4) - t;
ax = zeros(1,length(t));
subplot(3,1,3)
plot(t,tt,'b',t,ax,'k');
xlabel('x')
ylabel('f(x)')
```

This script should not be too difficult to read like text. Note the use of array arguments, for instance, in defining the array `tt` in terms of the array `t`. ■

Iterative methods for finding roots

It is not realistic to expect, except in special cases, to find a solution of a nonlinear equation by using a closed form formula or a procedure that has a finite, small number of steps. Indeed, it is enough to consider finding roots of polynomials to realize how rare closed formulas are: they practically exist only for very low order polynomials. Thus, one has to resort to *iterative* methods: starting with an initial iterate x_0, the method generates a sequence of iterates $x_1, x_2, \ldots, x_k, \ldots$ that (if all works well) converge to a root of the given, continuous function. In general, our methods will require rough knowledge of the root's location. To find more than one root, we can fire up the same method starting from different initial iterates x_0.

To find approximate locations of roots we can *roughly plot* the function, as done in Example 3.1. (Yes, it sounds a little naive, but sometimes complicated things can be made simple by one good picture.) Alternatively, we can *probe* the function, i.e., evaluate it at several points, looking for locations where $f(x)$ changes sign.

If $f(a) \cdot f(b) < 0$, i.e., f changes sign on the interval $[a,b]$, then by the Intermediate Value Theorem given on page 10 there is a number $c = x^*$ in this interval for which $f(c) = s = 0$. To see this intuitively, imagine trying to graph a scalar function from a positive to a negative value, without lifting the pen (because the function is continuous): somewhere the curve has to cross the x-axis!

Such simple procedures for roughly locating roots are unfortunately not easy to generalize to several equations in several unknowns.

Stopping an iterative procedure

Typically, an iterative procedure starts with an initial iterate x_0 and yields a sequence of iterates $x_1, x_2, \ldots, x_k, \ldots$. Note that in general we *do not* expect the iterative procedure to produce the exact solution x^* exactly. We would conclude that the series of iterates converges if the values of $|f(x_k)|$ and/or of $|x_k - x_{k-1}|$ decrease towards 0 sufficiently fast as the iteration counter k increases.

Correspondingly, one or more of the following general criteria are used to terminate such an iterative process successfully after n iterations:

$$|x_n - x_{n-1}| < \texttt{atol} \quad \text{and/or}$$
$$|x_n - x_{n-1}| < \texttt{rtol}|x_n| \quad \text{and/or}$$
$$|f(x_n)| < \texttt{ftol},$$

where `atol`, `rtol`, and `ftol` are user-specified constants.

Usually, but not always, the second, relative criterion is more robust than the first, absolute one. A favorite combination uses one tolerance value `tol`, which reads

$$|x_n - x_{n-1}| < \texttt{tol}(1 + |x_n|).$$

The third criterion is independent of the first two; it takes the function value into account. The function $f(x)$ may in general be very flat, or very steep, or neither, near its root.

Desired properties of root finding algorithms

When assessing the qualities of a given root finding algorithm, a key property is its efficiency. In determining an algorithm's efficiency it is convenient to concentrate on the number of function evaluations, i.e., the evaluations of $f(x)$ at the iterates $\{x_k\}$, required to achieve convergence to a given accuracy. Other details of the algorithm, which may be considered as overhead, are then generally neglected.[5] Now, if the function $f(x)$ is as simple to evaluate as those considered in Example 3.1, then it is hard to understand why we concentrate on this aspect alone. But in these circumstances any of the algorithms considered in this chapter is very fast indeed. What we really keep in the back of our minds is a possibility that the evaluation of $f(x)$ is rather costly. For instance, think of simulating a space shuttle returning to earth, with x being a control parameter that affects the distance $f(x)$ of the shuttle's landing spot from the location of the reception committee awaiting this event. The calculation of $f(x)$ for each value of x involves a precise simulation of the flight's trajectory for the given x, and it may then be very costly. An algorithm that does not require too many such function evaluations is then sought.

Desirable qualities of a root finding algorithm are the following:

- Efficient—requires a small number of function evaluations.

- Robust—fails rarely, if ever. Announces failure if it does fail.

- Requires a minimal amount of additional data such as the function's derivative.

- Requires f to satisfy only minimal smoothness properties.

- Generalizes easily and naturally to many equations in many unknowns.

No algorithm we are aware of satisfies all of these criteria. Moreover, which of these is more important to honor depends on the application. So we study several possibilities in the following sections.

---

[5] An exception is the number of evaluations of the derivative $f'(x)$ required, for instance, by Newton's method. See Section 3.3.

3.2 Bisection method

The method developed in this section is simple and safe and requires minimal assumptions on the function $f(x)$. However, it is also slow and hard to generalize to higher dimensions.

Suppose that for a given $f(x)$ we know an interval $[a,b]$ where f changes sign, i.e., $f(a) \cdot f(b) < 0$. The Intermediate Value Theorem given on page 10 then assures us that there is a root x^* such that $a \leq x^* \leq b$. Now evaluate $f(p)$, where $p = \frac{a+b}{2}$ is the midpoint, and check the sign of $f(a) \cdot f(p)$. If it is negative, then the root is in $[a,p]$ (by the same Intermediate Value Theorem), so we can set $b \leftarrow p$ and repeat; else $f(a) \cdot f(p)$ is positive, so the root must be in $[p,b]$, hence we can set $a \leftarrow p$ and repeat. (Of course, if $f(a) \cdot f(p) = 0$ exactly, then p is the root and we are done.)

In each such iteration the interval $[a,b]$ where x^* is trapped shrinks by a factor of 2; at the kth step, the point x_k is the midpoint p of the kth subinterval trapping the root. Thus, after a total of n iterations, $|x^* - x_n| \leq \frac{b-a}{2} \cdot 2^{-n}$. Therefore, the algorithm is guaranteed to converge. Moreover, if required to satisfy

$$|x^* - x_n| \leq \mathtt{atol}$$

for a given absolute error tolerance $\mathtt{atol} > 0$, we may determine the number of iterations n needed by demanding $\frac{b-a}{2} \cdot 2^{-n} \leq \mathtt{atol}$. Multiplying both sides by $2^n/\mathtt{atol}$ and taking log, we see that it takes

$$n = \left\lceil \log_2 \left(\frac{b-a}{2\,\mathtt{atol}} \right) \right\rceil$$

iterations to obtain a value $p = x_n$ that satisfies

$$|x^* - p| \leq \mathtt{atol}.$$

The following MATLAB function does the job:

```
function [p,n] = bisect(func,a,b,fa,fb,atol)
%
% function [p,n] = bisect(func,a,b,fa,fb,atol)
%
% Assuming fa = func(a), fb = func(b), and fa*fb < 0,
% there is a value root in (a,b) such that func(root) = 0.
% This function returns in p a value such that
%       | p - root | < atol
% and in n the number of iterations required.

% check input
if (a >= b) | (fa*fb >= 0) | (atol <= 0)
  disp('something wrong with the input: quitting');
  p = NaN; n=NaN;
  return
end

n = ceil ( log2 (b-a) - log2 (2*atol));
for k=1:n
  p = (a+b)/2;
  fp = feval(func,p);
  if fa * fp < 0
    b = p;
    fb = fp;
  else
    a = p;
```

```
        fa = fp;
    end
  end
  p = (a+b)/2;
```

Example 3.3. Using our MATLAB function `bisect` for two of the instances appearing in Example 3.1, we find

- for func$(x) = x^3 - 30x^2 + 2552$, starting from the interval $[0,20]$ with tolerance 1.e-8 gives $x^* \approx 11.86150151$ in 30 iterations;
- for func$(x) = 2.5\sinh(x/4) - 1$, starting from the interval $[-10,10]$ with tolerance 1.e-10 gives $x^* \approx 1.5601412791$ in 37 iterations.

Here is the MATLAB script for the second function instance:

```
format long g
[x,n] = bisect('fex3',-10,10,fex3(-10),fex3(10),1.e-10)
function f = fex3(x)
f = 2.5 * sinh (x/4) - 1;
```

Please make sure that you can produce the corresponding script and results for the first instance of this example. ∎

The stopping criterion in the above implementation of the bisection method is *absolute*, rather than *relative*, and it relates to the values of x rather than to those of f.

Note that in the function `bisect`, if an evaluated p happens to be an exact root, then the code can fail. Such an event would be rather rare in practice, unless we purposely, not to say maliciously, aim for it (e.g., by starting from $a = -1$, $b = 1$, for $f(x) = \sin(x)$). Adding the line

```
if abs(fp) < eps, n = k; return, end
```

just after evaluating fp would handle this exception.

We note here that the situation where a root is known to be bracketed so decisively as with the bisection method is not common.

Recursive implementation

The bisection method is a favorite example in elementary computer programming courses, because in addition to its conceptual simplicity it admits a natural presentation in *recursive* form. In MATLAB this can look as follows:

```
  function [p] = bisect_recursive (func,a,b,fa,fb,atol)
  p = (a+b)/2;
  if b-a < 2*atol
    return
  else
    fp = feval(func,p);
    if fa * fp < 0
      b = p;
      fb = fp;
    else
      a = p;
```

```
        fa = fp;
    end
    p = bisect_recursive (func,a,b,fa,fb,atol);
end
```

Here then is an incredibly short, yet complete, method implementation. However, what makes recursion unappealing for effective implementation in general is the fact that it is wasteful in terms of storage and potentially suboptimal in terms of CPU time. A precise characterization of the reasons for that is beyond the scope of our discussion, belonging more naturally in an introductory book on programming techniques.

Specific exercises for this section: Exercises 1–2.

3.3 Fixed point iteration

The methods discussed in the present section and in the next two, unlike the previous one, have direct extensions to more complicated problems, e.g., to systems of nonlinear equations and to more complex functional equations.

Our problem

$$f(x) = 0$$

can be written as

$$x = g(x).$$

We will discuss how this can be done in a moment. Given the latter formulation, we are looking for a *fixed point*, i.e., a point x^* satisfying

$$x^* = g(x^*).$$

In the *fixed point iteration* process we define a sequence of iterates $x_1, x_2, \ldots, x_k, \ldots$ by

$$x_{k+1} = g(x_k), \quad k = 0, 1, \ldots,$$

starting from an initial iterate x_0. If such a sequence converges, then the limit must be a fixed point.

The convergence properties of the fixed point iteration depend on the choice of the function g. Before we see how, it is important to understand that for a given problem $f(x) = 0$, we can define many functions g (not all of them "good," in a sense that will become clear soon). For instance, we can consider any of the following, and more:

- $g(x) = x - f(x)$,
- $g(x) = x + 2f(x)$,
- $g(x) = x - f(x)/f'(x)$ (assuming f' exists and $f'(x) \neq 0$).

Thus, we are really considering not one method but a *family of methods*. The *algorithm* of fixed point iteration for finding the roots of $f(x)$ that appears on the following page includes the selection of an appropriate $g(x)$.

> **Algorithm: Fixed Point Iteration.**
> Given a scalar continuous function in one variable, $f(x)$, select a function $g(x)$ such that x satisfies $f(x) = 0$ if and only if $g(x) = x$. Then:
>
> 1. Start from an *initial guess* x_0.
>
> 2. For $k = 0, 1, 2, \ldots$, set
>
> $$x_{k+1} = g(x_k)$$
>
> until x_{k+1} satisfies termination criteria.

Suppose that we have somehow determined the continuous function $g \in C[a,b]$, and let us consider the fixed point iteration. Obvious questions arise:

1. Is there a fixed point x^* in $[a,b]$?

2. If yes, is it unique?

3. Does the sequence of iterates converge to a fixed point x^*?

4. If yes, how fast?

5. If not, does this mean that no fixed point exists?

Fixed point theorem

To answer the first question above, suppose that there are two values $a < b$ such that $g(a) \geq a$ and $g(b) \leq b$. If $g(a) = a$ or $g(b) = b$, then a fixed point has been found, so now assume $g(a) > a$ and $g(b) < b$. Then for the continuous function

$$\phi(x) = g(x) - x$$

we have $\phi(a) > 0$, $\phi(b) < 0$. (Note that ϕ *does not* have to coincide with the function f that we have started with; there are lots of ϕ's for a given f.) Hence, by the Intermediate Value Theorem given on page 10, just as before, there is a root $a < x^* < b$ such that $\phi(x^*) = 0$. Thus, $g(x^*) = x^*$, so x^* is a fixed point. We have established the existence of a root: $f(x^*) = 0$.

Next, suppose that g is not only continuous but also differentiable and that there is a positive number $\rho < 1$ such that

$$|g'(x)| \leq \rho, \quad a < x < b.$$

Then the root x^* is unique in the interval $[a,b]$, for if there is also y^* which satisfies $y^* = g(y^*)$, then

$$|x^* - y^*| = |g(x^*) - g(y^*)| = |g'(\xi)(x^* - y^*)| \leq \rho |x^* - y^*|,$$

where ξ is an intermediate value between x^* and y^*. Obviously, this inequality can hold with $\rho < 1$ only if $y^* = x^*$.

We can summarize our findings so far as the *Fixed Point Theorem*.

3.3. Fixed point iteration

> **Theorem: Fixed Point.**
> If $g \in C[a,b]$, $g(a) \geq a$ and $g(b) \leq b$, then there is a fixed point x^* in the interval $[a,b]$.
>   If, in addition, the derivative g' exists and there is a constant $\rho < 1$ such that the derivative satisfies
> $$|g'(x)| \leq \rho \quad \forall x \in (a,b),$$
> then the fixed point x^* is unique in this interval.

Convergence of the fixed point iteration

Turning to the fixed point iteration and the third question on the preceding page, similar arguments establish convergence (now that we know that there is a unique solution): under the same assumptions we have

$$|x_{k+1} - x^*| = |g(x_k) - g(x^*)| \leq \rho |x_k - x^*|.$$

This is a **contraction** by the factor ρ. So

$$|x_{k+1} - x^*| \leq \rho |x_k - x^*| \leq \rho^2 |x_{k-1} - x^*| \leq \cdots \leq \rho^{k+1} |x_0 - x^*|.$$

Since $\rho < 1$, we have $\rho^k \to 0$ as $k \to \infty$. This establishes convergence of the fixed point iteration.

Example 3.4. For the function $g(x) = e^{-x}$ on $[0.2, 1.2]$, Figure 3.2 shows the progression of iterates towards the fixed point x^* satisfying $x^* = e^{-x^*}$.

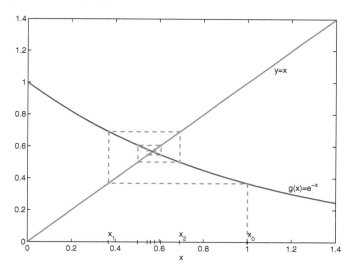

Figure 3.2. *Fixed point iteration for $x = e^{-x}$, starting from $x_0 = 1$. This yields $x_1 = e^{-1}$, $x_2 = e^{-e^{-1}}, \ldots$ Convergence is apparent.*

Note the contraction effect. ∎

To answer the question about the speed of convergence, it should be clear from the bounds derived before Example 3.4 that the smaller ρ is, the faster the iteration converges. In case of the bisection method, the speed of convergence does not depend on the function f (or any g, for that matter), and it is in fact identical to the speed of convergence obtained with $\rho \equiv 1/2$. In contrast, for the fixed point iteration, there is dependence on $|g'(x)|$. This, and the (negative) answer to our fifth question on page 46, are demonstrated next.

Example 3.5. Consider the function

$$f(x) = \alpha \cosh(x/4) - x,$$

where α is a parameter. For $\alpha = 10$ we saw in Example 3.1 that there is no root. But for $\alpha = 2$ there are actually two roots. Indeed, setting

$$g(x) = 2\cosh(x/4)$$

and plotting $g(x)$ and x as functions of x, we obtain Figure 3.3, which suggests not only that there are two fixed points (i.e., roots of f)—let's call them x_1^* and x_2^*—but also that we can bracket them, say, by

$$2 \leq x_1^* \leq 4, \quad 8 \leq x_2^* \leq 10.$$

Next we apply our trusted routine `bisect`, introduced in the previous section, to $f(x)$ with $\alpha = 2$. This yields

$$x_1^* \approx 2.35755106, \quad x_2^* \approx 8.50719958,$$

correct to an absolute tolerance 1.e-8, in 27 iterations for each root. (You should be able to explain why precisely the same number of iterations was required for each root.)

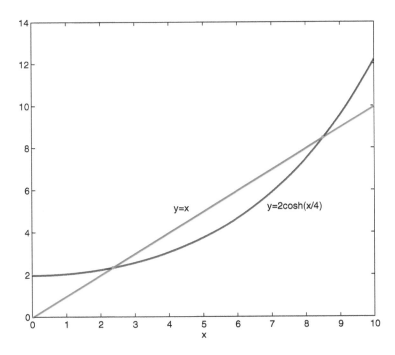

Figure 3.3. *The functions x and $2\cosh(x/4)$ meet at two locations.*

3.3. Fixed point iteration

Now, it is very tempting, and rather natural here, to use the same function g defined above in a fixed point iteration, thus defining

$$x_{k+1} = 2\cosh(x_k/4).$$

For the first root, on the interval $[2,4]$ we have the conditions of the Fixed Point Theorem holding. In fact, near x_1^*, $|g'(x)| < 0.4$, so we expect faster convergence than with the bisection method (why?). Indeed, starting from $x_0 = 2$ the method requires 16 iterations, and starting from $x_0 = 4$ it requires 18 iterations to get to within 1.e-8 of the root x_1^*.

For the second root, however, on the interval $[8,10]$ the conditions of the Fixed Point Theorem do not hold! In particular, $|g'(x_2^*)| > 1$. Thus, a fixed point iteration using this g *will not* converge to the root x_2^*. Indeed, starting with $x_0 = 10$ the iteration diverges quickly, and we obtain *overflow* after 3 iterations, whereas starting this fixed point iteration with $x_0 = 8$ we do obtain convergence, but to x_1^* rather than to x_2^*.

It is important to realize that the root x_2^* is there, even though the fixed point iteration with the "natural g" does not converge to it. Not everything natural is good for you! There are other choices of g for the purpose of fixed point iteration that perform better here. ∎

> **Note:** A discussion of *rates of convergence* similar to that appearing here is also given in Section 7.3. There it is more crucial, because here we will soon see methods that converge faster than a rate of convergence can suitably quantify.

Rate of convergence

Suppose now that at a particular root of a given fixed point iteration, $\rho = |g'(x^*)|$ satisfies $0 < \rho < 1$. Starting with x_0 sufficiently close to x^* we can write $x_k - x^* \approx g'(x^*)(x_{k-1} - x^*)$. Hence we get

$$|x_k - x^*| \approx \rho |x_{k-1} - x^*| \approx \cdots \approx \rho^k |x_0 - x^*|.$$

To quantify the speed of convergence in terms of ρ we can ask, how many iterations does it take to reduce the error by a fixed factor, say, 10?

To answer this we set $|x_k - x^*| \approx 0.1 |x_0 - x^*|$, obtaining

$$\rho^k \approx 0.1.$$

Taking $\log_{10}$ of both sides yields $k \log_{10} \rho \approx -1$. Let us define the *rate of convergence* as

$$rate = -\log_{10} \rho. \qquad (3.1)$$

Then it takes about $k = \lceil 1/rate \rceil$ iterations to reduce the error by more than an order of magnitude. Obviously, the smaller ρ the higher the convergence rate and correspondingly, the smaller the number of iterations required to achieve the same error reduction effect.

Example 3.6. The bisection method is not exactly a fixed point iteration, but it corresponds to an iterative method of a similar sort with $\rho = 0.5$. Thus its convergence rate according to (3.1) is

$$rate = -\log_{10} 0.5 \approx 0.301.$$

This yields $k = 4$, and indeed the error reduction factor with this many bisections is 16, which is more than 10.

For the root x_1^* of Example 3.5 we have $\rho \approx 0.3121$ and

$$rate = -\log_{10} 0.3121 \approx 0.506.$$

Here it takes only two iterations to roughly reduce the error by a factor of 10 if starting close to x^*.

For the root x_2^* of Example 3.5 we obtain from (3.1) a *negative* rate of convergence, as is appropriate for a case where the method does not converge. ∎

Of course, it makes no sense to calculate ρ or the convergence rate so precisely as in Example 3.6: we did this only so that you can easily follow the algebra with a calculator. Indeed, such accurate estimation of ρ would require knowing the solution first! Moreover, there is nothing magical about an error reduction by a particularly precise factor. The rate of convergence should be considered as a rough indicator only. It is of importance, however, to note that the rate becomes negative for $\rho > 1$, indicating no convergence of the fixed point iteration, and that it becomes infinite for $\rho = 0$, indicating that the error reduction per iteration is faster than by any constant factor. We encounter such methods next.

Specific exercises for this section: Exercises 3–4.

3.4 Newton's method and variants

The bisection method requires the function f to be merely continuous (which is good) and makes no further use of further information on f such as availability of its derivatives (which causes it to be painfully slow at times). At the other end of the scale is *Newton's method*, which requires more knowledge and smoothness of the function f but which converges much faster in appropriate circumstances.

Newton's method

Newton's method is the most basic fast method for root finding. The principle we use below to derive it can be directly extended to more general problems.

Assume that the first and second derivatives of f exist and are continuous: $f \in C^2[a,b]$. Assume also that f' can be evaluated with sufficient ease. Let x_k be a current iterate. By Taylor's expansion on page 5 we can write

$$f(x) = f(x_k) + f'(x_k)(x - x_k) + f''(\xi(x))(x - x_k)^2/2,$$

where $\xi(x)$ is some (unknown) point between x and x_k.

Now, set $x = x^*$, for which $f(x^*) = 0$. If f were linear, i.e., $f'' \equiv 0$, then we could find the root by solving $0 = f(x_k) + f'(x_k)(x^* - x_k)$, yielding $x^* = x_k - f(x_k)/f'(x_k)$. For a nonlinear function we therefore define the next iterate by the same formula, which gives

$$x_{k+1} = x_k - \frac{f(x_k)}{f'(x_k)}, \quad k = 0, 1, 2, \ldots.$$

This corresponds to neglecting the term $f''(\xi(x^*))(x^* - x_k)^2/2$ when defining the next iterate; however, if x_k is already close to x^*, then $(x^* - x_k)^2$ is very small, so we would expect x_{k+1} to be much closer to x^* than x_k is.

A *geometric interpretation* of Newton's method is that x_{k+1} is the x-intercept of the tangent line to f at x_k; see Figure 3.4.

3.4. Newton's method and variants

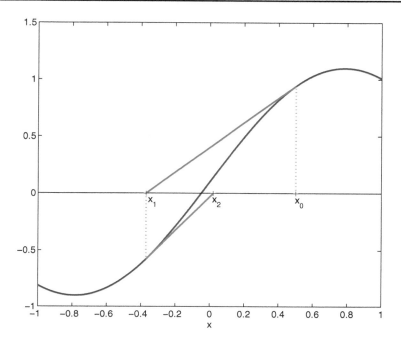

Figure 3.4. *Newton's method: the next iterate is the x-intercept of the tangent line to f at the current iterate.*

Algorithm: Newton's Method.
Given a scalar differentiable function in one variable, $f(x)$:

1. Start from an *initial guess* x_0.

2. For $k = 0, 1, 2, \ldots$, set
$$x_{k+1} = x_k - \frac{f(x_k)}{f'(x_k)},$$
until x_{k+1} satisfies termination criteria.

Example 3.7. Consider the same function as in Example 3.5, given by
$$f(x) = 2\cosh(x/4) - x.$$
The Newton iteration here is
$$x_{k+1} = x_k - \frac{2\cosh(x_k/4) - x_k}{0.5\sinh(x_k/4) - 1}.$$
We use the same absolute tolerance of 1.e-8 and the same four initial iterates as in Example 3.5.

- Starting from $x_0 = 2$ requires 4 iterations to reach x_1^* to within the given tolerance.
- Starting from $x_0 = 4$ requires 5 iterations to reach x_1^* to within the given tolerance.
- Starting from $x_0 = 8$ requires 5 iterations to reach x_2^* to within the given tolerance.
- Starting from $x_0 = 10$ requires 6 iterations to reach x_2^* to within the given tolerance.

The values of $f(x_k)$, starting from $x_0 = 8$, are displayed below:

k	0	1	2	3	4	5
$f(x_k)$	-4.76e-1	8.43e-2	1.56e-3	5.65e-7	7.28e-14	1.78e-15

This table suggests (indirectly) that the number of significant digits essentially doubles at each iteration. Of course, when roundoff level is reached, no meaningful improvement can be obtained upon heaping more floating point operations, and the improvement from the 4th to the 5th iteration in this example is minimal. ∎

Speed of convergence

How fast does a nonlinear iteration converge, if it does? Let us assume that the method indeed converges and that x_k is already "close enough" to the root x^*. We define *convergence orders* as follows. The method is said to be

- **linearly convergent** if there is a constant $\rho < 1$ such that

$$|x_{k+1} - x^*| \leq \rho |x_k - x^*|$$

  for all k sufficiently large;

- **quadratically convergent** if there is a constant M such that

$$|x_{k+1} - x^*| \leq M |x_k - x^*|^2$$

  for all k sufficiently large;

- **superlinearly convergent** if there is a sequence of constants $\rho_k \to 0$ such that

$$|x_{k+1} - x^*| \leq \rho_k |x_k - x^*|$$

  for all k sufficiently large.

Note that for quadratic convergence we can set $\rho_k = M|x_k - x^*| \to_{k \to \infty} 0$; thus the quadratic order implies superlinear order. Superlinear convergence may require less than this but more than linear convergence.

Higher convergence orders (e.g., cubic) may be defined similarly to the quadratic order. However, normally there is really no practical reason to consider such higher convergence orders: methods of this sort would typically require more work per iteration, and roundoff error sets in quickly even for quadratically convergent methods, as we have seen in Example 3.7.

For Newton's method we have already noted that in the derivation a quadratic term was dropped. Indeed, it turns out that if $f \in C^2[a,b]$ and there is a root x^* in $[a,b]$ such that $f(x^*) = 0$, $f'(x^*) \neq 0$, then there is a number δ such that, starting with x_0 from anywhere in the neighborhood $[x^* - \delta, x^* + \delta]$, Newton's method converges quadratically. See Exercise 9.

How is the general fixed point iteration $x_{k+1} = g(x_k)$, considered in the previous section, related to Newton's method? We have seen that the speed of convergence may strongly depend on the choice of g (recall Example 3.5, for instance). It is not difficult to see (Exercise 3) that if, in addition to the assumptions that guarantee convergence in the Fixed Point Theorem, also $g'(x^*) \neq 0$, then the method converges *linearly*. In this case the size of $|g'(x^*)|$ matters and this is quantified by the *rate of convergence*, defined in Section 3.3. The method may converge faster than linearly if $g'(x^*) = 0$. This is precisely the case with Newton's method. Here $g(x) = x - f(x)/f'(x)$, and

3.4. Newton's method and variants

hence $g' = 1 - \frac{(f')^2 - ff''}{(f')^2} = \frac{ff''}{(f')^2}$. Now, since $x = x^*$ is the root of f it follows immediately that $g'(x^*) = 0$.

Newton's method does have two disadvantages: (i) the need to know not only that the derivative f' exists but also how to evaluate it, and (ii) the local nature of the method's convergence. The first of these two aspects is addressed (or, rather, circumvented) in what follows.

Secant method

The requirement that not only the function f but also its derivative f' be supplied by the user of Newton's method is at times simple to satisfy, as in our Example 3.7. But at other instances it can be a drag. An extra effort is required by the user, and occasionally the evaluation of f' can be much more costly than the evaluation of f at a given argument value. Moreover, sometimes the derivative is simply not available, for example, in certain experimental setups where it is possible to evaluate only the function itself.

The secant method is a variant of Newton's method, where $f'(x_k)$ is replaced by its finite difference approximation based on the evaluated function values at x_k and at the previous iterate x_{k-1}. Assuming convergence, observe that near the root

$$f'(x_k) \approx \frac{f(x_k) - f(x_{k-1})}{x_k - x_{k-1}}.$$

(A similar formula was used for different purposes in Example 1.2.) Substitution of this approximation into the formula for Newton's method yields the *Secant method*, also referred to as *quasi-Newton*, given by

$$x_{k+1} = x_k - \frac{f(x_k)(x_k - x_{k-1})}{f(x_k) - f(x_{k-1})}, \quad k = 1, 2, \ldots.$$

Algorithm: Secant Method.
Given a scalar differentiable function in one variable, $f(x)$:

1. Start from two *initial guesses* x_0 and x_1.

2. For $k = 1, 2, \ldots$, set

$$x_{k+1} = x_k - \frac{f(x_k)(x_k - x_{k-1})}{f(x_k) - f(x_{k-1})}$$

until x_{k+1} satisfies termination criteria.

It is worthwhile to compare at this point the Secant algorithm to the Newton algorithm on page 51.

Example 3.8. Consider again the same function as in Example 3.5, given by

$$f(x) = 2\cosh(x/4) - x.$$

We apply the secant method with the same stopping criterion and the same tolerance as in Examples 3.5 and 3.7. Note that the secant method requires two initial iterates, x_0 and x_1.

- Starting from $x_0 = 2$ and $x_1 = 4$ requires 7 iterations to reach x_1^* to within the given tolerance.
- Starting from $x_0 = 10$ and $x_1 = 8$ requires 7 iterations to reach x_2^* to within the given tolerance.

The values of $f(x_k)$, starting from $x_0 = 10$ and $x_1 = 8$, are displayed below:

k	0	1	2	3	4	5	6
$f(x_k)$	2.26	−4.76e-1	−1.64e-1	2.45e-2	−9.93e-4	−5.62e-6	1.30e-9

This table suggests that the number of significant digits increases more rapidly as we get nearer to the root (i.e., better than a linear order), but not as fast as with Newton's method. Thus we observe indeed a demonstration of superlinear convergence. ∎

> **Theorem: Convergence of the Newton and Secant Methods.**
> If $f \in C^2[a,b]$ and there is a root x^* in $[a,b]$ such that $f(x^*) = 0$, $f'(x^*) \neq 0$, then there is a number δ such that, starting with x_0 (and also x_1 in case of the secant method) from anywhere in the neighborhood $[x^* - \delta, x^* + \delta]$, Newton's method converges quadratically and the secant method converges superlinearly.

This theorem specifies the conditions under which Newton's method is guaranteed to converge quadratically, while the secant method is guaranteed to converge superlinearly.

Constructing secant-like modifications for Newton's method for the case of systems of equations is more involved but possible and practically useful; we will discuss this in Section 9.2.

The case of a multiple root

The fast local convergence rate of both Newton and secant methods is predicated on the assumption that $f'(x^*) \neq 0$. What if $f'(x^*) = 0$? This is the case of a *multiple root*. In general, if $f(x)$ can be written as

$$f(x) = (x - x^*)^m q(x),$$

where $q(x^*) \neq 0$, then x^* is a root of multiplicity m. If $m > 1$, then obviously $f'(x^*) = 0$. As it turns out, Newton's method may still converge, but only at a linear (rather than quadratic) rate.

Example 3.9. For the monomial $f(x) = x^m$, $m > 1$, we get

$$\frac{f(x)}{f'(x)} = \frac{x}{m}.$$

Newton's method reads

$$x_{k+1} = x_k - \frac{x_k}{m} = \frac{m-1}{m} x_k.$$

Since the root is $x^* = 0$, we can somewhat trivially write

$$|x_{k+1} - x^*| = \frac{m-1}{m} |x_k - x^*|,$$

so the method is linearly convergent, with the contraction factor $\rho = \frac{m-1}{m}$. The corresponding rate of convergence is $rate = -\log_{10} \frac{m-1}{m}$, and it is not difficult to calculate these values and see that they deteriorate as m grows. If $f''(x^*) \neq 0$, i.e., $m = 2$, then $\rho = 0.5$. See Example 3.6. ∎

3.5. Minimizing a function in one variable

Globalizing Newton's method

Both Newton's method and the secant method are guaranteed to converge under appropriate conditions *only locally*: there is the assumption in the theorem statement that the initial iterate x_0 is already "close enough" to an isolated root. This is in contrast to the situation for the bisection method, which is guaranteed to converge provided only that a bracketing interval is found on which f changes sign. The δ-neighborhood for which Newton's method is guaranteed to converge may be large for some applications, as is the case for Example 3.7, but it may be small for others, and it is more involved to assess ahead of time which is the case for a given problem.

A natural idea, then, is to combine two methods into a hybrid one. We start with a rough plot or probe of the given function f in order to bracket roots. Next, starting from a given bracket with $f(x_0) \cdot f(x_1) < 0$, initiate a Newton or a secant method, monitoring the iterative process by requiring sufficient improvement at each iteration. This "sufficient improvement" can, for instance, be that $|f(x_{k+1})| < 0.5|f(x_k)|$ or the bisection guaranteed improvement ratio $|x_{k+1} - x_k| < 0.5|x_k - x_{k-1}|$. If a Newton or secant iteration is applied but there is no sufficient improvement, then we conclude that the current iterate must be far from where the superior fire power of the fast method can be utilized, and we therefore return to x_0 and x_1, apply a few bisection iterations to obtain new x_0 and x_1 which bracket the root x^* more tightly, and restart the Newton or secant routine.

This hybrid strategy yields a robust solver for scalar nonlinear equations; see Exercise 13.

Convergence and roundoff errors

Thus far in this chapter we have scarcely noticed the possible existence of roundoff errors. But rest assured that roundoff errors are there, as promised in Chapter 2. They are simply dominated by the convergence errors, so long as the termination criteria of our iterative methods are well above rounding unit, and so long as the problem is *well-conditioned*. In the last entry of the table in Example 3.7, the convergence error is so small that roundoff error is no longer negligible. The result is certainly not bad in terms of error magnitude, but the quadratic convergence pattern (which does not hold for roundoff errors, barring a miracle) is destroyed.

Our essential question here is, given that the error tolerances are well above rounding unit in magnitude, do we ever have to worry about roundoff errors when solving a nonlinear equation? The answer is that we might, if the *problem* is such that roundoff error accumulates significantly.

When could ill-conditioning occur for the problem of finding roots of a scalar nonlinear function? This can happen if the function $f(x)$ is very flat near its root, because small changes to f may then significantly affect the precise location of the root. Exercise 2 illustrates this point. The general case is that of a *multiple root*.

Specific exercises for this section: Exercises 5–12.

3.5 Minimizing a function in one variable

A major source of applications giving rise to root finding is *optimization*. In its one-variable version we are required to find an argument $x = \hat{x}$ that minimizes a given *objective function* $\phi(x)$.[6] For a simple instance, we have already briefly examined in Example 3.1 the minimization of the function

$$\phi(x) = 10\cosh(x/4) - x$$

over the real line.

---

[6] We arbitrarily concentrate on finding minimum rather than maximum points. If a given problem is naturally formulated as finding a maximum of a function $\psi(x)$, say, then define $\phi(x) = -\psi(x)$ and consider minimizing ϕ.

Another, general, instance is obtained from any root finding or fixed point problem, $g(x) = x$, by setting
$$\phi(x) = [g(x) - x]^2.$$
But here we concentrate on cases where the given objective is not necessarily that of root finding. Note that in general $\phi(x^*) \neq 0$.

Conditions for a minimum point

Assume that $\phi \in C^2[a,b]$, i.e., that ϕ and its first two derivatives are continuous on a given interval $[a,b]$. Denote
$$f(x) = \phi'(x).$$
An argument x^* satisfying $a < x^* < b$ is called a **critical point** if
$$f(x^*) = 0.$$
Now, for a parameter h small enough so that $x^* + h \in [a,b]$ we can expand in Taylor's series on page 5 and write
$$\phi(x^* + h) = \phi(x^*) + h\phi'(x^*) + \frac{h^2}{2}\phi''(x^*) + \cdots$$
$$= \phi(x^*) + \frac{h^2}{2}[\phi''(x^*) + \mathcal{O}(h)].$$

(See the discussion on page 7 regarding order $\mathcal{O}$ notation.) Since $|h|$ can be taken arbitrarily small, it is now clear that at a critical point:

- If $\phi''(x^*) > 0$, then $\hat{x} = x^*$ is a *local minimizer* of $\phi(x)$. This means that ϕ attains minimum at $\hat{x} = x^*$ in some neighborhood which includes x^*.

- If $\phi''(x^*) < 0$, then x^* is a *local maximizer* of $\phi(x)$. This means that ϕ attains maximum at $\hat{x} = x^*$ in some neighborhood which includes x^*.

- If $\phi''(x^*) = 0$, then a further investigation at x^* is required.

Algorithms for function minimization

It is not difficult to see that if $\phi(x)$ attains a minimum (or maximum) at a point $\hat{x}$, then this point must be critical, i.e., $f(\hat{x}) = 0$. Thus, the problem of finding all minima of a given function $\phi(x)$ can be solved by finding all the critical roots[7] and then checking for each if it is a minimum by examining the sign of the second derivative of ϕ.

Example 3.10. For the function $\phi(x) = 10\cosh(x/4) - x$ we have
$$\phi'(x) = f(x) = \frac{10}{4}\sinh(x/4) - 1,$$
$$\phi''(x) = f'(x) = \frac{10}{16}\cosh(x/4).$$

---

[7]Note that for quadratic convergence of Newton's method, $\phi(x)$ must have three continuous derivatives.

3.5. Minimizing a function in one variable

Since $\phi''(x) > 0$ for all x, we can have only one minimum point. (Think of how the curve must look like if we had more than one minimum.) Applying any of the algorithms described in Sections 3.2–3.4, we obtain the unique minimum point as the root of $f(x)$, given by

$$\hat{x} \approx 1.5601412791.$$

At this point, $\phi(\hat{x}) \approx 9.21018833518615$. ∎

Example 3.11. For an example with many critical points, consider $\phi(x) = \cos(x)$.
The critical points are where $f(x) = \phi'(x) = -\sin(x) = 0$; thus

$$x^* = j\pi, \quad j = 0, \pm 1, \pm 2, \ldots.$$

Since

$$\phi''(x^*) = -\cos(j\pi) = (-1)^{j+1}, \quad j = 0, \pm 1, \pm 2, \ldots,$$

we have minimum points where j is odd, i.e., at every second critical point: the local minimizers are

$$\hat{x} = (2l+1)\pi, \quad l = 0, \pm 1, \pm 2, \ldots.$$

The other critical points are maximum points. ∎

For simple problems this could mark the end of this section. However, for problems where function evaluations are really dear we note that there are two imperfections in the simple procedure outlined above:

- We are investing a considerable effort into finding all critical points, even those that end up being local maxima. Is there a way to avoid finding such unwanted points?

- No advantage is taken of the fact that we are minimizing $\phi(x)$. Can't we gauge how good an iterate x_k is by checking $\phi(x_k)$?

These objections become more of an issue when considering the minimization of functions of several variables. Nonetheless, let us proceed to quickly outline some general ideas with Newton's method on page 51 for the scalar case.

The kth iteration of Newton's method applied to $f(x) = \phi'(x)$ can be written here as

$$x_{k+1} = x_k + \alpha_k \xi_k, \quad \text{where}$$

$$\xi_k = -\frac{\phi'(x_k)}{\phi''(x_k)}, \quad \alpha_k = 1.$$

Now, if $\phi''(x_k) > 0$ and the *step size* $\alpha_k > 0$ is small enough, then

$$\phi(x_{k+1}) \approx \phi(x_k) + \alpha_k \xi_k \phi'(x_k) = \phi(x_k) - \alpha_k \phi'(x_k)^2 / \phi''(x_k) < \phi(x_k).$$

Therefore, we have a decrease in ϕ, i.e., a step in the right direction.

Thus, it is possible to design a procedure where we insist that $\phi(x)$ decrease in each iteration. This would prevent convergence to critical points which are local maxima rather than minima. In such a procedure we monitor $\phi''(x_k)$, replacing it by $\max\{\phi''(x_k), \epsilon\}$, where $\epsilon > 0$ is a parameter chosen not large. We then select a step size α_k by trying $\alpha_k = 1$ first and decreasing it if necessary until a sufficient decrease is obtained in $\phi(x_{k+1})$ relatively to $\phi(x_k)$. We shall have more to say about this sort of procedure when considering minimizing functions in several variables.

Local and global minimum

The procedures outlined above find *local* minima. The problem of finding a *global* minimizer, i.e., a point $\hat{x}$ such that $\phi(\hat{x}) \leq \phi(x)$ for all $x \in [a,b]$, is significantly more difficult in general. In the worst situations, where not much is known about the objective function ϕ, a suitable algorithm may involve a complete enumeration, i.e., finding all local minima and then comparing them. At the other end of the scale, if the objective function is known to be *convex*, which means that

$$\phi(\theta x + (1-\theta)y) \leq \theta \phi(x) + (1-\theta)\phi(y) \quad \forall \, 0 \leq \theta \leq 1$$

for any two points $x, y \in [a,b]$, then there is a unique minimum and the problems of finding local and global minima coincide. Such is the instance considered in Example 3.10, but not that considered in Example 3.11.

Specific exercises for this section: Exercises 21–23.

3.6 Exercises

0. **Review questions**

    (a) What is a nonlinear equation?

    (b) Is the bisection method (i) efficient? (ii) robust? Does it (iii) require a minimal amount of additional knowledge? (iv) require f to satisfy only minimum smoothness properties? (v) generalize easily to several functions in several variables?

    (c) Answer similar questions for the Newton and secant methods.

    (d) State at least one advantage and one disadvantage of the recursive implementation of the bisection method over the iterative nonrecursive implementation.

    (e) In what way is the fixed point iteration a *family* of methods, rather than just one method like bisection or secant?

    (f) What is the basic condition for convergence of the fixed point iteration, and how does the speed of convergence relate to the derivative of the iteration function g?

    (g) Suppose a given fixed point iteration does not converge: does this mean that there is no root in the relevant interval? Answer a similar question for the Newton and secant methods.

    (h) State at least two advantages and two disadvantages of Newton's method.

    (i) What are order of convergence and rate of convergence, and how do they relate?

    (j) State at least one advantage and one disadvantage of the secant method over Newton's.

    (k) In what situation does Newton's method converge only linearly?

    (l) Explain the role that roundoff errors play in the convergence of solvers for nonlinear equations, and explain their relationship with convergence errors.

    (m) State a similarity and a difference between the problem of minimizing a function $\phi(x)$ and that of solving the nonlinear equation $\phi'(x) = 0$.

    (n) State what a convex function is, and explain what happens if an objective function is convex.

1. Apply the bisection routine `bisect` to find the root of the function

    $$f(x) = \sqrt{x} - 1.1$$

    starting from the interval $[0,2]$ (that is, $a = 0$ and $b = 2$), with `atol = 1.e-8`.

3.6. Exercises

   (a) How many iterations are required? Does the iteration count match the expectations, based on our convergence analysis?

   (b) What is the resulting absolute error? Could this absolute error be predicted by our convergence analysis?

2. Consider the polynomial function[8]
$$f(x) = (x-2)^9$$
$$= x^9 - 18x^8 + 144x^7 - 672x^6 + 2016x^5 - 4032x^4 + 5376x^3 - 4608x^2 + 2304x - 512.$$

   (a) Write a MATLAB script which evaluates this function at 161 equidistant points in the interval $[1.92, 2.08]$ using two methods:

      i. Apply nested evaluation (cf. Example 1.4) for evaluating the polynomial in the expanded form $x^9 - 18x^8 + \cdots$.

      ii. Calculate $(x-2)^9$ directly.

      Plot the results in two separate figures.

   (b) Explain the difference between the two graphs.

   (c) Suppose you were to apply the bisection routine from Section 3.2 to find a root of this function, starting from the interval $[1.92, 2.08]$ and using the nested evaluation method, to an absolute tolerance 10^{-6}. *Without computing anything*, select the correct outcome:

      i. The routine will terminate with a root p satisfying $|p-2| \leq 10^{-6}$.

      ii. The routine will terminate with a root p *not* satisfying $|p-2| \leq 10^{-6}$.

      iii. The routine will not find a root.

      Justify your choice in one short sentence.

3. Consider the fixed point iteration $x_{k+1} = g(x_k)$, $k = 0, 1, \ldots$, and let all the assumptions of the Fixed Point Theorem hold. Use a Taylor's series expansion to show that the order of convergence depends on how many of the derivatives of g vanish at $x = x^*$. Use your result to state how fast (at least) a fixed point iteration is expected to converge if $g'(x^*) = \cdots = g^{(r)}(x^*) = 0$, where the integer $r \geq 1$ is given.

4. Consider the function $g(x) = x^2 + \frac{3}{16}$.

   (a) This function has two fixed points. What are they?

   (b) Consider the fixed point iteration $x_{k+1} = g(x_k)$ for this g. For which of the points you have found in (a) can you be sure that the iterations will converge to that fixed point? Briefly justify your answer. You may assume that the initial guess is sufficiently close to the fixed point.

   (c) For the point or points you found in (b), roughly how many iterations will be required to reduce the convergence error by a factor of 10?

5. Write a MATLAB script for computing the cube root of a number, $x = \sqrt[3]{a}$, with only basic arithmetic operations using Newton's method, by finding a root of the function $f(x) = x^3 - a$. Run your program for $a = 0, 2, 10$. For each of these cases, start with an initial guess reasonably close to the solution. As a stopping criterion, require the function value whose root you are searching to be smaller than 10^{-8}. Print out the values of x_k and $f(x_k)$ in each iteration. Comment on the convergence rates and explain how they match your expectations.

---

[8] This beautiful example is inspired by Demmel [21]. We gave a modified version of it as an exam question once. Unfortunately, not everyone thought it was beautiful.

6. (a) Derive a third order method for solving $f(x) = 0$ in a way similar to the derivation of Newton's method, using evaluations of $f(x_n)$, $f'(x_n)$, and $f''(x_n)$. The following remarks may be helpful in constructing the algorithm:
    - Use the Taylor expansion with three terms plus a remainder term.
    - Show that in the course of derivation a quadratic equation arises, and therefore *two* distinct schemes can be derived.

    (b) Show that the order of convergence (under the appropriate conditions) is cubic.

    (c) Estimate the number of iterations and the cost needed to reduce the initial error by a factor of 10^m.

    (d) Write a script for solving the problem of Exercise 5. To guarantee that your program does not generate complex roots, make sure to start sufficiently close to a real root.

    (e) Can you speculate what makes this method less popular than Newton's method, despite its cubic convergence? Give two reasons.

7. Consider *Steffensen's method*

$$x_{k+1} = x_k - \frac{f(x_k)}{g(x_k)}, \quad k = 0, 1, \ldots,$$

    where

$$g(x) = \frac{f(x + f(x)) - f(x)}{f(x)}.$$

    (a) Show that in general the method converges quadratically to a root of $f(x)$.

    (b) Compare the method's efficiency to the efficiency of the secant method.

8. It is known that the order of convergence of the secant method is $p = \frac{1+\sqrt{5}}{2} = 1.618\ldots$ and that of Newton's method is $p = 2$. Suppose that evaluating f' costs approximately α times the cost of approximating f. Determine approximately for what values of α Newton's method is more efficient (in terms of number of function evaluations) than the secant method. You may neglect the asymptotic error constants in your calculations. Assume that both methods are starting with initial guesses of a similar quality.

9. This exercise essentially utilizes various forms of Taylor's expansion and relies on expertise in calculus.

    (a) Prove that if $f \in C^2[a,b]$ and there is a root x^* in $[a,b]$ such that $f(x^*) = 0$, $f'(x^*) \neq 0$, then there is a number δ such that, starting with x_0 from anywhere in the neighborhood $[x^* - \delta, x^* + \delta]$, Newton's method converges quadratically.

    (b) This is more challenging: prove the same conclusions under the same assumptions, except that now f is only assumed to have a first Lipschitz continuous derivative. Thus, while f'' may not exist everywhere in $[a,b]$, there is a constant γ such that for any $x, y \in [a,b]$

$$|f'(x) - f'(y)| \leq \gamma |x - y|.$$

10. The function

$$f(x) = (x-1)^2 e^x$$

    has a double root at $x = 1$.

3.6. Exercises

    (a) Derive Newton's iteration for this function. Show that the iteration is well-defined so long as $x_k \neq -1$ and that the convergence rate is expected to be similar to that of the bisection method (and certainly not quadratic).

    (b) Implement Newton's method and observe its performance starting from $x_0 = 2$.

    (c) How easy would it be to apply the bisection method? Explain.

11. In general, when a smooth function $f(x)$ has a multiple root at x^*, the function

$$\psi(x) = \frac{f(x)}{f'(x)}$$

has a simple root there.

    (a) Show that a Newton method for $\psi(x)$ yields the iteration

$$x_{k+1} = x_k - \frac{f(x_k)f'(x_k)}{[f'(x_k)]^2 - f(x_k)f''(x_k)}.$$

    (b) Give one potential advantage and two potential disadvantages for this iteration as compared to Newton's method for $f(x)$.

    (c) Try this method on the problem of Exercise 10. What are your observations?

12. Given $a > 0$, we wish to compute $x = \ln a$ using addition, subtraction, multiplication, division, and the exponential function, e^x.

    (a) Suggest an iterative formula based on Newton's method, and write it in a way suitable for numerical computation.

    (b) Show that your formula converges quadratically.

    (c) Write down an iterative formula based on the secant method.

    (d) State which of the secant and Newton's methods is expected to perform better in this case in terms of overall number of exponential function evaluations. Assume a fair comparison, i.e., same floating point system, "same quality" initial guesses, and identical convergence criterion.

13. Write a MATLAB program to find all the roots of a given, twice continuously differentiable, function $f \in C^2[a,b]$.

Your program should first probe the function $f(x)$ on the given interval to find out where it changes sign. (Thus, the program has, in addition to f itself, four other input arguments: a, b, the number *nprobe* of equidistant values between a and b at which f is probed, and a tolerance `tol`.)

For each subinterval $[a_i, b_i]$ over which the function changes sign, your program should then find a root as follows. Use either Newton's method or the secant method to find the root, monitoring decrease in $|f(x_k)|$. If an iterate is reached for which there is no sufficient decrease (e.g., if $|f(x_k)| \geq 0.5|f(x_{k-1})|$), then revert back to $[a_i, b_i]$, apply three bisection steps and restart the Newton or secant method.

The ith root is deemed "found" as x_k if both

$$|x_k - x_{k-1}| < \mathtt{tol}(1 + |x_k|) \quad \text{and} \quad |f(x_k)| < \mathtt{tol}$$

hold.

Verify your program by finding the two roots (given in Example 3.5) of the function
$$f(x) = 2\cosh(x/4) - x,$$
starting your search with $[a,b] = [0,10]$ and $nprobe = 10$. Then use your program to answer other questions in this section where relevant.

14. Find all the roots of the function
$$f(x) = \begin{cases} \frac{\sin(x)}{x}, & x \neq 0, \\ 1, & x = 0, \end{cases}$$
in the interval $[-10, 10]$ for `tol` $= 10^{-7}$.

    [This function is special enough to have a name. It is called the `sinc` function.]

15. For $x > 0$ consider the equation
$$x + \ln x = 0.$$
It is a reformulation of the equation of Example 3.4.

    (a) Show analytically that there is exactly one root, $0 < x^* < \infty$.
    (b) Plot a graph of the function on the interval $[0.1, 1]$.
    (c) As you can see from the graph, the root is between 0.5 and 0.6. Write MATLAB routines for finding the root, using the following:
        i. The bisection method, with the initial interval $[0.5, 0.6]$. Explain why this choice of the initial interval is valid.
        ii. A linearly convergent fixed point iteration, with $x_0 = 0.5$. Show that the conditions of the Fixed Point Theorem (for the function g you have selected) are satisfied.
        iii. Newton's method, with $x_0 = 0.5$.
        iv. The secant method, with $x_0 = 0.5$ and $x_1 = 0.6$.

        For each of the methods:
        - Use $|x_k - x_{k-1}| < 10^{-10}$ as a convergence criterion.
        - Print out the iterates and show the progress in the number of correct decimal digits throughout the iteration.
        - Explain the convergence behavior and how it matches theoretical expectations.

16. We saw in Example 3.5 that the function
$$f(x) = \alpha \cosh(x/4) - x$$
has two roots for $\alpha = 2$ and none for $\alpha = 10$. Is there an α for which there is precisely one root? If yes, then find such an α and the corresponding root; if not, then justify.

17. The derivative of the sinc function is given by
$$f(x) = \frac{x\cos(x) - \sin(x)}{x^2}.$$

    (a) Show that near $x = 0$, this function can be approximated by
    $$f(x) \approx -x/3.$$
    The error in this approximation gets smaller as x approaches 0.
    (b) Find all the roots of f in the interval $[-10, 10]$ for `tol` $= 10^{-8}$.

3.6. Exercises 63

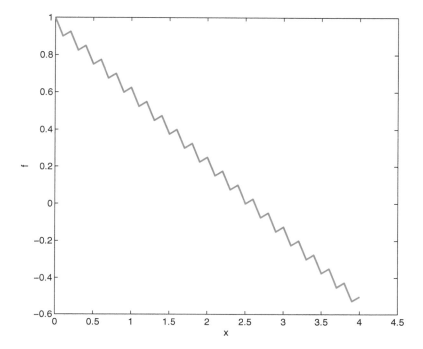

Figure 3.5. *Graph of an anonymous function; see Exercise* 18.

18. Consider finding the root of a given nonlinear function $f(x)$, known to exist in a given interval $[a,b]$, using one of the following three methods: *bisection, Newton,* and *secant.* For each of the following instances, one of these methods has a distinct advantage over the other two. Match problems and methods and justify briefly.

    (a) $f(x) = x - 1$ on the interval $[0, 2.5]$.

    (b) $f(x)$ is given in Figure 3.5 on $[0, 4]$.

    (c) $f \in C^5[0.1, 0.2]$, the derivatives of f are all bounded in magnitude by 1, and $f'(x)$ is hard to specify explicitly or evaluate.

19. You are required by a computer manufacturer to write a library function for a given floating point system to find the cube root $y^{1/3}$ of any given positive number y. Any such relevant floating point number can be represented as $y = a \times 2^e$, where a is a normalized fraction ($0.5 \le a < 1$) and e is an integer exponent. This library function must be very efficient and it should always work. For efficiency purposes it makes sense to store some useful constants ahead of computation time, e.g., the constants $2^{1/3}$, $\frac{2}{3}$, and $a/3$, should these prove useful.

    (a) Show how $y^{1/3}$ can be obtained, once $a^{1/3}$ has been calculated for the corresponding fraction, in at most five additional flops.

    (b) Derive the corresponding Newton iteration. What is the flop count per iteration?

    (c) How would you choose an initial approximation? Roughly how many iterations are needed? (The machine rounding unit is 2^{-52}.)

    [You might find this exercise challenging.]

20. Suppose that the division button of your calculator has stopped working, and you have addition, subtraction, and multiplication only. Given a real number $b \neq 0$, suggest a quadratically convergent iterative formula to compute $\frac{1}{b}$, correct to a user-specified tolerance. Write a MATLAB routine that implements your algorithm, using $|x_k - x_{k-1}| < 10^{-10}$ as a convergence criterion, and apply your algorithm to $b = \pi$ (that is, we compute $\frac{1}{\pi}$), with two different initial guesses: (a) $x_0 = 1$; and (b) $x_0 = 0.1$.

    Explain your results.

21. Write a MATLAB program to minimize a smooth, scalar function in one variable $\phi(x)$ over a given interval. The specifications are similar to those in Exercise 13. Your program should first find and display all critical points, i.e., zeros of $\phi'(x)$. Then it should determine which of these correspond to a local minimum by checking the sign of ϕ'', and display the local minimum points. Finally, it should determine the global minimum point(s), value, and arguments by minimizing over the local minimum values.

22. Apply your program from Exercise 21 to find the (unique) *global* minimum of the function $\phi(x) = \cos(x) + x/10$ over the interval $[-10, 10]$. What are $\hat{x}$ and $\phi(\hat{x})$?

23. Use your program from Exercise 21 to minimize

    (a) $\phi(x) = \sin(x)$ and

    (b) $\phi(x) = -\text{sinc}(x) = -\frac{\sin(x)}{x}$

    over the interval $[-10, 10]$ with `tol` $= 10^{-8}$.

    You might find that determining the global minimum of the second function is significantly more challenging, even though the global minimum for the first one is not unique. Explain the source of difficulties and how you have addressed them.

    [Exercises 17 and 14 are relevant here.]

3.7 Additional notes

Just about any introductory text on numerical methods or numerical analysis covers the topics introduced in this chapter, although the goals and the manner of these treatments may vary significantly. For different flavors, see Heath [39], Cheney and Kincaid [12], and Burden and Faires [11], as well as the classics Conte and de Boor [13] and Dahlquist and Björck [15].

For us this chapter is mainly an opportunity to introduce and demonstrate, within a simple context, several basic notions and issues arising also in much more complex numerical computations. We have thus avoided methods that do not seem to shed much light on other problems as well.

There is a (somewhat more complicated) analogue to the bisection method for function minimization. This classical algorithm is called **golden section search** and is described, together with extensions, in Press et al. [59]; alternatively, see the crisper Heath [39].

Chapter 4
Linear Algebra Background

Systems of linear equations and algebraic eigenvalue (or singular value) problems arise very frequently in large numerical computations, often as a subproblem. Computational tasks associated with models arising in computer science, mathematics, statistics, the natural sciences, social sciences, various engineering disciplines, and business and economics all require the solution of such linear algebra systems. The simplest techniques for solving these problems have been known for centuries, and yet their efficient implementation and the investigation of modern methods in large, complex situations is an active area of research.

Numerical methods that relate directly to linear algebra problems are covered in several chapters following this one. Here we set up the mathematical preliminaries and tools used later on, and we provide some typical examples of simple applications that give rise to common systems of this sort. What this chapter barely contains are actual algorithms or their analysis. We lightly touch upon some of those in Section 4.5 but leave the more systematic treatment of linear algebra algorithms for later chapters.

> **Note:** A large portion of this chapter is a review of material that can be found in an introductory linear algebra text. We do, however, make a point of spicing things up by giving a preliminary taste of numerical issues and by highlighting notions that are particularly important for computations. Readers who feel comfortable with the basic concepts and are familiar with applications that yield such problems may skip this chapter. But we have a feeling that those who do stick around will be able to learn a new thing or two, particularly in the later parts of our discussion.

In Section 4.1 we quickly review basic concepts that are covered in any text on linear algebra. Tools for measuring sizes of vectors and matrices and distances between them are provided in Section 4.2. In Section 4.3 we describe two very important special classes of matrices that you will encounter again and again: **symmetric positive definite** and **orthogonal** matrices. Section 4.4 introduces the singular value decomposition. Finally, the examples in Section 4.5 provide "teasers" for what may arise in applications.

4.1 Review of basic concepts

In this section we provide a brief (and by no means comprehensive) review of basic concepts in linear algebra. The two fundamental problems that we consider are linear systems and eigenvalue problems.

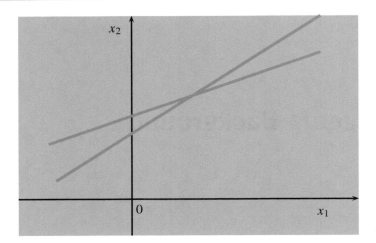

Figure 4.1. *Intersection of two straight lines: $a_{11}x_1 + a_{12}x_2 = b_1$ and $a_{21}x_1 + a_{22}x_2 = b_2$.*

Basics of linear system solution

Example 4.1. Consider the simple example of finding the point of intersection of two straight lines in a plane; see Figure 4.1. If one line is defined by $a_{11}x_1 + a_{12}x_2 = b_1$ and the other by $a_{21}x_1 + a_{22}x_2 = b_2$, then the point of intersection, lying on both lines, must satisfy both equations

$$a_{11}x_1 + a_{12}x_2 = b_1,$$
$$a_{21}x_1 + a_{22}x_2 = b_2.$$

For instance, let $a_{11} = 1$, $a_{12} = -4$, $a_{21} = .5$, $a_{22} = -1$, $b_1 = -10$, $b_2 = -2$. Then the point of intersection (x_1, x_2) is the solution of the 2×2 system

$$x_1 - 4x_2 = -10,$$
$$.5x_1 - x_2 = -2.$$

How would we go about finding a solution $\mathbf{x} = \begin{pmatrix} x_1 \\ x_2 \end{pmatrix}$ that satisfies these equations? Recall that we can multiply any row in such a system by a scalar and add to another row. Here, if we multiply the first row by -0.5 and add to the second, then x_1 is eliminated and we obtain

$$0 \cdot x_1 + x_2 \equiv x_2 = 3.$$

Substituting in the first equation then yields $x_1 = -10 + 12 = 2$. Hence, the point of intersection is $(2, 3)$. ∎

Singularity and linear independence

It is surprising how directly Example 4.1, which is trivial by all accounts, extends in many ways. Let us stay with it for another brief moment and ask whether there is always a unique intersection point for the two lines in the plane. Geometrically, the answer is that there is a unique solution if and only if the lines are not parallel. Here the term "parallel" includes also the case of coinciding lines. In the case of parallel lines we have either no intersection point, i.e., no solution to the linear

4.1. Review of basic concepts

system of equations, as for the system

$$x_1 + x_2 = 5,$$
$$3x_1 + 3x_2 = 16,$$

or infinitely many "intersection points" if the two straight lines coincide, as for the system

$$x_1 + x_2 = 5,$$
$$3x_1 + 3x_2 = 15.$$

When are the lines parallel? When their slopes are equal, expressed algebraically by the condition $-\frac{a_{11}}{a_{12}} = -\frac{a_{21}}{a_{22}}$, or

$$\det(A) = a_{11}a_{22} - a_{21}a_{12} = 0.$$

When the determinant[9] of a square matrix A equals 0 we say that A is **singular**. Equivalently, A is nonsingular if and only if its columns (or rows) are **linearly independent**.[10,11]

Range space and nullspace

Generally, we can write $A\mathbf{x} = \mathbf{b}$ as[12]

$$\mathbf{a}_1 x_1 + \mathbf{a}_2 x_2 + \cdots + \mathbf{a}_n x_n = \mathbf{b},$$

where $\mathbf{a}_j$, or $A(:, j)$ in MATLAB-speak, is the jth column of A. Pictorially, we can write

$$\begin{pmatrix} \vdots \\ \mathbf{a}_1 \\ \vdots \end{pmatrix} x_1 + \begin{pmatrix} \vdots \\ \mathbf{a}_2 \\ \vdots \end{pmatrix} x_2 + \cdots + \begin{pmatrix} \vdots \\ \mathbf{a}_n \\ \vdots \end{pmatrix} x_n = \begin{pmatrix} \vdots \\ \mathbf{b} \\ \vdots \end{pmatrix} \quad \text{and} \quad \mathbf{x} = \begin{pmatrix} x_1 \\ x_2 \\ \vdots \\ x_{n-1} \\ x_n \end{pmatrix}.$$

If A is nonsingular, then the columns $\mathbf{a}_j$ are linearly independent. Then there is a unique way of writing any vector $\mathbf{b}$ as a linear combination of these columns. Hence, if A is nonsingular, then there is a unique solution for the system $A\mathbf{x} = \mathbf{b}$.

---

[9] Please refer to the nearest text on linear algebra for the definition of a *determinant*. Our algorithms will not use determinants.

[10] Recall that a set of vectors $\{\mathbf{v}_i\}$ are linearly independent if $\sum_i \alpha_i \mathbf{v}_i = \mathbf{0}$ necessarily implies that all coefficients α_i are equal to zero.

[11] For a given set $\{\mathbf{v}_i\}_{i=1}^m$ of m real, linearly independent vectors $\mathbf{v}_i \in \mathcal{R}^n$ (where necessarily $m \leq n$), the **linear space** $V = \text{span}\{\mathbf{v}_1, \ldots, \mathbf{v}_m\}$ consists of all linear combinations $\sum_{i=1}^m \alpha_i \mathbf{v}_i$. Obviously V is a **subspace** of $\mathcal{R}^n$, i.e., the space V is contained in the space $\mathcal{R}^n$, because any $\mathbf{v} \in V$ also satisfies $\mathbf{v} \in \mathcal{R}^n$.

[12] Throughout this text a given matrix A is by default square and real, i.e., it can be written as

$$A = \begin{pmatrix} a_{11} & a_{12} & \cdots & a_{1n} \\ a_{21} & a_{22} & \cdots & a_{2n} \\ \vdots & \vdots & \ddots & \vdots \\ a_{n1} & a_{n2} & \cdots & a_{nn} \end{pmatrix},$$

and the elements $a_{i,j}$ are real numbers. Exceptions where A is rectangular or complex are noted in context.

On the other hand, if A is singular, or if it actually has more rows than columns, then it may or may not be possible to write **b** as a linear combination of the columns of A. Let the matrix generally have m rows and n columns, and denote

- range$(A) \equiv$ all vectors that can be written as a linear combination of the columns of A, i.e., all vectors **y** that can be written as $\mathbf{y} = A\mathbf{x}$, for some vector **x** (note that $\mathbf{y} \in \mathcal{R}^m$ while $\mathbf{x} \in \mathcal{R}^n$) ;

- null$(A) \equiv$ *nullspace* of $A \equiv$ all vectors $\mathbf{z} \in \mathcal{R}^n$ for which $A\mathbf{z} = \mathbf{0}$, where **0** denotes a vector of zeros.[13]

If A is $n \times n$ and singular, then there are vectors in $\mathcal{R}^n$ which do not belong to range(A), and there are nonzero vectors which belong to A's nullspace, null(A). So, in this case, for a particular given right-hand-side vector **b** the equation $A\mathbf{x} = \mathbf{b}$ will not have a solution if $\mathbf{b} \notin$ range(A). On the other hand, if A is singular and $\mathbf{b} \in$ range(A), then there are infinitely many solutions. This is because if some particular $\hat{\mathbf{x}}$ satisfies the linear system of equations $A\hat{\mathbf{x}} = \mathbf{b}$, then $\mathbf{x} = \hat{\mathbf{x}} + \tilde{\mathbf{x}}$ is also a solution for any $\tilde{\mathbf{x}} \in$ null(A): $A(\hat{\mathbf{x}} + \tilde{\mathbf{x}}) = \mathbf{b} + \mathbf{0} = \mathbf{b}$.

To summarize, the following statements are equivalent:

- A is nonsingular.
- $\det(A) \neq 0$.
- The columns of A are linearly independent.
- The rows of A are linearly independent.
- There exists a matrix which we denote by A^{-1}, which satisfies $A^{-1}A = I = AA^{-1}$.
- range$(A) = \mathcal{R}^n$.
- null$(A) = \{\mathbf{0}\}$.

Example 4.2. The matrix $A = \begin{pmatrix} 1 & 1 \\ 3 & 3 \end{pmatrix}$ is singular, as noted earlier. The nullspace of A can be found in a straightforward way by considering a linear system of the form $A\mathbf{x} = \mathbf{0}$. It readily follows that the elements of **x** are characterized by the relation $x_1 + x_2 = 0$; hence the nullspace is given by

$$\text{null}(A) = \left\{ \alpha \begin{pmatrix} 1 \\ -1 \end{pmatrix} \quad \forall \alpha \in \mathcal{R} \right\}.$$

Next, observe that if $\mathbf{b} \in$ range(A), then it can be written as a linear combination of the columns of A; hence

$$\text{range}(A) = \left\{ \beta \begin{pmatrix} 1 \\ 3 \end{pmatrix} \quad \forall \beta \in \mathcal{R} \right\}.$$

For the case $\mathbf{b} = \begin{pmatrix} 2 \\ 3 \end{pmatrix}$ there is clearly no solution because $\mathbf{b} \notin$ range(A). By the same token, for the case $\mathbf{b} = \begin{pmatrix} 2 \\ 6 \end{pmatrix} = 2\begin{pmatrix} 1 \\ 3 \end{pmatrix}$ we can be sure that $\mathbf{b} \in$ range(A). By inspection there is a solution $\hat{\mathbf{x}} = \begin{pmatrix} 1 \\ 1 \end{pmatrix}$. Now, any vector of the form $\mathbf{x} = \begin{pmatrix} 1+\alpha \\ 1-\alpha \end{pmatrix}$ for any scalar α is also a solution. In fact, for this particular choice of **b** any vector **x** whose elements satisfy $x_1 + x_2 = 2$ is a solution. For instance, $\mathbf{x} = \begin{pmatrix} 1001 \\ -999 \end{pmatrix}$ is a solution, although intuitively it is quite far from $\hat{\mathbf{x}}$. ∎

---

[13]Mathematicians often prefer the term *kernel* over nullspace and use ker(A) rather than null(A) to denote it.

4.1. Review of basic concepts

Almost singularity

In practice we often (though certainly not always) encounter situations where the matrix A for a given linear system is known to be nonsingular. However, what if A is "almost singular"?

Example 4.3. Consider the previous example, slightly perturbed to read

$$\begin{pmatrix} 1+\varepsilon & 1 \\ 3 & 3 \end{pmatrix} \begin{pmatrix} x_1 \\ x_2 \end{pmatrix} = \begin{pmatrix} 2+\varepsilon \\ 6 \end{pmatrix},$$

where ε is a small but positive number, $0 < \varepsilon \ll 1$. (You can think of $\varepsilon = 10^{-10}$, for instance.)

The matrix in this example is nonsingular, so there is a unique solution, $\hat{\mathbf{x}} = \begin{pmatrix} 1 \\ 1 \end{pmatrix}$. However, upon a slight perturbation (removing ε) the matrix can become singular and there are many solutions, some of which are rather far from $\hat{\mathbf{x}}$; for instance, set $\alpha = 1000$ in Example 4.2. ∎

The simple Examples 4.2 and 4.3 indicate that if the matrix A is "almost singular," in the sense that a slight perturbation makes it singular, then a small change in the problem of solving linear systems with this matrix may cause a large change in the solution. Such problems are **ill-conditioned**.

Basics of eigenvalue problems

The algebraic eigenvalue problem

$$A\mathbf{x} = \lambda \mathbf{x}$$

is fundamental. Basically, one wants to characterize the action of the matrix A in simple terms. Here the product of A and the **eigenvector x** equals the product of the scalar **eigenvalue** λ and the same vector. We require $\mathbf{x} \neq \mathbf{0}$ in order not to say something trivial. Together, $(\lambda, \mathbf{x})$ is an **eigenpair**, and the set of eigenvalues forms the **spectrum** of A.[14] If there are n eigenpairs $(\lambda_j, \mathbf{x}_j)$ and if the eigenvectors are linearly independent, then any other vector $\mathbf{y} \in \mathcal{R}^n$ can be written as their linear combination, i.e., there are coefficients α_j such that

$$\mathbf{y} = \sum_{j=1}^{n} \alpha_j \mathbf{x}_j,$$

and hence the operation of our matrix A on this arbitrary vector can be characterized in terms of the eigenpairs as

$$A\mathbf{y} = \sum_{j=1}^{n} \alpha_j A \mathbf{x}_j = \sum_{j=1}^{n} \alpha_j \lambda_j \mathbf{x}_j.$$

The coefficient α_j in $\mathbf{y}$ is thus replaced by $\alpha_j \lambda_j$ in $A\mathbf{y}$ and a powerful decomposition is obtained.

The n eigenvalues of an $n \times n$ matrix

Does every real $n \times n$ matrix have n eigenvalues? We can write the eigenvalue problem as a homogeneous linear system

$$(\lambda I - A)\mathbf{x} = \mathbf{0}.$$

---

[14] The eigenvectors $\mathbf{x}$ are sometimes referred to as *right* eigenvectors, to distinguish between them and *left* eigenvectors, which are given by

$$\mathbf{w}^T A = \lambda \mathbf{w}^T.$$

Since we want a nontrivial **x**, this means that $\lambda I - A$ must be singular. Therefore we can find λ by forming the **characteristic polynomial** and finding its roots. That is, in principle we solve

$$\det(\lambda I - A) = 0.$$

Example 4.4. For a 2×2 matrix we have

$$\det(\lambda I - A) = \det \begin{pmatrix} \lambda - a_{11} & -a_{12} \\ -a_{21} & \lambda - a_{22} \end{pmatrix} = (\lambda - a_{11})(\lambda - a_{22}) - a_{12}a_{21}$$
$$= \lambda^2 - (a_{11} + a_{22})\lambda + (a_{11}a_{22} - a_{12}a_{21}).$$

We therefore have to solve a quadratic equation of the form

$$\lambda^2 - b\lambda + c = 0.$$

From high school we know that to solve this we should form the discriminant $\Delta = b^2 - 4c$. A short calculation shows that here $\Delta = (a_{11} - a_{22})^2 + 4a_{12}a_{21}$. If $\Delta > 0$, then there are two real solutions given by

$$\lambda_{1,2} = \frac{1}{2}(b \pm \sqrt{\Delta}) = \frac{1}{2}(a_{11} + a_{22} \pm \sqrt{\Delta});$$

if $\Delta = 0$, then there is one real solution; and if $\Delta < 0$, then there are no real solutions. This is also the spirit of Chapter 3, where only distinct real solutions are sought for nonlinear equations.

Here, however, there is no reason to restrict ourselves to real numbers: we want two eigenvalues rain or shine, and this means we must be ready for a situation where the eigenvalues may be *complex numbers*.[15]

In the complex plane we now have that in case of $\Delta < 0$ the eigenvalues are

$$\lambda_{1,2} = \frac{1}{2}(a_{11} + a_{22} \pm \imath \sqrt{-\Delta}),$$

and if $\Delta = 0$, then there is the double root $\frac{1}{2}(a_{11} + a_{22})$.

Thus, for the matrix $A = \begin{pmatrix} 1 & 2 \\ 2 & 4 \end{pmatrix}$ we have $\Delta = 3^2 + 4^2 = 25$; hence $\lambda_1 = (5+5)/2 = 5$, $\lambda_2 = (5-5)/2 = 0$. For the matrix $A = \begin{pmatrix} 0 & 1 \\ -1 & 0 \end{pmatrix}$ we have $\Delta = -4$, and it follows that $\lambda_1 = \imath$, $\lambda_2 = -\imath$. ∎

The lessons of Example 4.4, if not its simplicity, generalize. Clearly, the degree of the characteristic polynomial is equal to the dimension of the matrix. A polynomial of degree n has n generally complex and not necessarily distinct roots, and we can write

$$\det(\lambda I - A) = (\lambda - \lambda_1)(\lambda - \lambda_2) \cdot \ldots \cdot (\lambda - \lambda_n) = 0,$$

where the roots $\lambda_1, \lambda_2, \ldots, \lambda_n$ are the eigenvalues of A. If there are k roots that are equal to each other, call their value λ_j; then we say that λ_j is an eigenvalue with **algebraic multiplicity** k. The sum of the algebraic multiplicities of all the eigenvalues of A must be n, the dimension of A.

---

[15]Complex numbers make a rare appearance in our text, but when they do their role can be critical. This is so in eigenvalue problems (Chapter 8), Fourier transform (Chapter 13), and differential equations (Chapter 16). Their basic properties are summarized on the facing page.

4.1. Review of basic concepts

Complex Numbers.
A complex number z can be written as

$$z = x + \iota y,$$

where x and y are two real scalars and $\iota = \sqrt{-1}$. We write $x = \Re(z)$, $y = \Im(z)$. (In MATLAB, x = real(z) and y = imag(z).)

It is customary to view z as a point (a vector from the origin) on the **complex plane** with $\Re(z)$ on the x-axis and $\Im(z)$ on the y-axis. The Euclidean length of the vector z is then the **magnitude**

$$|z| = \sqrt{x^2 + y^2} = \sqrt{(x + \iota y)(x - \iota y)} = \sqrt{z \bar{z}},$$

where $\bar{z}$ is the **conjugate** of z. Complex eigenvalues, for instance, appear in conjugate pairs.

The fundamental **Euler identity** reads

$$e^{\iota \theta} = \cos(\theta) + \iota \sin(\theta) \tag{4.1}$$

for any real angle θ (in radians). In polar coordinates (r, θ) we can then write a complex number z as

$$z = r e^{\iota \theta} = r \cos(\theta) + \iota r \sin(\theta) = x + \iota y,$$

where $r = |z|$ and $\tan(\theta) = y/x$.

Example 4.5. The matrix

$$A = \begin{pmatrix} -0.9880 & 1.8000 & -0.8793 & -0.5977 & -0.7819 \\ -1.9417 & -0.5835 & -0.1846 & -0.7250 & 1.0422 \\ 0.6003 & -0.0287 & -0.5446 & -2.0667 & -0.3961 \\ 0.8222 & 1.4453 & 1.3369 & -0.6069 & 0.8043 \\ -0.4187 & -0.2939 & 1.4814 & -0.2119 & -1.2771 \end{pmatrix}$$

has eigenvalues given approximately by $\lambda_1 = -2$, $\lambda_2 = -1 + 2.5\iota$, $\lambda_3 = -1 - 2.5\iota$, $\lambda_4 = 2\iota$, and $\lambda_5 = -2\iota$. These eigenvalues are plotted in Figure 4.2. ∎

The connection between eigenvalues and the roots of the characteristic polynomial has further implications. It is known that closed form formulas for the roots of a polynomial do not generally exist if the polynomial is of degree 5 or higher. Thus, unlike the solution of a linear system of equations, we may not expect to be able to solve the eigenvalue problem to infinite precision, even in the absence of roundoff errors.

Diagonalizability and spectral decomposition

Note that for a given eigenvalue λ_j the corresponding eigenvector $\mathbf{x}_j$ may be complex valued if λ_j is not real. Moreover, the eigenvector is fixed only up to a constant.

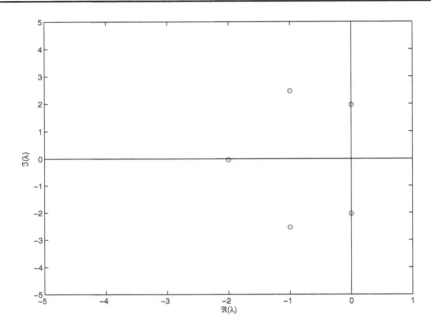

Figure 4.2. *Eigenvalues in the complex plane, Example 4.5. Note that the complex ones arrive in pairs: if λ is an eigenvalue, then so is $\bar{\lambda}$. Also, here all eigenvalues have nonpositive real parts.*

Let us mention a few basic properties of eigenvalues:

1. If $A\mathbf{x} = \lambda\mathbf{x}$, then for any complex scalar α

$$(A + \alpha I)\mathbf{x} = (\lambda + \alpha)\mathbf{x}.$$

2. For any positive integer k

$$A^k\mathbf{x} = \lambda^k\mathbf{x}.$$

3. If $B = S^{-1}AS$ and $A\mathbf{x} = \lambda\mathbf{x}$, then $B\mathbf{y} = \lambda\mathbf{y}$, where $\mathbf{x} = S\mathbf{y}$. In other words, B has the same eigenvalues as A. We say that B is **similar** to A and call the transformation $A \to S^{-1}AS$ a **similarity transformation** of A.

4. **Spectral decomposition:** Let X be the matrix whose columns are eigenvectors of A, and suppose that X is square and nonsingular. Then

$$\begin{aligned} AX &= A[\mathbf{x}_1, \mathbf{x}_2, \ldots, \mathbf{x}_n] \\ &= [\lambda_1\mathbf{x}_1, \lambda_2\mathbf{x}_2, \ldots, \lambda_n\mathbf{x}_n] \\ &= X\Lambda, \end{aligned}$$

where Λ is a diagonal matrix with the eigenvalues on its diagonal, $\Lambda = \text{diag}(\lambda_1, \lambda_2, \ldots, \lambda_n)$. Therefore, we can write

$$A = X\Lambda X^{-1}.$$

4.2. Vector and matrix norms

This decomposition is the *spectral decomposition* of A, and any matrix A that can be decomposed in this fashion is **diagonalizable**.

Unfortunately, the last property above does not apply to just any matrix; there are matrices that do not have a complete, linearly independent set of eigenvectors. Such matrices are not diagonalizable and are labeled by less than flattering adjectives, such as **defective** matrices. Given an eigenvalue, its **geometric multiplicity** is defined as the dimension of the space that its associated eigenvectors span; in other words, it is the number of linearly independent eigenvectors associated with this eigenvalue. It is enough to have one eigenvalue whose geometric multiplicity is smaller than its algebraic multiplicity for the matrix to lose its diagonalizability. In the sense described earlier, eigenvalues of a defective matrix capture less of its action than in the diagonalizable case.

Example 4.6. The eigenvalues of any upper triangular or lower triangular matrix can be read off its main diagonal. For example, the matrix $A = \begin{pmatrix} 4 & 1 \\ 0 & 4 \end{pmatrix}$ is 2×2 and has the eigenvalue 4 with algebraic multiplicity 2 and eigenvector $(1,0)^T$. But no matter how hard you look, you will not be able to find another (linearly independent) eigenvector. The geometric multiplicity of the eigenvalue 4 is, then, only 1.

Take next the matrix $A = \begin{pmatrix} 4 & 0 \\ 0 & 4 \end{pmatrix}$. Trivially, this is just the identity matrix multiplied by 4. Here we can easily find two linearly independent eigenvectors, the simplest choice being $(1,0)^T$ and $(0,1)^T$, so the geometric multiplicity of the eigenvalue 4 equals 2. ∎

A question that we may ask ourselves is in which circumstances we should be concerned with potential difficulties when solving an eigenvalue problem. It turns out that nondiagonalizable matrices are such a source for trouble, as illustrated below in a very simple setting.

Example 4.7. Take the first matrix from Example 4.6, and perturb its off-diagonal elements slightly so that now $A = \begin{pmatrix} 4 & 1.01 \\ 0.01 & 4 \end{pmatrix}$. Using the MATLAB function `eig` we find that the eigenvalues of our (diagonalizable) matrix are approximately 4.1005 and 3.8995. This result is somewhat alarming: a perturbation of magnitude 0.01 has produced a change of magnitude 0.1005 in the eigenvalues! Such a magnification is precisely the type of ill-conditioning mentioned previously for linear systems, and it can cause a considerable headache if we do not detect it early enough.

On the other hand, applying the same perturbation to the second, diagonal matrix from Example 4.6 produces the eigenvalues 4.01 and 3.99, so the matrix perturbation has not been magnified in the eigenvalues. The eigenvalue 4 in this case is well-conditioned. ∎

Specific exercises for this section: Exercises 1–3.

4.2 Vector and matrix norms

In previous chapters we have defined relative and absolute errors for scalars and have seen their use for assessing the quality of an algorithm. Errors measure the distance between the true solution and the computed solutions, so for solutions of linear systems and other problems in linear algebra we need to equip ourselves with the notion of a distance between vectors and between matrices. This gives rise to the notion of a **norm**, generalizing the concept of absolute value (or magnitude) of a scalar.

Vector norms

For a vector $\mathbf{x} \in \mathcal{R}^n$ the Euclidean length is defined as

$$\|\mathbf{x}\|_2 = \sqrt{\mathbf{x}^T \mathbf{x}}$$
$$= (x_1^2 + x_2^2 + \cdots + x_n^2)^{1/2}$$
$$= \left(\sum_{i=1}^n x_i^2\right)^{1/2}.$$

This is called the ℓ_2-*norm*, and it is the usual way for measuring a vector's length. But there are others:

- The ℓ_∞-*norm* is the magnitude of the largest element in $\mathbf{x}$, defined by

$$\|\mathbf{x}\|_\infty = \max_{1 \leq i \leq n} |x_i|.$$

- The ℓ_1-*norm* is the sum of magnitudes of the elements of $\mathbf{x}$, defined by

$$\|\mathbf{x}\|_1 = \sum_{i=1}^n |x_i|.$$

But what exactly *is* a norm? It is a function with a range of scalar values, denoted $\|\cdot\|$, that satisfies the essential properties of the magnitude of a scalar:

1. $\|\mathbf{x}\| \geq 0$; $\|\mathbf{x}\| = 0$ if and only if $\mathbf{x} = \mathbf{0}$,
2. $\|\alpha \mathbf{x}\| = |\alpha| \|\mathbf{x}\| \quad \forall \alpha \in \mathcal{R}$,
3. $\|\mathbf{x} + \mathbf{y}\| \leq \|\mathbf{x}\| + \|\mathbf{y}\| \quad \forall \mathbf{x}, \mathbf{y} \in \mathcal{R}^n$.

The three norms introduced above (ℓ_1, ℓ_2, and ℓ_∞) are all special cases of a family of ℓ_p-norms defined by

$$\|\mathbf{x}\|_p = \left(\sum_{i=1}^n |x_i|^p\right)^{1/p}, \quad 1 \leq p \leq \infty.$$

It is possible to verify that ℓ_p-norm is a norm by confirming that it satisfies the three properties listed above that define a norm function. (The third property, however, may be nontrivial to show.) Incidentally, this definition of an ℓ_p-norm is good also for complex-valued vectors (such as eigenvectors can be). The maximum norm can be viewed as a limiting case for $p \to \infty$.

Example 4.8. Suppose we want to measure the distance between the two vectors

$$\mathbf{x} = \begin{pmatrix} 11 \\ 12 \\ 13 \end{pmatrix} \quad \text{and} \quad \mathbf{y} = \begin{pmatrix} 12 \\ 14 \\ 16 \end{pmatrix}.$$

Let

$$\mathbf{z} = \mathbf{y} - \mathbf{x} = \begin{pmatrix} 1 \\ 2 \\ 3 \end{pmatrix}.$$

4.2. Vector and matrix norms

Then

$$\|z\|_1 = 1+2+3 = 6,$$
$$\|z\|_2 = \sqrt{1+4+9} \approx 3.7417,$$
$$\|z\|_\infty = 3.$$

In MATLAB we get the above results by typing

```
zz = [12,14,16]' - [11,12,13]';
norm(zz,1)
norm(zz)
norm(zz,inf)
```

Although the values of the different norms are different, they have the same order of magnitude. ∎

We see that in Example 4.8, $\|z\|_1 \geq \|z\|_2 \geq \|z\|_\infty$. This is true in general. Moreover, it is easy to see (and worth remembering) that for any vector $x \in \mathcal{R}^n$

$$\|x\|_\infty \leq \|x\|_2 \leq \sqrt{n}\|x\|_\infty,$$
$$\|x\|_\infty \leq \|x\|_1 \leq n\|x\|_\infty.$$

Example 4.9. Figure 4.3 displays plots of all vectors in the plane $\mathcal{R}^2$ whose norm equals 1, according to each of the three norms we have seen.

Figure 4.3. *The "unit circle" according to the three norms, ℓ_1, ℓ_2, and ℓ_∞. Note that the diamond is contained in the circle, which in turn is contained in the square.*

Our usual geometric notion of a circle is captured by the ℓ_2-norm, less so by the other two.[16] Indeed, even though the ℓ_2-norm is harder to compute (especially for matrices, as becomes clear below) we will use it in the sequel, unless otherwise specifically noted. ∎

Matrix norms

To be able to answer questions such as how far a matrix is from being singular, and to be able to examine quantitatively the stability of our algorithms, we need matrix norms.

---

[16]The ℓ_1-norm is occasionally called the "Manhattan norm." Figure 4.3 should help explain why.

Matrix norms (i.e., functions which take a matrix as an argument, produce a scalar value, and satisfy the above three norm conditions) can be obtained in many different ways. But of particular interest to us is the **induced** or *natural* norm associated with each vector norm. Given an $m \times n$ matrix A and a vector norm $\|\cdot\|$, define

$$\|A\| = \max_{\mathbf{x} \neq 0} \frac{\|A\mathbf{x}\|}{\|\mathbf{x}\|}$$
$$= \max_{\|\mathbf{x}\|=1} \|A\mathbf{x}\|.$$

Note that for all vectors $\mathbf{x}$ which are not identically equal to the zero-vector, $\|\mathbf{x}\| > 0$. Also, for each such vector, dividing by its norm we can normalize it. Thus, the above definition makes sense.

It is not difficult to show that induced matrix norms satisfy the three norm conditions, as well as the *consistency* condition $\|AB\| \leq \|A\|\|B\|$. In particular, for any vector $\mathbf{x}$ of an appropriate dimension, we have

$$\|A\mathbf{x}\| \leq \|A\|\|\mathbf{x}\|.$$

Here, then, are the four properties that our matrix norms satisfy:

1. $\|A\| \geq 0$; $\|A\| = 0$ if and only if $A = 0$ (elementwise);
2. $\|\alpha A\| = |\alpha|\|A\|$ $\forall \alpha \in \mathcal{R}$;
3. $\|A + B\| \leq \|A\| + \|B\|$ $\forall A, B \in \mathcal{R}^{m \times n}$;
4. $\|A \cdot B\| \leq \|A\|\|B\|$ $\forall A \in \mathcal{R}^{m \times n}$, $B \in \mathcal{R}^{n \times l}$.

When referring to the usual induced norms, a common convention is to omit the ℓ and refer to the analogues of the ℓ_1-, ℓ_2-, ℓ_∞-norms for vectors as the 1-norm, 2-norm, and ∞-norm, respectively, for matrices.

Calculating induced matrix norms

Calculating $\|A\|_\infty$ is simple. Letting $\mathbf{x}$ be any vector with $\|\mathbf{x}\|_\infty = 1$ we have $|x_j| \leq 1$, $j = 1,\ldots,n$, and hence

$$\|A\mathbf{x}\|_\infty = \max_i \left| \sum_{j=1}^n a_{ij} x_j \right|$$
$$\leq \max_i \sum_{j=1}^n |a_{ij}||x_j| \leq \max_i \sum_{j=1}^n |a_{ij}|,$$

so

$$\|A\|_\infty = \max_{\|\mathbf{x}\|_\infty=1} \|A\mathbf{x}\|_\infty \leq \max_i \sum_{j=1}^n |a_{ij}|.$$

But equality can be achieved by a judicious choice of $\mathbf{x}$: if p is the maximizing row in the above expression, i.e., $\sum_{j=1}^n |a_{pj}| = \max_i \sum_{j=1}^n |a_{ij}|$, then choose $\tilde{x}_j = sign(a_{pj})$ for each j. This yields that $a_{pj} \tilde{x}_j = |a_{pj}|$, so for this particular vector $\mathbf{x} = \tilde{\mathbf{x}}$ we have

$$\|A\tilde{\mathbf{x}}\|_\infty = \max_i \left| \sum_{j=1}^n a_{ij} \tilde{x}_j \right| \geq \left| \sum_{j=1}^n a_{pj} \tilde{x}_j \right| = \sum_{j=1}^n |a_{pj}|.$$

4.2. Vector and matrix norms

Combining the two estimates, we obtain

$$\sum_{j=1}^{n} |a_{pj}| \leq \|A\tilde{\mathbf{x}}\|_\infty \leq \|A\|_\infty \leq \sum_{j=1}^{n} |a_{pj}|,$$

and this implies that there is equality throughout in the derivation. We conclude that in general

$$\|A\|_\infty = \max_{1 \leq i \leq m} \sum_{j=1}^{n} |a_{ij}|.$$

An expression of similar simplicity is available also for the 1-norm and is given by

$$\|A\|_1 = \max_{1 \leq j \leq n} \sum_{i=1}^{m} |a_{ij}|.$$

No such simple expression is available for the 2-norm, though!

Example 4.10. Consider

$$A = \begin{pmatrix} 1 & 3 & 7 \\ -4 & 1.2725 & -2 \end{pmatrix}.$$

Then

$$\|A\|_\infty = \max\{11, 7.2725\} = 11,$$
$$\|A\|_1 = \max\{5, 4.2725, 9\} = 9,$$
$$\|A\|_2 = \max_{\mathbf{x}^T\mathbf{x}=1} \|A\mathbf{x}\|_2 = \max_{\mathbf{x}^T\mathbf{x}=1} \sqrt{(A\mathbf{x})^T A\mathbf{x}} = \max_{\mathbf{x}^T\mathbf{x}=1} \sqrt{\mathbf{x}^T(A^T A)\mathbf{x}}.$$

We can calculate

$$A^T A = \begin{pmatrix} 17 & -2.09 & 15 \\ -2.09 & 10.6193 & 18.455 \\ 15 & 18.455 & 53 \end{pmatrix},$$

but we still need to figure out how to maximize the quadratic form $\mathbf{x}^T(A^T A)\mathbf{x}$. Please stay tuned. ■

The 2-norm and spectral radius

To calculate the 2-norm of a matrix, we need to consider the eigenvalues of $A^T A$. Let B be a square, $n \times n$ real matrix. From our discussion in the previous section, recall that B has n *eigenvalues*. We define the **spectral radius** of B as

$$\rho(B) = \max\{|\lambda|; \ \lambda \text{ is an eigenvalue of } B\}.$$

Obviously for any induced matrix norm, if $(\tilde{\lambda}, \tilde{\mathbf{x}})$ is the eigenvalue-eigenvector pair which maximizes $|\lambda|$ for B, with $\tilde{\mathbf{x}}$ normalized to satisfy $\|\tilde{\mathbf{x}}\| = 1$, then

$$\rho(B) = |\tilde{\lambda}| = \|\tilde{\lambda}\tilde{\mathbf{x}}\| = \|B\tilde{\mathbf{x}}\| \leq \|B\|.$$

So, any induced norm is always bounded below by the spectral radius! Note that the spectral radius of a general square matrix is *not* a norm. For example, for $C = \begin{pmatrix} 0 & 1 \\ 0 & 0 \end{pmatrix}$ we have $\rho(C) = 0$ even though $C \neq 0$.

The spectral radius is often a pretty good lower bound for a norm. Further, it can be shown that even for rectangular (i.e., not necessarily square) matrices, we have

$$\|A\|_2 = \sqrt{\rho(A^T A)}.$$

Example 4.11. Continuing with Example 4.10 the eigenvalues of the matrix $A^T A$, found by the MATLAB command `eig(A'*A)`, are (approximately) 0, 16.8475, and 63.7718. Thus, we finally get

$$\|A\|_2 \approx \sqrt{63.7718} \approx 7.9857.$$

This approximate value is in fact much more accurate than we would ever want for a norm in practice. ∎

The square roots of the eigenvalues of $B = A^T A$ (which are all real and nonnegative) are called the **singular values** of A. The 2-norm of a matrix is therefore equal to its largest singular value. We return to singular values in full force in Section 4.4; see especially page 81. For now, let us just say that for general square matrices (not symmetric positive definite ones) the singular values can be significantly more friendly than eigenvalues: the singular values are real and positive, and they can be computed without forming $A^T A$.

Specific exercises for this section: Exercises 4–6.

4.3 Special classes of matrices

There are several important classes of matrices for which the algorithms for solution of linear systems or eigenvalue problems can be simplified or made significantly more efficient than in the general case. We will see such matrices in detail in the next few chapters, where their importance and treatment become relevant in a local context. In this section we define two particularly important classes of matrices: symmetric positive definite matrices and orthogonal matrices.

Why single out these particular two classes? Because of their fundamental nature. Symmetric positive definite matrices extend the notion of a positive scalar, arise in many real-world applications, and virtually divide numerical linear algebra techniques into two groups—those that are designed for these specific matrices, and those that are not. Orthogonal matrices form a particularly useful class of matrix transformations which are the basis of many modern methods, in a sense that will become clearer in Chapters 6–8.

Symmetric positive definite matrices

The square matrix A is **symmetric** if $A^T = A$. Furthermore, a matrix A is **positive definite** if

$$\mathbf{x}^T A \mathbf{x} > 0 \quad \forall \, \mathbf{x} \neq \mathbf{0}.$$

Thus, for any column vector $\mathbf{x} = (x_1, \ldots, x_n)^T$ we require $\sum_{i,j=1}^{n} a_{i,j} x_i x_j > 0$, provided that at least one component $x_j \neq 0$.

Symmetric positive definite matrices may be considered an extension of positive scalars to n dimensions. They arise often in practice.

4.3. Special classes of matrices

Example 4.12. Suppose we wish to find a minimum of a nonlinear C^2 function of n variables, $\phi(\mathbf{x}) = \phi(x_1, \ldots, x_n)$. This task extends minimization in one variable, considered in Section 3.4. Recall that for one variable the necessary condition for a minimum at $x = x^*$ is $\phi'(x^*) = 0$, and a sufficient condition is that also $\phi''(x^*) > 0$.

In n variables the necessary condition for a minimum of ϕ at a point $\mathbf{x}^*$ is that the vector of first derivatives $\frac{\partial \phi}{\partial x_i}$, called the **gradient**, vanish at $\mathbf{x}^*$; a sufficient condition that such a point actually yields a minimum (and not a maximum or a saddle point) is that the matrix A of second derivatives of ϕ, having the elements $a_{i,j} = \frac{\partial^2 \phi}{\partial x_i \partial x_j}$, be symmetric positive definite at $\mathbf{x}^*$. We study such problems in Section 9.2. ∎

A symmetric $n \times n$ matrix has n *real* eigenpairs. If A is also positive definite, then for any eigenvalue $\lambda = \lambda_j$ and its associated eigenvector $\mathbf{x} = \mathbf{x}_j$ we can write

$$0 < \mathbf{x}^T A \mathbf{x} = \lambda \mathbf{x}^T \mathbf{x} = \lambda \sum_{i=1}^{n} x_i^2,$$

implying that $\lambda_j > 0$. Hence all eigenvalues are positive, and we can order them as

$$\lambda_1 \geq \lambda_2 \geq \cdots \geq \lambda_n > 0.$$

It is easy to see that if A is symmetric positive definite, then all principal minors $A_{[j:k,j:k]}$ are also symmetric positive definite (simply consider all vectors $\mathbf{x}$ that have zeros in the first $j-1$ and the last $n-k$ places). The determinants of these principal minors are all positive as well.

Note that for any rectangular matrix B with linearly independent columns the matrix $A = B^T B$ is symmetric positive definite, because for any nontrivial vector $\mathbf{x}$ we can write

$$\mathbf{x}^T A \mathbf{x} = \mathbf{x}^T B^T B \mathbf{x} = \|B\mathbf{x}\|_2^2 > 0.$$

Recall that

$$\|B\|_2 = \sqrt{\rho(A)}.$$

Orthogonal vectors and matrices

Two vectors $\mathbf{u}$ and $\mathbf{v}$ of the same length are **orthogonal** to each other if their inner product vanishes, i.e.,

$$\mathbf{u}^T \mathbf{v} = 0.$$

If in addition each vector has a unit length, i.e., $\|\mathbf{u}\|_2 = \|\mathbf{v}\|_2 = 1$, then we say that the vectors are **orthonormal**.

In geometric terms, orthogonality of two vectors simply means that the vectors form a right angle between them. Of course, the notion of orthogonality extends to more than two vectors: a set of vectors are orthogonal (orthonormal) if each pair among them satisfies the above-mentioned orthogonality (orthonormality) relation.

Example 4.13. The simplest example of a set of orthogonal vectors is those which form the axes of the standard coordinate system. In three dimensions we have $\mathbf{e}_1 = (1,0,0)^T$, $\mathbf{e}_2 = (0,1,0)^T$, and $\mathbf{e}_3 = (0,0,1)^T$, scaling the vectors so they are orthonormal; see Figure 4.4.

In this system, any vector $\mathbf{w} = (x,y,z)^T$ can be expressed as $\mathbf{w} = x\mathbf{e}_1 + y\mathbf{e}_2 + z\mathbf{e}_3$. We then say that x, y, and z are the coordinates of $\mathbf{w}$ with respect to the orthogonal basis $(\mathbf{e}_1, \mathbf{e}_2, \mathbf{e}_3)$. All this may seem trivial, because these basis vectors $(\mathbf{e}_1, \mathbf{e}_2, \mathbf{e}_3)$ form an identity matrix I. But in fact, expressing a vector in terms of its coordinates with respect to a given orthogonal basis is quite useful in general for more complicated settings. ∎

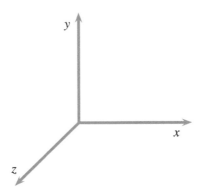

Figure 4.4. *The Cartesian coordinate system as a set of orthonormal vectors.*

Orthogonal matrices

A real square matrix Q is *orthogonal* if its columns are orthonormal, i.e.,

$$Q^T Q = I.$$

It is straightforward to verify that for any vector $\mathbf{x}$ we have in the ℓ_2-norm, $\|Q\mathbf{x}\| = \|\mathbf{x}\|$, because

$$\|Q\mathbf{x}\|^2 = (Q\mathbf{x})^T(Q\mathbf{x}) = \mathbf{x}^T Q^T Q \mathbf{x} = \mathbf{x}^T \mathbf{x} = \|\mathbf{x}\|^2.$$

This property of norm preservation is important in numerical computations. For one thing it means that applying an orthogonal transformation (read, "multiplying by Q") in a floating point environment does not (in principle) amplify errors. Moreover, this mathematical property can be conveniently used for transforming matrices in certain situations, for example, to devise effective solution methods for least squares problems. Details on this are provided in Section 6.2.

In Example 4.13, where a standard coordinate system is considered, if we line up $\mathbf{e}_1$, $\mathbf{e}_2$, and $\mathbf{e}_3$ as columns of a matrix, then we get the identity matrix. The latter is the simplest example of an orthogonal matrix. It is straightforward to observe that orthogonal matrices remain orthogonal under row and column permutations.

Example 4.14. Rigid rotations are another example of an orthogonal linear operator. The matrix

$$\begin{pmatrix} \cos(\theta) & \sin(\theta) \\ -\sin(\theta) & \cos(\theta) \end{pmatrix}$$

is orthogonal for any θ and corresponds simply to a rotation of a two-dimensional vector by an angle θ. In other words, multiplying a given vector by this matrix results in a vector whose length is the same but whose angle with the axes is shifted by θ compared to the original vector. ∎

Specific exercises for this section: Exercises 7–13.

4.4 Singular values

The **singular value decomposition** (SVD) is a basic matrix decomposition that is particularly stable under all weather conditions. The definition is given on the next page.

4.4. Singular values

Singular Value Decomposition.
Given a rectangular, $m \times n$ matrix A, the SVD reads

$$A = U \Sigma V^T,$$

where U is an $m \times m$ orthogonal matrix, V is an $n \times n$ orthogonal matrix, and Σ is an $m \times n$ matrix given by

$$\Sigma = \begin{pmatrix} \sigma_1 & & & & \\ & \sigma_2 & & & \\ & & \ddots & & \\ & & & \sigma_r & \\ & & & & \\ & & & & \end{pmatrix},$$

with the *singular values* $\sigma_1 \geq \sigma_2 \geq \cdots \geq \sigma_r > 0$. Everything else in Σ is zero.

Note the generality of the SVD in terms of matrix dimensions: no relationship between the positive integers m and n is imposed, and the rank of A satisfies $\text{rank}(A) = r \leq \min(m,n)$. To get a feeling for the dimensions of the matrices involved in such a decomposition, see Figure 4.5.

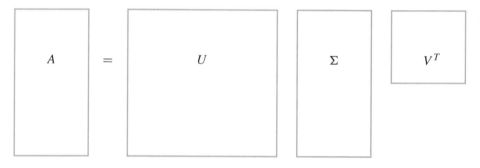

Figure 4.5. *Singular value decomposition:* $A_{m \times n} = U_{m \times m} \Sigma_{m \times n} V^T_{n \times n}$.

In MATLAB the relevant command is (you guessed correctly) svd. The instruction [U,Sigma,V]=svd(A) yields the three matrices appearing in the SVD definition. Somewhat confusingly, though usefully, the instruction s=svd(A) gives just the singular values in a one-dimensional array of length $\min(m,n)$. The first r positive singular values are padded with zeros if $r < \min(m,n)$.

Example 4.15. The SVD of the matrix

$$A = \begin{pmatrix} 1 & 2 \\ 3 & 4 \\ 5 & 6 \end{pmatrix}$$

is given by

$$U = \begin{pmatrix} 0.229847696400071 & 0.883461017698525 & -0.408248290463864 \\ 0.524744818760294 & 0.240782492132547 & 0.816496580927726 \\ 0.819641941120516 & -0.401896033433433 & -0.408248290463863 \end{pmatrix};$$

$$\Sigma = \begin{pmatrix} 9.525518091565107 & 0 \\ 0 & 0.514300580658642 \\ 0 & 0 \end{pmatrix};$$

$$V = \begin{pmatrix} 0.619629483829340 & -0.784894453267053 \\ 0.784894453267053 & 0.619629483829340 \end{pmatrix}.$$

The singular values of A are extracted from the diagonal of Σ and hence are $\sigma_1 = 9.525518091565107$ and $\sigma_2 = 0.514300580658642$. ∎

For a square $n \times n$ matrix A all three matrices U, V, and Σ are also $n \times n$. Then we can write the diagonal matrix Σ as

$$\Sigma = \begin{pmatrix} \sigma_1 & & & & & \\ & \ddots & & & & \\ & & \sigma_r & & & \\ & & & 0 & & \\ & & & & \ddots & \\ & & & & & 0 \end{pmatrix}.$$

Next, let $r = n \le m$. Then $C = A^T A$ is a symmetric positive definite matrix, as we have seen in Section 4.3. Denote its eigenvalues, in decreasing order, by λ_i, $i = 1,\ldots,n$. We can write

$$C = A^T A = V \Sigma^T U^T U \Sigma V^T = V \Sigma^T \Sigma V^T.$$

The matrix $\Sigma^T \Sigma$ is diagonal with $\sigma_1^2,\ldots,\sigma_n^2$ on its main diagonal, and V^T provides a similarity transformation for C. Thus, $\sigma_i = \sqrt{\lambda_i}$ and

$$\|A\|_2 = \sigma_1.$$

So, the 2-norm of A is simply given by its largest singular value. This applies to any nonsquare matrix.

> **Note:** In many ways singular values are to a general matrix what eigenvalues are to a symmetric positive definite one. On the other hand, make no mistake: for a general square matrix, singular values correspond to eigenvalue magnitudes and as such *do not* simply replace eigenvalues in terms of information content.

Specific exercises for this section: Exercises 14–16.

4.5 Examples

Linear systems and eigenvalue problems arise in numerous applications and come in many flavors. In this section we present three examples that illustrate the diversity of applications that lead to such linear algebra problems.

The following descriptions are relatively long and may be considered as setting the stage for three case studies, to be returned to repeatedly in later chapters.

Example 4.16 (data fitting). Consider the problem of linear *data fitting*. Given measurements, or observations, $(t_1,b_1),(t_2,b_2),\ldots,(t_m,b_m) = \{(t_i,b_i)\}_{i=1}^m$, we want to fit to this data a function $v(t)$ which is described as a linear combination of n known, linearly independent functions of t, $\phi_1(t),\phi_2(t),\ldots,\phi_n(t)$. The points t_i where the observations b_i are made are assumed distinct, $t_1 < t_2 < \cdots < t_m$. The functions $\phi_i(t)$ are called **basis functions**.

Thus, we write

$$v(t) = \sum_{j=1}^n x_j \phi_j(t)$$

and seek coefficients $x_1,\ldots,x_n$ such that $v(t_i) = b_i, i = 1,\ldots,m$. Since the sought function v obviously depends on the choice of coefficients, let us denote this explicitly as $v(t;\mathbf{x})$.

We can define

$$a_{ij} = \phi_j(t_i)$$

and write these **interpolation** requirements as

$$\begin{aligned}
a_{11}x_1 + a_{12}x_2 + \cdots + a_{1n}x_n &= b_1, \\
a_{21}x_1 + a_{22}x_2 + \cdots + a_{2n}x_n &= b_2, \\
\vdots &= \vdots \\
a_{m1}x_1 + a_{m2}x_2 + \cdots + a_{mn}x_n &= b_m.
\end{aligned}$$

Let us further assume that $m = n$ (i.e., the same number of data and unknown coefficients) and rewrite the requirements in matrix form $A\mathbf{x} = \mathbf{b}$, where

$$A = \begin{pmatrix} a_{11} & a_{12} & \cdots & a_{1n} \\ a_{21} & a_{22} & \cdots & a_{2n} \\ \vdots & \vdots & \ddots & \vdots \\ a_{n1} & a_{n2} & \cdots & a_{nn} \end{pmatrix}, \quad \mathbf{x} = \begin{pmatrix} x_1 \\ x_2 \\ \vdots \\ x_n \end{pmatrix}, \quad \mathbf{b} = \begin{pmatrix} b_1 \\ b_2 \\ \vdots \\ b_n \end{pmatrix}.$$

If the points t_i are chosen so that the columns of A are linearly independent then there is a unique solution to the above matrix problem, and the interpolant $v(t;\mathbf{x})$ is thus uniquely determined.

For the special case where we choose $\phi_j(t) = t^{j-1}$ we obtain that $v(t;\mathbf{x})$ is the interpolating polynomial of degree $< n$. Our matrix A then becomes

$$A = \begin{pmatrix} 1 & t_1 & \cdots & t_1^{n-1} \\ 1 & t_2 & \cdots & t_2^{n-1} \\ \vdots & \vdots & \ddots & \vdots \\ 1 & t_n & \cdots & t_n^{n-1} \end{pmatrix}.$$

This form of a matrix is called *Vandermonde*. The polynomial interpolation problem always has a solution, but as it turns out, under certain conditions there are numerical difficulties in solving the Vandermonde system.

As a specific instance suppose we have four observations, $(0,1)$, $(0.1,-0.9)$, $(0.8,10)$, and $(1,9)$. The following MATLAB script generates the corresponding 4×4 matrix A and data vector **b**, and then it solves the system $A\mathbf{x} = \mathbf{b}$ by the "backslash" command. The plotting instructions which follow this produce Figure 4.6.

```
t = [0,0.1,0.8,1]'; b = [1,-0.9,10,9]';
A = zeros(4,4); % This tells Matlab we have a 4 by 4 matrix in mind
powers = 0:3;
for j=1:4
  A(:,j) = t.^powers(j);
end

x = A \ b; % This solves the system Ax = b

tt = -0.1:.01:1.1;
pt = x(1) + x(2).*tt + x(3).*tt.^2 + x(4).*tt.^3;
plot(tt,pt)
hold on
plot(t',b','ro','LineWidth',2)
xlabel('t')
ylabel('v')
```

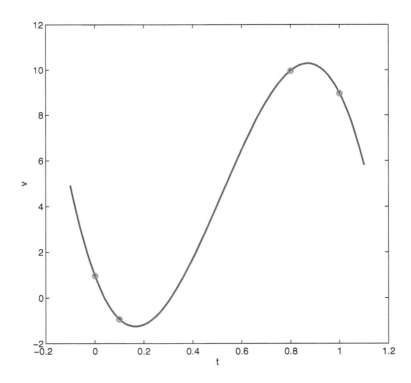

Figure 4.6. *Data fitting by a cubic polynomial using the solution of a linear system for the polynomial's coefficients.*

4.5. Examples

In real-world applications it is often not advisable or even feasible to have as many basis functions as data points. If laboratory measurements are the source for the data points, then there could very well be some inaccurate measurements, and hence redundancy (in the form of extra data points) would be a blessing. In any case it makes little sense to interpolate noisy measurements. Another instance is where we are merely trying to pick up a trend in the data: we consider a list of average temperatures in each of the past 10 years, say, not to carefully approximate what happened between those years but to obtain an indication of a slope, i.e., whether it has generally been warming up or cooling down.

Thus, we now want to find $\mathbf{x} \in \mathcal{R}^n$ such that $v(t_i; \mathbf{x}) \approx b_i$, $i = 1, \ldots, m$, where $m \geq n$, but usually $m > n$. When the latter occurs, we say that the problem is **overdetermined**. It is convenient to imagine values such as $n = 1, 2$, or 3 and $m = 100$, say, when you read the following.

An obvious question is in what sense we want to do the data fitting. It is often natural (or at least easy enough) to use the ℓ_2-norm, which defines a general least squares problem

$$\min_{\mathbf{x}} \sum_{i=1}^{m} [b_i - v(t_i; \mathbf{x})]^2.$$

This in turn can be written as determining the parameters $\mathbf{x}$ from

$$\min_{\mathbf{x}} \sum_{i=1}^{m} \left[b_i - \sum_{j=1}^{n} x_j \phi_j(t_i) \right]^2,$$

and further, with the rectangular matrix A having elements $a_{i,j} = \phi_j(t_i)$ the linear least squares problem can be formulated as

$$\min_{\mathbf{x}} \|\mathbf{b} - A\mathbf{x}\|_2,$$

where the dimensions of the matrix and vectors are

$$A \in \mathcal{R}^{m \times n}, \ \mathbf{x} \in \mathcal{R}^n, \ \mathbf{b} \in \mathcal{R}^m, \ m \geq n.$$

(Note that squaring $\|\mathbf{b} - A\mathbf{x}\|_2$ yields the same minimizer $\mathbf{x}$.)

We describe several methods for solving the least squares problem in Chapter 6. ∎

Example 4.17 (differential equations). Consider the recovery of a function $v(t)$ from its given second derivative $-g(t)$ on the interval $[0, 1]$. Such a problem is typical for many applications. An example is the celebrated heat equation in one dimension: if the intensity of a heat source is given and we wish to recover the steady-state temperature of the body that is being heated, then under certain simplifying assumptions about the heat conductivity properties of the material we obtain a problem of the same type.

Obviously the process as described above is not unique: for any v which satisfies $-v''(t) = g(t)$, also $w(t) = v(t) + \alpha + \beta t$ has the same second derivative for any constants α and β. To fix these constants we need two additional constraints, and we consider the following possibilities:

1. $v(0) = v(1) = 0$, or

2. $v(0) = 0$, $v'(1) = 0$.

This yields a (fortunately simple) *differential equation*, $-v'' = g$, with boundary conditions. To solve it numerically we approximate the differential equation on a discrete mesh, as in Figure 4.7. We subdivide the interval $[0, 1]$ into subintervals of size h each (say, $h = 0.01$), define $t_i = ih$,

Figure 4.7. *A discrete mesh for approximating a function and its second derivative in Example* 4.17.

$i = 0, 1, \ldots, N$, where $Nh = 1$, and approximate $-v'' = g$ by[17]

$$-\frac{v_{i+1} - 2v_i + v_{i-1}}{h^2} = g(t_i), \quad i = 1, 2, \ldots, N-1.$$

For the first case of **boundary conditions** $v(0) = 0$, $v(1) = 0$, we write correspondingly $v_0 = 0$, $v_N = 0$, to close the system. Collecting our unknowns into a vector and the difference equations into a matrix, we get $A\mathbf{v} = \mathbf{g}$, where

$$\mathbf{v} = \begin{pmatrix} v_1 \\ v_2 \\ \vdots \\ v_{N-2} \\ v_{N-1} \end{pmatrix}, \quad \mathbf{g} = \begin{pmatrix} g(t_1) \\ g(t_2) \\ \vdots \\ g(t_{N-2}) \\ g(t_{N-1}) \end{pmatrix}, \quad A = \frac{1}{h^2} \begin{pmatrix} 2 & -1 & & & \\ -1 & 2 & -1 & & \\ & \ddots & \ddots & \ddots & \\ & & -1 & 2 & -1 \\ & & & -1 & 2 \end{pmatrix}.$$

For instance, if $N = 5$, then $h = .2$ and

$$A = 25 \begin{pmatrix} 2 & -1 & 0 & 0 \\ -1 & 2 & -1 & 0 \\ 0 & -1 & 2 & -1 \\ 0 & 0 & -1 & 2 \end{pmatrix}.$$

The matrix A is symmetric positive definite. (Indeed, this is why we have attached the minus sign to v''; without it A would have been negative definite.) It is also very **sparse** when N is large: $a_{i,j} = 0$ if $i > j+1$ or $j > i+1$, so out of $(N-1)^2$ elements of A only fewer than $3(N-1)$ are nonzero. A typical result is displayed in Figure 4.8.

Turning to the other set of boundary conditions, we still set $v_0 = 0$, but now at the other end we have to discretize the condition $v'(1) = 0$. We can do this by adding a *ghost unknown*, v_{N+1}, and

---

[17]For any function $f(t) \in C^4[0,1]$, the second derivative f'' has the finite difference approximation

$$f''(t_i) = \frac{f(t_{i+1}) - 2f(t_i) + f(t_{i-1})}{h^2} + \mathcal{O}(h^2).$$

This can be readily seen by manipulating Taylor expansions of $f(t_{i+1})$ and $f(t_{i-1})$. Thus, for h small, $-(v(t_{i+1}) - 2v(t_i) + v(t_{i-1}))/h^2 \approx -v''(t_i) = g(t_i)$. Solving the specified linear system for the v_i's we then expect that $v_i \approx v(t_i)$. Indeed, as it turns out, $v_i = v(t_i) + \mathcal{O}(h^2)$, but showing this is well beyond the scope of this chapter. All we are interested in here is the resulting linear system for these v_i's.

4.5. Examples

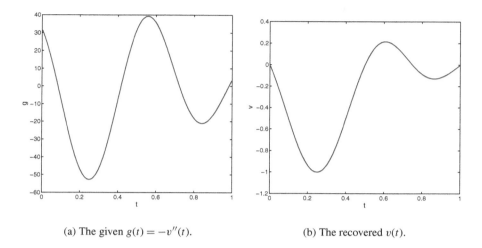

(a) The given $g(t) = -v''(t)$.

(b) The recovered $v(t)$.

Figure 4.8. *Recovering a function v satisfying $v(0) = v(1) = 0$ from its second derivative. Here $N = 507$.*

writing
$$\frac{v_{N+1} - v_{N-1}}{2h} = 0,$$
$$\frac{-v_{N+1} + 2v_N - v_{N-1}}{h^2} = g(t_N),$$

thus extending the differential equation domain all the way to the boundary $t = 1$. The "ghost unknown" corresponds to a location beyond the boundary, but you really don't need to understand this in depth for the present purpose. Thus, the first of these equations yields $v_{N+1} = v_{N-1}$, and the second then reads $\frac{2}{h^2}(v_N - v_{N-1}) = g(t_N)$. So we get the linear system $A\mathbf{v} = \mathbf{g}$ of size N, rather than $N - 1$, because now v_N qualifies as an unknown, as opposed to the situation previously discussed, where we had $v(1) = 0$. The system is

$$\mathbf{v} = \begin{pmatrix} v_1 \\ v_2 \\ \vdots \\ v_{N-1} \\ v_N \end{pmatrix}, \; \mathbf{g} = \begin{pmatrix} g(t_1) \\ g(t_2) \\ \vdots \\ g(t_{N-1}) \\ g(t_N) \end{pmatrix}, \; A = \frac{1}{h^2} \begin{pmatrix} 2 & -1 & & & \\ -1 & 2 & -1 & & \\ & \ddots & \ddots & \ddots & \\ & & -1 & 2 & -1 \\ & & & -2 & 2 \end{pmatrix}.$$

This matrix is no longer symmetric. But it is still very sparse and nonsingular.

What about the boundary conditions $v'(0) = v'(1) = 0$? The above derivation yields a *singular* matrix A! Its nullspace consists of all constant vectors (i.e., vectors whose components all have one and the same value). Indeed, the differential solution $v(t)$ is also determined only up to a constant in this case. ∎

Example 4.18 (principal component analysis). A veteran technique for data analysis that at its core is nothing but an SVD of a rectangular data matrix is principal component analysis (PCA). Below we first briefly discuss some generalities and then settle into a specific instance of a PCA application. We have no intention of reproducing a text on data analysis and statistics here, and if you are not familiar with some of the terminology below, then in our restricted context this should not stop you.

Let A be an $m \times n$ matrix, where each column corresponds to a different experiment of the same type and m is the dimension. We can think of each such column as a point (vector) in $\mathcal{R}^m$. Further, we must assume that the empirical mean has already been subtracted from these columns, so that A has zero mean.

Assuming a normal (Gaussian) probability distribution for the data errors leads to considerations in Euclidean space; i.e., we are thus justified in using the ℓ_2-norm. Hence orthogonal transformations naturally arise. The PCA is such a transformation. It is really a coordinate rotation that aligns the transformed axes with the directions of maximum variance. Thus, writing the SVD as

$$A = U \Sigma V^T,$$

the largest variance is in the direction of the first column of U (the first **principal component**), the largest variance on the subspace orthogonal to the first principal component is in the direction of the second column of U, and so on. The matrix

$$B = U^T A = \Sigma V^T$$

therefore represents a better alignment than the given A in terms of variance differentiation, and this can occasionally be of use.

Note that since the mean is zero, the **covariance matrix** is given by

$$C = A A^T = U \Sigma \Sigma^T U^T.$$

This matrix is positive semidefinite and its eigenvectors are the columns of U, namely, the singular vectors which are the principal components here.

A way to make this process more practical is dimensionality reduction. Letting U_r consist of the first r columns of U, $r < n$, we represent the data by the smaller matrix $B_r = U_r^T A$. Then $B_r = \Sigma_r V_r^T$ with Σ_r and V_r defined correspondingly. See examples in Section 8.2.

Let us emphasize that the assumptions that make PCA work are not always applicable. For statisticians this is a linear, nonparametric analysis that cannot incorporate prior knowledge and assumes some sublime importance of the variance as the way to differentiate or imply similarity between entities. Nonetheless, on many occasions this approach performs well.

Moving on to show a particular application, consider Figure 4.9(a). Graphics objects in $\mathcal{R}^3$ are often represented or approximated by a **point cloud**, a scattered data set of points that together presumably form a coherent surface. Such representations arise in numerous applications including laser scanning, medical imaging, and visualization. Relevant tasks are "cleaning" the data set from noise and outliers, resampling, downsampling, or upsampling it, and displaying the surface that gave rise to this cloud in a pleasing way (see Figure 11.6 on page 359). For such applications we need the normal directions at the points with respect to the implied surface.

Finding these normals (without forming the surface first!) can be done using PCA. For a fixed point $\mathbf{p}$ in the cloud, we define a neighborhood $\mathcal{N}_p$ of nearby points. A simple choice would be those points whose distance in ℓ_2-norm from $\mathbf{p}$ is below some value, but there are more sophisticated ways for determining just who is a neighbor of the point $\mathbf{p}$ under focus. For our present purposes, let us assume it has been done and there are $n = n_p$ such neighbors. Next, we calculate the mean of these neighbors to find the centroid of $\mathcal{N}_p$, call it $\bar{\mathbf{p}}$. Our $3 \times n_p$ data matrix A therefore has $\mathbf{p}_{i_p} - \bar{\mathbf{p}}$ for its ith column, where $\mathbf{p}_{i_p}$ is the ith neighbor of $\mathbf{p}$.

Next, we find the three singular vectors of A (also known as the eigenvectors of the covariance matrix). The first two principal vectors span the **tangent plane** at $\mathbf{p}$. The third one, orthogonal to the first two and hence to the tangent plane, is our desired unsigned normal direction.

4.6. Exercises

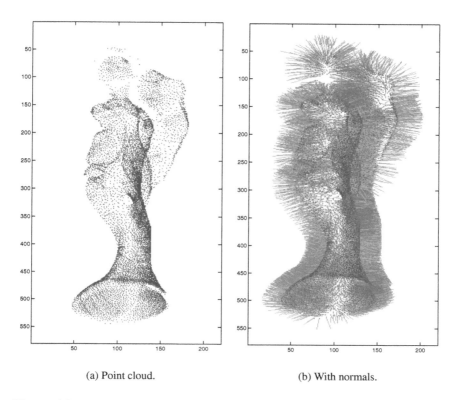

(a) Point cloud.

(b) With normals.

Figure 4.9. *Example* 4.18: *a point cloud representing* (a) *a surface in three-dimensional space, and* (b) *together with its unsigned normals.*

An additional, global, pass over the points is then required to determine what *sign* our normal vectors should have to point outward rather than inward of the implied surface, but this is beyond the present scope. The results of applying this approach to Figure 4.9(a) are depicted in Figure 4.9(b). ∎

Specific exercises for this section: Exercise 17.

4.6 Exercises

0. **Review questions**

    (a) What is a singular matrix?

    (b) Suppose $A\mathbf{x} = \mathbf{b} \neq \mathbf{0}$ is a linear system and A is a square, singular matrix. How many solutions is it possible for the system to have?

    (c) Suppose we know that a matrix is nonsingular. How many solutions are there for the linear system $A\mathbf{x} = \mathbf{0}$?

    (d) What is the spectrum of a matrix?

    (e) Define algebraic multiplicity and geometric multiplicity of eigenvalues.

    (f) What is a diagonalizable matrix?

(g) How is the norm of a vector related to the absolute value of a scalar?

(h) Prove that for any given vector, $\|\mathbf{x}\|_1 \geq \|\mathbf{x}\|_2$.

(i) What is an induced matrix norm?

(j) What is the spectral radius of a matrix? Is it a norm?

(k) Suppose we define a function of a matrix to be the maximum absolute value among all its entries. Is it a norm? If not, give a counterexample.

(l) Is it true that every symmetric positive definite matrix is necessarily nonsingular?

(m) Give an example of a 3×3 orthogonal matrix not equal to the identity.

(n) What is the fundamental difference (from a linear algebra point of view) between a data fitting problem, where the number of data points is larger than the number of coefficients to be determined, and the case where these quantities are equal?

1. Show that if all row-sums of a square matrix are equal to zero, then the matrix is singular.

2. Determine whether the following is true or false and justify your answer: if $A\mathbf{x} = \lambda \mathbf{x}$ and A is nonsingular, then $\frac{1}{\lambda}$ is an eigenvalue of A^{-1}.

3. This exercise requires a fresher knowledge of complex arithmetic than what is assumed elsewhere in this chapter; consult page 71.

    (a) Show that the eigenvalues of a symmetric matrix are real.

    (b) Show that if the eigenvalues of a real matrix are not real, then the matrix cannot be symmetric.

    (c) A real square matrix A is called **skew-symmetric** if $A^T = -A$.
    Show that the eigenvalues of a skew-symmetric matrix are either purely imaginary or zero.

4. (a) Consider the ℓ_p-norm of a vector, defined on page 74. Show that the first two properties of a norm hold. (The third, triangle inequality property of a norm holds as well but is tougher to show in general.)

    (b) Find the 1-norm, 2-norm, 3-norm, and maximum norm of the vectors
    $$\mathbf{x} = (1, 2, 3, 4)^T \quad \text{and} \quad \mathbf{y} = (1000, -2000, 3000, -4000)^T.$$

5. Determine whether the following statement is true or false and give a justification: If A is nonsingular, then for any induced norm, $\|A^{-1}\| = \|A\|^{-1}$.

6. For an $m \times n$ matrix A show that
$$\|A\|_1 = \max_{1 \leq j \leq n} \sum_{i=1}^{m} |a_{ij}|.$$

7. Let A be symmetric positive definite. Show that the so-called *energy norm*
$$\|\mathbf{x}\|_A = \sqrt{\mathbf{x}^T A \mathbf{x}}$$
is indeed a (vector) norm.

[Hint: Show first that it suffices to consider only diagonal matrices A with positive entries on the main diagonal.]

4.6. Exercises

8. Which of the following matrices are necessarily orthogonal?

    (a) Permutation matrices, which are obtained by permuting rows or columns of the identity matrix, so that in each row and each column we still have precisely one value equal to 1;

    (b) symmetric positive definite matrices;

    (c) nonsingular matrices;

    (d) diagonal matrices.

9. Find all the values of a and b for which the matrix

$$\begin{pmatrix} a & 1 & 1+b \\ 1 & a & 1 \\ 1-b^2 & 1 & a \end{pmatrix}$$

   is symmetric positive definite.

10. (a) Show that the matrix

$$\begin{pmatrix} c & s \\ -s & c \end{pmatrix}$$

   is orthogonal if $c^2 + s^2 = 1$.

    (b) *Givens rotations* are based on rotation operations of the form

$$\begin{pmatrix} c & s \\ -s & c \end{pmatrix} \begin{pmatrix} a_1 \\ a_2 \end{pmatrix} = \begin{pmatrix} \alpha \\ 0 \end{pmatrix}.$$

   In other words, given a vector we rotate it (while preserving its length) so that one of its components is zeroed out.

      i. Use orthogonality to express α in terms of a_1 and a_2.
      ii. Find c and s that do the job.

11. (a) Show that the eigenvalues of an orthogonal matrix are equal to 1 in modulus.

    (b) A *projector* is a square matrix P that satisfies $P^2 = P$. Find the eigenvalues of a projector.

12. The *condition number* of an eigenvalue λ of a given matrix A is defined as

$$s(\lambda) = \frac{1}{\mathbf{x}^T \mathbf{w}},$$

where $\mathbf{x}$ is a (right) eigenvector of the matrix, satisfying $A\mathbf{x} = \lambda \mathbf{x}$, and $\mathbf{w}$ is a left eigenvector, satisfying $\mathbf{w}^T A = \lambda \mathbf{w}^T$. Both $\mathbf{x}$ and $\mathbf{w}$ are assumed to have a unit ℓ_2-norm. Loosely speaking, the condition number determines the difficulty of computing the eigenvalue in question accurately; the smaller $s(\lambda)$ is, the more numerically stable the computation is expected to be.

Determine the condition numbers of the eigenvalue 4 for the two matrices discussed in Example 4.7. Explain the meaning of your results and how they are related to the observations made in the example.

13. Let A be a real $m \times n$ matrix and denote by S the subspace of $\mathcal{R}^n$ consisting of all those vectors that are orthogonal to any vector in null(A^T). Show that $S = \text{range}(A)$.

14. Let A be real and symmetric, and denote its eigenvalues by λ_i, $i = 1,\ldots,n$. Find an expression for the singular values of A in terms of its eigenvalues.

15. Let A be skew-symmetric, and denote its singular values by $\sigma_1 \geq \sigma_2 \geq \cdots \sigma_n \geq 0$. Show that

    (a) If n is even, then $\sigma_{2k} = \sigma_{2k-1} \geq 0$, $k = 1,2,\ldots,n/2$. If n is odd, then the same relationship holds up to $k = (n-1)/2$ and also $\sigma_n = 0$.

    (b) The eigenvalues λ_j of A can be written as
    $$\lambda_j = (-1)^j \iota \sigma_j, \quad j = 1,\ldots,n.$$

16. (a) Suppose that A is an orthogonal matrix. What are its singular values?

    (b) Is the SVD of a given matrix A unique in general?

17. Obtain the matrix A for Example 4.17 in the case where $v'(0) = v'(1) = 0$, and show that it is singular.

4.7 Additional notes

This chapter is different from all others in our book in that it aims to provide background and as such does not propose or discuss algorithms except in the case studies of Section 4.5.

We have found it necessary to include it (while not including a similar chapter on calculus) because there appears to be more variation in students' backgrounds when it comes to linear algebra. The more theoretical concepts of linear algebra such as rings and ideals are not necessary for our present purposes, but the notions of norms, orthogonality, and symmetric positive definiteness, to mention a few, are fundamental to numerical computing.

If this book were written without an eye on practical computing issues, a review chapter of this sort would likely have had a bit more on determinants. They are interesting creatures indeed, but their computational usefulness has greatly declined in recent decades.

Needless to say, our chapter is not meant as a replacement for a linear algebra textbook. One of many good books is Strang [64].

The behavior of a vector norm function $\|\cdot\|$ near zero is interesting. Let $\mathbf{x}_\varepsilon$ be a family of vectors depending on a parameter $\varepsilon \geq 0$ such that $\|\mathbf{x}_\varepsilon\| = \varepsilon$. Then so long as $\varepsilon > 0$ we have no further information on the components of $\mathbf{x}_\varepsilon$. But for $\varepsilon = 0$ we know that each and every component of $\mathbf{x}_0$ is zero.

Example 4.16 is a primer for problems of data fitting, further considered in several later chapters; see in particular Sections 6.1 and 10.1. Similarly, Example 4.17 is a primer for boundary value problems in ordinary differential equations; see in particular Section 16.7. Finally, Example 4.18 relates directly to Chapter 8 and elsewhere. Figure 4.9 was produced by H. Huang, and further details and references for point cloud treatment can be found in Huang et al. [45].

Chapter 5
Linear Systems: Direct Methods

In this chapter we consider numerical methods for solving a system of linear equations $A\mathbf{x} = \mathbf{b}$. We assume that the given matrix A is real, $n \times n$, and nonsingular and that $\mathbf{b}$ is a given real vector in $\mathcal{R}^n$, and we seek a solution $\mathbf{x}$ that is necessarily also a vector in $\mathcal{R}^n$. Such problems arise frequently in virtually any branch of science, engineering, economics, or finance.

There is really no single technique that is best for all cases. Nonetheless, the many available numerical methods can generally be divided into two classes: *direct methods* and *iterative methods*. The present chapter is devoted to methods of the first type. In the absence of roundoff error, such methods would yield the exact solution within a finite number of steps.

The basic direct method for solving linear systems of equations is **Gaussian elimination**, and its various aspects and variants occupy the first seven sections of this chapter. Section 5.1 presents the method in simple terms.

The bulk of the algorithm involves only the matrix A and amounts to its decomposition into a product of two matrices that have a simpler form. This is called an **LU decomposition**, developed in Section 5.2. Such an alternative view is useful, for instance, when there are several right-hand-side vectors $\mathbf{b}$ each requiring a solution, as the LU decomposition can then be shared.

The simple algorithm of Section 5.1 is not guaranteed to be stable or even well-defined in general. In Section 5.3 we modify it using **pivoting** strategies to make the algorithm of Gaussian elimination with pivoting practically stable in general. Then in Section 5.4 we discuss efficient methods for implementing these algorithms.

The following three sections are devoted to variants of the same basic algorithm for special classes of matrices. Important simplifications can be made for symmetric positive definite matrices, as discussed in Section 5.5. If the matrix A contains only a few nonzero elements, then it is called **sparse**. Techniques for storing such matrices and for solving linear systems for the special case where all elements are zero outside a narrow band along the main diagonal are presented in Section 5.6. Direct methods for handling more general sparse matrices are briefly considered in Section 5.7.

In Section 5.8 we consider not the numerical solution of $A\mathbf{x} = \mathbf{b}$ but rather its assessment. Since there are roundoff errors, the result of any of our algorithms is approximate, and so the question of how close the approximate solution is to the exact one arises. This in turn depends on the **condition number** of the matrix, a concept that is defined and further discussed here.

5.1 Gaussian elimination and backward substitution

In this section we show the following:

- How to solve linear equations when A is in upper triangular form. The algorithm is called *backward substitution*.

- How to transform a general system of linear equations into an upper triangular form, to which backward substitution can be applied. The algorithm is called *Gaussian elimination*.

> **Note:** Experience suggests that many of our students have already seen much of the material in Section 5.1. As the section count in this chapter increases, though, there may be fewer people to whom this is all old news, so pick your spot for jumping in.

Backward substitution

Occasionally, A has a special structure that makes the solution process simple. For instance, if A is *diagonal*, written as

$$A = \begin{pmatrix} a_{11} & & & \\ & a_{22} & & \\ & & \ddots & \\ & & & a_{nn} \end{pmatrix}$$

(by this notation we mean that the off-diagonal elements are all 0), then the linear equations are uncoupled, reading $a_{ii} x_i = b_i$, and the solution is obviously

$$x_i = \frac{b_i}{a_{ii}}, \quad i = 1, 2, \ldots, n.$$

A more involved instance of a special structure is an **upper triangular** matrix

$$A = \begin{pmatrix} a_{11} & a_{12} & \cdots & a_{1n} \\ & a_{22} & \ddots & \vdots \\ & & \ddots & \vdots \\ & & & a_{nn} \end{pmatrix},$$

where all elements below the main diagonal are zero: $a_{ij} = 0$ for all $i > j$.

In this case each equation is possibly coupled only to those following it but not to those preceding it. Thus, we can solve an upper triangular system of equations *backwards*. The last row reads $a_{nn} x_n = b_n$, so $x_n = \frac{b_n}{a_{nn}}$. Next, now that we know x_n, the row before last can be written as $a_{n-1,n-1} x_{n-1} = b_{n-1} - a_{n-1,n} x_n$, so $x_{n-1} = \frac{b_{n-1} - a_{n-1,n} x_n}{a_{n-1,n-1}}$. Next the previous row can be dealt with, yielding x_{n-2}, etc. We obtain the *backward substitution* algorithm given on the next page. In MATLAB this can be written as

```
x(n) = b(n) / A(n,n);
for k = n-1:-1:1
    x(k) = ( b(k) - A(k,k+1:n)*x(k+1:n) ) / A(k,k);
end
```

5.1. Gaussian elimination and backward substitution

> **Algorithm: Backward Substitution.**
> Given an upper triangular matrix A and a right-hand-side $\mathbf{b}$,
> $$\text{for } k = n : -1 : 1$$
> $$x_k = \frac{b_k - \sum_{j=k+1}^{n} a_{kj} x_j}{a_{kk}}$$
> end

In Example 4.1 we performed one step of this backward substitution algorithm when obtaining x_1 once we knew x_2.

It is worth noting that the successful completion of the algorithm is possible only if the diagonal elements of A, a_{ii}, are nonzero. If any of them is zero, then the algorithm experiences what is known as a *breakdown*. However, for an upper triangular matrix such as A, a zero diagonal term will occur only if the matrix is singular, which violates our assumption from the beginning of this chapter. So we do not need to worry about breakdowns here.

Cost of backward substitution

What is the cost of this algorithm? In a simplistic way we just count each floating point operation (such as $+$ and $*$) as a *flop*. The number of flops required here is

$$1 + \sum_{k=1}^{n-1} ((n-k) + (n-k) + 1) = \sum_{k=1}^{n} (2(n-k) + 1) = n^2.$$

In general, this way of evaluating complexity of algorithms considers only part of the picture. It does not take into account data movement between elements of the computer's memory hierarchy. In fact, concerns of data locality can be crucial to the execution of an algorithm. The situation is even more complex on multiprocessor machines. In MATLAB, the same backward substitution algorithm implemented as

```
for k = n:-1:1
  x(k) = b(k);
  for j = k+1:n
    x(k) = x(k) - A(k,j)*x(j);
  end
  x(k) = x(k) / A(k,k);
end
```

may take much longer to execute compared to the *vectorized* script presented before, especially for large matrices, because no advantage is taken of the way in which the matrix and vector elements are stored. A general rule for generating efficient MATLAB implementations is to avoid for- and while-loops whenever possible.

Still, noting that the run time of this algorithm is $\mathcal{O}(n^2)$ and not $\mathcal{O}(n^3)$ is often meaningful in practice.

Forward substitution

Another instance of a special matrix structure is a **lower triangular** matrix

$$A = \begin{pmatrix} a_{11} & & & \\ a_{21} & a_{22} & & \\ \vdots & \ddots & \ddots & \\ a_{n1} & a_{n2} & \cdots & a_{nn} \end{pmatrix},$$

where all elements above the main diagonal are zero: $a_{ij} = 0$ for all $i < j$. The situation is similar to the upper triangular case, except that we apply *forward substitution* to compute a solution. The algorithm is given on this page.

Algorithm: Forward Substitution.
Given a lower triangular matrix A and a right-hand-side $\mathbf{b}$,

for $k = 1 : n$
$$x_k = \frac{b_k - \sum_{j=1}^{k-1} a_{kj} x_j}{a_{kk}}$$
end

Example 5.1. Consider solving the equations

$$5x_1 = 15,$$
$$x_1 + 2x_2 = 7,$$
$$-x_1 + 3x_2 + 2x_3 = 5.$$

The matrix A is lower triangular, and in our general notation we have

$$A = \begin{pmatrix} 5 & 0 & 0 \\ 1 & 2 & 0 \\ -1 & 3 & 2 \end{pmatrix}, \quad \mathbf{b} = \begin{pmatrix} 15 \\ 7 \\ 5 \end{pmatrix}.$$

Applying the forward substitution algorithm we get $x_1 = \frac{15}{5} = 3$, then $x_2 = \frac{7-3}{2} = 2$, then $x_3 = \frac{5+3-6}{2} = 1$. ∎

Gaussian elimination

Next, assume that the matrix A has no special zero-structure. *Gaussian elimination* is a generalization of the procedure used in Example 4.1. Thus, we use elementary row transformations to reduce the given linear system to an equivalent upper triangular form. Then we solve the resulting system using backward substitution.

The reason this works is that the solution to our problem $A\mathbf{x} = \mathbf{b}$ is *invariant* to

- multiplication of a row by a constant,
- subtraction of a multiple of one row from another, and
- row interchanges.

5.1. Gaussian elimination and backward substitution

These claims can be easily verified directly. Simply recall that $A\mathbf{x} = \mathbf{b}$ is a concise form for the system of equations

$$a_{11}x_1 + a_{12}x_2 + \cdots + a_{1n}x_n = b_1,$$
$$a_{21}x_1 + a_{22}x_2 + \cdots + a_{2n}x_n = b_2,$$
$$\vdots \quad = \vdots$$
$$a_{n1}x_1 + a_{n2}x_2 + \cdots + a_{nn}x_n = b_n,$$

and check what each of these operations amounts to.

Performing each of these operations can be recorded directly on the *augmented matrix*

$$(A|\mathbf{b}) = \begin{pmatrix} a_{11} & a_{12} & \cdots & a_{1n} & b_1 \\ a_{21} & a_{22} & \cdots & a_{2n} & b_2 \\ \vdots & \vdots & \ddots & \vdots & \vdots \\ a_{n1} & a_{n2} & \cdots & a_{nn} & b_n \end{pmatrix}.$$

Eliminating one column at a time

1. Eliminate the first column elements below the main diagonal:

   - Subtract $\frac{a_{21}}{a_{11}} \times$ the first row from the second row.
   - Subtract $\frac{a_{31}}{a_{11}} \times$ the first row from the third row.

   $\vdots$

   - Subtract $\frac{a_{n1}}{a_{11}} \times$ the first row from the nth row.

   This produces

   $$(A^{(1)}|\mathbf{b}^{(1)}) = \begin{pmatrix} a_{11} & a_{12} & \cdots & a_{1n} & b_1 \\ 0 & a_{22}^{(1)} & \cdots & a_{2n}^{(1)} & b_2^{(1)} \\ \vdots & \vdots & \ddots & \vdots & \vdots \\ 0 & a_{n2}^{(1)} & \cdots & a_{nn}^{(1)} & b_n^{(1)} \end{pmatrix}.$$

2. Next, consider the $(n-1) \times n$ submatrix of $(A^{(1)}|\mathbf{b}^{(1)})$ obtained by ignoring the first row and column. These are the elements that have been modified by the first stage of the process, described above, but not set to 0. Apply exactly the same step as before to this submatrix; i.e., subtract $\frac{a_{32}}{a_{22}^{(1)}} \times$ the second row from the third row, etc.

   The reason we are "allowed" to do this is that the first row remains untouched by these operations, and the first column for the other rows is all zeros, so nothing changes in the first column during subsequent operations.

After this second stage we have an augmented matrix of the form

$$(A^{(2)}|\mathbf{b}^{(2)}) = \begin{pmatrix} a_{11} & a_{12} & \cdots & & a_{1n} & b_1 \\ 0 & a_{22}^{(1)} & a_{23}^{(1)} & \cdots & a_{2n}^{(1)} & b_2^{(1)} \\ \vdots & 0 & a_{33}^{(2)} & \cdots & a_{3n}^{(2)} & b_3^{(2)} \\ \vdots & \vdots & \vdots & \ddots & \vdots & \vdots \\ 0 & 0 & a_{n3}^{(2)} & \cdots & a_{nn}^{(2)} & b_n^{(2)} \end{pmatrix}.$$

The superscripts in the above expression show for each element at which stage it has last been modified.

3. Repeat the process. After $n-1$ such stages we obtain an upper triangular system

$$(A^{(n-1)}|\mathbf{b}^{(n-1)}) = \begin{pmatrix} a_{11} & a_{12} & \cdots & & a_{1n} & b_1 \\ 0 & a_{22}^{(1)} & a_{23}^{(1)} & \cdots & a_{2n}^{(1)} & b_2^{(1)} \\ \vdots & 0 & a_{33}^{(2)} & \cdots & a_{3n}^{(2)} & b_3^{(2)} \\ \vdots & \vdots & \ddots & \ddots & \vdots & \vdots \\ 0 & 0 & \cdots & 0 & a_{nn}^{(n-1)} & b_n^{(n-1)} \end{pmatrix}.$$

This procedure leads to the Gaussian elimination algorithm in its simplest form given on the next page. It is depicted pictorially for $n = 4$ in Figure 5.1.

In stating the Gaussian elimination algorithm we have not bothered to zero out the elements below the main diagonal in the modified matrix A: this part of the matrix is not considered in the ensuing backward substitution.

Of course, for this algorithm to work we need to assume that all those pivotal elements $a_{kk}^{(k-1)}$ by which we divide are nonzero and in fact not very close to 0 either. We deal with the question of ensuring this later, in Section 5.3. For now, let's assume there is simply no such issue.

Example 5.2. For Example 4.1 we have $n = 2$; hence the loops in the general algorithm collapse into defining just one update for $k = 1, i = 2, j = 2$:

$$(A|\mathbf{b}) = \begin{pmatrix} 1 & -4 & -10 \\ 0.5 & -1 & -2 \end{pmatrix},$$

$$l_{21} = \frac{a_{21}}{a_{11}} = 0.5,$$

$$a_{22}^{(1)} = -1 - 0.5 \cdot (-4) = 1,$$

$$b_2^{(1)} = -2 - 0.5 \cdot (-10) = 3,$$

$$(A^{1)}|\mathbf{b}^{(1)}) = \begin{pmatrix} 1 & -4 & -10 \\ 0 & 1 & 3 \end{pmatrix}.$$

Note that in the algorithm we never actually define $a_{22}^{(1)}$ or $b_2^{(1)}$. Instead, we simply overwrite the original value $a_{22} = -1$ with the new value $a_{22} = 1$ and the original value $b_2 = -2$ with the new value $b_2 = 3$. ∎

5.1. Gaussian elimination and backward substitution

Algorithm: Gaussian Elimination.
Given a real, nonsingular $n \times n$ matrix A and a vector $\mathbf{b}$ of size n, first transform into upper triangular form,

$$\begin{aligned}
&\text{for } k = 1 : n-1 \\
&\quad \text{for } i = k+1 : n \\
&\quad\quad l_{ik} = \frac{a_{ik}}{a_{kk}} \\
&\quad\quad \text{for } j = k+1 : n \\
&\quad\quad\quad a_{ij} = a_{ij} - l_{ik} a_{kj} \\
&\quad\quad \text{end} \\
&\quad\quad b_i = b_i - l_{ik} b_k \\
&\quad \text{end} \\
&\text{end}
\end{aligned}$$

Next, apply the algorithm of *backward substitution*.

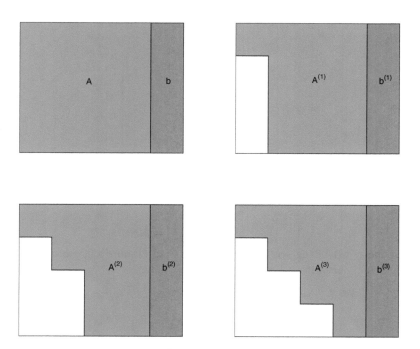

Figure 5.1. *Gaussian elimination for the case $n = 4$. Only areas of potentially nonzero entries are shown.*

The cost of the Gaussian elimination algorithm in terms of operation count is approximately

$$\sum_{k=1}^{n-1}((n-k) + 2(n-k)(n-k+1)) \approx 2\sum_{k=1}^{n-1}(n-k)^2$$

$$= 2((n-1)^2 + \cdots + 1) = \frac{2}{3}n^3 + \mathcal{O}(n^2)$$

flops.[18] In particular, as the size of the matrix A increases the cost of Gaussian elimination rises cubically. Comparing this to the cost of backward substitution we see that the cost of the elimination phase dominates for all but very small problems.

Specific exercises for this section: Exercises 1–2.

5.2 LU decomposition

In this section we show that the process of Gaussian elimination in fact decomposes the matrix A into a product $L \times U$ of a lower triangular matrix L and an upper triangular matrix U.

Let us continue to assume that the elements $a_{kk}^{(k-1)}$ encountered in the Gaussian elimination process are all bounded safely away from 0. We will remove this assumption in the next section.

Elementary lower triangular matrices

Consider the first stage of Gaussian elimination described above. These are elementary row operations applied to the matrix A to zero out its first column below the $(1,1)$ entry. These operations are also applied to the right-hand-side vector $\mathbf{b}$. Note, however, that the operations on $\mathbf{b}$ can actually be done at a later time because they do not affect those done on A.

We can capture the effect of this first stage by defining the elementary $n \times n$ lower triangular matrix

$$M^{(1)} = \begin{pmatrix} 1 & & & & \\ -l_{21} & 1 & & & \\ -l_{31} & & 1 & & \\ \vdots & & & \ddots & \\ -l_{n1} & & & & 1 \end{pmatrix}.$$

(This matrix has zeros everywhere except in the main diagonal and in the first column: when describing matrices, blank does not mean nothing—it means zero.) Then the effect of this first stage is seen to produce

$$A^{(1)} = M^{(1)} A \quad \text{and} \quad \mathbf{b}^{(1)} = M^{(1)} \mathbf{b}.$$

Likewise, the effect of the second stage of Gaussian elimination, which is to zero out the second column of $A^{(1)}$ below the main diagonal, can be written as

$$A^{(2)} = M^{(2)} A^{(1)} = M^{(2)} M^{(1)} A \quad \text{and} \quad \mathbf{b}^{(2)} = M^{(2)} \mathbf{b}^{(1)} = M^{(2)} M^{(1)} \mathbf{b},$$

where

$$M^{(2)} = \begin{pmatrix} 1 & & & & \\ & 1 & & & \\ & -l_{32} & \ddots & & \\ & \vdots & & \ddots & \\ & -l_{n2} & & & 1 \end{pmatrix}.$$

---

[18] For those readers who remember and appreciate the usefulness of Riemann sums, the final result can be approximated by recognizing that for large enough n this is nothing but an approximation of the integral $\int_0^n x^2 dx$.

5.2. LU decomposition

(This matrix has zeros everywhere except in the main diagonal and in the second column below the main diagonal.)

Obtaining the LU decomposition

This procedure is continued, and after $n-1$ such stages it transforms the matrix A into an upper triangular matrix, which we call U, by

$$U = A^{(n-1)} = M^{(n-1)} \cdots M^{(2)} M^{(1)} A.$$

Likewise, $\mathbf{b}^{(n-1)} = M^{(n-1)} \cdots M^{(2)} M^{(1)} \mathbf{b}$. Multiplying U by $[M^{(n-1)}]^{-1}$, then by $[M^{(n-2)}]^{-1}$, etc., we obtain

$$A = LU,$$

where $L = [M^{(1)}]^{-1} \cdots [M^{(n-2)}]^{-1} [M^{(n-1)}]^{-1}$.

To see what L looks like, note first that $[M^{(k)}]^{-1}$ has exactly the same form as $M^{(k)}$, i.e., it's the identity matrix plus nonzero elements in the kth column below the main diagonal, except that these elements are without the minus sign, viz., $l_{k+1,k}, l_{k+2,k}, \ldots, l_{n,k}$. For instance, you can verify directly that the matrix $M^{(2)}$ defined above, multiplied by

$$[M^{(2)}]^{-1} = \begin{pmatrix} 1 & & & & \\ & 1 & & & \\ & l_{32} & \ddots & & \\ & \vdots & & \ddots & \\ & l_{n2} & & & 1 \end{pmatrix},$$

yields the identity matrix.

So, $[M^{(k)}]^{-1}$ are also elementary lower triangular matrices. Now, you can verify directly that the product of such matrices yields

$$L = \begin{pmatrix} 1 & & & & \\ l_{21} & 1 & & & \\ l_{31} & l_{32} & 1 & & \\ \vdots & \vdots & \ddots & \ddots & \\ l_{n1} & l_{n2} & \cdots & l_{n,n-1} & 1 \end{pmatrix}.$$

The elements of L are therefore obtained during the Gaussian elimination process!

The LU decomposition is depicted pictorially for $n = 10$ in Figure 5.2.

Example 5.3. Let us verify the above claims for the matrix

$$A = \begin{pmatrix} 1 & -1 & 3 \\ 1 & 1 & 0 \\ 3 & -2 & 1 \end{pmatrix}.$$

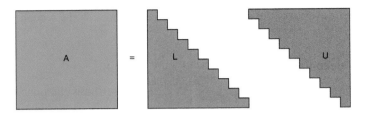

Figure 5.2. *LU decomposition for the case $n = 10$. Only areas of potentially nonzero entries in the square matrices L and U are shown.*

1. The Gaussian elimination process for the first column yields $l_{21} = \frac{1}{1} = 1$, $l_{31} = \frac{3}{1} = 3$, so

$$M^{(1)} = \begin{pmatrix} 1 & 0 & 0 \\ -1 & 1 & 0 \\ -3 & 0 & 1 \end{pmatrix}, \quad A^{(1)} = M^{(1)}A = \begin{pmatrix} 1 & -1 & 3 \\ 0 & 2 & -3 \\ 0 & 1 & -8 \end{pmatrix}.$$

2. The Gaussian elimination process for the second column yields $l_{32} = \frac{1}{2}$, so

$$M^{(2)} = \begin{pmatrix} 1 & 0 & 0 \\ 0 & 1 & 0 \\ 0 & -0.5 & 1 \end{pmatrix}, \quad U = A^{(2)} = M^{(2)}A^{(1)} = \begin{pmatrix} 1 & -1 & 3 \\ 0 & 2 & -3 \\ 0 & 0 & -6.5 \end{pmatrix}.$$

3. Note that

$$[M^{(1)}]^{-1} = \begin{pmatrix} 1 & 0 & 0 \\ 1 & 1 & 0 \\ 3 & 0 & 1 \end{pmatrix}, \quad [M^{(2)}]^{-1} = \begin{pmatrix} 1 & 0 & 0 \\ 0 & 1 & 0 \\ 0 & 0.5 & 1 \end{pmatrix},$$

$$[M^{(1)}]^{-1}[M^{(2)}]^{-1} = \begin{pmatrix} 1 & 0 & 0 \\ 1 & 1 & 0 \\ 3 & 0.5 & 1 \end{pmatrix} = L,$$

and

$$LU = \begin{pmatrix} 1 & 0 & 0 \\ 1 & 1 & 0 \\ 3 & 0.5 & 1 \end{pmatrix} \begin{pmatrix} 1 & -1 & 3 \\ 0 & 2 & -3 \\ 0 & 0 & -6.5 \end{pmatrix} = \begin{pmatrix} 1 & -1 & 3 \\ 1 & 1 & 0 \\ 3 & -2 & 1 \end{pmatrix} = A.$$

This explicitly specifies the LU decomposition of the given matrix A. ∎

We therefore conclude that the Gaussian elimination procedure of the previous section *decomposes* A into a product of a unit lower triangular[19] matrix L and an upper triangular matrix U.

---

[19] A *unit* lower triangular matrix is a lower triangular matrix with 1's on the main diagonal.

5.2. LU decomposition

This is the famous **LU decomposition**. Together with the ensuing backward substitution the entire solution algorithm for $A\mathbf{x} = \mathbf{b}$ can therefore be described in three steps, as depicted in the algorithm given on this page.

Algorithm: Solving $A\mathbf{x} = \mathbf{b}$ by LU Decomposition.
Given a real nonsingular matrix A, apply LU decomposition first:

$$A = LU.$$

Given also a right-hand-side vector $\mathbf{b}$:

1. *Forward substitution*: solve

$$L\mathbf{y} = \mathbf{b}.$$

2. *Backward substitution*: solve

$$U\mathbf{x} = \mathbf{y}.$$

Separating the decomposition step

Note that decomposing A and solving $L\mathbf{y} = \mathbf{b}$ simultaneously is identical to applying the Gaussian elimination algorithm given on page 99. Note also that the decomposition step costs $\mathcal{O}(n^3)$ operations, whereas the forward and backward substitutions each cost about n^2 operations. Thus, the decomposition step dominates the total cost. Moreover, since we do not need $\mathbf{b}$ at all to perform the decomposition step, we may perform the LU decomposition once and then solve different linear systems of equations for different right-hand-sides $\mathbf{b}$ at an $\mathcal{O}(n^2)$ cost for each system solve.

Example 5.4. Continuing Example 5.3, suppose we are to solve $A\mathbf{x} = \mathbf{b}$ for a new right-hand-side vector

$$\mathbf{b} = \begin{pmatrix} 2 \\ 4 \\ 1 \end{pmatrix}.$$

The LU decomposition phase having already been accomplished, we proceed with the forward substitution phase, solving $L\mathbf{y} = \mathbf{b}$ and obtaining $y_1 = 2$, $y_2 = 4 - y_1 = 2$, $y_3 = 1 - 3y_1 - 0.5y_2 = -6$. Then solving $U\mathbf{x} = \mathbf{y}$ yields $x_3 = \frac{6}{6.5} = \frac{12}{13}$, $x_2 = \frac{1}{2}(2 + 3\frac{12}{13}) = \frac{31}{13}$, $x_1 = 2 + \frac{31}{13} - 3\frac{12}{13} = \frac{21}{13}$, so

$$\mathbf{x} = \frac{1}{13}\begin{pmatrix} 21 \\ 31 \\ 12 \end{pmatrix}$$

is the sought solution. ∎

Example 5.5. Let us denote the inverse of a nonsingular matrix A by $G = A^{-1}$. If we are given G, then the solution of a linear system $A\mathbf{x} = \mathbf{b}$ can be calculated as $G\mathbf{b}$, which takes $\mathcal{O}(n^2)$ flops to evaluate in general.

But first we have to find G! One way, as good as any, is to use the LU decomposition of A. Thus, we calculate L and U first. Now, the identity matrix is composed of the *unit vectors*

$$\mathbf{e}_i = \begin{pmatrix} 0 \\ \vdots \\ 0 \\ 1 \\ 0 \\ \vdots \\ 0 \end{pmatrix}$$

(with the value 1 in the ith location, $i = 1,\ldots,n$). For each of these unit vectors we then solve $A\mathbf{x}_i = \mathbf{e}_i$ by writing $L(U\mathbf{x}_i) = \mathbf{e}_i$ and performing forward and then backward substitutions. Thus $\mathbf{x}_i$ is the ith column of G and

$$G = A^{-1} = [\mathbf{x}_1,\ldots,\mathbf{x}_n].$$

Using again a simple operation count, each forward or backward substitution costs about n^2 flops, so for n right-hand sides we have $n \cdot 2n^2$ flops added to the cost of the LU decomposition. Discounting lower order terms in n this approximately yields

$$\frac{2}{3}n^3 + 2n^3 = \frac{8}{3}n^3$$

flops. ∎

Forming A^{-1} explicitly and multiplying by $\mathbf{b}$ is generally not a recommended way to solve linear systems of equations for several reasons. For one, it can be wasteful in storage, a point that will become much clearer when we discuss banded matrices in Section 5.6. Moreover, it is more computationally expensive than going by the LU decomposition route, though by less than an order of magnitude. Also, it may give rise to a more pronounced presence of roundoff errors. Finally, it simply has no advantage over the methods presented here to offset the disadvantages mentioned above.

Example 5.6. Another simple, yet general, by-product of LU decomposition is a tool for computing the *determinant* of a matrix. Recall that the determinant of a product of matrices equals the product of the matrix determinants. Moreover, the determinant of an upper triangular or a lower triangular matrix is the product of the elements on the main diagonal of such a matrix. Thus, for a general square matrix $A = LU$, we have

$$\det(A) = \det(L) \cdot \det(U) = 1 \cdots 1 \cdot u_{11}u_{22}\cdots u_{nn}$$
$$= u_{11}u_{22}\cdots u_{nn} = \Pi_{k=1}^{n} u_{kk}.$$

In particular, A is nonsingular if and only if all u_{kk} are nonzero, $k = 1,\ldots,n$. ∎

Note that storage can be handled very efficiently for the LU decomposition. Assuming that the input matrix A can be discarded, its storage locations can be reused to store both U and L in the form

$$\begin{pmatrix} u_{11} & u_{12} & u_{13} & \cdots & u_{1n} \\ l_{21} & u_{22} & u_{23} & \ddots & u_{2n} \\ l_{31} & l_{32} & u_{33} & \ddots & u_{3n} \\ \vdots & \ddots & \ddots & \ddots & \vdots \\ l_{n1} & l_{n2} & \cdots & l_{n,n-1} & u_{nn} \end{pmatrix}.$$

The main diagonal of L need not be stored since all its values are equal to 1.

Specific exercises for this section: Exercises 3–4.

5.3 Pivoting strategies

In this section we modify the algorithm of Gaussian elimination by pivoting, thus making it generally stable.

The process of Gaussian elimination (or LU decomposition) described in Sections 5.1 and 5.2 cannot be applied unmodified to all nonsingular matrices. For instance, the matrix

$$A = \begin{pmatrix} 0 & 1 \\ 1 & 0 \end{pmatrix}$$

is clearly nonsingular, and yet $a_{11} = 0$, so the algorithm breaks down immediately. Here is another example, slightly more subtle.

Example 5.7. Consider the system

$$\begin{aligned} x_1 + x_2 + x_3 &= 1, \\ x_1 + x_2 + 2x_3 &= 2, \\ x_1 + 2x_2 + 2x_3 &= 1. \end{aligned}$$

The matrix A defined by these equations is nonsingular (indeed, the unique solution is $\mathbf{x} = (1, -1, 1)^T$), also all elements of A are nonzero, and yet Gaussian elimination yields after the first stage

$$\left(\begin{array}{ccc|c} 1 & 1 & 1 & 1 \\ 1 & 1 & 2 & 2 \\ 1 & 2 & 2 & 1 \end{array}\right) \Rightarrow \left(\begin{array}{ccc|c} 1 & 1 & 1 & 1 \\ 0 & 0 & 1 & 1 \\ 0 & 1 & 1 & 0 \end{array}\right).$$

Next, $a_{22}^{(1)} = 0$ and we cannot proceed further.

In this case there is an obvious remedy: simply interchange the second and the third rows! This yields

$$\left(\begin{array}{ccc|c} 1 & 1 & 1 & 1 \\ 1 & 2 & 2 & 1 \\ 1 & 1 & 2 & 2 \end{array}\right) \Rightarrow \left(\begin{array}{ccc|c} 1 & 1 & 1 & 1 \\ 0 & 1 & 1 & 0 \\ 0 & 0 & 1 & 1 \end{array}\right),$$

and backward substitution proceeds as usual. ∎

In the above examples the process broke down because a zero pivotal element $a_{kk}^{(k-1)}$ was encountered. But $a_{kk}^{(k-1)}$ does not have to exactly equal 0 for trouble to arise: undesirable accumulation of roundoff error may arise also when $a_{kk}^{(k-1)}$ is near 0.

Example 5.8. Consider a perturbation of Example 5.7 that reads

$$x_1 + x_2 + x_3 = 1,$$
$$x_1 + 1.0001 x_2 + 2x_3 = 2,$$
$$x_1 + 2x_2 + 2x_3 = 1.$$

The exact solution, correct to 5 digits, is $\mathbf{x} \approx (1, -1.0001, 1.0001)^T$. Now, Gaussian elimination in exact arithmetic can be completed and yields

$$\left(\begin{array}{ccc|c} 1 & 1 & 1 & 1 \\ 1 & 1.0001 & 2 & 2 \\ 1 & 2 & 2 & 1 \end{array} \right) \Rightarrow \left(\begin{array}{ccc|c} 1 & 1 & 1 & 1 \\ 0 & 0.0001 & 1 & 1 \\ 0 & 1 & 1 & 0 \end{array} \right) \Rightarrow \left(\begin{array}{ccc|c} 1 & 1 & 1 & 1 \\ 0 & 0.0001 & 1 & 1 \\ 0 & 0 & -9999.0 & -10000 \end{array} \right).$$

Assume next that we are using floating point arithmetic with base $\beta = 10$ and precision $t = 3$. Then backward substitution gives

$$x_3 = 1, \quad x_2 = 0, \quad x_1 = 0.$$

On the other hand, if we interchange the second and third rows we obtain

$$\left(\begin{array}{ccc|c} 1 & 1 & 1 & 1 \\ 1 & 2 & 2 & 1 \\ 1 & 1.0001 & 2 & 2 \end{array} \right) \Rightarrow \left(\begin{array}{ccc|c} 1 & 1 & 1 & 1 \\ 0 & 1 & 1 & 0 \\ 0 & 0.0001 & 1 & 1 \end{array} \right) \Rightarrow \left(\begin{array}{ccc|c} 1 & 1 & 1 & 1 \\ 0 & 1 & 1 & 0 \\ 0 & 0 & 0.9999 & 1 \end{array} \right).$$

Now backward substitution with 3 decimal digits gives

$$x_3 = 1.000, \quad x_2 = -1.000, \quad x_1 = 1.000,$$

which is correct, in that the difference between this solution and the exact solution is less than the rounding unit. ∎

Partial pivoting

The problem highlighted in these simple examples is real and general. Recall from Chapter 2 that roundoff errors may be unduly magnified if divided by a small value. Here, if a pivotal element $a_{kk}^{(k-1)}$ is small in magnitude, then roundoff errors are amplified, both in subsequent stages of the Gaussian elimination process and (even more apparently) in the backward substitution phase.

These simple examples also suggest a strategy to resolve the difficulty, called *partial pivoting*: as the elimination proceeds for $k = 1, \ldots, n-1$, at each stage k choose $q = q(k)$ as the smallest integer for which

$$|a_{qk}^{(k-1)}| = \max_{k \leq i \leq n} |a_{ik}^{(k-1)}|,$$

and interchange rows k and q. (Recall from page 97 that this operation does not change the solution, provided that the corresponding entries of $\mathbf{b}^{(k-1)}$ are exchanged as well.) Then proceed with the elimination process.

5.3. Pivoting strategies

This Gaussian elimination with partial pivoting (GEPP) strategy certainly resolves the difficulty in all the above examples. In Examples 5.7 and 5.8, in particular, for $k=1$ we have $q=1$, i.e., no row interchange is necessary, but for $k=2$ we get $q=3$, so the second and third rows are interchanged.

How does partial pivoting affect the LU decomposition? Note first that for the matrix of Example 5.7 it really is not possible to write $A = LU$ with L unit lower triangular and U nonsingular upper triangular (try it!). However, the operation of row interchange is captured in general by a *permutation matrix* P. We start with $P = I$ and proceed with GEPP; at each stage k where a row interchange is mandated, we record this fact by interchanging rows k and q of P. For Examples 5.7 and 5.8 this yields

$$PA = LU, \quad \text{where}$$
$$P = \begin{pmatrix} 1 & 0 & 0 \\ 0 & 0 & 1 \\ 0 & 1 & 0 \end{pmatrix}.$$

Forming $PA = LU$

But does this really work in general? Recall from Section 5.2 that L is obtained as the inverse of a product of elementary matrices, $L^{-1} = M^{(n-1)} \cdots M^{(2)} M^{(1)}$. Now the process is captured instead by the sequence

$$B = M^{(n-1)} P^{(n-1)} \cdots M^{(2)} P^{(2)} M^{(1)} P^{(1)},$$

where each $P^{(i)}$ is an elementary permutation matrix encoding one row interchange. What is implied, then, is that we can pull all the permutations together and write

$$B = L^{-1} P, \quad \text{where } P = P^{(n-1)} \cdots P^{(2)} P^{(1)}.$$

Such a claim is no longer elementary, dear Watson! Indeed, it is by far the most subtle point in Sections 5.1–5.3. But it turns out to hold in general, namely, the straightforward process of partial pivoting that results in $U = BA$ being upper triangular can be written as $PA = LU$ for an appropriate unit lower triangular matrix L.

Let us show that this is indeed so. Suppose we have a 4×4 matrix, and thus the process terminates after three steps. (Everything we discuss here straightforwardly extends to an $n \times n$ matrix, and the only reason we opt for the specific case of $n = 4$ is for ease of exposition.)

We have

$$U = M^{(3)} P^{(3)} M^{(2)} P^{(2)} M^{(1)} P^{(1)} A.$$

Define $\tilde{M}^{(3)} = M^{(3)}$, $\tilde{M}^{(2)} = P^{(3)} M^{(2)} (P^{(3)})^T$, and $\tilde{M}^{(1)} = P^{(3)} P^{(2)} M^{(1)} (P^{(2)})^T (P^{(3)})^T$. Then, by using the fact that the inverse of an elementary permutation matrix is its transpose, we have

$$U = \underbrace{\tilde{M}^{(3)} \tilde{M}^{(2)} \tilde{M}^{(1)}}_{\tilde{M}} \underbrace{P^{(3)} P^{(2)} P^{(1)}}_{P} A.$$

It is tempting to think that we are done, because we have managed to have the $\tilde{M}^{(j)}$ and the $P^{(j)}$ comfortably bundled with their own specimen. Not so fast! The order is right, and P is certainly a valid permutation matrix, but to declare victory we need to be convinced that $\tilde{M}$ (and hence its inverse) is indeed a unit lower triangular matrix.

In general, a symmetric permutation of a lower triangular matrix does not maintain this structure. The only reason it works in this case is because for a given j, the unit lower triangular matrix $M^{(j)}$ is basically an identity matrix with the conveniently limited "correction" of having only the

$n - j$ entries in the jth column below the main diagonal possibly nonzero. A symmetric permutation of the form we are discussing corresponds to multiplying on the left by an elementary permutation matrix $P^{(i)}$ with $i > j$ and on the right by its transpose, and amounts to swapping rows and columns in a way that merely affects nonzero off-diagonal entries in the strictly lower triangular part of $M^{(j)}$. Indeed, zeros that change places are of no interest, and neither are swapped diagonal entries, which are all equal to 1 to begin with! In Exercises 7 and 8 you are given an opportunity to see all this up close, by working on a 4×4 matrix with actual numbers and by looking into a similar procedure for symmetric indefinite matrices.

In general, note that the permutation matrix P is nonsingular, being just a permutation of the rows of the identity matrix. Moreover, P is *orthogonal*, i.e., $P^{-1} = P^T$, so we can write

$$A = (P^T L) U,$$

where the rows and columns of P^T are just unit vectors.

For GEPP it follows immediately that the multipliers l_{ik} (the elements of the unit lower triangular matrix L in the decomposition $PA = LU$) are all bounded in magnitude by 1, written explicitly as

$$|l_{ik}| \leq 1, \quad 1 \leq i, k \leq n.$$

In an actual implementation we will not store a matrix P, most of which consists of zeros, nor do we have to physically interchange matrix rows in memory (although we may end up doing the latter for reasons of easier memory access). We can simply keep track of what row interchanges are implied by the algorithm in a one-dimensional array **p**. The elements of **p** are used for indexing the correct rows, as specified in algorithmic form later on. For example, the array **p** that corresponds to the matrix P in Examples 5.7 and 5.8 is $\mathbf{p} = [1, 3, 2]$.

GEPP stability

Does the incorporation of this partial pivoting procedure guarantee *stability* of the Gaussian elimination algorithm, in the sense that roundoff errors do not get amplified by factors that grow unboundedly as the matrix size n increases? As far as $P^T L$ is concerned the answer is affirmative, but we have to check U, too. We need to be assured that

$$g_n(A) = \max_{i,j,k} |a_{i,j}^{(k)}|$$

(in words, the maximum magnitude of all elements that arise in the course of the elimination process) does not grow too fast, i.e., does not grow exponentially as a function of n.

Unfortunately, it is not possible in general to guarantee stability of GEPP. Indeed, this algorithm is not even invariant to scaling of the rows: in Example 5.8 if the third row were originally multiplied by 10^{-5}, then no row interchanges would have resulted upon applying the GEPP strategy. But the difficulty highlighted in that example would persist.

We could modify the partial pivoting strategy into one called **scaled partial pivoting**. Thus, define initially for each row i of A a *size*

$$s_i = \max_{1 \leq j \leq n} |a_{ij}|.$$

Then, at each stage k of the Gaussian elimination procedure, choose $q = q(k)$ as the smallest integer for which

$$\frac{|a_{qk}^{(k-1)}|}{s_q} = \max_{k \leq i \leq n} \frac{|a_{ik}^{(k-1)}|}{s_i},$$

and interchange rows k and q.

5.3. Pivoting strategies

This latter strategy entails a computational overhead and cannot completely eliminate potential instability, as we see below in Example 5.9. Still, we should not underestimate scaling as an important tool for enhancing the numerical stability of the process in general. At the very least, we should ensure that the linear system to be solved has rows that all have comparable norm. In certain situations it may be worth paying the extra computational cost of ensuring reasonable scaling throughout the whole elimination process. See Example 5.11 for a case in point.

Example 5.9. Let us give a rare instance where Gaussian elimination with scaled partial pivoting is unstable. Applying it to the matrix

$$A = \begin{pmatrix} 1 & 0 & \cdots & \cdots & 0 & 1 \\ -1 & 1 & 0 & \cdots & 0 & 1 \\ -1 & -1 & 1 & \ddots & \vdots & 1 \\ \vdots & \ddots & -1 & \ddots & 0 & \vdots \\ \vdots & \ddots & \ddots & \ddots & \ddots & 1 \\ -1 & \cdots & \cdots & \cdots & -1 & 1 \end{pmatrix}$$

produces no row interchanges, and yet after the kth step, $k = 1, 2, \ldots, n-1$, the elements in the last column satisfy $a_{i,n}^{(k)} = 2^k$, $i = k+1, \ldots, n$ (please check!), so $g_n(A) = a_{n,n}^{(n-1)} = 2^{n-1}$. Such a growth rate of g_n is unacceptable. ∎

A strategy that does guarantee stability is **complete pivoting**: at each stage choose q and r as the smallest integers for which

$$|a_{qr}^{(k-1)}| = \max_{k \leq i, j \leq n} |a_{ij}^{(k-1)}|,$$

and interchange both row q and column r with row i and column j, respectively. However, this strategy is significantly more expensive than partial pivoting, and instances in which (scaled) partial pivoting really fails in a big way appear to be extremely rare in practice. For this reason, most of the commonly used computer codes for directly solving linear systems use a GEPP approach.

Matrices requiring no pivoting

We end this section by mentioning that there are certain families of matrices for which it is known that pivoting is not required (at least theoretically). One such family is symmetric positive matrices, to the decomposition of which we will devote a whole section soon. Another class are **diagonally dominant** matrices. An $n \times n$ matrix A is diagonally dominant if

$$|a_{ii}| \geq \sum_{\substack{j=1 \\ j \neq i}}^{n} |a_{ij}|.$$

In words, for each row the diagonal element is at least as large in magnitude as the sum of magnitudes of the rest of the elements combined. It is possible to show that if a nonsingular matrix is diagonally dominant, Gaussian elimination does not require pivoting.

Specific exercises for this section: Exercises 5–9.

5.4 Efficient implementation

In general, the performance of an algorithm depends not only on the number of arithmetic operations that are carried out but also on the frequency of memory access. So, if the algorithms we have seen thus far in this chapter can be arranged to work on chunks of vectors and matrices, then their efficiency increases. In this section we explore this important practical issue. Specifically, we briefly discuss the efficient implementation of LU decomposition and other basic linear algebra operations.

Vectorization

MATLAB, in particular, used to implement *if-*, *for-*, and *while*-loops in a relatively inefficient way, while doing operations on chunks of matrices very efficiently. MathWorks, the commercial company that owns the MATLAB product, has put considerable effort in the last few years into closing the gaps in performance, and these are now much smaller compared to the past. In any case, different variants of the algorithms we have seen in this chapter which have precisely the same operation count may perform very differently, depending on the implementation. We have already had a glimpse of this in Section 5.1, where the two for-loops which the backward substitution algorithm uses were implemented in a MATLAB script using only one for-loop, with the other replaced by an inner product of the form A(k,k+1:n)*x(k+1:n).

Even more important, this also can be done for the LU decomposition algorithm. Let us assume no pivoting at first, and consider the first algorithm we saw on page 99, ignoring the operations on the right-hand-side **b**. Of the three loops involving k, i, and j, the loop on k appears to be particularly sequential (i.e., it is not easy to couple together or perform more than one stage at a time). The results of the altered matrix at stage k are used at stage $k+1$, and later on. But this is not so for the other two loops. The determination of the multipliers l_{ik}, done in the i-loop for a given k, can be done all at once, as a vector

$$\mathbf{l}_k = \begin{pmatrix} l_{k+1,k} \\ l_{k+2,k} \\ \vdots \\ l_{n,k} \end{pmatrix}.$$

Similarly for the subsequent update (the j-loop) of the rows of the submatrix of A: each row i is updated by a multiple l_{ik} of the row $\mathbf{a}_{k,k+1:n}$.

For two vectors **y** and **z** of the same length, the *inner product* is defined to recall as the scalar

$$\mathbf{y}^T\mathbf{z} = \sum_{i=1}^{m} y_i z_i.$$

This is used in the forward and backward substitution algorithms. The *outer product*, on the other hand, is defined as

$$\mathbf{y}\mathbf{z}^T,$$

i.e., it is the matrix whose (i,j)th element is $y_i z_j$.

Example 5.10. For the two vectors

$$\mathbf{y} = \begin{pmatrix} 3 \\ 2 \end{pmatrix}, \quad \mathbf{z} = \begin{pmatrix} 1 \\ 2 \\ 3 \end{pmatrix},$$

5.4. Efficient implementation

there is no inner product because the lengths don't match, but there are two outer products, given by

$$\mathbf{y}\mathbf{z}^T = \begin{pmatrix} 3 & 6 & 9 \\ 2 & 4 & 6 \end{pmatrix} \quad \text{and} \quad \mathbf{z}\mathbf{y}^T = \begin{pmatrix} 3 & 2 \\ 6 & 4 \\ 9 & 6 \end{pmatrix}.$$

MATLAB carries out these operations quietly, when issuing instructions such as y * z'. So you have to be careful about whether the array you have there corresponds to a row or a column vector, as unwelcome surprises may otherwise result. ∎

Now, the i and j for-loops in the LU decomposition algorithm can be expressed in terms of an outer product:

$$\begin{aligned} &\text{for } k = 1:n-1 \\ &\quad \mathbf{l}_k = \left(\frac{a_{k+1,k}}{a_{kk}}, \ldots, \frac{a_{n,k}}{a_{kk}} \right)^T \\ &\quad A_{k+1:n,k+1:n} = A_{k+1:n,k+1:n} - \mathbf{l}_k * \mathbf{a}_{k,k+1:n} \\ &\text{end} \end{aligned}$$

This outer product form allows us to express the bulk of Gaussian elimination as operations on consecutive elements in chunks of matrices. If we now incorporate partial pivoting, then to retain the ability to relate to chunks of A we will actually execute row interchanges if necessary, rather than use indirect indexing.

Our code for solving linear systems of equations

We are now in position to write an efficient linear system solver. It is not as robust, general and efficient as MATLAB's backslash instruction x = A\b, but on the other hand we can see what is going on. Our code uses the algorithm given on page 103, and you should be able to just read it.

```
function x = ainvb(A,b)
%
% function x = ainvb(A,b)
%
% solve Ax = b

[p,LU] = plu (A);
y = forsub (LU,b,p);
x = backsub (LU,y);

function [p,A] = plu (A)
%
% function [p,A] = plu (A)
%
% Perform LU decomposition with partial pivoting.
% Upon return the coefficients of L and U replace those
% of the input n by n nonsingular matrix A. The row interchanges
% performed are recorded in the 1D array p.

n = size(A,1);
```

```
% initialize permutation vector p
p = 1:n;

% LU decomposition with partial pivoting
for k = 1:n-1

  % find row index of relative maximum in column k
  [val,q] = max(abs(A(k:n,k)));
  q = q + k-1;

  % interchange rows k and q and record this in p
  A([k,q],:)=A([q,k],:);
  p([k,q])=p([q,k]);

  % compute the corresponding column of L
  J=k+1:n;
  A(J,k) = A(J,k) / A(k,k);

  % update submatrix by outer product
  A(J,J) =  A(J,J) - A(J,k) * A(k,J);
end

function y = forsub (A,b,p)
%
% function y = forsub (A,b,p)
%
% Given a unit lower triangular, nonsingular n by n matrix A,
% an n-vector b, and a permutation p,
% return vector y which solves Ay = Pb

n = length(b);

% permute b according to p
b=b(p);

% forward substitution
y(1) = b(1);
for k = 2:n
  y(k) = b(k) - A(k,1:k-1) * y(1:k-1);
end

function x = backsub (A,b)
%
% function x = backsub (A,b)
%
% Given an upper triangular, nonsingular n by n matrix A and
% an n-vector b, return vector x which solves Ax = b

n = length(b); x = b;
x(n) = b(n) / A(n,n);
for k = n-1:-1:1
  x(k) = ( b(k) - A(k,k+1:n)*x(k+1:n) ) / A(k,k);
end
```

5.4. Efficient implementation

In the MATLAB version we have used a function like `plu` which actually utilizes the three for-loops on i, j, and k would take about three times longer to obtain the LU decomposition of a 100×100 matrix. Note that these two implementations have the same operation (or flop) count. Other examples can be easily constructed where two algorithms require the same operation count to achieve the same output, differing only by the amount of memory access, with vastly different CPU times. How can they differ so drastically in execution time, then?

Fast memory access and BLAS

Indeed, the simple flop count that was given in Section 5.1 for Gaussian elimination and for forward and backward substitutions is useful for a coarse ("first cut") indication of which algorithm is expected to be faster. However, these operation counts completely ignore the relative cost of logical operations and the important question of memory access.

Computer memories are built as *hierarchies*, ranging from very fast, expensive, and small to slow, cheap, and large. In decreasing order of speed, such a hierarchy would typically include

- registers,
- cache,
- memory, and
- disk.

Arithmetic and logical operations are typically done only at the registers, at the top of the hierarchy, so data stored elsewhere have to be moved into the registers first. Such data movement, especially if it does not start at the cache, can be much more expensive than the cost of an arithmetic operation. Moving data in chunks, rather than using many independent data item moves, makes a big difference in overall performance.

Computer manufacturers, especially of high-performance machines, have standardized basic matrix operations such as matrix-matrix and matrix-vector multiplications into basic linear algebra subroutines (BLAS).[20]

The actual implementation is machine-dependent, but the user can see a library of BLAS which are machine-independent, so one is not required to modify code because the laptop battery has died. The BLAS are organized into levels: the higher the level, the larger the number of flops per memory reference that can be achieved. The first level (BLAS1) includes operations such as multiplying a vector by a scalar in single-precision and adding another vector, as in

$$\mathbf{y} \leftarrow a \ast \mathbf{x} + \mathbf{y} \quad (\text{SAXPY}),$$

as well as inner products. The mysterious word *SAXPY* is actually a widely used term in numerical linear algebra, standing for "single-precision a·x plus y". The second level (BLAS2) are matrix-vector multiplications. These include triangular systems and matrix updates by vector outer products. Thus, the LU decomposition utilizing outer products uses level-2 rather than level-1 BLAS. The third-level routines (BLAS3) are for matrix-matrix multiplications and include triangular systems with many right-hand sides. The details become complex, involving, for instance, the possibility of LU decomposition by blocks. This may be important in special circumstances for high-performance computing, but it goes well beyond the scope of this chapter.

---

[20]The S in this acronym is a remnant of days when the Fortran language reigned, to the fury of computer scientists.

Scaling the rows of the given problem

Now that we have introduced the function `plu` let us return to the question of scaling. Note that the partial pivoting strategy implemented here does not take scaling into account, so we should ensure ahead of time that the matrix A is not scaled too poorly, i.e., replace the problem $A\mathbf{x} = \mathbf{b}$ if necessary by a scaled version $SA\mathbf{x} = S\mathbf{b}$, with S a diagonal matrix.

Example 5.11. Let us define a linear system to be solved by the script

```
n = 100; h = 1/(n-1); K = 100;
A = zeros(n,n);
for i = 2:n-1
  A(i,i) = -2/h^2 - K;
  A(i,i-1) = -1/h^2; A(i,i+1) = -1/h^2;
end
A(1,1) = 1; A(n,n) = 1;   % end definition of A

xe = ones(n,1);           % exact solution of 1's
b = A*xe;                 % corresponding right hand side
```

Having defined the exact solution to be $\mathbf{x} = (1, 1, \ldots, 1)^T$, we can calculate the errors that result upon applying our linear system solver introduced above and those for MATLAB's staple "backslash" command. The obtained errors are 1.92e-11 and 5.68e-11, respectively.

Such errors can often be attributed to the conditioning of the problem; see Section 5.8. However, here the source for the loss of accuracy is different. Dividing each b_i and row i of A by $s_i = \max_{1 \le j \le n} |a_{ij}|$ (notice that $s_1 = s_n = 1$ and $s_i \approx 2h^{-2}$ otherwise), and solving the scaled systems by the same two codes, the obtained errors are 1.52e-14 and 1.38e-14, respectively. This is an improvement by three orders of magnitude!

The scaling does induce a different pivoting strategy in `plu`. Without it the first row is used for pivoting only in the 76th step, whereas with scaling no row permutation is found necessary. ∎

Specific exercises for this section: Exercises 10–11.

5.5 The Cholesky decomposition

Recall from Chapter 4 that a matrix A is symmetric if $A^T = A$ and positive definite if

$$\mathbf{x}^T A \mathbf{x} > 0 \quad \forall \, \mathbf{x} \neq \mathbf{0}.$$

Such matrices extend in important ways the concept of a positive scalar. In this section we see how solving linear systems for such matrices through the LU decomposition correspondingly simplifies, leading to the Cholesky decomposition.

No need for pivoting

It is easy to see that if A is symmetric positive definite, then all principal submatrices $A_{[j:k,j:k]}$ are also symmetric positive definite (simply consider all vectors $\mathbf{x}$ that have zeros in the first $j-1$ and the last $n-k$ places). The determinants of these principal minors are all positive as well, and it transpires

5.5. The Cholesky decomposition

that the pivots u_{kk} resulting from LU decomposition are also all positive. Indeed, it can be shown that Gaussian elimination without pivoting is a stable algorithm for symmetric positive definite matrices.

Symmetrizing the LU decomposition

Consider the LU decomposition of a symmetric positive definite matrix, written as

$$A = LU.$$

Recall that L is unit lower triangular, whereas U is upper triangular with diagonal elements $u_{kk} > 0$. We can write

$$U = \begin{pmatrix} u_{11} & & & & \\ & u_{22} & & & \\ & & \ddots & & \\ & & & \ddots & \\ & & & & u_{nn} \end{pmatrix} \begin{pmatrix} 1 & \frac{u_{12}}{u_{11}} & \cdots & \cdots & \frac{u_{1n}}{u_{11}} \\ & 1 & \frac{u_{23}}{u_{22}} & \cdots & \frac{u_{2n}}{u_{22}} \\ & & \ddots & & \vdots \\ & & & \ddots & \vdots \\ & & & & 1 \end{pmatrix} = D\tilde{U},$$

so the LU decomposition reads

$$A = LD\tilde{U}.$$

But transposing it we have

$$A = A^T = \tilde{U}^T D L^T.$$

Since this decomposition is unique we must have $\tilde{U}^T = L$, i.e., the decomposition reads

$$A = LDL^T,$$

where L is unit lower triangular and D is diagonal with positive elements u_{kk}.

It is also possible to write $D = D^{1/2} D^{1/2}$ with

$$D^{1/2} = \text{diag}\{\sqrt{u_{11}}, \ldots, \sqrt{u_{nn}}\},$$

whence the LU decomposition is written as

$$A = GG^T$$

with $G = LD^{1/2}$ a lower triangular matrix. This is called the *Cholesky decomposition*. It is possible to derive an algorithm for its implementation which requires not only roughly half the space (due to symmetry, of course) but also half the flops of the general LU decomposition algorithm, viz., $\frac{1}{3}n^3 + \mathcal{O}(n^2)$. This can be done by imposing the equality $A = GG^T$ elementwise. As naive as this may sound at first, it is something that one cannot do for Gaussian elimination algorithms for general matrices without being potentially forced to employ complex arithmetic.

Example 5.12. In the 2×2 case we can require

$$A = \begin{pmatrix} a_{11} & a_{12} \\ a_{12} & a_{22} \end{pmatrix} = \begin{pmatrix} g_{11} & 0 \\ g_{21} & g_{22} \end{pmatrix} \begin{pmatrix} g_{11} & g_{21} \\ 0 & g_{22} \end{pmatrix}.$$

Now we have precisely three conditions for the three unknown quantities g_{11}, g_{12}, and g_{22}. For a_{11} we get $a_{11} = g_{11}^2$, from which it follows that $g_{11} = \sqrt{a_{11}}$. Next, we have for a_{12} the condition $g_{21} = \frac{a_{12}}{g_{11}}$, where g_{11} is already known. Finally, using the equation for a_{22} we get $g_{22} = \sqrt{a_{22} - g_{21}^2}$. ∎

This algorithm, given below, can be straightforwardly applied to larger matrices in the same fashion. In this algorithm the entries of A are overwritten by the entries of G. A key here, and the reason attempting to follow the same strategy for general matrices (that is, not necessarily symmetric positive definite) will fail, is that the arguments of the square root are guaranteed to be positive if A is symmetric positive definite. (This is not entirely trivial to show, but please trust us: it is true.)

Algorithm: Cholesky Decomposition.
Given a symmetric positive definite $n \times n$ matrix A, this algorithm overwrites its lower part with its Cholesky factor.

$$\text{for } k = 1 : n-1$$
$$a_{kk} = \sqrt{a_{kk}}$$
$$\text{for } i = k+1 : n$$
$$a_{ik} = \frac{a_{ik}}{a_{kk}}$$
$$\text{end}$$
$$\text{for } j = k+1 : n$$
$$\text{for } i = j : n$$
$$a_{ij} = a_{ij} - a_{ik}a_{jk}$$
$$\text{end}$$
$$\text{end}$$
$$\text{end}$$
$$a_{nn} = \sqrt{a_{nn}}$$

In MATLAB, the reliable built-in function `chol` calculates the so-called *Cholesky factor* of a matrix. Notice, though, that `R=chol(A)` gives the matrix $R = G^T$ such that $A = R^T R$. It does not make a difference whether R or G is used, so long as it is done consistently.

Example 5.13. Let us create a symmetric positive definite matrix as follows. For a given n, fill an $n \times n$ array C by random numbers. With a little luck, the corresponding matrix C will be nonsingular, and thus $A = C^T C$ will be symmetric positive definite (check this!).

We now create also a random exact solution **x** and define the right-hand-side **b** = A**x**. Next we "forget" that we know **x** and try to find it, once through the Cholesky decomposition (involving no pivoting) and once through the general LU decomposition procedure with pivoting. The following script carries all this out and records relative errors:

```
C = randn(n,n); A = C'*C;
xe = randn(n,1);            % the exact solution
b = A*xe;                   % generate right-hand-side data

R = chol(A);                % Cholesky factor
% the following line is for compatibility with our forsub
```

```
D = diag(diag(R));  L = D \ R';  bb = D \ b;  p = 1:n;
y = forsub(L,bb,p);          % forward substitution R'y = b
x = backsub(R,y);            % backward substitution Rx = y
rerx = norm(x-xe)/norm(xe)   % error by Cholesky

xd = ainvb(A,b);             % ignores spd use partial pivoting
rerxd = norm(xd-xe)/norm(xe) % error by general routine
```

Running this script with $n = 500$, say, the routine `ainvb` uses very few (and, as it turns out, unnecessary!) row permutations. When using the more efficient Cholesky decomposition, errors of at least a similar quality and occasionally better are produced. Of course, no partial pivoting is utilized with Cholesky.

We return to this script in Example 5.20. ∎

Specific exercises for this section: Exercises 12–14.

5.6 Sparse matrices

There are many situations in practice where the matrix is large and *sparse*. What this means is that most of its elements are equal to zero. For an $n \times n$ matrix, it is convenient to think of a sparse matrix as having $\mathcal{O}(n)$ nonzero elements.

Frequently, the sparsity of a matrix originates from the nature of the underlying operator. The latter often represents *local* operations, where only a few components of a given grid or network of nodes "interact" with each other. For instance, the second tridiagonal matrix that we have seen in Example 4.17 in the context of discretization of a one-dimensional second derivative operator is $N \times N$, but of its N^2 elements only $3N - 2$ are nonzero. For $h = 0.01$ we have $N = 100$, so this translates into 298 out of 10,000 elements of A being nonzero. The discretized second derivative is indeed an example of a local operator: its value for a certain point on a grid depends only on the point itself and its two immediate neighbors.

For very large sparse matrices, simply ignoring the sparsity property is not a practical option, because memory requirements may become prohibitive. For instance, think of the above-mentioned tridiagonal matrix, with $N = 1,000,000$. Storing the approximately 3 million nonzero elements in double precision, and other data such as row and column indices, consumes a fraction of computer memory in today's terms: a mere few tens of megabytes. But if we try to store this matrix on all of its 1 trillion entries in memory, then we would be seriously testing (or, in fact, wildly exceeding) the limits of our computer's fast memory resources, and for no good reason at all: there is no need to waste space on storing zero entries—we just need to know where they are.

Storing sparse matrices

So, a problem has been identified, but how do we solve it? When a matrix is sparse it is not necessarily best to store it as a two-dimensional array. Rather, we could store it in the form of three nonzero *one-dimensional* arrays, which contain row indices, column indices, and the corresponding numerical values. Notice that the first two are integer arrays.

The *triplet form* (or *coordinate form*) is the simplest format: an $n \times n$ matrix that has m nonzero entries is represented by vectors $\{\mathbf{i}, \mathbf{j}, \mathbf{v}\}$, which are m-long each. For an integer $1 \leq k \leq m$, the corresponding kth entries of i_k, j_k, v_k represent $v_k = a_{i_k, j_k}$. This is the format used in MATLAB's sparse matrix representation.

Other commonly used formats are the *compressed sparse row* form and the *compressed sparse column* form. These forms are in fact slightly more economical to work with (in terms of performing matrix operations) than the triplet form. The idea here is that if we go in an orderly fashion by rows or by columns, one of the integer arrays could simply contain pointers to the start of every row or column in the other integer array, saving some storage along the way. For instance, in sparse row form, the vector **i** has $n+1$ entries, whereas **j** and **v** have m entries. Column indices of the entries in row k are stored in j_{i_k} through $j_{i_{k+1}-1}$.

Example 5.14. The matrix
$$A = \begin{pmatrix} 2.4 & -3 & 0 \\ 0 & 0 & 1.3 \\ 0 & 2 & -6 \end{pmatrix}$$
is represented in MATLAB's triplet form as follows:

```
(1,1)        2.4000
(1,2)       -3.0000
(3,2)        2.0000
(2,3)        1.3000
(3,3)       -6.0000
```

Note that the entries are presented column by column.
Alternatively, Figure 5.3 illustrates the compressed sparse row form of the matrix.

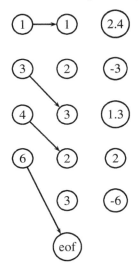

Figure 5.3. *Compressed row form of the matrix in Example 5.14. Shown are the vectors* **i, j, v**. *The arrows and the "eof" node are intended merely for illustration of where the elements of the vector* **i** *point to in* **j** *and correspondingly in* **v**.

It is impossible to appreciate the merits of sparse storage schemes for such a small matrix, but these concepts become useful for larger sparse matrices. ∎

We also mention data structures that are based on storing matrices by their diagonals. The MATLAB command `spdiags` does so. This often comes in handy, as many matrices in applications can be characterized in terms of a few nonzero diagonals. The matrices of Example 4.17 are in such a form.

5.6. Sparse matrices

Check out the commands `sparse` and `full`. The help documentation for these commands provides the names of a few additional MATLAB commands that deal with sparse matrices.

When representing matrices in sparse form, part of the programming effort is related to writing functions for matrix operations. For example, matrix addition typically requires inserting new nonzero entries, unless the matrices share identical nonzero patterns. It is not difficult to program, but is not as trivial and seamless as matrix addition in the dense case. Instances of basic operations that entail writing specialized routines are the transpose operation, matrix-vector product, matrix multiply, and several other operations.

Banded matrices and their LU decomposition

When solving a linear system of equations, $A\mathbf{x} = \mathbf{b}$, the effectiveness of Gaussian elimination and its variants greatly depends on the particular sparsity pattern of the matrix A. In Section 5.7 we briefly mention ways of increasing the efficiency by performing effective permutations. In Chapter 7 we discuss in detail alternatives to direct methods for situations where forming the LU decomposition results in a prohibitive amount of extra floating point operations and extra storage. But there are special cases of sparse matrices for which the LU decomposition works very effectively. One such instance is the class of banded matrices.

The matrix A is *banded* if all its elements are zero outside a band of diagonals around the main diagonal. Thus, there are positive integers p and q, $1 \leq p, q \leq n$, such that

$$a_{ij} = 0 \text{ if } i \geq j + p \text{ or if } j \geq i + q.$$

The **bandwidth** is defined as the number of consecutive diagonals that may contain nonzero elements, i.e., $p + q - 1$. The *upper bandwidth* is q and the *lower bandwidth* is p.

A particular banded matrix is depicted pictorially for $n = 10$, $p = 2$, and $q = 3$ in Figure 5.4. Elementwise we have

$$A = \begin{pmatrix} a_{11} & \cdots & a_{1q} & & & & & \\ \vdots & \ddots & \ddots & \ddots & & & & \\ a_{p1} & \ddots & \ddots & \ddots & \ddots & & & \\ & \ddots & \ddots & \ddots & \ddots & \ddots & & \\ & & \ddots & \ddots & \ddots & \ddots & a_{n-q+1,n} & \\ & & & \ddots & \ddots & \ddots & \ddots & \vdots \\ & & & & a_{n,n-p+1} & \cdots & & a_{nn} \end{pmatrix}.$$

For a diagonal matrix, $p = q = 1$; for a full (or dense) matrix, $p = q = n$. More interestingly, in Example 4.17 we see *tridiagonal* matrices, where $p = q = 2$.[21]

---

[21] If $p = q$, then $p - 1 = q - 1$ is often referred to as the *semibandwidth* of the matrix.

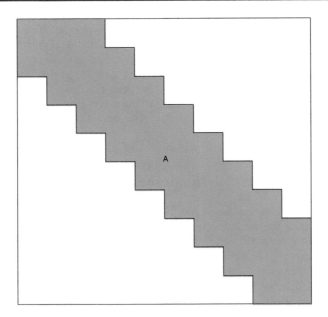

Figure 5.4. *A banded matrix for the case $n = 10$, $p = 2$, $q = 3$. Only areas of potentially nonzero entries are shaded in green.*

LU decomposition for banded matrices

Let us consider the LU decomposition without pivoting first. In the loops on i and on j we simply don't need to zero out elements that are already zero or to update rows with that part of the pivot's row which is zero. For a banded matrix the algorithm is therefore directly modified and is given on the current page.

Algorithm: LU Decomposition for Banded Matrices.
Given a real, banded, $n \times n$ matrix A which requires no pivoting, the following overwrites the upper triangular part of A by U:

$$
\begin{aligned}
&\text{for } k = 1 : n - 1 \\
&\quad \text{for } i = k+1 : \min(k + p - 1, n) \\
&\quad\quad l_{ik} = \frac{a_{ik}}{a_{kk}} \\
&\quad\quad \text{for } j = k+1 : \min(k + q - 1, n) \\
&\quad\quad\quad a_{ij} = a_{ij} - l_{ik} a_{kj} \\
&\quad\quad \text{end} \\
&\quad \text{end} \\
&\text{end}
\end{aligned}
$$

5.6. Sparse matrices

The resulting L and U factors then inherit the band form and can be written as

$$L = \begin{pmatrix} 1 & & & & & & & \\ \vdots & 1 & & & & & & \\ l_{p1} & \ddots & 1 & & & & & \\ & \ddots & \ddots & \ddots & & & & \\ & & \ddots & \ddots & \ddots & & & \\ & & & \ddots & \ddots & \ddots & & \\ & & & & l_{n,n-p+1} & \cdots & 1 \end{pmatrix},$$

$$U = \begin{pmatrix} u_{11} & \cdots & u_{1q} & & & & \\ & \ddots & \ddots & \ddots & & & \\ & & \ddots & \ddots & \ddots & & \\ & & & \ddots & \ddots & \ddots & \\ & & & & \ddots & \ddots & u_{n-q+1,n} \\ & & & & & \ddots & \vdots \\ & & & & & & u_{nn} \end{pmatrix}.$$

(Combine Figures 5.2 and 5.4 to see it in color.)

Of course, none of these zero triangles is ever stored. If the band sizes p and q remain small and fixed as the matrix size n grows, then we obtain an algorithm with $\mathcal{O}(n)$ storage and $\mathcal{O}(n)$ execution time—a great improvement. Applying higher level BLAS is more complicated, though.

For the case of a tridiagonal matrix, in particular, the i and j loops simplify into single statements. The resulting algorithm is occasionally referred to as the *Thomas algorithm*; see Exercise 17.

Pivoting

Just because a matrix is banded does not mean that it no longer requires pivoting! If the usual row partial pivoting is applied, then the band structure may be altered due to row interchanges. Using such pivoting, the upper bandwidth q of U may increase to $q + p - 1$. Moreover, although the number of nonzeros in each column of the unit lower triangular L remains as without pivoting, they may no longer be tucked next to the main diagonal in a neat band. (Please check this statement, e.g., by constructing a suitable example and running MATLAB's routine `lu`.)

In the above discussion we did not assume anything about the sparsity pattern *within* the band. In general, those locations in A where an original zero is overwritten by a nonzero element during LU decomposition are referred to as a **fill-in**. Banded matrices for which no pivoting is required and which do not have zeros within the band yield no fill-in, as we have seen. Row permutations used during partial pivoting do introduce some fill-in into L, but not overwhelmingly so when the band is narrow.

However, significant sparsity within a not-very-narrow band is harder to take care of. Banded matrices that have many zeros also within the band serve as a motivating example for the class of *iterative* solution methods, discussed in Chapter 7. Techniques for minimizing the damage caused by fill-in when using direct methods, with which the present chapter is concerned, are discussed next.

Specific exercises for this section: Exercises 15–19.

5.7 Permutations and ordering strategies

We have already noticed that the effectiveness of Gaussian elimination may strongly depend on the sparsity pattern of the matrix; banded matrices introduced in the previous section provide a convincing example that great advances can be made with LU decomposition if no crippling fill-in results. In this section we discuss ways of making the Gaussian elimination procedure for more general sparse matrices as effective as possible by permutations and reordering strategies.

> **Note:** The topic of Section 5.7 is more advanced and more current than those of other sections in this chapter.

The effect of row and column permutations on fill-in

Let us start with a long but motivating example which shows that in fact reordering may result in a significant performance difference in operation count and storage requirements.

Example 5.15. Consider an $n \times n$ matrix A whose elements satisfy

$$a_{ij} = \begin{cases} \neq 0 & \text{if } i = j \text{ or } i = 1 \text{ or } j = 1, \\ 0 & \text{otherwise.} \end{cases}$$

An illustration of A for a 5×5 case is given below, where $\times$ stands for a possibly nonzero matrix element.[22] Consider

$$A = \begin{pmatrix} \times & \times & \times & \times & \times \\ \times & \times & 0 & 0 & 0 \\ \times & 0 & \times & 0 & 0 \\ \times & 0 & 0 & \times & 0 \\ \times & 0 & 0 & 0 & \times \end{pmatrix}.$$

Suppose now that C is a matrix obtained by swapping the first and the last rows of A, so

$$C = \begin{pmatrix} \times & 0 & 0 & 0 & \times \\ \times & \times & 0 & 0 & 0 \\ \times & 0 & \times & 0 & 0 \\ \times & 0 & 0 & \times & 0 \\ \times & \times & \times & \times & \times \end{pmatrix}.$$

---

[22] To fully appreciate the effect of sparsity and fill-in you really should think of a large matrix, say, $n = 500$ or $n = 5000$.

5.7. Permutations and ordering strategies

If A represents a matrix associated with a linear system $A\mathbf{x} = \mathbf{b}$, then the first and the last elements of $\mathbf{b}$ are also exchanged. This corresponds to simply writing down the equations of the system in a slightly different order, just like we did when applying row partial pivoting in Section 5.3.

However, if A is symmetric, which can be important, as we saw in Section 5.5, then we lose this property in C. Therefore, we apply the same swapping also to the columns of A. This corresponds to swapping the first and last unknowns and yields a matrix B whose elements are given by

$$b_{ij} = \begin{cases} \neq 0 & \text{if } i = j \text{ or } i = n \text{ or } j = n, \\ 0 & \text{otherwise.} \end{cases}$$

An illustration of the sparsity pattern of B for the 5×5 case is given by

$$B = \begin{pmatrix} \times & 0 & 0 & 0 & \times \\ 0 & \times & 0 & 0 & \times \\ 0 & 0 & \times & 0 & \times \\ 0 & 0 & 0 & \times & \times \\ \times & \times & \times & \times & \times \end{pmatrix}.$$

The matrices A and B are often referred to as *arrow matrices*, for obvious reasons. If A is symmetric, then so is B.

Suppose that no pivoting is required (for example, if A and hence B are diagonally dominant), and let us denote by $A = L_A U_A$ the LU decomposition of A and by $B = L_B U_B$ the LU decomposition of B. What are the nonzero patterns of L_A, U_A, L_B, U_B and the overall computational costs entailed in solving the system with A and in solving the system with B?

For A the first stage of Gaussian elimination produces a disastrous fill-in: when zeroing out the first column below the $(1,1)$ element, we subtract scalar multiples of the first row from each of the other rows, and this yields

$$A^{(1)} = \begin{pmatrix} \times & \times & \times & \times & \times \\ 0 & \times & \times & \times & \times \\ 0 & \times & \times & \times & \times \\ 0 & \times & \times & \times & \times \\ 0 & \times & \times & \times & \times \end{pmatrix}.$$

At this point the original, attractive sparsity pattern of the matrix is gone! The resulting $A^{(1)}$ looks like what one would get by applying LU decomposition to any generic dense matrix, and subsequently the factors L_A and U_A are fully dense.

The situation with the matrix B is much more encouraging. It is straightforward to see that thanks to the fact that the first row of B has only nonzero elements in the first and last columns, we get no fill-in, so

$$B^{(1)} = \begin{pmatrix} \times & 0 & 0 & 0 & \times \\ 0 & \times & 0 & 0 & \times \\ 0 & 0 & \times & 0 & \times \\ 0 & 0 & 0 & \times & \times \\ 0 & \times & \times & \times & \times \end{pmatrix}.$$

The same happens in the next stages of the LU decomposition, yielding

$$L_B = \begin{pmatrix} \times & 0 & 0 & 0 & 0 \\ 0 & \times & 0 & 0 & 0 \\ 0 & 0 & \times & 0 & 0 \\ 0 & 0 & 0 & \times & 0 \\ \times & \times & \times & \times & \times \end{pmatrix}, \quad U_B = \begin{pmatrix} \times & 0 & 0 & 0 & \times \\ 0 & \times & 0 & 0 & \times \\ 0 & 0 & \times & 0 & \times \\ 0 & 0 & 0 & \times & \times \\ 0 & 0 & 0 & 0 & \times \end{pmatrix}.$$

Thus, the overall storage requirements as well as the overall computational work are very modest for B.

There is nothing special about the size $n = 5$; everything above is applicable to a general n. It is worth your time doing Exercise 21: the difference in storage and flop count estimates between handling A and B is intriguing, showing that swapping equations and unknowns in the original linear system results in a substantially improved situation. ∎

Permutation strategies

To understand things in a more general way, let us revisit the notion of *permutation matrices*. We have defined those in the context of partial pivoting in Section 5.3 and have observed that a permutation matrix P satisfies $P^T = P^{-1}$. Now, suppose that P is a given permutation matrix, and p is the corresponding one-dimensional array that represents the same permutation. Then $A\mathbf{x} = \mathbf{b}$ and $PA\mathbf{x} = P\mathbf{b}$ have the same solution. Using $P^T P = I$, let us write

$$(PAP^T)(P\mathbf{x}) = P\mathbf{b}.$$

Doing this incorporates mathematically the fact that we are now looking at $P\mathbf{x}$ rather than $\mathbf{x}$, as per the performed permutation. Note that if A is symmetric, then so is $B = PAP^T$. We can rewrite the linear system as

$$B\mathbf{y} = \mathbf{c},$$

where $\mathbf{y} = P\mathbf{x}$ and $\mathbf{c} = P\mathbf{b}$. In Example 5.15, P is a permutation matrix associated with the array $\mathrm{p} = [n, 2, 3, 4, \ldots, n-2, n-1, 1]$.

In modern direct solvers much attention is devoted to the question of what permutation matrix should be selected. Based on the discussion so far, we may intuitively think of two possible ways of aiming to reduce the storage and computational work:

- Reduce the bandwidth of the matrix.

- Reduce the expected fill-in in the decomposition stage.

The question is how to accomplish the above mentioned goals in practice. One danger to be aware of is the combinatorial nature of the issue at hand. Suffice it to observe that for a linear system with n unknowns there are $n!$ different orderings. We certainly do not want to sort through such a number of possibilities when the solution algorithm itself involves $\mathcal{O}(n^3)$ operations; the latter all of a sudden seems small in comparison. Therefore, we need to be assured that the amount of work for determining the ordering does not dominate the computation, and hence seeking an "optimal" graph may turn out to be too costly an adventure. The art of finding effective orderings within a reasonable cost must rely on good heuristics and does not always have a solid theory to back it up.

Commonly used approaches for deriving effective ordering strategies utilize the *matrix graph*, discussed next.

5.7. Permutations and ordering strategies

Matrix graph

Suppose a matrix is *structurally symmetric*. This means that if $a_{ij} = 0$, then also $a_{ji} = 0$. For instance, the matrices A and B in Example 5.15 are structurally symmetric, whereas C is not. The graph of our matrix A comprises *vertices* and *edges* as follows: a row/column index is represented as a vertex, and if $a_{ij} \neq 0$, then there is an edge that connects vertex i to vertex j. Since we assume structural symmetry we need not worry about distinguishing between a_{ij} and a_{ji}.

The graphs of the matrices A and B of Example 5.15 are depicted in Figure 5.5. The graphs of any matrix A and a symmetric permutation PAP^T are identical in terms of the edges; the only difference is in the labeling of the vertices. In particular, vertex i is replaced by vertex p_i, where p is the permutation array.

Figure 5.5. *Graphs of the matrices A (left) and B (right) of Example 5.15.*

In the process of Gaussian elimination, each stage can be described in terms of the matrix graph as follows: upon zeroing out the (i, j) entry of the matrix with entry (j, j) being the current pivot, the vertices that are the neighbors of vertex j will cause the creation of an edge connecting them to vertex i, if such an edge does not already exist. For instance, in Example 5.15, when attempting to zero out entry $(5,1)$ using entry $(1,1)$ as pivot, in the graph of $A^{(1)}$ all vertices j connected to vertex 1 generate an edge $(5, j)$, and hence new edges $(2,5)$, $(3,5)$, and $(4,5)$ appear. In contrast, for B no new edge is generated because there are no edges connected to vertex 1 other than vertex 5 itself.

The above underlines the importance of the *degree* of a vertex, i.e., the number of edges emanating from it. Generally speaking, we should postpone dealing with vertices of a high degree as much as possible. For the matrix A in Example 5.15 all the vertices except vertex 1 have degree 1, but vertex 1 has degree 4 and we start off the Gaussian elimination by eliminating it; this results in disastrous fill-in. On the other hand, for the matrix B all vertices except vertex 5 have degree 1, and vertex 5 is the last vertex we deal with. Until we hit vertex 5 there is no fill because the latter is the only vertex that is connected to the other vertices. When we deal with vertex 5 we identify vertices that should hypothetically be generated, but they are already in existence to begin with, so we end up with no fill whatsoever.

Two vertex labeling strategies

Ordering strategies in abundance have been derived throughout the years. As tempting as it is, we will not describe them in detail.

Many of these algorithms rely on heuristics related to the degrees of the vertices of the matrix graphs and use a good amount of auxiliary objects from graph theory. The **reverse Cuthill Mc-Kee** (RCM) algorithm is one that is effective at reducing bandwidth, and the *minimum degree* or **approximate minimum degree** (AMD) algorithm aims at minimizing the expected fill-in.

Example 5.16. Let us illustrate how intriguing the nonzero structure of these matrices could be.

The leftmost plot in Figure 5.6 shows the sparsity pattern of a certain well-known symmetric positive definite matrix A. This matrix is derived in Example 7.1 on page 168, but all that concerns us here is its zero pattern.

The middle plot of Figure 5.6 shows the effect of RCM ordering. It can be believed from the picture that the goal was to minimize the bandwidth. The pattern is pretty elegant, at least in the eyes of those who appreciate the contemporary art of matrix sparsity plots (or, alternatively, wind surfing). The rightmost plot depicts the result of applying the AMD algorithm. Here it is harder to identify a unifying theme. But note that, as expected, rows with more elements (corresponding to vertices with a higher degree) appear closer to the bottom of the matrix, i.e., their index or label value is higher.

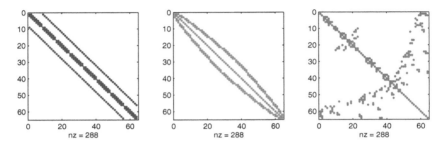

Figure 5.6. *Sparsity pattern of a certain symmetric positive definite matrix A (left), its RCM ordering (middle), and approximate minimum degree ordering (right).*

Notice the numbers `nz` at the bottom of each plot. This stands for "nonzeros" and states how many nonzero entries the matrix has. Since we only permute rows and columns, `nz` does not change here. But it will change following the procedure in Example 5.17.

The MATLAB code for generating the plots of Figure 5.6 (omitting the generation of A and a couple of output commands) is

```
p=symrcm(A); spy(A(p,p));
p=symamd(A); spy(A(p,p));
```

The actual work is encoded in the two built-in functions with the suggestive names (other than `spy`). ■

Example 5.17. Figure 5.7 shows the sparsity patterns of the Cholesky factors of the same matrices as in Example 5.16. There is fill in the original matrix as well as for the RCM matrix, but since the latter has a lower bandwidth the overall number of nonzeros is smaller, so we see an improvement. The fill-in is yet smaller for the AMD algorithm, which as expected (by its name, at least) produces the Cholesky factor with approximately the least number of nonzeros.

Try to figure out the corresponding MATLAB instructions for Figure 5.7 that are additional to those displayed in Example 5.16. ■

Specific exercises for this section: Exercises 20–21.

5.8. Estimating errors and the condition number

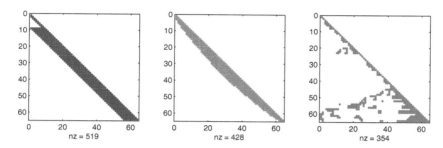

Figure 5.7. *Sparsity pattern of the Cholesky factors of A (left), its RCM ordering (middle), and approximate minimum degree ordering (right).*

5.8 Estimating errors and the condition number

Thus far in this chapter we have discussed at length solution methods for solving

$$A\mathbf{x} = \mathbf{b},$$

and Chapter 7 will add several more. In this section we take a short break from methods and ponder the question of how accurate we may expect our numerical solutions to be.

The error in the numerical solution

Suppose that, using some algorithm, we have computed an approximate solution $\hat{\mathbf{x}}$. With $\mathbf{x}$ denoting as usual the exact, unknown solution, we would like to be able to evaluate the absolute error $\|\mathbf{x} - \hat{\mathbf{x}}\|$, or the relative error

$$\frac{\|\mathbf{x} - \hat{\mathbf{x}}\|}{\|\mathbf{x}\|},$$

in some norm such as those defined in Section 4.2. Needless to say, an exact expression for these errors is not something we can get our hands on: if the error were known, then the error-free solution $\mathbf{x}$ would also be known! So, we seek an upper bound on the error and rely on quantities that we can directly calculate.

One such quantity is the *residual*

$$\hat{\mathbf{r}} = \mathbf{b} - A\hat{\mathbf{x}}.$$

In the ideal world we have $\mathbf{r} = \mathbf{b} - A\mathbf{x} = \mathbf{0}$, but in practice there will always be some nonzero residual $\hat{\mathbf{r}}$ because of roundoff errors. When using iterative methods in Chapter 7 there is also convergence errors to reckon with, as in Chapter 3, but let's not worry about that now. A stable Gaussian elimination algorithm variant will deliver a residual with a small norm. The question is, what can be concluded from this about the error in $\hat{\mathbf{x}}$?

Example 5.18. Let

$$A = \begin{pmatrix} 1.2969 & .8648 \\ .2161 & .1441 \end{pmatrix}, \quad \mathbf{b} = \begin{pmatrix} .8642 \\ .1440 \end{pmatrix}.$$

Suppose that, using some algorithm, the approximate solution

$$\hat{\mathbf{x}} = \begin{pmatrix} .9911 \\ -.4870 \end{pmatrix}$$

was calculated. Then

$$\hat{\mathbf{r}} = \mathbf{b} - A\hat{\mathbf{x}} = \begin{pmatrix} -10^{-8} \\ 10^{-8} \end{pmatrix},$$

so $\|\hat{\mathbf{r}}\|_\infty = 10^{-8}$. This is a sensibly small residual for many practical purposes. However, the exact solution is

$$\mathbf{x} = \begin{pmatrix} 2 \\ -2 \end{pmatrix}, \quad \text{so } \|\mathbf{x} - \hat{\mathbf{x}}\|_\infty = 1.513.$$

Thus, the error is of the same size as the solution itself, roughly 10^8 times that of the residual! ∎

It is important to understand that we cannot expect our algorithms to deliver much more than a small residual. We return to this in Section 7.4.

Condition number and a relative error estimate

How does the residual $\hat{\mathbf{r}}$ generally relate to the error in $\hat{\mathbf{x}}$? We can write

$$\hat{\mathbf{r}} = \mathbf{b} - A\hat{\mathbf{x}} = A\mathbf{x} - A\hat{\mathbf{x}} = A(\mathbf{x} - \hat{\mathbf{x}}),$$

so

$$\mathbf{x} - \hat{\mathbf{x}} = A^{-1}\hat{\mathbf{r}}.$$

Let us choose a vector norm (say, the maximum norm), and its corresponding matrix norm.[23] Then

$$\|\mathbf{x} - \hat{\mathbf{x}}\| = \|A^{-1}\hat{\mathbf{r}}\| \leq \|A^{-1}\|\|\hat{\mathbf{r}}\|.$$

This gives a bound on the absolute error in $\hat{\mathbf{x}}$ in terms of $\|A^{-1}\|$. But usually the relative error is more meaningful. Since $\|\mathbf{b}\| \leq \|A\|\|\mathbf{x}\|$ implies $\frac{1}{\|\mathbf{x}\|} \leq \frac{\|A\|}{\|\mathbf{b}\|}$, we have

$$\frac{\|\mathbf{x} - \hat{\mathbf{x}}\|}{\|\mathbf{x}\|} \leq \|A^{-1}\|\|\hat{\mathbf{r}}\|\frac{\|A\|}{\|\mathbf{b}\|}.$$

We therefore define the *condition number* of the matrix A as

$$\kappa(A) = \|A\|\|A^{-1}\|$$

and write the bound obtained on the relative error as

$$\frac{\|\mathbf{x} - \hat{\mathbf{x}}\|}{\|\mathbf{x}\|} \leq \kappa(A)\frac{\|\hat{\mathbf{r}}\|}{\|\mathbf{b}\|}. \tag{5.1}$$

---

[23]Whereas the choice of which vector norm to use is less important for our current purposes, it does fix the choice of the matrix norm as the corresponding induced matrix norm.

5.8. Estimating errors and the condition number

In words, the relative error in the solution is bounded by the condition number of the matrix A times the relative error in the residual. Memorizing this estimate would be worth your while.

The notion of **conditioning** of a problem is fundamental and has been briefly discussed in Section 1.3. Here it can be expressed in terms of a quantity relating to the matrix A alone. Note that $\kappa(A)$ does depend on the norm in which it is defined. To make sure it is clear what norm is used (when it is important to make the distinction) we will use notation such as $\kappa_2(A)$, $\kappa_\infty(A)$, and so on. Regardless of norm, we have

$$1 = \|I\| = \|A^{-1}A\| \le \kappa(A),$$

at one end of the scale, so a matrix is ideally conditioned if its condition number equals 1, and $\kappa(A) = \infty$ for a singular matrix at the other end of the scale. In fact, it can be shown that if A is nonsingular and E is the matrix of minimum ℓ_2-norm such that $A + E$ is singular, then

$$\frac{\|E\|_2}{\|A\|_2} = \frac{1}{\kappa_2(A)}.$$

So, the condition number measures how close to singularity in the relative sense a matrix is.

Example 5.19. Continuing with Example 5.18, we have

$$A^{-1} = 10^8 \begin{pmatrix} .1441 & -.8648 \\ -.2161 & 1.2969 \end{pmatrix}.$$

This yields

$$\|A^{-1}\|_\infty = 1.513 \times 10^8, \quad \text{hence } \kappa_\infty(A) = 2.1617 \cdot 1.513 \times 10^8 \approx 3.27 \times 10^8.$$

Regarding the actual computed $\hat{\mathbf{x}}$, note that $\frac{1.513}{2} < \frac{3.27}{.8642}$. So, indeed

$$\frac{\|\mathbf{x}-\hat{\mathbf{x}}\|_\infty}{\|\mathbf{x}\|_\infty} < \kappa_\infty(A) \frac{\|\hat{\mathbf{r}}\|_\infty}{\|\mathbf{b}\|_\infty}.$$

However, the mere knowledge that the bound (5.1) holds would not make the computed solution necessarily satisfactory for all purposes.

Finally, if we modify $a_{22} \leftarrow \frac{.8648 \cdot .2161}{1.2969}$, then obviously the matrix becomes singular, because the columns will be a scalar multiple of one another. So, the matrix $A + E$ is singular, where

$$E = \begin{pmatrix} 0 & 0 \\ 0 & \frac{.8648 \cdot .2161}{1.2969} - .1441 \end{pmatrix} \approx \begin{pmatrix} 0 & 0 \\ 0 & -7.7 \times 10^{-9} \end{pmatrix}.$$

Indeed, for this particular perturbation we have

$$\frac{\|E\|}{\|A\|} \approx \frac{1}{\kappa(A)}$$

in both ℓ_2- and ℓ_∞-norms. ∎

In MATLAB use cond(A,inf), or cond(A,2), etc., to find the condition number in the specified norm. But usually there is really no reason, other than for deductive purposes, why we would want to find $\kappa(A)$ exactly. Typically, we want just a cheap estimate, especially when the condition number may be large. A cheap estimate of a lower bound for $\kappa_1(A)$ is obtained by the instruction `condest(A)`. Assuming we have such an estimate, let us now tie this to an actual computation.

The error when using a direct method

There are two separate issues. The more practical one is the actual estimation, or bounding, of the relative error in the result, given a calculated residual $\hat{\mathbf{r}} = \mathbf{b} - A\hat{\mathbf{x}}$. This has already been answered in Equation (5.1), regardless of how the residual $\hat{\mathbf{r}}$ arises.[24]

The other issue is tying the above to our direct methods for solving linear equations. Consider Gaussian elimination with pivoting, and assume no error other than the accumulation of roundoff errors. The accumulated error in the LU decomposition can be viewed as the exact decomposition of a slightly perturbed matrix $\tilde{A} = A + \delta A$. Then the forward and backward substitutions add more small perturbations, each of the order of the rounding unit times n, so the entire algorithm may be considered as producing the *exact solution of a perturbed problem*

$$(A + \delta A)\hat{\mathbf{x}} = \mathbf{b} + \delta \mathbf{b}.$$

This is called **backward error analysis**.

A simple substitution shows that

$$\hat{\mathbf{r}} = \mathbf{b} - A\hat{\mathbf{x}} = (\delta A)\hat{\mathbf{x}} - \delta \mathbf{b}.$$

How large are δA and $\delta \mathbf{b}$, and thus $\hat{\mathbf{r}}$? An LU decomposition procedure with pivoting can be shown to yield a perturbation bounded by

$$\|\delta A\|_\infty \leq \eta \phi(n) g_n(A),$$

where $\phi(n)$ is a low order polynomial in n (cubic at most) and η is the rounding unit. The extra perturbations which the forward and backward substitutions yield, in particular the bound on $\|\delta \mathbf{b}\|_\infty$, are significantly smaller. Thus, as long as the pivoting keeps $g_n(A)$ growing only moderately in n (like a low order polynomial, but *not* exponentially), and assuming n not to be gigantic, the overall perturbations δA and $\delta \mathbf{b}$ are not larger than a few orders of magnitude times η.

Plugging the expression for $\hat{\mathbf{r}}$ into Equation (5.1) we finally obtain

$$\frac{\|\mathbf{x} - \hat{\mathbf{x}}\|}{\|\mathbf{x}\|} \leq \kappa(A) \frac{\|\delta \mathbf{b}\| + \|\delta A\| \|\hat{\mathbf{x}}\|}{\|\mathbf{b}\|}.$$

This can be further beautified, provided $\|\delta A\| < 1/\|A^{-1}\|$, to read

$$\frac{\|\mathbf{x} - \hat{\mathbf{x}}\|}{\|\mathbf{x}\|} \leq \frac{\kappa(A)}{1 - \kappa(A)(\|\delta A\|/\|A\|)} \left(\frac{\|\delta \mathbf{b}\|}{\|\mathbf{b}\|} + \frac{\|\delta A\|}{\|A\|} \right).$$

In summary, a *stable* algorithm is responsible for producing a small residual. This yields acceptably small error in the solution if the problem is *well-conditioned*, i.e., if $\kappa(A)$ is not too large. Just what is "small" and what is "too large" may depend on the particular circumstances, but the qualitative concept is general.

---

[24] The residual could actually arise as a result of measurement errors in data, for instance.

5.8. Estimating errors and the condition number

Example 5.20. Let us return to Example 5.13. If you invoke its script with $n = 500$ several times, you'll get different results, as befits an unseeded random number generator. But a typical run (unless you are incredibly lucky) would yield errors that are well above the rounding unit.

Using the Cholesky decomposition, in one instance tried we had a relative error of 2.6×10^{-11} and in another the relative error was 8.0×10^{-10}. The corresponding errors using `ainvb` (which required two and three theoretically unnecessary row permutations) were fairly close and slightly worse. Recall that the default rounding unit is $\eta \approx 10^{-16}$.

In search for the lost digits of accuracy we computed the condition numbers κ_2 (A) of the constructed random matrices A and these were, respectively, 8.8×10^5 and 3.9×10^8. These condition numbers, then, account for the loss of accuracy in the computed solutions.

Indeed, the relative residuals were 8.1×10^{-16} and 9.2×10^{-16}, respectively, which shows not only that both our variants of Gaussian elimination performed very well but also that the bound (5.1) is respected and indicative of actual error behavior. ∎

More on the condition number

The condition number is of fundamental importance in numerical computing. Note that we *cannot* use the determinant of A to indicate closeness to singularity, as the following example demonstrates.

Example 5.21. Consider $A = 0.1I$, explicitly written as

$$A = \begin{pmatrix} 0.1 & & & \\ & 0.1 & & \\ & & \ddots & \\ & & & 0.1 \end{pmatrix}.$$

This matrix is perfectly conditioned: $\kappa(A) = 1$. Yet, $\det(A) = 0.1^n$. For $n = 16$ the determinant is already of the order of our standard rounding unit.

The opposite can also hold, namely, a matrix A can be nearly singular and yet its determinant is not small. Such is the case for

$$A = \begin{pmatrix} 1 & -1 & \cdots & -1 \\ & \ddots & \ddots & \vdots \\ & & \ddots & -1 \\ & & & 1 \end{pmatrix},$$

whose inverse has the element 2^{n-2} in the upper right corner. The condition number grows exponentially in n while $\det(A) = 1$. ∎

Conditioning of orthogonal matrices

An important class of matrices that are perfectly conditioned in the induced ℓ_2-norm are *orthogonal matrices*, introduced in Section 4.3. Recall that an $n \times n$, real matrix Q is orthogonal if $Q^T = Q^{-1}$. (For instance, a permutation matrix is orthogonal.) Then for any vector $\mathbf{x}$ satisfying $\|\mathbf{x}\|_2 = 1$ we have

$$\|Q\mathbf{x}\|_2^2 = \mathbf{x}^T Q^T Q \mathbf{x} = \mathbf{x}^T \mathbf{x} = 1,$$

so $\|Q\|_2 = 1$. Likewise, $\|Q^{-1}\|_2 = \|Q^T\|_2 = 1$, hence $\kappa_2(Q) = 1$. Orthogonal transformations are therefore particularly stable and thus a long-time favorite in numerical linear algebra.

> **Note:** Orthogonal matrices are ideally conditioned. On the other hand, there is nothing in principle to prevent a symmetric positive definite matrix from having a large condition number.

Conditioning of symmetric positive definite matrices

Let us consider another important class of matrices, the symmetric positive definite ones. Recall from Section 4.3 that the eigenvalues of such matrices are all real and positive, so let's write them as

$$\lambda_1 \geq \lambda_2 \geq \cdots \geq \lambda_n > 0.$$

Thus, $\|A\|_2 = \sqrt{\rho(A^T A)} = \sqrt{\lambda_1^2} = \lambda_1$. Also, the inverse of A is symmetric positive definite with eigenvalues

$$0 < \frac{1}{\lambda_1} \leq \frac{1}{\lambda_2} \leq \cdots \leq \frac{1}{\lambda_n}.$$

Thus, $\|A^{-1}\|_2 = \frac{1}{\lambda_n}$, and this yields the compact expression

$$\kappa_2(A) = \frac{\lambda_1}{\lambda_n}.$$

Example 5.22. Recall from Example 4.9 that the "unit circle", i.e., all the vectors with unit norm, actually corresponds to our usual notion of a circle in $\mathcal{R}^2$ when using the ℓ_2-norm. In Figure 5.8 we plot the mapping of this unit circle by the symmetric positive definite matrix

$$A = \begin{pmatrix} 1 & .75 \\ .75 & 1 \end{pmatrix}.$$

Producing such a plot is very simple in MATLAB:

```
d = .75;
A = [1,d;d,1];
t = 0:.01:10; m = length(t);
x(1,1:m) = sin(t);
x(2,1:m) = cos(t);
y = A*x;
plot (y(1,:),y(2,:),'b', x(1,:),x(2,:),'r:')
xlabel('x_1')
ylabel('x_2')
```

The eigenvalues of A are .25 and 1.75, so $\kappa(A) = \frac{1.75}{0.25} = 7$. These eigenvalues are the lengths of the semi-axes of the ellipse in Figure 5.8. The larger the difference between the eigenvalues (i.e., the larger the condition number), the more "long and skinny" the unit circle becomes under this transformation. ∎

Specific exercises for this section: Exercises 22–24.

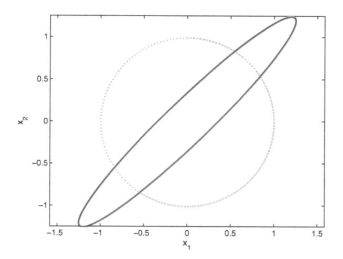

Figure 5.8. *Stretching the unit circle by a symmetric positive definite transformation.*

5.9 Exercises

0. **Review questions**

    (a) What is the overall cost in flops of applying a forward or backward solve?

    (b) What is the overall cost in flops of decomposing a matrix using LU?

    (c) During the course of Gaussian elimination without pivoting a zero pivot has been encountered. Is the matrix singular?

    (d) State three disadvantages of computing the inverse of a matrix to solve a linear system rather than using the LU decomposition approach.

    (e) The complete pivoting strategy is numerically stable, whereas partial pivoting is not always stable. Why is the latter approach preferred in practice in spite of this?

    (f) Show that for any symmetric positive definite matrix A there is a nonsingular matrix H such that $A = H^T H$.

    (g) Give three examples of data structures for storing a sparse matrix. Why aren't they used for full (dense) matrices, too?

    (h) How are the storage requirements and overall computational work different for banded matrices compared with general dense matrices?

    (i) State two possible strategies for reordering the equations and unknowns in a way that improves the efficiency of Gaussian elimination for sparse matrices.

    (j) What is the fundamental difference between error and residual in terms of computability?

    (k) Suppose we compute an approximate solution $\tilde{\mathbf{x}}$ for $A\mathbf{x} = \mathbf{b}$ and get $\mathbf{r} = \mathbf{b} - A\tilde{\mathbf{x}}$ whose norm is very small. Can we conclude in this case that the error $\mathbf{x} - \tilde{\mathbf{x}}$ must also be small?

    (l) The condition number is an important concept, but it is rarely computed exactly. Why?

1. Consider the problem
$$x_1 - x_2 + 3x_3 = 2,$$
$$x_1 + x_2 = 4,$$
$$3x_1 - 2x_2 + x_3 = 1.$$

   Carry out Gaussian elimination in its simplest form for this problem. What is the resulting upper triangular matrix?

   Proceed to find the solution by backward substitution.

2. The *Gauss–Jordan method* used to solve the prototype linear system can be described as follows. Augment A by the right-hand-side vector $\mathbf{b}$ and proceed as in Gaussian elimination, except use the pivot element $a_{kk}^{(k-1)}$ to eliminate not only $a_{ik}^{(k-1)}$ for $i = k+1, \ldots, n$ but also the elements $a_{ik}^{(k-1)}$ for $i = 1, \ldots, k-1$, i.e., all elements in the kth column other than the pivot. Upon reducing $(A|\mathbf{b})$ into

$$\begin{bmatrix} a_{11}^{(n-1)} & 0 & \cdots & 0 & \bigg| & b_1^{(n-1)} \\ 0 & a_{22}^{(n-1)} & \ddots & \vdots & \bigg| & b_2^{(n-1)} \\ \vdots & \ddots & \ddots & 0 & \bigg| & \vdots \\ 0 & \cdots & 0 & a_{nn}^{(n-1)} & \bigg| & b_n^{(n-1)} \end{bmatrix},$$

   the solution is obtained by setting

$$x_k = \frac{b_k^{(n-1)}}{a_{kk}^{(n-1)}}, \quad k = 1, \ldots, n.$$

   This procedure circumvents the backward substitution part necessary for the Gaussian elimination algorithm.

   (a) Write a pseudocode for this Gauss–Jordan procedure using, e.g., the same format as for the one appearing in Section 5.2 for Gaussian elimination. You may assume that no pivoting (i.e., no row interchanging) is required.

   (b) Show that the Gauss–Jordan method requires $n^3 + \mathcal{O}(n^2)$ floating point operations for one right-hand-side vector $\mathbf{b}$—roughly 50% more than what's needed for Gaussian elimination.

3. Let A and T be two nonsingular, $n \times n$ real matrices. Furthermore, suppose we are given two matrices L and U such that L is unit lower triangular, U is upper triangular, and

$$TA = LU.$$

   Write an algorithm that will solve the problem

$$A\mathbf{x} = \mathbf{b}$$

   for any given vector $\mathbf{b}$ in $\mathcal{O}(n^2)$ complexity. First explain briefly yet clearly why your algorithm requires only $\mathcal{O}(n^2)$ flops (you may assume without proof that solving an upper triangular or a lower triangular system requires only $\mathcal{O}(n^2)$ flops). Then specify your algorithm in detail (including the details for lower and upper triangular systems) using pseudocode or a MATLAB script.

4. The classical way to invert a matrix A in a basic linear algebra course augments A by the $n \times n$ identity matrix I and applies the Gauss–Jordan algorithm of Exercise 2 to this augmented matrix (including the solution part, i.e., the division by the pivots $a_{kk}^{(n-1)}$). Then A^{-1} shows up where I initially was.

   How many floating point operations are required for this method? Compare this to the operation count of $\frac{8}{3}n^3 + \mathcal{O}(n^2)$ required for the same task using LU decomposition (see Example 5.5).

5. Let
$$A = \begin{pmatrix} 5 & 6 & 7 & 8 \\ 0 & 4 & 3 & 2 \\ 0 & 0 & 0 & 1 \\ 0 & 0 & -1 & -2 \end{pmatrix}.$$

   (a) The matrix A can be decomposed using partial pivoting as
$$PA = LU,$$
   where U is upper triangular, L is unit lower triangular, and P is a permutation matrix. Find the 4×4 matrices U, L, and P.

   (b) Given the right-hand-side vector $\mathbf{b} = (26, 9, 1, -3)^T$, find $\mathbf{x}$ that satisfies $A\mathbf{x} = \mathbf{b}$. (Show your method: do not just guess.)

6. Let B be any real, nonsingular $n \times n$ matrix, where n is even, and set $A = B - B^T$. Show that A does not admit an LU decomposition (i.e., some pivoting *must* be applied, even if A is nonsingular).

7. Let
$$A = \begin{pmatrix} 0 & 1 & 0 & 0 \\ 1 & 0 & 3 & 0 \\ -.5 & 0 & -.2 & 1 \\ -.5 & -.3 & 1 & 0 \end{pmatrix}.$$

   Decompose A using partial pivoting as
$$PA = LU,$$
   where U is upper triangular, L is unit lower triangular and P is a permutation matrix. Record all the elementary matrices, $P^{(i)}$, and $M^{(i)}$ along the way. (You can use MATLAB to help you with these small, annoying calculations.)

   In Section 5.3 we define matrices $\tilde{M}^{(i)}$ and relate them to $P^{(i)}$ and $M^{(i)}$. Find $\tilde{M}^{(1)}$, $\tilde{M}^{(2)}$, and $\tilde{M}^{(3)}$, and show that $L = [\tilde{M}^{(1)}]^{-1}[\tilde{M}^{(2)}]^{-1}[\tilde{M}^{(3)}]^{-1}$.

8. Symmetric indefinite matrices require pivoting, and it can be shown that a numerically stable Gaussian elimination process has the form
$$PAP^T = LDL^T,$$

where P is a permutation matrix and D is a block diagonal matrix of a certain form. The MATLAB command `ldl` produces such a decomposition.

For simplicity, suppose that D is diagonal (rather than block diagonal), and suppose we are given a sequence of elementary permutation matrices interleaved with unit lower triangular matrices. Show that, similarly to GEPP in the general case, we can collect all elementary permutation matrices together into a single permutation matrix, bringing about the above defined LDL^T decomposition.

9. Can you find an example other than the one given on page 109 where scaled partial pivoting is unstable while complete pivoting does well?

   [If you have difficulties finding such a example, then the one in the text may look contrived.]

10. The MATLAB code given in Section 5.4 for solving linear systems of equations, using LU decomposition in outer form with partial pivoting, works well if the matrix A is nonsingular to a working precision. But if A is singular, then the exit is not graceful.

    Please fix this by modifying the functions `ainvb` and `plu` to include checks so that `ainvb+plu` will always return as follows:

    (i) In case of a nonsingular system, return silently with the solution **x**.

    (ii) In case of a singular system, display a message regarding the singularity and return with **x** assigned the value `NaN`.

    Assume that MATLAB will complain if a division by a number smaller than $\text{eps} = 2.22 \times 10^{-16}$ is attempted. (You want to avoid precisely this sort of complaint.)

11. Apply the modified solver obtained in Exercise 10 to the following systems. In each case, check the difference between the computed solution **x** and the result of MATLAB's built-in solver $A \setminus \mathbf{b}$.

    (a)
    $$x_1 + x_2 + x_4 = 2,$$
    $$2x_1 + x_2 - x_3 + x_4 = 1,$$
    $$4x_1 - x_2 - 2x_3 + 2x_4 = 0,$$
    $$3x_1 - x_2 - x_3 + x_4 = -3.$$

    (b) Same as the previous system, but with the coefficient of x_4 in the last equation set to $a_{4,4} = 2$.

    (c)
    $$x_1 + x_2 + x_3 = 1,$$
    $$x_1 + (1 + 10^{-15})x_2 + 2x_3 = 2,$$
    $$x_1 + 2x_2 + 2x_3 = 1.$$

    (d) Same as the previous system, but with the second equation multiplied by 10^{20}.

    (e)
    $$A = \begin{pmatrix} 1 & 2 & 3 \\ 1 & 2 & 3 \\ 1 & 2 & 3 \end{pmatrix}, \quad \mathbf{b} = \begin{pmatrix} 1 \\ 2 \\ 1 \end{pmatrix}.$$

5.9. Exercises

(f)
$$A = \begin{pmatrix} 1 & 2 & 3 \\ 0 & 0 & 0 \\ 3 & 2 & 1 \end{pmatrix}$$

with the same right-hand side as before.

12. The Cholesky algorithm given on page 116 has all those wretched loops as in the Gaussian elimination algorithm in its simplest form. In view of Section 5.4 and the program `ainvb` we should be able to achieve also the Cholesky decomposition effect more efficiently.

    Write a code implementing the Cholesky decomposition with only one loop (on k), utilizing outer products.

13. The script in Example 5.13 will give different answers each time you run it, because the random number generator starts at different spots.

    Please fix this, making the program having the option to run on the same input when reinvoked. (Check `help randn` to see how to do this.)

    Run the program for $n = 200$ and $n = 500$. What are your observations?

14. This exercise is for lovers of complex arithmetic.

    Denote by $\mathbf{x}^H$ the conjugate transpose of a given complex-valued vector $\mathbf{x}$, and likewise for a matrix. The ℓ_2-norm is naturally extended to complex-valued vectors by

    $$\|\mathbf{x}\|_2^2 = \mathbf{x}^H \mathbf{x} = \sum_{i=1}^{n} |x_i|^2,$$

    which in particular is real, possibly unlike $\mathbf{x}^T \mathbf{x}$.

    A complex-valued $n \times n$ matrix A is called **Hermitian** if $A^H = A$. Further, A is positive definite if
    $$\mathbf{x}^H A \mathbf{x} > 0$$
    for all complex-valued vectors $\mathbf{x} \neq \mathbf{0}$.

    (a) Show that the 1×1 matrix $A = \iota$ is symmetric but not Hermitian, whereas $B = \begin{pmatrix} 2 & \iota \\ -\iota & 2 \end{pmatrix}$ is Hermitian positive definite but not symmetric. Find the eigenvalues of A and of B.
    [In general, all eigenvalues of a Hermitian positive definite matrix are real and positive.]

    (b) Show that an $n \times n$ Hermitian positive definite matrix A can be decomposed as $A = LDL^H$, where D is a diagonal matrix with positive entries and L is a complex-valued unit lower triangular matrix.

    (c) Extend the Cholesky decomposition algorithm to Hermitian positive definite matrices. Justify.

15. (a) Write a MATLAB script that stores a matrix in compressed column form.

    (b) Write a symmetric version of your program, that is, a script for symmetric matrices, where if $a_{i,j}$ is stored ($i \neq j$), then $a_{j,i}$ is not.

    Try your programs on a 5×5 example of your choice.

16. Continuing Exercise 15, write a program that gets a symmetric positive definite matrix in compressed column form as input and performs a *sparse Cholesky* factorization. The program should generate the Cholesky factor in compressed column form too. To increase modularity, you may want to write a few separate routines that perform operations such as inserting a new nonzero entry into a matrix.

    Test your program using the matrix $A = B^T B$, where B is the matrix from Exercise 15. Invent additional data as needed for your test.

17. Write a MATLAB function that solves tridiagonal systems of equations of size n. Assume that no pivoting is needed, but do not assume that the tridiagonal matrix A is symmetric. Your program should expect as input four vectors of size n (or $n-1$): one right-hand-side **b** and the three nonzero diagonals of A. It should calculate and return $\mathbf{x} = A^{-1}\mathbf{b}$ using a Gaussian elimination variant that requires $\mathcal{O}(n)$ flops and consumes no additional space as a function of n (i.e., in total $5n$ storage locations are required).

    Try your program on the matrix defined by $n = 10$, $a_{i-1,i} = a_{i+1,i} = -i$, and $a_{i,i} = 3i$ for all i such that the relevant indices fall in the range 1 to n. Invent a right-hand-side vector **b**.

18. Apply your program from Exercise 17 to the problem described in Example 4.17 using the second set of boundary conditions, $v(0) = v'(1) = 0$, for $g(t) = (\frac{\pi}{2})^2 \sin(\frac{\pi}{2}t)$ and $N = 100$. Compare the results to the vector **u** composed of $u(ih) = \sin(\frac{\pi}{2}ih)$, $i = 1,\ldots,N$, by recording $\|\mathbf{v} - \mathbf{u}\|_\infty$.

19. Suppose that we rewrite the problem of Example 4.17 and Exercise 18 as the system of two simple differential equations,

    $$v'(t) = w(t), \quad w'(t) = -g(t), \quad v(0) = 0, \; w(1) = 0.$$

    From this it is clear that we can solve first for $w(t)$ and then for $v(t)$, but let us ignore this and solve for the two unknown functions simultaneously.

    On the same mesh as in Example 4.17, define the discretization

    $$\frac{v_{i+1} - v_i}{h} = \frac{w_{i+1} + w_i}{2}, \quad i = 0,\ldots,N-1, \; v_0 = 0,$$

    $$\frac{w_{i+1} - w_i}{h} = -g(ih + h/2), \quad i = 0,\ldots,N-1, \; w_N = 0.$$

    (The resulting values **v** in the notation of Exercise 18 should have comparable accuracy as an approximation to **u**, but let us ignore that, too.)

    Defining the vector of $n = 2N$ unknowns $\mathbf{x} = (w_0, v_1, w_1, v_2, \ldots, v_{N-1}, w_{N-1}, v_N)^T$, construct the corresponding matrix problem. Show that the resulting matrix is banded: what is the bandwidth? How would you go about finding **x**?

20. Consider the LU decomposition of an upper Hessenberg (no, it's not a place in Germany) matrix, defined on the facing page, assuming that no pivoting is needed: $A = LU$.

    (a) Provide an efficient algorithm for this LU decomposition (do not worry about questions of memory access and vectorization).

    (b) What is the sparsity structure of the resulting matrix L (i.e., where are its nonzeros)?

    (c) How many operations (to a leading order) does it take to solve a linear system $A\mathbf{x} = \mathbf{b}$, where A is upper Hessenberg?

5.10. Additional notes

(d) Suppose now that partial pivoting is applied. What are the sparsity patterns of the factors of A?

21. For the arrow matrices of Example 5.15 determine the overall storage and flop count requirements for solving the systems with A and with B in the general $n \times n$ case.

22. Given that a and b are two real positive numbers, the eigenvalues of the symmetric tridiagonal matrix A=tri [b,a,b] of size $n \times n$ are $\lambda_j = a + 2b\cos(\frac{\pi j}{n+1})$, $j = 1, \ldots, n$. (A nonsparse version of this matrix can be obtained in MATLAB with the instruction

```
A = diag(a*ones(n,1),0) + diag(b*ones(n-1,1),1)
    + diag(b*ones(n-1,1),-1)
```

Type help spdiags for the scoop on the sparse version.)

(a) Find $\|A\|_\infty$.

(b) Show that if A is strictly diagonally dominant, then it is symmetric positive definite.

(c) Suppose $a > 0$ and $b > 0$ are such that A is symmetric positive definite. Find the condition number $\kappa_2(A)$. (Assuming that n is large, an approximate value would suffice. You may also assume that $a \neq 2b$.)

23. Let $\mathbf{b} + \delta\mathbf{b}$ be a perturbation of a vector $\mathbf{b}$ ($\mathbf{b} \neq \mathbf{0}$), and let $\mathbf{x}$ and $\delta\mathbf{x}$ be such that $A\mathbf{x} = \mathbf{b}$ and $A(\mathbf{x} + \delta\mathbf{x}) = \mathbf{b} + \delta\mathbf{b}$, where A is a given nonsingular matrix. Show that

$$\frac{\|\delta\mathbf{x}\|}{\|\mathbf{x}\|} \leq \kappa(A)\frac{\|\delta\mathbf{b}\|}{\|\mathbf{b}\|}.$$

24. Run the MATLAB script of Example 5.22, plugging in different values for d. In particular, try $d = .25$, $d = .85$, and $d = -.75$. What do you observe? What will happen as $d \downarrow (-1)$?

Hessenberg matrix
An $n \times n$ matrix A is said to be in *Hessenberg* or *upper Hessenberg* form if all its elements below the first subdiagonal are zero, so that

$$a_{ij} = 0, \quad i > j+1.$$

5.10 Additional notes

This chapter is the first in our text to address and describe some important and nontrivial numerical methods, or at least major components thereof, that see an enormous amount of use in real-world applications. It is not for naught that the MATLAB command "\" is both the shortest in the language's repertoire and one that invokes a program rumored to contain over 100,000 lines of code.

Just about any introductory text on numerical methods addresses in various ways the material in Sections 5.1–5.3 and usually also Section 5.5, part of Section 5.6, and Section 5.8.

The analysis of stability for the various variants of Gaussian elimination with pivoting was done by J. Wilkinson in the 1960s and is considered a classic of numerical analysis. Several modern texts describe it in detail, e.g., Trefethen and Bau [70], Golub and van Loan [30], Watkins [74], and Demmel [21].

The Cholesky decomposition is loved for much more than the relatively modest gains in CPU time and storage that it offers over LU decomposition. It is a very clean decomposition that in essence corresponds to taking the square root of a positive scalar.

A much more thorough discussion than what is contained in Section 5.4 can be found in [21]. A modern reference discussing the important topic of direct methods for sparse matrices is Davis [19].

Occasionally, solution of many linear systems $A\mathbf{x}_j = \mathbf{b}_j$, $j = 1, 2, 3, \ldots$, is required, where A is fixed, large, and sparse, and the given right-hand-side vectors $\mathbf{b}_j$ are not particularly close to one another. For such problems, the iterative methods discussed in Chapter 7 lose some of their attraction, because they must deal with each right-hand side essentially from scratch. Direct methods may be preferred then, because the LU decomposition of A is carried out just once and then only a pair of forward-backward substitutions is required for each j. Lots of fast access memory may be required for the successful utilization of such direct methods, though, because of the fill-in that results from the matrix factorization even when the best ordering strategies are utilized. This consideration is central enough to influence in a major way the sort of computing equipment purchased for carrying out simulations for certain real-world applications.

Chapter 6
Linear Least Squares Problems

This chapter is concerned with methods for solving the algebraic problem

$$\min_{\mathbf{x}} \|\mathbf{b} - A\mathbf{x}\|_2,$$

where the dimensions of the real matrix and vectors are

$$A \in \mathcal{R}^{m \times n}, \ \mathbf{x} \in \mathcal{R}^n, \ \mathbf{b} \in \mathcal{R}^m, \ m \geq n.$$

We assume that A has full column rank. Notice that in the overdetermined case, $m > n$, there is typically no $\mathbf{x}$ satisfying $A\mathbf{x} = \mathbf{b}$ exactly, even in the absence of roundoff error.

The least squares problem arises often in many diverse application fields, especially where **data fitting** is required. Instances arise in machine learning, computer vision, and computer graphics applications, to name but a few. In computer vision people may want to match local invariant features of cluttered images under arbitrary rotations, scalings, change of brightness and contrast, and so on. Such actions are parametrized and the best values for the parameters are then sought to match the data. In *computer graphics* a parametrization of a surface mesh in three dimensions may be sought that yields a *morphing* of one animal into another, say. In such applications the question of just how to parametrize, or how to generate an efficient predictive model A in the present terminology, is often the crux of the matter and is far from trivial. But once the method of parametrization is determined, what follows is an instance of the problem considered here. Other applications seek to find an approximating function $v(t, \mathbf{x})$ depending on a continuous variable t that fits data pairs $(t_i, b_i), i = 1, \ldots, m$, as in Example 4.16. We will apply our solution techniques to such instances in Section 6.1.

In Section 6.1 we formulate an $n \times n$ system of linear equations called the **normal equations** that yield the solution for the stated minimization problem. Solving the normal equations is a fast and straightforward approach. However, the resulting algorithm is not as stable as can be in general. **Orthogonal transformations** provide a more stable way to proceed, and this is described in Section 6.2. Algorithms to actually carry out the necessary **QR decomposition** are described in Section 6.3.

6.1 Least squares and the normal equations

In this section we first solve the least squares minimization problem, obtaining the *normal equations*. Then we show examples related to *data fitting*.

The space spanned by the columns of our $m \times n$ matrix A (i.e., all vectors $\mathbf{z}$ of the form $\mathbf{z} = A\mathbf{y}$) is generally of dimension at most n, and we further assume that the dimension equals n, so that A has full column rank. In other words, we assume that its columns are linearly independent. Notice also that in the overdetermined case, $m > n$, $\mathbf{b}$ generally does not lie in the range space of A.

Deriving the normal equations

Let us rewrite the problem that we aim to solve as

$$\min_{\mathbf{x}} \frac{1}{2} \| \mathbf{b} - A\mathbf{x} \|^2.$$

We have squared the normed expression, thus getting rid of the square root sign,[25] and multiplied it by $1/2$: this will not change the minimizer, in the sense that the same solution coefficients x_j will be obtained. Notice that we have dropped the subscript 2 in the norm notation: there is no other norm here to get confused by. Finally, we define the residual vector as usual by

$$\mathbf{r} = \mathbf{b} - A\mathbf{x}.$$

Writing these matrix and vectors explicitly we have

$$\mathbf{x} = \begin{pmatrix} x_1 \\ \vdots \\ x_n \end{pmatrix}, \; A = \begin{pmatrix} a_{1,1} & \cdots & a_{1,n} \\ \vdots & & \vdots \\ \vdots & \cdots & \vdots \\ a_{m,1} & \cdots & a_{m,n} \end{pmatrix}, \; \mathbf{b} = \begin{pmatrix} b_1 \\ \vdots \\ \vdots \\ b_m \end{pmatrix}, \; \mathbf{r} = \begin{pmatrix} r_1 \\ \vdots \\ \vdots \\ r_m \end{pmatrix}.$$

Note that the matrix A is $m \times n$, with $m > n$ and perhaps $m \gg n$, so it is "long and skinny," and correspondingly we do not expect $\mathbf{r}$ to vanish at the optimum. See Figure 6.1.

We have a minimization problem for a smooth scalar function in several variables, given by

$$\min_{\mathbf{x}} \psi(\mathbf{x}), \quad \text{where} \quad \psi(\mathbf{x}) = \frac{1}{2} \| \mathbf{r} \|^2.$$

The *necessary conditions* for a minimum are obtained by setting the derivatives of ψ with respect to each unknown x_k to zero, yielding

$$\frac{\partial}{\partial x_k} \psi(\mathbf{x}) = 0, \quad k = 1, \ldots, n.$$

Since

$$\psi(\mathbf{x}) = \frac{1}{2} \| \mathbf{r} \|^2 = \frac{1}{2} \sum_{i=1}^{m} \left(b_i - \sum_{j=1}^{n} a_{i,j} x_j \right)^2,$$

---

[25] If you are not sure you remember everything you need to remember about norms, Section 4.2 may come to your aid.

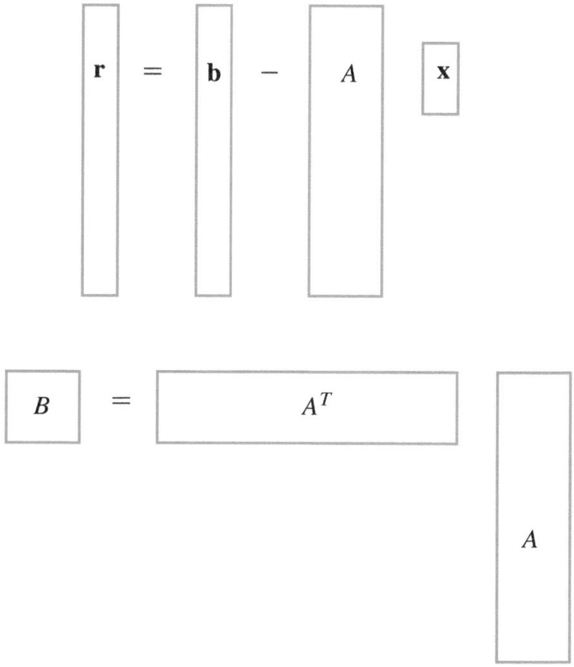

6.1. Least squares and the normal equations

Figure 6.1. *Matrices and vectors and their dimensions in ℓ_2 data fitting.*

the conditions for a minimum yield

$$\frac{\partial}{\partial x_k}\psi(\mathbf{x}) = \sum_{i=1}^{m}\left[\left(b_i - \sum_{j=1}^{n}a_{i,j}x_j\right)(-a_{i,k})\right] = 0$$

for $k = 1, 2, \ldots, n$. The latter expression can be rewritten as

$$\sum_{i=1}^{m}a_{i,k}\sum_{j=1}^{n}a_{i,j}x_j = \sum_{i=1}^{m}a_{i,k}b_i \quad \text{for } k = 1, \ldots, n.$$

In matrix-vector form this expression looks much simpler; it reads

$$A^T A\mathbf{x} = A^T \mathbf{b}.$$

This system of n linear equations in n unknowns is called the *normal equations*. Note that $B = A^T A$ can be much smaller in size than A; see Figure 6.1. The matrix B is symmetric positive definite given that A has full column rank; see Section 4.3.

Least squares solution uniqueness

Is the solution of the normal equations really a minimizer (and not, say, a maximizer) of the least squares norm? and if yes, is it the *global* minimum? The Least Squares Theorem given on the next page says it is. The answer is affirmative because the matrix B is positive definite.

To see this (completing the proof of the theorem), note first that our objective function is a quadratic in n variables. Indeed, we can write

$$\psi(\mathbf{x}) = \frac{1}{2}(\mathbf{b} - A\mathbf{x})^T(\mathbf{b} - A\mathbf{x}) = \frac{1}{2}\mathbf{x}^T B\mathbf{x} - \mathbf{b}^T A\mathbf{x} + \frac{1}{2}\|\mathbf{b}\|^2.$$

> **Theorem: Least Squares.**
> The least squares problem
> $$\min_{\mathbf{x}} \|A\mathbf{x} - \mathbf{b}\|_2,$$
> where A has full column rank, has a unique solution that satisfies the normal equations
> $$\left(A^T A\right)\mathbf{x} = A^T \mathbf{b}.$$

Expanding this quadratic function in a Taylor series (see page 259) around the alleged minimizer amounts to the usual Taylor expansion in each variable: for any nonzero vector increment $\Delta \mathbf{x}$ we write

$$\psi(\mathbf{x} + \Delta \mathbf{x}) = \psi(\mathbf{x}) + \sum_{j=1}^{n} \Delta x_j \underbrace{\frac{\partial \psi}{\partial x_j}(\mathbf{x})}_{=0} + \frac{1}{2} \sum_{j=1}^{n} \sum_{k=1}^{n} \Delta x_j \Delta x_k \underbrace{\frac{\partial^2 \psi}{\partial x_j \partial x_k}(\mathbf{x})}_{=B_{j,k}}.$$

Hence, in vector notation we obtain

$$\psi(\mathbf{x} + \Delta \mathbf{x}) = \psi(\mathbf{x}) + (\Delta \mathbf{x})^T B \Delta \mathbf{x} > \psi(\mathbf{x}),$$

completing the proof. ◆

Solving via the normal equations

It is important to realize that $\mathbf{x}$, which has been our vector argument above, is now specified as the solution for the normal equations, which is indeed the solution of the least squares minimization problem.

Furthermore, if A has full column rank, then $B = A^T A$ is symmetric positive definite. The least squares problem has therefore been reduced, at least in principle, to that of solving an $n \times n$ system of linear equations of the form described in Section 5.5.

The beauty does not end here, geometrically speaking. Notice that for the corresponding residual at optimum we have

$$A^T(\mathbf{b} - A\mathbf{x}) = A^T \mathbf{r} = \mathbf{0}.$$

Hence we seek a solution satisfying that *the residual is orthogonal to the column space of A*. Since a picture is better than a thousand words (or at least a thousand bytes), Figure 6.2 is provided to illustrate the projection.

Our solution is given by $\mathbf{x} = (A^T A)^{-1} A^T \mathbf{b}$. The matrix multiplying $\mathbf{b}$ is important enough to have a name and special notation: it is called the **pseudo-inverse** of A, denoted by

$$A^\dagger = (A^T A)^{-1} A^T.$$

The solution via the normal equations using direct solvers amounts to the four steps laid out in the algorithm given on the facing page.

6.1. Least squares and the normal equations

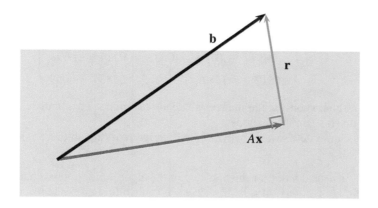

Figure 6.2. *Discrete least squares approximation.*

Algorithm: Least Squares via Normal Equations.

1. Form $B = A^T A$ and $\mathbf{y} = A^T \mathbf{b}$.

2. Compute the Cholesky Factor, i.e., the lower triangular matrix G satisfying $B = GG^T$.

3. Solve the lower triangular system $G\mathbf{z} = \mathbf{y}$ for $\mathbf{z}$.

4. Solve the upper triangular system $G^T \mathbf{x} = \mathbf{z}$ for $\mathbf{x}$.

Example 6.1. Let's see some numbers. Choose $m = 5$, $n = 3$, and specify

$$A = \begin{pmatrix} 1 & 0 & 1 \\ 2 & 3 & 5 \\ 5 & 3 & -2 \\ 3 & 5 & 4 \\ -1 & 6 & 3 \end{pmatrix}, \quad \mathbf{b} = \begin{pmatrix} 4 \\ -2 \\ 5 \\ -2 \\ 1 \end{pmatrix}.$$

We wish to solve the least squares problem $\min_\mathbf{x} \|\mathbf{b} - A\mathbf{x}\|$.

Following the normal equations route we calculate

$$B = A^T A = \begin{pmatrix} 40 & 30 & 10 \\ 30 & 79 & 47 \\ 10 & 47 & 55 \end{pmatrix}, \quad \mathbf{y} = A^T \mathbf{b} = \begin{pmatrix} 18 \\ 5 \\ -21 \end{pmatrix}.$$

Now solving the square system $B\mathbf{x} = \mathbf{y}$ yields the final result $\mathbf{x} = (.3472, .3990, -.7859)^T$, correct to the number of digits shown.

Here are more numbers. The optimal residual (rounded) is

$$\mathbf{r} = \mathbf{b} - A\mathbf{x} = (4.4387, .0381, .495, -1.893, 1.311)^T.$$

This vector is orthogonal to each column of A. Finally, the pseudo-inverse of A (rounded) is

$$A^\dagger = B^{-1}A^T = \begin{pmatrix} .049 & .0681 & .1049 & .0531 & -.1308 \\ -.0491 & -.0704 & .0633 & .0114 & .1607 \\ .0512 & .1387 & -.1095 & .0533 & -.059 \end{pmatrix}.$$

As with the inverse of a square nonsingular matrix, we do not recommend explicitly calculating such a matrix, and will not do so again in this book.

Please see if you can write a seven-line MATLAB script verifying all these calculations. ∎

The overall computational work for the first step of the algorithm is approximately mn^2 floating point operations (flops); for step 2 it is $n^3/3 + \mathcal{O}(n^2)$ flops, while steps 3 and 4 cost $\mathcal{O}(n^2)$ flops. This is another case where although operation counts are in general unsatisfactory for measuring true performance, they do deliver the essential result that the main cost here, especially when $m \gg n$, is in *forming* the matrix $B = A^T A$.

Data fitting

Generally, data fitting problems arise as follows. We have *observed data* **b** and a model function that for any candidate model **x** provides *predicted data*. The task is to find **x** such that the predicted data match the observed data to the extent possible, by minimizing their difference in the least squares sense. In the linear case which we study here, the predicted data are given by $A\mathbf{x}$. In this context the assumption that A has full column rank does not impose a serious restriction: it just implies that there is no redundancy in the representation of the predicted data, so that for any vector $\hat{\mathbf{x}} \in \mathcal{R}^n$ there is no other vector $\tilde{\mathbf{x}} \in \mathcal{R}^n$ such that $A\tilde{\mathbf{x}} = A\hat{\mathbf{x}}$.

Example 6.2 (linear regression). Consider fitting a given data set of m pairs (t_i, b_i) by a straight line. Thus, we want to find the coefficients x_1 and x_2 of

$$v(t) = x_1 + x_2 t,$$

such that $v(t_i) \approx b_i$, $i = 1, \ldots, m$. So $n = 2$ here, and

$$A = \begin{pmatrix} 1 & t_1 \\ 1 & t_2 \\ \vdots & \vdots \\ 1 & t_m \end{pmatrix}.$$

The components of the normal equations are

$$B_{1,1} = \sum_{i=1}^m 1 = m, \quad B_{1,2} = \sum_{i=1}^m 1 \cdot t_i = \sum_{i=1}^m t_i,$$

$$B_{2,1} = B_{1,2}, \quad B_{2,2} = \sum_{i=1}^m t_i \cdot t_i = \sum_{i=1}^m t_i^2,$$

$$y_1 = \sum_{i=1}^m b_i, \quad y_2 = \sum_{i=1}^m t_i b_i.$$

6.1. Least squares and the normal equations

This leads to a system of just two equations given by

$$mx_1 + \left(\sum_{i=1}^m t_i\right) x_2 = \sum_{i=1}^m b_i,$$

$$\left(\sum_{i=1}^m t_i\right) x_1 + \left(\sum_{i=1}^m t_i^2\right) x_2 = \sum_{i=1}^m t_i b_i.$$

The solution is written explicitly as the famous formula

$$x_1 = \frac{\sum_{i=1}^m t_i^2 \sum_{i=1}^m b_i - \sum_{i=1}^m t_i b_i \sum_{i=1}^m t_i}{m \sum_{i=1}^m t_i^2 - \left(\sum_{i=1}^m t_i\right)^2},$$

$$x_2 = \frac{m \sum_{i=1}^m t_i b_i - \sum_{i=1}^m t_i \sum_{i=1}^m b_i}{m \sum_{i=1}^m t_i^2 - \left(\sum_{i=1}^m t_i\right)^2}.$$

For readers who like to see examples with numbers, consider the data

i	1	2	3
t_i	0.0	1.0	2.0
b_i	0.1	0.9	2.0

Then $m = 3$, $\sum_{i=1}^m t_i = 0.0 + 1.0 + 2.0 = 3.0$, $\sum_{i=1}^m b_i = 0.1 + 0.9 + 2.0 = 3.0$, $\sum_{i=1}^m t_i^2 = 5.0$, $\sum_{i=1}^m t_i b_i = 0.9 + 4.0 = 4.9$. This yields (please verify) the solution

$$x_1 = 0.05, \quad x_2 = 0.95.$$

One purpose of this example is to convince you of the beauty of matrix-vector notation and computer programs as an alternative to the above exposition.

A more common regression curve would of course be constructed for many more data points. See Figure 6.3 for a plot of a regression curve obtained with these formulas where $m = 25$. Here $t_i = i < t_{i+1}$ denotes years since 1980, and b_i is the average change in temperature off the base of 1 in a certain secret location. Although it does not always hold that $b_i < b_{i+1}$, the straight line fit has a positive slope, which indicates a clear warming trend. ∎

While in a typical course on statistics for the social sciences the regression formulas of Example 6.2 appear as if by magic, here they are a simple by-product of a general treatment—albeit, without the statistical significance.

Polynomial data fitting

Extending the linear regression formulas to a higher degree polynomial fit, $v(t) \equiv p_{n-1}(t) = x_1 + x_2 t + \cdots + x_n t^{n-1}$, is straightforward too. Writing for each data point $v(t_i) = x_1 + x_2 t_i + \cdots + x_n t_i^{n-1} = (1, t_i, \ldots, t_i^{n-1})\,\mathbf{x}$, the matrix A is the extension of the previously encountered **Vander-**

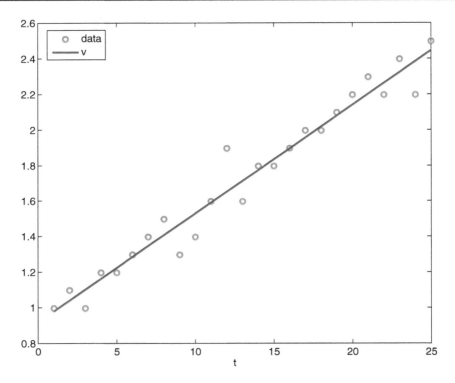

Figure 6.3. *Linear regression curve (in blue) through green data. Here, $m = 25$, $n = 2$.*

monde matrix given by

$$A = \begin{pmatrix} 1 & t_1 & t_1^2 & \cdots & t_1^{n-1} \\ 1 & t_2 & t_2^2 & \cdots & t_2^{n-1} \\ \vdots & \vdots & \vdots & & \vdots \\ \vdots & \vdots & \vdots & & \vdots \\ 1 & t_{m-1} & t_{m-1}^2 & \cdots & t_{m-1}^{n-1} \\ 1 & t_m & t_m^2 & \cdots & t_m^{n-1} \end{pmatrix}.$$

Note that the structure of the matrix A depends on our choice of the basis functions used to describe polynomials. More on this is provided in Chapter 10. For now let us just say that the simple basis used here, called *monomial basis*, is good only for polynomials of a really low degree. But then these are the ones we may want in the present context anyway.

Here is a MATLAB function for best approximation by low order polynomials, using the normal equations:

```
function coefs = lsfit (t, b, n)
%
% function coefs = lsfit (t, b, n)
%
```

6.1. Least squares and the normal equations

```
% Construct coefficients of the polynomial of
% degree at most n-1 that best fits data (t,b)

t = t(:); b = b(:); % make sure t and b are column vectors
m = length(t);

% long and skinny A
A = ones(m,n);
for j=1:n-1
  A(:,j+1) = A(:,j).*t;
end

% normal equations and solution
B = A'*A;   y = A'*b;
coefs = B \ y;
```

Example 6.3. Sample the function $f(t) = \cos(2\pi t)$ at 21 equidistant points on the interval $[0, 1]$ and construct best fits by polynomials of degree at most $n - 1$ for each $n = 1, 2, 3, 4, 5$.

Following is an appropriate MATLAB script:

```
% data
m = 21;
tt = 0:1/(m-1):1;
bb = cos(2*pi*tt);

% find polynomial coefficients
for n=1:5
  coefs{n} = lsfit(tt,bb,n);
end

% Evaluate and plot
t = 0:.01:1;
z = ones(5,101);
for n=1:5
  z(n,:) = z(n,:) * coefs{n}(n);
  for j=n-1:-1:1
      z(n,:) = z(n,:).*t + coefs{n}(j);
  end
end
plot(t,z,tt,bb,'ro')
xlabel('t')
ylabel('p_{n-1}')
```

The resulting approximants $p_{n-1}(t)$ are plotted in Figure 6.4. Note that here, due to symmetry, $p_1(t) \equiv p_0(t)$ and $p_3(t) \equiv p_2(t)$. So, the degree "at most $n - 1$" turns out to be equal to $n - 2$ rather than to $n - 1$ for odd values of $n - 1$. ∎

Data fitting vs. interpolation

Reflecting on Examples 6.3 and 4.16 (page 85), there is a seeming paradox hidden in our arguments, namely, that we attempt to minimize the residual for a fixed n, $n < m$, but refuse to simply increase n until $n = m$. Indeed, this would drive the residual to zero, the resulting scheme being a polynomial

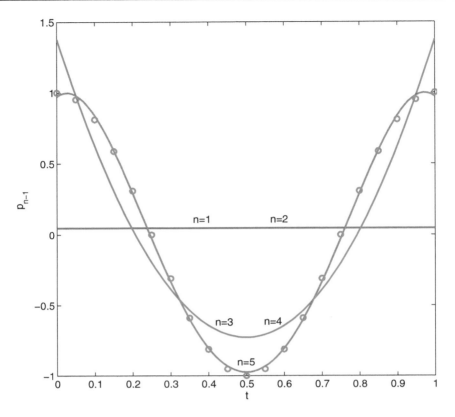

Figure 6.4. *The first 5 best polynomial approximations to $f(t) = \cos(2\pi t)$ sampled at $0 : .05 : 1$. The data values appear as red circles. Clearly, p_4 fits the data better than p_2, which in turn is a better approximation than p_0. Note $p_{2j+1} = p_{2j}$.*

that interpolates the data. Why not just interpolate?! (Interpolation techniques are discussed at length in Chapter 10.)

The simple answer is that choosing n is part of our *modeling* efforts, and the ensuing least squares minimization problem is part of the *solution process* with n already fixed. But there are reasons for choosing n small in the first place. One is hinted at in Example 6.2, namely, that we are trying to find the **trend** in the data on a long time scale.

An additional reason for not interpolating the data values is that they may contain measurement errors. Moreover, we may want a model function that depends only on a few parameters x_j for ease of manipulation, although we determine them based on all the given data. For a poignant example we refer the reader to Exercise 3.

The solution of the least squares data fitting problem through solving the normal equations has the advantage of being straightforward and efficient. A linear least squares solver is implemented in the MATLAB backslash operator as well as in the MATLAB command `polyfit`. Replacing the last three lines in our function `lsfit` by the line

```
coefs = A \ b;
```

would implement for the same purpose, albeit more enigmatically, an algorithm which in terms of roundoff error accumulation is at least as good. See Section 6.2. The routine `polyfit` is even easier to use, as it does not require forming the matrix and the right-hand-side vector; the input consists of the data points and the degree of the required polynomial.

Data fitting in other norms

Before moving on let us also mention data fitting in other norms.

1. Using ℓ_1 we consider
$$\min_{\mathbf{x}} \|\mathbf{b} - A\mathbf{x}\|_1.$$

   Here we would be looking at finding coefficients $\mathbf{x}$ such that the sum of absolute values of the deviations is minimized. This norm is particularly useful if we need to automatically get rid of an undue influence of *outliers* in the data, which are data values that conspicuously deviate from the rest due to measurement error.

2. Using ℓ_∞ we consider
$$\min_{\mathbf{x}} \|\mathbf{b} - A\mathbf{x}\|_\infty.$$

   This is a *min-max* problem of finding the minimum over $x_1, x_2, \ldots, x_n$ of the maximum data deviation. This norm is useful if the worst-case error in the approximation is important and must be kept in check.

Both ℓ_1 and ℓ_∞ best approximations lead to **linear programming** problems, briefly discussed in the more advanced Section 9.3. They are significantly more complex than the problem that we are faced with using least squares yet are very important in many modern areas of application.

Note that if $n = m$, then we have one and the same solution regardless of the norm used, because with all three norms the minimum is obtained with $\mathbf{x} = A^{-1}\mathbf{b}$, which yields a zero residual. But when $n < m$ these different norms usually yield significantly different best approximations.

The least squares approximation (which is the approximation associated with the ℓ_2-norm) is not only the simplest to calculate, it also has a variety of beautiful mathematical properties. We continue to concentrate on it in the remainder of this chapter.

Specific exercises for this section: Exercises 1–4.

6.2 Orthogonal transformations and QR

The main drawback of the normal equations approach for solving linear least squares problems is accuracy in the presence of large condition numbers. Information may be lost when forming $A^T A$ when $\kappa(A^T A)$ is very large. In this section we derive better algorithms in this respect.

But first let us investigate the source of difficulty.

Example 6.4. This example demonstrates the possible inaccuracy that arises when the normal equations are formed. Given

$$A = \begin{pmatrix} 1 & 1 & 1 \\ \varepsilon & 0 & 0 \\ 0 & \varepsilon & 0 \\ 0 & 0 & \varepsilon \end{pmatrix},$$

we have

$$A^T A = \begin{pmatrix} 1+\varepsilon^2 & 1 & 1 \\ 1 & 1+\varepsilon^2 & 1 \\ 1 & 1 & 1+\varepsilon^2 \end{pmatrix}.$$

Now, if $\eta < \varepsilon \leq \sqrt{\eta}$, where η is the rounding unit, then the evaluated $A^T A$ is numerically singular even though A has numerically full column rank. ∎

Troubling as this little example may appear to be, it is yet unclear who is to blame: is the *problem* to be solved simply more ill-conditioned, or is it the *algorithm* that insists on forming $A^T A$ that is responsible for the effective singularity? To see things better, we now consider a perturbation analysis as in Section 5.8, sticking to the 2-norm for vectors and matrices alike.

Condition number by SVD

Let us assume that $\mathbf{b}$ is in the range space of A, so for the exact least squares solution $\mathbf{x}$ we have

$$\mathbf{b} = A\mathbf{x}, \quad \mathbf{r} = \mathbf{0}.$$

Note that we will not use this to define a new algorithm (that would be cheating); rather, we seek an error estimate that is good also when $\|\mathbf{r}\|$ is small, and we will only use the bound

$$\|\mathbf{b}\| \leq \|A\| \|\mathbf{x}\|.$$

Next, for a given approximate solution $\hat{\mathbf{x}}$ we can calculate $\hat{\mathbf{b}} = A\hat{\mathbf{x}}$ and

$$\hat{\mathbf{r}} = \mathbf{b} - A\hat{\mathbf{x}} = \mathbf{b} - \hat{\mathbf{b}} = A(\mathbf{x} - \hat{\mathbf{x}}).$$

Thus, $\|\mathbf{x} - \hat{\mathbf{x}}\| \leq \|A^\dagger\| \|\hat{\mathbf{r}}\|$ and, corresponding to the bound (5.1) on page 128, we get

$$\frac{\|\mathbf{x} - \hat{\mathbf{x}}\|}{\|\mathbf{x}\|} \leq \kappa(A) \frac{\|\hat{\mathbf{r}}\|}{\|\mathbf{b}\|}, \quad \kappa(A) = \|A\| \|A^\dagger\|.$$

To obtain an expression for the condition number $\kappa(A) = \kappa_2(A)$, recall from Section 4.4 (page 81) that A enjoys having an SVD, so there are orthogonal matrices U and V such that

$$A = U \Sigma V^T.$$

In the present case (with rank$(A) = n \leq m$) we can write

$$\Sigma = \begin{pmatrix} S \\ 0 \end{pmatrix},$$

with S a square diagonal matrix having the singular values $\sigma_1 \geq \sigma_2 \geq \cdots \geq \sigma_n > 0$ on its main diagonal.

Now, recall that a square matrix Q is *orthogonal* if its columns are orthonormal, i.e., $Q^T Q = I$. It is straightforward to verify that for any vector $\mathbf{y}$ we have in the ℓ_2-norm $\|Q\mathbf{y}\| = \|\mathbf{y}\|$, because

$$\|Q\mathbf{y}\|^2 = (Q\mathbf{y})^T (Q\mathbf{y}) = \mathbf{y}^T Q^T Q \mathbf{y} = \mathbf{y}^T \mathbf{y} = \|\mathbf{y}\|^2.$$

6.2. Orthogonal transformations and QR

(If the last couple of sentences look terribly familiar to you, it may be because we have mentioned this in Section 4.3.) The important point here is that multiplication by an orthogonal matrix does not change the norm. Thus, we simply have

$$\|A\| = \|\Sigma\| = \sigma_1.$$

Furthermore, $A^T A = V \Sigma^T U^T U \Sigma V^T = V S^2 V^T$, so for the pseudo-inverse we get

$$A^\dagger = (A^T A)^{-1} A^T = V S^{-2} V^T V \Sigma^T U^T = V(S^{-1}\ 0) U^T.$$

This yields $\|A^\dagger\| = \|(S^{-1}\ 0)\| = \frac{1}{\sigma_n}$, and hence the *condition number* is given by

$$\kappa(A) = \kappa_2(A) = \frac{\sigma_1}{\sigma_n}. \tag{6.1}$$

The error bound (5.1) is therefore directly generalized to the least squares solution for overdetermined systems, with the condition number given by (6.1) in terms of the largest over the smallest singular values.

What can go wrong with the normal equations algorithm

Turning to the algorithm that requires solving a linear system with $B = A^T A$, the condition number of this symmetric positive definite matrix is the ratio of its largest to smallest eigenvalues (see page 132), which in turn is given by

$$\kappa_2(B) = \frac{\lambda_1}{\lambda_n} = \frac{\sigma_1^2}{\sigma_n^2} = \kappa_2(A)^2.$$

Thus, the relative error in the solution using the normal equations is bounded by $\kappa(A)^2$, which is square what it could be using a more stable algorithm. In practice, this is a problem when $\kappa(A)$ is "large but not incredibly large," as in Example 6.4.

Next we seek algorithms that allow for the error to be bounded in terms of $\kappa(A)$ and not its square.

The QR decomposition

We next develop solution methods via the QR *orthogonal decomposition*. An SVD approach also yields an algorithm that is discussed in Section 8.2.

Let us emphasize again the essential point that for any orthogonal matrix P and vector $\mathbf{w}$ of appropriate sizes we can write

$$\|\mathbf{b} - A\mathbf{w}\| = \|P\mathbf{b} - PA\mathbf{w}\|.$$

We can therefore design an **orthogonal transformation** such that PA would have a more yielding form than the given matrix A. For the least squares problem it could be beneficial to transform our given A into an upper triangular form in this way.

Fortunately, this is possible using the celebrated **QR decomposition**. For a matrix A of size $m \times n$ with full column rank, there exists an orthogonal matrix Q of size $m \times m$ and an $m \times n$ matrix $\binom{R}{0}$, where R is an upper triangular $n \times n$ matrix, such that

$$A = Q \begin{pmatrix} R \\ 0 \end{pmatrix}.$$

Exercise 6 shows that R is nonsingular if and only if A has full column rank.

The details on how to construct such a decomposition are given in Section 6.3; see in particular the description of *Householder reflections*. Assuming the decomposition is available let us now see how to utilize it for solving our problem.

It is possible to multiply a residual vector by Q^T, still retaining the same least squares problem. This yields

$$\|\mathbf{b} - A\mathbf{x}\| = \left\|\mathbf{b} - Q\begin{pmatrix} R \\ 0 \end{pmatrix}\mathbf{x}\right\| = \left\|Q^T\mathbf{b} - \begin{pmatrix} R \\ 0 \end{pmatrix}\mathbf{x}\right\|.$$

So we can now minimize the right-hand term. But at this point it is clear how to do it! Partitioning the vector $Q^T\mathbf{b}$ to its first n components $\mathbf{c}$ and its last $m - n$ components $\mathbf{d}$, so

$$Q^T\mathbf{b} = \begin{pmatrix} \mathbf{c} \\ \mathbf{d} \end{pmatrix},$$

we see that

$$\|\mathbf{r}\|^2 = \|\mathbf{b} - A\mathbf{x}\|^2 = \|\mathbf{c} - R\mathbf{x}\|^2 + \|\mathbf{d}\|^2.$$

We have no control over $\|\mathbf{d}\|^2$. But the first term can be set to its minimal value of zero by solving the upper triangular system $R\mathbf{x} = \mathbf{c}$. This yields the solution $\mathbf{x}$, and we also get $\|\mathbf{r}\| = \|\mathbf{d}\|$.

Example 6.5. Let

$$A = \begin{pmatrix} 1 & 0 \\ 1 & 1 \\ 1 & 2 \end{pmatrix}, \quad \mathbf{b} = \begin{pmatrix} 0.1 \\ 0.9 \\ 2.0 \end{pmatrix}.$$

The data are the same as in Example 6.2. Note that this example can be happily solved by a simple method; but we use it to demonstrate a simple instance of QR decomposition.

The MATLAB command [Q,T] = qr(A) produces an orthogonal $m \times m$ matrix Q and an upper triangular $m \times n$ matrix T such that $A = QT$. Applying it for the above data we get

$$Q = \begin{pmatrix} -.5774 & .7071 & .4082 \\ -.5774 & 0 & -.8165 \\ -.5774 & -.7071 & .4082 \end{pmatrix}, \quad T = \begin{pmatrix} -1.7321 & -1.7321 \\ 0 & -1.4142 \\ 0 & 0 \end{pmatrix}.$$

Corresponding to our notation we see that

$$R = \begin{pmatrix} -1.7321 & -1.7321 \\ 0 & -1.4142 \end{pmatrix}.$$

Multiplying A and $\mathbf{b}$ by Q^T we have the problem

$$\min_{\mathbf{x}} \left\| \begin{pmatrix} -1.7321 \\ -1.3435 \\ 0.1225 \end{pmatrix} - \begin{pmatrix} -1.7321 & -1.7321 \\ 0 & -1.4142 \\ 0 & 0 \end{pmatrix} \mathbf{x} \right\|.$$

Hence $\mathbf{c} = (-1.7321, -1.3435)^T$, $\mathbf{d} = .1225$. Solving $R\mathbf{x} = \mathbf{c}$ yields the solution $\mathbf{x} = (.05, .95)^T$, while the norm of the residual is $\|\mathbf{r}\| = \|\mathbf{d}\| = .1225$. ∎

6.2. Orthogonal transformations and QR

Note that the QR decomposition obtained in Example 6.5 is not unique: $-Q$ and $-R$ provide perfectly valid QR factors, too.

> **Note:** Standard methods for solving the linear least squares problem include the following:
>
> 1. Normal equations: fast, simple, intuitive, but less robust in ill-conditioned situations.
>
> 2. QR decomposition: this is the "standard" approach implemented in general-purpose software. It is more computationally expensive than the normal equations approach if $m \gg n$ but is more robust.
>
> 3. SVD: used mostly when A is rank deficient or nearly rank deficient (in which case the QR approach may not be sufficiently robust). The SVD-based approach is very robust but is significantly more expensive in general and cannot be adapted to deal efficiently with sparse matrices. We will discuss this further in Section 8.2.

Economy size QR decomposition

An *economy size* QR decomposition also exists, with Q now being $m \times n$ with orthonormal columns and R remaining an $n \times n$ upper triangular matrix. Here we may write

$$A = QR.$$

Note that Q has the dimensions of A; in particular, it is rectangular and thus has no inverse if $m > n$.

Such a decomposition may be derived because if we look at the original QR decomposition, we see that the last $m - n$ columns of the $m \times m$ matrix Q actually multiply rows of zeros in the right-hand factor (those zero rows that appear right below the upper triangular matrix R).

Another way to get the same least squares algorithm is by looking directly at the normal equations and applying the decomposition. To see this, it is notationally convenient to use the economy size QR, obtaining

$$\mathbf{x} = A^\dagger \mathbf{b} = (A^T A)^{-1} A^T \mathbf{b} = (R^T Q^T Q R)^{-1} R^T Q^T \mathbf{b}$$
$$= (R^T R)^{-1} R^T Q^T \mathbf{b} = R^{-1} Q^T \mathbf{b}.$$

See Exercise 5.

The least squares solution via the QR decomposition thus comprises the three steps specified in the algorithm given on the next page.

The dominant cost factor in this algorithm is the computation of the QR decomposition. Ways to compute it will be discussed soon, in Section 6.3.

Algorithm efficiency

If done efficiently, the cost of forming the QR decomposition is approximately $2mn^2 - 2n^3/3$ flops: it is the same as for the normal equations method if $m \approx n$ and is twice as expensive if $m \gg n$. The QR decomposition method takes some advantage of sparsity and bandedness, similarly to the LU decomposition if $n = m$, but twice as slow. It is more robust than the normal equations method: while for the latter the relative error is proportional to $[\kappa(A)]^2$ and breaks down already if $\kappa(A) \approx \frac{1}{\sqrt{\eta}}$, a robust version of QR yields a relative error proportional to $\kappa(A)$, so it breaks down only if $\kappa(A) \approx \frac{1}{\eta}$, with η the rounding unit fondly remembered from Chapter 2.

Algorithm: Least Squares via the QR Decomposition (full and economy versions).

1. Decompose

   (a) $A = QR$, with R upper triangular and Q $m \times n$ with orthonormal columns; or

   (b) in the full version, $A = Q \begin{pmatrix} R \\ 0 \end{pmatrix}$, with Q $m \times m$ orthogonal.

2. Compute

   (a) $\mathbf{c} = Q^T \mathbf{b}$; or

   (b) in the full version, $\begin{pmatrix} \mathbf{c} \\ \mathbf{d} \end{pmatrix} = Q^T \mathbf{b}$.

3. Solve the upper triangular system $R\mathbf{x} = \mathbf{c}$.

Example 6.6. A lot of special effort has gone into optimizing the backslash operation in MATLAB. It is therefore of interest to see how the different options for solving overdetermined systems play out in terms of computing time.

Thus, we fill a matrix A and a vector $\mathbf{b}$ with random numbers and measure CPU times for solving the overdetermined system using the QR and normal equations options. We assume that no singularity or severe ill-conditioning arise, which is reasonable: as important as singular matrices are, they are not encountered at random. Here is the code:

```
for n = 300:100:1000
   % fill a rectangular matrix A and a vector b with random numbers
   % hoping that A'*A is nonsingular
   m = n+1; % or m= 3*n+1, or something else
   A = randn(m,n); b = randn(m,1);

   % solve and find execution times; first, Matlab way using QR
   t0 = cputime;
   xqr = A \ b;
   temp = cputime;
   tqr(n/100-2) = temp - t0;

   % next use normal equations
   t0 = temp;
   B = A'*A; y = A'*b;
   xne = B \ y;
   temp = cputime;
   tne(n/100-2) = temp - t0;
end

ratio = tqr./tne;
plot(300:100:1000,ratio)
```

Note that the backslash operator appears in both methods. But its meaning is different. In particular, the system for B involves a square matrix.

From Figure 6.5 we see that as n gets large enough the normal equations method is about twice more efficient when $m \approx n$ and about 4 times more efficient when $m \approx 3n$. These results do depend on the computing environment and should not be taken as more than a local, rough indication. Regarding accuracy of the results, even for $n = 1000$ the maximum difference between the two obtained solutions was still very small in this experiment. But then again, random matrices do not give rise to particularly ill-conditioned matrices, and when the going gets tough in terms of conditioning, the differences may be much more significant. ■

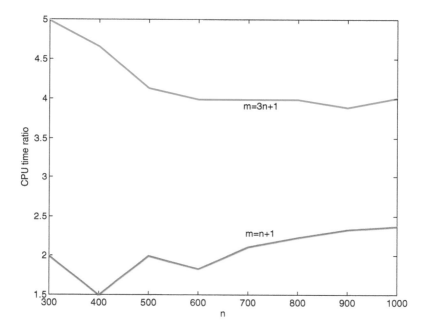

Figure 6.5. *Ratio of execution times using QR vs. normal equations. The number of rows for each n is $3n + 1$ for the upper curve and $n + 1$ for the lower one.*

Note: That the MATLAB designers decided to have QR as the default least squares solver is a tribute to the cause of robustness.

Specific exercises for this section: Exercises 5–6.

6.3 Householder transformations and Gram–Schmidt orthogonalization

In the previous section we have discussed the QR decomposition and explained its merits. But how does one actually compute it? That is the subject of the present section.

There are two intuitive ways of going about this: either by constructing R bit by bit until we obtain an orthogonal matrix Q, or by applying a sequence of orthogonal transformations that bring the given matrix A into an upper triangular matrix R. Throughout this section we continue to use the shorthand notation $\|\mathbf{v}\|$ for $\|\mathbf{v}\|_2$.

Note: A particularly robust QR decomposition method for the least squares problem is through **Householder reflections** (also known as *Householder transformations*). Below we first describe other, conceptually important methods that may be easier to grasp, but programs and a numerical example are only provided for this method of choice. This section is more technical than the previous ones in the present chapter.

Gram–Schmidt orthogonalization

The first of these two approaches is probably the more intuitive one, although not necessarily the better way to go computationally, and it yields the *Gram–Schmidt* process. Let us illustrate it for a simple example and then generalize.

Example 6.7. Consider a 3×2 instance. We write

$$\begin{pmatrix} a_{11} & a_{12} \\ a_{21} & a_{22} \\ a_{31} & a_{32} \end{pmatrix} = \begin{pmatrix} q_{11} & q_{12} \\ q_{21} & q_{22} \\ q_{31} & q_{32} \end{pmatrix} \begin{pmatrix} r_{11} & r_{12} \\ 0 & r_{22} \end{pmatrix}.$$

For notational convenience, denote inner products by

$$\langle \mathbf{z}, \mathbf{y} \rangle \equiv \mathbf{z}^T \mathbf{y}.$$

Writing the above column by column we have

$$(\mathbf{a}_1, \mathbf{a}_2) = (r_{11}\mathbf{q}_1,\ r_{12}\mathbf{q}_1 + r_{22}\mathbf{q}_2).$$

Requiring orthonormal columns yields the three conditions $\|\mathbf{q}_1\| = 1$, $\|\mathbf{q}_2\| = 1$, and $\langle \mathbf{q}_1, \mathbf{q}_2 \rangle = 0$. We thus proceed as follows: for the first column, use $\|\mathbf{q}_1\| = 1$ to obtain

$$r_{11} = \|\mathbf{a}_1\|\ ;\quad \mathbf{q}_1 = \mathbf{a}_1/r_{11}.$$

For the second column we have $\mathbf{a}_2 = r_{12}\mathbf{q}_1 + r_{22}\mathbf{q}_2$, and applying an inner product with $\mathbf{q}_1$ yields

$$r_{12} = \langle \mathbf{a}_2, \mathbf{q}_1 \rangle.$$

Next, observe that once r_{12} is known we can compute $\tilde{\mathbf{q}}_2 = \mathbf{a}_2 - r_{12}\mathbf{q}_1$ and then set $r_{22} = \|\tilde{\mathbf{q}}_2\|$ and $\mathbf{q}_2 = \tilde{\mathbf{q}}_2/r_{22}$. This completes the procedure; we now have the matrices Q and R in the economy size version. ∎

Example 6.7 should convince you that it is possible to obtain a QR decomposition by an orthogonalization process that constructs the required factors column by column. By directly extending the above procedure we obtain the Gram–Schmidt orthogonalization procedure, which computes $r_{ij} = \langle \mathbf{a}_j, \mathbf{q}_i \rangle$ such that

$$\mathbf{a}_j = \sum_{i=1}^{j} r_{ij} \mathbf{q}_i.$$

For any given j, once $\mathbf{q}_1$ through $\mathbf{q}_{j-1}$ are known, $\mathbf{q}_j$ (and hence r_{jj}) can be computed in a straightforward manner.

It is assumed that the columns of A are linearly independent, or in other words that A has full column rank. If there is linear dependence, say, $\mathbf{a}_j$ is linearly dependent on $\mathbf{a}_1$ through $\mathbf{a}_{j-1}$, then at the jth stage of the algorithm we will get (ignoring roundoff errors) $r_{jj} = 0$. Thus, robust codes have a condition for gracefully exiting if r_{jj} is zero or approximately zero.

Modified Gram–Schmidt

The procedure we just described is called *classical Gram–Schmidt*. For all its endearing simplicity, it is numerically unstable if the columns of A are nearly linearly dependent.

Stability can be improved by a simple fix: instead of using $\mathbf{a}_1$ through $\mathbf{a}_{j-1}$ for constructing $\mathbf{q}_j$, we employ the already computed $\mathbf{q}_1$ through $\mathbf{q}_{j-1}$, which by construction are orthogonal to one another and thus less prone to damaging effects of roundoff errors. The resulting algorithm, given on this page, is identical to the classical one except for the fourth line and is called the *modified Gram–Schmidt algorithm*. It is the preferred version in all numerical codes.

> **Algorithm: Modified Gram–Schmidt Orthogonalization.**
> Input: matrix A of size $m \times n$.
>
> $\quad$ for $j = 1 : n$
> $\quad\quad \mathbf{q}_j = \mathbf{a}_j$
> $\quad\quad$ for $i = 1 : j - 1$
> $\quad\quad\quad r_{ij} = \langle \mathbf{q}_j, \mathbf{q}_i \rangle$
> $\quad\quad\quad \mathbf{q}_j = \mathbf{q}_j - r_{ij} \mathbf{q}_i$
> $\quad\quad$ end
> $\quad\quad r_{jj} = \|\mathbf{q}_j\|$
> $\quad\quad \mathbf{q}_j = \mathbf{q}_j / r_{jj}$
> $\quad$ end

Orthogonal transformations

The Gram–Schmidt procedure just described is a process of orthogonalization: we use the elements of a triangular matrix as coefficients while turning the given A into an orthogonal matrix Q. A natural alternative is to use orthogonal *transformations* to turn A into an upper triangular matrix R. In many ways, the concept is reminiscent of Gaussian elimination, at least in terms of the "zeroing out" aspect of it.

A slow but occasionally useful way of accomplishing this goal is by a surgical, pointed algorithm that takes aim at one entry of A at a time. Such are **Givens rotations**. They are transformations that successively rotate vectors in a manner that turns 2×2 submatrices into upper triangular 2×2 ones. Givens rotations are worth considering in cases where the original matrix has a special structure. A famous instance here are the Hessenberg matrices (page 139); see Exercise 5.20. Such matrices arise in abundance in advanced techniques for solving linear systems and eigenvalue problems, and Givens rotations thus form an integral part of many of the relevant algorithms. They are easy to parallelize and are useful in a variety of other situations as well.

Householder reflections

The technique employing *Householder transformations* is the most suitable for general-purpose QR decomposition and the most reminiscent of LU decomposition.

Turning a column into a unit vector, orthogonally

Suppose we are given a vector $\mathbf{z}$ and are looking for an orthogonal transformation that zeros out all its entries except the first one. How can we find such a transformation?

Consider the matrix

$$P = I - 2\mathbf{u}\mathbf{u}^T,$$

where $\mathbf{u}$ is a given unit vector, $\|\mathbf{u}\| = 1$, and $I = I_m$ is the $m \times m$ identity matrix. Clearly, $P\mathbf{u} = -\mathbf{u}$, so we can say that P is a *reflector*. We could proceed as follows: for the given vector $\mathbf{z}$, find a vector $\mathbf{u}$ such that P, defined as above, has the property that $P\mathbf{z}$ transforms $\mathbf{z}$ into the vector $(\alpha, 0, \ldots, 0)^T = \alpha \mathbf{e}_1$. Denoting $\beta = 2\mathbf{u}^T\mathbf{z}$, we want

$$P\mathbf{z} = \mathbf{z} - 2\mathbf{u}\mathbf{u}^T\mathbf{z} = \mathbf{z} - \beta\mathbf{u} = \alpha \mathbf{e}_1.$$

From this it follows that $\mathbf{u}$ is a vector in the direction $\mathbf{z} - \alpha\mathbf{e}_1$. Now, since P is an orthogonal transformation (Exercise 9) we have that $\|\mathbf{z}\| = \|P\mathbf{z}\| = |\alpha|$; hence it follows that $\mathbf{u}$ is a unit vector in the direction of $\mathbf{z} \pm \|\mathbf{z}\|\mathbf{e}_1$. We finally select the sign in the latter expression in accordance with that of z_1 (the first element of $\mathbf{z}$), because this reduces the possibility of cancellation error.

Extending to QR decomposition

The procedure outlined above can be applied and extended in an analogous way to the derivation of LU decomposition in Section 5.2. To compute the QR decomposition we proceed as follows. Start by applying a reflection, call it $P^{(1)}$, that turns the first column of A into a multiple of $\mathbf{e}_1$ as described above, using the first column of A for $\mathbf{z}$, so the matrix $P^{(1)}A$ now has all zeros in its first column except the $(1,1)$ entry. Next, apply a similar transformation based on the second column of $P^{(1)}A$, zeroing it below the diagonal $(2,2)$ entry, as when using $M^{(2)}$ in the formation of the LU decomposition. After completing this stage, the matrix $P^{(2)}P^{(1)}A$ has all zero entries below the $(1,1)$ and $(2,2)$ elements.

In a general step of the algorithm, $P^{(k)}$ is a concatenation of the $(k-1) \times (k-1)$ identity matrix with an $(m-k+1) \times (m-k+1)$ reflector which is responsible for zeroing out all entries below the (k,k) entry of $P^{(k-1)}P^{(k-2)}\cdots P^{(1)}A$. See Figure 6.6. Carrying this through until the end (n steps in total when $n < m$), we obtain the desired QR decomposition. The fact that the product of orthogonal matrices is also an orthogonal matrix comes in handy.

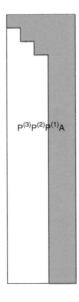

Figure 6.6. *Householder reflection. Depicted is a 20×5 matrix A after 3 reflections.*

6.3. Householder transformations and Gram–Schmidt orthogonalization

The Householder reflection is not only elegant, it is also very robust. It has better numerical stability properties than modified Gram–Schmidt, and its storage requirements are modest: there is no need to store the $P^{(k)}$ matrices, since each of them is defined by a single vector, say, $\mathbf{u}_k$. Similar to the LU decomposition, when Householder reflections are used the vectors $\mathbf{u}_k$ can overwrite A column by column. So, this QR decomposition, although not in the economy-size version, is actually rather economical! The following script demonstrates how it can all be done:

```
function [A,p] = house(A)
%
% function [A,p] = house(A)
%
% perform QR decomposition using Householder reflections
% Transformations are of the form P_k = I - 2u_k(u_k^T), so
% store effecting vector u_k in p(k) + A(k+1:m,k). Assume m > n.

[m,n]=size(A); p = zeros(1,n);
for k = 1:n
   % define u of length = m-k+1
   z = A(k:m,k);
   e1 = [1; zeros(m-k,1)];
   u = z+sign(z(1))*norm(z)*e1; u = u/norm(u);
   % update nonzero part of A by I-2uu^T
   A(k:m,k:n) = A(k:m,k:n)-2*u*(u'*A(k:m,k:n));
   % store u
   p(k) = u(1);
   A(k+1:m,k) = u(2:m-k+1);
end
```

Solving least squares problems with house

Now, suppose we wish to compute the solution $\mathbf{x}$ of the least squares problem $\min_\mathbf{x} \|\mathbf{b} - A\mathbf{x}\|$. For that, we will need to compute $Q^T \mathbf{b}$. If the QR decomposition of A is already given, having executed the above script, all we need to do is the following:

```
function [x,nr] = lsol(b,AA,p)
%
% function [x,r] = lsol(b,AA,p)
%
% Given the output AA and p of house, containing Q and R of matrix A,
% and given a right-hand-side vector b,  solve min || b - Ax ||.
% Return also the norm of the residual, nr = ||r|| = ||b - Ax||.

y = b(:); [m,n] = size(AA);
% transform b
for k=1:n
   u = [p(k);AA(k+1:m,k)];
   y(k:m) = y(k:m) - 2*u *(u'*y(k:m));
end
% form upper triangular R and solve
R = triu(AA(1:n,:));
x = R \ y(1:n); nr = norm(y(n+1:m));
```

Notice that these programs are remarkably short, especially given the length of our description leading to them. This reflects our acquired stage of knowing what we're doing!

Example 6.8. We show for a small matrix how Householder transformations can be used to construct the QR decomposition and solve a least squares problem. For this, consider the data of Example 6.1.

The first column of A is $\mathbf{a}_1 = (1, 2, 5, 3 - 1)^T$, and the sign of its first element is positive. Thus, compute $\mathbf{a}_1 + \|\mathbf{a}_1\|(1, 0, 0, 0, 0)^T = (7.3246, 2, 5, 3, -1)^T$ and normalize it, obtaining the reflection vector $\mathbf{u} = (0.7610, 0.2078, 0.5195, 0.3117, -0.1039)^T$. This reflection vector defines the transformation

$$P^{(1)} = I_5 - 2\mathbf{u}\mathbf{u}^T = \begin{pmatrix} -0.1581 & -0.3162 & -0.7906 & -0.4743 & 0.1581 \\ -0.3162 & 0.9137 & -0.2159 & -0.1295 & 0.0432 \\ -0.7906 & -0.2159 & 0.4603 & -0.3238 & 0.1079 \\ -0.4743 & -0.1295 & -0.3238 & 0.8057 & 0.0648 \\ 0.1581 & 0.0432 & 0.1079 & 0.0648 & 0.9784 \end{pmatrix}$$

and

$$P^{(1)}A = \begin{pmatrix} -6.3246 & -4.7434 & -1.5811 \\ 0 & 1.7048 & 4.2952 \\ 0 & -0.2380 & -3.7620 \\ 0 & 3.0572 & 2.9428 \\ 0 & 6.6476 & 3.3524 \end{pmatrix} \; ; \quad P^{(1)}\mathbf{b} = \begin{pmatrix} -2.8460 \\ -3.8693 \\ 0.3266 \\ -4.8040 \\ 1.9347 \end{pmatrix}.$$

Note, again, that $P^{(1)}$ need not be constructed explicitly; all we need to store in this step is the vector $\mathbf{u}$. Next, we work on zeroing out the elements of $P^{(1)}A$ below the (2,2) entry. Thus, we define $\mathbf{z} = (1.7048, -0.2380, 3.0572, 6.6476)^T$, and our reflector is $\mathbf{u} = \mathbf{z} + \text{sign}(\mathbf{z}(1))\|\mathbf{z}\|\mathbf{e}_1 = (0.7832, -0.0202, 0.2597, 0.5646)^T$. Notice that now $\mathbf{z}$ and $\mathbf{e}_1$ are only four elements in length.

The matrix $P^{(2)}$ is a concatenation of the 1×1 identity matrix with $I_4 - 2\mathbf{u}\mathbf{u}^T$, and now the second column has been taken care of. Finally, we work on the third column. Please verify that, setting $\mathbf{z} = (-3.5155, -0.2234, -3.5322)^T$ and $\mathbf{u} = (-0.9232, -0.0243, -0.3835)^T$. After completing this step we have $A = Q\binom{R}{0}$ where

$$Q = (P^{(3)}P^{(2)}P^{(1)})^T = \begin{pmatrix} -0.1581 & 0.0998 & 0.2555 & -0.3087 & 0.8969 \\ -0.3162 & -0.1996 & 0.6919 & -0.4755 & -0.3942 \\ -0.7906 & 0.0998 & -0.5464 & -0.2454 & -0.0793 \\ -0.4743 & -0.3659 & 0.2661 & 0.7427 & 0.1369 \\ 0.1581 & -0.8980 & -0.2945 & -0.2585 & 0.1227 \end{pmatrix}$$

and

$$R = \begin{pmatrix} -6.3246 & -4.7434 & -1.5811 \\ 0 & -7.5166 & -5.2550 \\ 0 & 0 & 4.9885 \end{pmatrix}.$$

6.4. Exercises

For the right-hand-side vector given in Example 6.1, $Q^T\mathbf{b} = (-2.8460, 1.1308, -3.9205, -3.2545, 3.8287)^T$. The solution of the least squares problem is the product of R^{-1} with the first $n = 3$ elements of $Q^T\mathbf{b}$, yielding again $\mathbf{x} = (0.3472, 0.3990, -0.7859)^T$. ∎

Specific exercises for this section: Exercises 7–9.

6.4 Exercises

0. **Review questions**

    (a) What are the normal equations? What orthogonality property of least squares solutions do they reflect?

    (b) Let A be an $m \times n$ matrix, with $m > n$. Under what conditions on the matrix A is the matrix $A^T A$:

    i. symmetric?

    ii. nonsingular?

    iii. positive definite?

    (c) Suppose a given rectangular matrix A is sparse but has one dense row. What is the sparsity pattern of $A^T A$? Explain the meaning of this for solving least squares problems using normal equations.

    (d) Define the pseudo-inverse of a rectangular matrix A that has more rows than columns, and explain how it connects to a matrix inverse and what is "pseudo" about it.

    (e) What is the floating point operation count for solving the least squares problem using the normal equations?

    (f) Can the regression curve of Example 6.2 be obtained using the program lsfit? If yes, then how?

    (g) Why do data fitting problems typically yield overdetermined systems and not systems of the form treated in Chapter 5?

    (h) In what circumstances is the QR decomposition expected to yield better least squares fitting results than the normal equations?

    (i) What is the difference between QR decomposition and the economy size QR decomposition?

    (j) What is the difference between the QR decomposition produced by modified Gram–Schmidt and that produced by Householder transformations?

    (k) What is the floating point operation count for solving the least squares problem using the QR decomposition?

1. The following data were obtained in connection with the values of a certain fictitious material property:

t	0.0	0.1	0.2	0.3	0.4	0.5	0.6	0.7	0.8	0.9	1.0
b	0.9	1.01	1.05	0.97	0.98	0.95	0.01	−0.1	0.02	−0.1	0.0

It was then hypothesized that the underlying material property is a piecewise constant function with one break point (i.e., two constant pieces).

   (a) Plot the data points and decide where (approximately) the break point should be.

   (b) Find the piecewise constant function which best fits the data by least squares, and plot it, too.

2. (a) Evaluate the function

$$f(t) = .05\sin(1000t) + .5\cos(\pi t) - .4\sin(10t)$$

   at the 101 points given by 0:.01:1. Plot the resulting broken line interpolant.

   (b) In order to study the slow scale trend of this function, we wish to find a low degree polynomial (degree at most 6) that best approximates f in the least squares norm at the above 101 data points. By studying the figure from part (a) find out the smallest n that would offer a good fit in this sense. (Try to do this without further computing.)

   (c) Find the best approximating polynomial v of degree n and plot it together with f. What are your observations?

3. Let us synthesize data in the following way. We start with the cubic polynomial

$$q(t) = -11 + \frac{55}{3}t - \frac{17}{2}t^2 + \frac{7}{6}t^3.$$

Thus, $n = 4$. This is sampled at 33 equidistant points between 0.9 and 4.1. Then we add to these values 30% noise using the random number generator `randn` in MATLAB, to obtain "the data" which the approximations that you will construct "see." From here on, no knowledge of $q(t)$, its properties, or its values anywhere is assumed: we pretend we don't know where the data came from!

Your programming task is to pass through these data three approximations:

   (a) An interpolating polynomial of degree 32. This can be done using the MATLAB function `polyfit`. You don't need to know how such interpolation works for this exercise, although details are given in Chapter 10.

   (b) An interpolating cubic spline using the MATLAB function `spline`. The corresponding method is described in Section 11.3, but again you don't need to rush and study that right now.

   (c) A cubic polynomial which best fits the data in the ℓ_2 sense, obtained by our function `lsfit`.

Plot the data and the obtained approximations as in Figures 6.3 and 6.4. Which of these approximations make sense? Discuss.

4. Often in practice, an approximation of the form

$$u(t) = \gamma_1 e^{\gamma_2 t}$$

is sought for a data fitting problem, where γ_1 and γ_2 are constants. Assume given data $(t_1, z_1), (t_2, z_2), \ldots, (t_m, z_m)$, where $z_i > 0$, $i = 1, 2, \ldots, m$, and $m > 0$.

   (a) Explain in one brief sentence why the techniques introduced in the present chapter cannot be directly applied to find this $u(t)$.

6.4. Exercises

(b) Considering instead
$$v(t) = \ln u(t) = (\ln \gamma_1) + \gamma_2 t,$$
it makes sense to define $b_i = \ln z_i$, $i = 1, 2, \ldots, m$, and then find coefficients x_1 and x_2 such that $v(t) = x_1 + x_2 t$ is the best least squares fit for the data
$$(t_1, b_1), (t_2, b_2), \ldots, (t_m, b_m).$$
Using this method, find $u(t)$ for the data

i	1	2	3
t_i	0.0	1.0	2.0
z_i	$e^{0.1}$	$e^{0.9}$	e^2

5. (a) Why can't one directly extend the LU decomposition to a long and skinny matrix in order to solve least squares problems?

   (b) When writing $\mathbf{x} = A^\dagger \mathbf{b} = \cdots = R^{-1} Q^T \mathbf{b}$ we have somehow moved from the conditioning $[\kappa(A)]^2$ to the conditioning $\kappa(A)$ through mathematical equalities. Where is the improvement step hidden? Explain.

6. (a) Let Q be an orthogonal $m \times m$ matrix and R an $n \times n$ upper triangular matrix, $m > n$, such that
$$A = Q \begin{pmatrix} R \\ 0 \end{pmatrix}.$$
Show that the diagonal elements of R all satisfy $r_{ii} \neq 0$, $i = 1, \ldots, n$, if and only if A has full column rank.

   (b) Next, let Q be $m \times n$ with orthonormal columns (so $Q^T Q = I$, but Q does not have an inverse) such that
$$A = QR.$$
Prove the same claim as in part (a) for this economy size decomposition.

7. Find the QR factorization of the general 2×2 matrix
$$A = \begin{pmatrix} a_{11} & a_{12} \\ a_{21} & a_{22} \end{pmatrix}.$$

8. (a) Explain what may happen during the course of the Gram–Schmidt process if the matrix A is rank deficient.

   (b) Show that classical Gram–Schmidt and the modified version are mathematically equivalent.

   (c) Construct a 3×2 example, using a decimal floating point system with a 2-digit fraction, in which modified Gram–Schmidt proves to be more numerically stable than the classical version.

9. (a) For a given real vector $\mathbf{u}$ satisfying $\|\mathbf{u}\|_2 = 1$, show that the matrix $P = I - 2\mathbf{u}\mathbf{u}^T$ is orthogonal.

   (b) Suppose A is a complex-valued matrix. Construct a complex analogue of Householder transformations, with the reflector given by $P = I - 2\mathbf{u}\mathbf{u}^*$, where $*$ denotes a complex conjugate transpose and $\mathbf{u}^*\mathbf{u} = 1$. (The matrix P is now *unitary*, meaning that $P^* P = I$.)

6.5 Additional notes

The material of this chapter elegantly extends the approaches of Chapter 5 to handle over-determined systems, and unlike Chapter 8 it does not require any serious change in our state of mind in the sense that exact direct algorithms are considered. Yet its range of application is ubiquitous. No wonder it is popular and central among fans of numerical linear algebra and users alike. A variety of issues not discussed or just skimmed over here are exposed and dealt with in books on numerical linear algebra; see, e.g., Demmel [21], Watkins [74], or the reference book of Golub and van Loan [30].

As mentioned in Section 6.1 the problems of data fitting in ℓ_1 and in ℓ_∞ can be posed as special instances of *linear programming*. There are many undergraduate textbooks on the latter topic. We mention instead the higher level but concise expositions in the true classics Luenberger [51] and Fletcher [25] and the more modern Nocedal and Wright [57]. In Section 9.3 we quickly address linear programming as well.

The same books are also excellent references for treatments of *nonlinear least squares* problems. Such problems are treated in Section 9.2. A very simple example is provided in Exercise 4. Nonlinear data fitting problems arise frequently when solving inverse problems. The books by Tikhonov and Arsenin [67] and Engl, Hanke, and Neubauer [24] are relevant, though admittedly not always very accessible.

Chapter 7

Linear Systems: Iterative Methods

In this chapter we return to the basic problem of Chapter 5, namely, solving a linear system of equations $A\mathbf{x} = \mathbf{b}$. The difference is that here we consider iterative methods for this purpose. The basic premise is that the nonsingular matrix A is large, and yet calculating the product $A\mathbf{v}$ for any given vector $\mathbf{v}$ of appropriate length is relatively inexpensive. Such a situation arises in many applications and typically happens when A is sparse, even if it is not tightly banded as in Section 5.6. In Section 7.1 we further justify the need for considering such problems, giving rise to iterative methods.

In Section 7.2 we describe **stationary** iterative methods, concentrating on some simple instances. The basic, so-called **relaxation** methods described in Section 7.2 are, well, basic. They are usually too slow to compete with more advanced methods in terms of efficiency but are often used instead as building blocks in more complex schemes. We analyze the performance of these methods in Section 7.3.

In Section 7.4 a solution technique for the special case where the matrix A is symmetric positive definite is described. This is the **conjugate gradient** (CG) method, and along the way we also introduce **gradient descent** methods such as steepest descent. These methods often require enhancement in the form of **preconditioning**, and we briefly consider this as well.

Extensions of the CG method to more general cases where the matrix A is not necessarily symmetric positive definite are considered in the more advanced Section 7.5.

Stationary methods do not have to be slow or simple, as is evidenced by the **multigrid method** considered in the more advanced Section 7.6. This family of methods is more context specific, and the exposition here is more demanding, but these methods can be very efficient in practice. The methods of Sections 7.4–7.6 all enjoy heavy, frequent use in practice today.

7.1 The need for iterative methods

> **Note:** The Gaussian elimination algorithm and its variations such as the LU decomposition, the Cholesky method, adaptation to banded systems, etc., is the approach of choice for many problems. Please do not allow anything said in the present chapter to make you forget this.

There are situations that require a treatment that is different from the methods of Chapter 5. Here are a few drawbacks of direct methods:

- As observed in Chapter 5, the Gaussian elimination (or LU decomposition) process may introduce **fill-in**, i.e., L and U may have nonzero elements in locations where the original matrix A has zeros. If the amount of fill-in is significant, then applying the direct method may become costly. This in fact occurs often, in particular when a banded matrix is sparse within its band.

- Sometimes we do not really need to solve the system exactly. For example, in Chapter 9 we will discuss methods for nonlinear systems of equations in which each iteration involves a linear system solve. Frequently, it is sufficient to solve the linear system within the nonlinear iteration only to a low degree of accuracy. Direct methods cannot accomplish this because, by definition, to obtain a solution the process must be completed; there is no notion of an early termination or an inexact (yet acceptable) solution.

- Sometimes we have a pretty good idea of an approximate guess for the solution. For example, in time-dependent problems we may solve a linear system at a certain time level and then move on to the next time level. Often the solution for the previous time level is quite close to the solution in the current time level, and it is definitely worth using it as an initial guess. This is called a **warm start**. Direct methods cannot make good use of such information, especially when A also changes slightly from one time instance to the next.

- Sometimes only matrix-vector products are given. In other words, the matrix is not available explicitly or is very expensive to compute. For example, in digital signal processing applications it is often the case that only input and output signals are given, without the transformation itself being explicitly formulated and available.

Concrete instances leading to iterative methods

We next discuss some relevant instances of a matrix A. As mentioned above, a situation of fill-in arises often. Below we describe one major source of such problems. Many linear systems arising in practice involve values defined on a grid, or a mesh, in two or even three space dimensions. For instance, consider the brightness values at each pixel of your two-dimensional computer monitor screen. The collection of such brightness values forms a two-dimensional *grid function*, $\{u_{i,j}, \ i = 1, \ldots, N, \ j = 1, \ldots, M\}$. As another instance, we endow a geophysical domain in three space dimensions with a discretization grid,[26] as depicted in Figure 7.1, before commencing with a computational analysis of its properties (such as conductivity σ). Imagine some function value associated with each of the grid nodes.

Often there are linear, algebraic relationships between values at neighboring grid nodes.[27] But neighboring locations in a two- or three-dimensional array do not necessarily correspond to consecutive locations when such a structure is reshaped into a one-dimensional array. Let us get more specific. The following case study is long but fundamental.

Example 7.1. An example of such a relationship as described above, arising in many applications, is the famous **Poisson equation**. This is a partial differential equation that in its simplest form is defined on the open *unit square*, $0 < x, y < 1$, and reads

$$-\left(\frac{\partial^2 u}{\partial x^2} + \frac{\partial^2 u}{\partial y^2}\right) = g(x, y).$$

---

[26] People also use the word *mesh* in place of grid. For us these words are synonymous.

[27] *Finite element methods* and *finite difference methods* for partial differential equations give rise to such local algebraic relationships.

7.1. The need for iterative methods

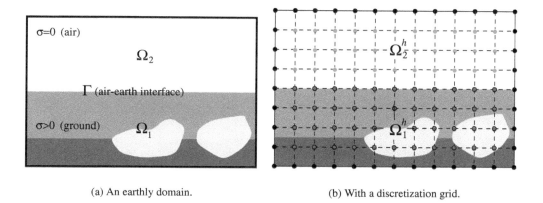

(a) An earthly domain.   (b) With a discretization grid.

Figure 7.1. *A two-dimensional cross section of a three-dimensional domain with a square grid added. (Reprinted from [2] with permission from World Scientific Publishing Co. Pte. Ltd.)*

Here $u(x, y)$ is the unknown function sought and $g(x, y)$ is a given *source*. Further to satisfying the differential equation, the sought function $u(x, y)$ is known to satisfy *homogeneous Dirichlet boundary conditions* along the entire boundary of the unit square, written as

$$u(x, 0) = u(x, 1) = u(0, y) = u(1, y) = 0.$$

In Example 4.17 we saw a one-dimensional version of this problem, but the present one involves many more issues, both mathematically and computationally. Nonetheless, we hasten to point out that you don't need to be an expert in partial differential equations to follow the exposition below.

Discretizing using centered differences for the partial derivatives we obtain the equations

$$4u_{i,j} - u_{i+1,j} - u_{i-1,j} - u_{i,j+1} - u_{i,j-1} = b_{i,j}, \quad 1 \leq i, j \leq N, \tag{7.1}$$

$$u_{i,j} = 0 \quad \text{otherwise,}$$

where $u_{i,j}$ is the value at the (i, j)th node of a square (with $M = N$) planar grid, and $b_{i,j} = h^2 g(ih, jh)$ are given values at the same grid locations. Here $h = 1/(N + 1)$ is the *grid width*; see Figure 7.2. The value of N can be fairly large, say, $N = 127$ or even $N = 1023$.

Obviously, these are linear relations, so we can express them as a system of linear equations

$$A\mathbf{u} = \mathbf{b},$$

where $\mathbf{u}$ consists of the $n = N^2$ unknowns $\{u_{i,j}\}$ somehow organized as a vector, and $\mathbf{b}$ is composed likewise from the values $\{b_{i,j}\}$. Notice that the resulting matrix A could be quite large. For example, for $N = 1023$ we have $n = N^2 > 1{,}000{,}000(!)$. But what does the matrix A look like? In general this depends on the way we choose to order the grid values, as we see next.

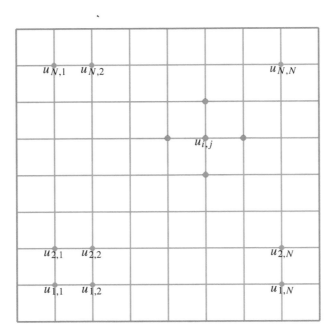

Figure 7.2. *A two-dimensional grid with grid function values at its nodes, discretizing the unit square. The length of each edge of the small squares is the grid width h, while their union is a square with edge length 1. The locations of $u_{i,j}$ and those of its neighbors that are participating in the difference formula (7.1) are marked in red.*

How should we order the grid unknowns $\{u_{i,j}\}_{i,j=1}^{N}$ into a vector **u**? We can do this in many ways. A simple way is lexicographically, say, by columns, which yields

$$\mathbf{u} = \begin{pmatrix} u_{1,1} \\ u_{2,1} \\ \vdots \\ u_{N,1} \\ u_{1,2} \\ u_{2,2} \\ \vdots \\ u_{N,2} \\ u_{1,3} \\ \vdots \\ u_{N,N} \end{pmatrix}, \quad \mathbf{b} = \begin{pmatrix} b_{1,1} \\ b_{2,1} \\ \vdots \\ b_{N,1} \\ b_{1,2} \\ b_{2,2} \\ \vdots \\ b_{N,2} \\ b_{1,3} \\ \vdots \\ b_{N,N} \end{pmatrix}.$$

In MATLAB this can be achieved by the instruction `reshape (b,n,1)`, where b is an $N \times N$

7.1. The need for iterative methods

array and $n = N^2$. This ordering of an array of values in two dimensions into a vector (that is to say, a one-dimensional array), as well as any other ordering of this sort, separates neighbors to some extent. The resulting $n \times n$ matrix A (with $n = N^2$) has the form

$$A = \begin{pmatrix} J & -I & & & \\ -I & J & -I & & \\ & \ddots & \ddots & \ddots & \\ & & -I & J & -I \\ & & & -I & J \end{pmatrix},$$

where J is the tridiagonal $N \times N$ matrix

$$J = \begin{pmatrix} 4 & -1 & & & \\ -1 & 4 & -1 & & \\ & \ddots & \ddots & \ddots & \\ & & -1 & 4 & -1 \\ & & & -1 & 4 \end{pmatrix},$$

and I denotes the identity matrix of size N. For instance, if $N = 3$, then

$$A = \left(\begin{array}{ccc|ccc|ccc} 4 & -1 & 0 & -1 & 0 & 0 & 0 & 0 & 0 \\ -1 & 4 & -1 & 0 & -1 & 0 & 0 & 0 & 0 \\ 0 & -1 & 4 & 0 & 0 & -1 & 0 & 0 & 0 \\ \hline -1 & 0 & 0 & 4 & -1 & 0 & -1 & 0 & 0 \\ 0 & -1 & 0 & -1 & 4 & -1 & 0 & -1 & 0 \\ 0 & 0 & -1 & 0 & -1 & 4 & 0 & 0 & -1 \\ \hline 0 & 0 & 0 & -1 & 0 & 0 & 4 & -1 & 0 \\ 0 & 0 & 0 & 0 & -1 & 0 & -1 & 4 & -1 \\ 0 & 0 & 0 & 0 & 0 & -1 & 0 & -1 & 4 \end{array} \right).$$

We can see that for any size N the matrix A is diagonally dominant and nonsingular. It can be verified directly that the $n = N^2$ eigenvalues of A are given by

$$\lambda_{l,m} = 4 - 2(\cos(l\pi h) + \cos(m\pi h)), \quad 1 \leq l, m \leq N$$

(recall $(N+1)h = 1$). Thus $\lambda_{l,m} > 0$ for all $1 \leq l, m \leq N$, and we see that the matrix A is also positive definite.

The MATLAB command spy(A) plots the nonzero structure of a matrix A. For the current example and $N = 10$, we obtain Figure 7.3.

We can see that the matrix remains banded, as we have seen before in Example 4.17, but now there are also $N - 2$ zero diagonals *between* the nonzero diagonals. Gaussian elimination without pivoting can be stably applied to this diagonally dominant system, retaining the sparsity

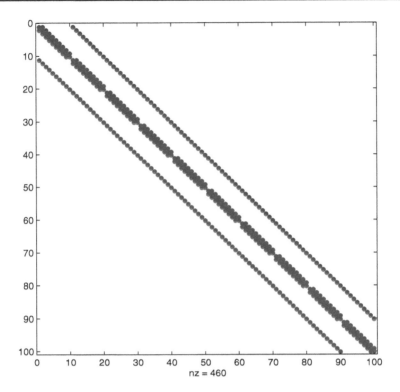

Figure 7.3. *The sparsity pattern of A for Example 7.1 with N = 10. Note that there are* nz = 460 *nonzeros out of 10,000 matrix locations.*

structure of the outer band. However, the inner zero diagonals get replaced, in general, by nonzeros. The leftmost plots in Figures 5.6 and 5.7 (page 126) depict this. It can be seen that the Cholesky decomposition costs $\mathcal{O}(N^4)$ flops (because the matrix is banded, with semibandwidth N) and, perhaps worse, requires $\mathcal{O}(N^3)$ storage locations instead of the original $5N^2$. For $N = 1,000$, say, the storage requirement $N^3 = 10^9$ is often prohibitive for a fast memory, whereas $N^2 = 10^6$ is acceptable. ∎

The situation for three-dimensional grids (arising, for instance, from discretizing the Poisson equation in three space variables) is even worse for the direct methods of Chapter 5, because for those problems the semibandwidth is larger (compared to the dimensions of the matrix) and within the band there is a much higher degree of sparsity. Hence we consider here alternative, *iterative methods*. These methods in their basic form do not require the storage of A. They are suitable mainly for special matrices, but such special matrices often arise in practice.

Fixed point iteration

Several methods considered in this chapter can be written in the form of a *fixed point iteration*, which is a vector version of the approach considered in Section 3.3. Thus, for a given linear system of equations

$$A\mathbf{x} = \mathbf{b},$$

7.2. Stationary iteration and relaxation methods

rewritten as a *vector equation* $\mathbf{f}(\mathbf{x}) = \mathbf{b} - A\mathbf{x} = \mathbf{0}$, we seek an equivalent form $\mathbf{g}(\mathbf{x}) = \mathbf{x}$, and define the iteration

$$\mathbf{x}_{k+1} = \mathbf{g}(\mathbf{x}_k), \quad k = 0, 1, \ldots,$$

starting from an initial guess $\mathbf{x}_0$. Moreover, the convergence of all these methods is *linear*, and the difference between them in terms of speed of convergence, although it can be significant, is only in the *rate of convergence*.

Specific exercises for this section: Exercises 1–2.

7.2 Stationary iteration and relaxation methods

In this section we consider the solution of our prototype system of linear equations by *stationary* methods, defined below.

Stationary methods using splitting

For a given *splitting* of our matrix A, $A = M - N$, where obviously $N = M - A$, we can write $M\mathbf{x} = N\mathbf{x} + \mathbf{b}$. This leads to a fixed point iteration of the form

$$\mathbf{x}_{k+1} = M^{-1}N\mathbf{x}_k + M^{-1}\mathbf{b} = \mathbf{x}_k + M^{-1}(\mathbf{b} - A\mathbf{x}_k).$$

We can write the iteration as $\mathbf{x}_{k+1} = \mathbf{g}(\mathbf{x}_k)$, where $\mathbf{g}(\mathbf{x}) = \mathbf{x} + M^{-1}(\mathbf{b} - A\mathbf{x})$. It is called stationary because the matrix multiplying $\mathbf{x}$ and the fixed vector appearing in the definition of the function $\mathbf{g}(\mathbf{x})$ are constant and do not depend on the iteration.

In Section 3.3 we use the notation $\mathbf{x}^*$ for the exact solution where $\mathbf{x}^* = \mathbf{g}(\mathbf{x}^*)$, hence $A\mathbf{x}^* = \mathbf{b}$ here. In this chapter, as in the previous three, we drop the superscript * for notational convenience and agree that the notation $\mathbf{x}$, which in the definition of $\mathbf{g}(\mathbf{x})$ above is the independent variable, doubles up as the exact (unknown) solution of the given linear system.

Clearly, the question of how to choose M is central. But before diving into it, let us spend a moment on figuring out how the error behaves in this general setting and how it is related to the residual. We start our iteration with an initial guess $\mathbf{x}_0$. The error is $\mathbf{e}_0 = \mathbf{x} - \mathbf{x}_0$, and the residual equation is $A\mathbf{e}_0 = \mathbf{b} - A\mathbf{x}_0 = \mathbf{r}_0$. Thus, $\mathbf{x} = \mathbf{x}_0 + A^{-1}\mathbf{r}_0$. Obviously, getting $\mathbf{x}$ this way is equivalent to solving $A\mathbf{x} = \mathbf{b}$, so we have nothing to write home about yet. But now we *approximate this error equation*. Instead of solving $A\mathbf{e}_0 = \mathbf{r}_0$, we solve $M\mathbf{p}_0 = \mathbf{r}_0$, where hopefully $\mathbf{p}_0 \approx \mathbf{e}_0$. We can thus write our iteration as calculating for each k the residual $\mathbf{r}_k = \mathbf{b} - A\mathbf{x}_k$ and then computing

$$\mathbf{x}_{k+1} = \mathbf{x}_k + M^{-1}\mathbf{r}_k.$$

This gives us precisely the fixed point iteration written above. It is the form in which we cast all our stationary iterations when implementing them.

Choosing the splitting matrix M

The matrix M should be chosen so that on one hand $\mathbf{x}_{k+1}$ can be easily found and on the other hand M^{-1} is "close to A^{-1}." These two seemingly contradictory requirements allow lots of room to play. Different choices of M lead to a variety of methods, from the simple iterative methods described below to very complicated multiresolution ones. Furthermore, by following this strategy we are giving up computing the exact solution $\mathbf{x}$. The matrix M^{-1} is sometimes referred to as an *approximate inverse* of A. Note that we do not have to form M^{-1} in order to apply the iteration— only the solutions of problems of the form $M\mathbf{p}_k = \mathbf{b} - A\mathbf{x}_k$ for $\mathbf{p}_k$ are required (followed by setting $\mathbf{x}_{k+1} = \mathbf{x}_k + \mathbf{p}_k$). For instance, we may design M such that solving for $\mathbf{p}_k$ can be done efficiently using a direct solution method. This will become obvious below, when we discuss specific choices of M.

Let us denote by D the diagonal matrix consisting of the diagonal elements of A and by E the $n \times n$ lower triangular matrix consisting of corresponding elements of A and zeros elsewhere. In MATLAB you can define these by `D = diag(diag(A)); E = tril(A)`. (Note that in cases where A is not available explicitly, it may not always be possible to actually evaluate these matrices.) Next, we describe methods based on specific choices of M that are related to D and E.

Two basic relaxation methods

The following iterative methods are often referred to as *relaxation methods*. To refresh our memory, we rewrite the system $A\mathbf{x} = \mathbf{b}$ in component form as

$$a_{11}x_1 + a_{12}x_2 + \cdots + a_{1n}x_n = b_1,$$
$$a_{21}x_1 + a_{22}x_2 + \cdots + a_{2n}x_n = b_2,$$
$$\vdots \qquad \qquad = \vdots$$
$$a_{n1}x_1 + a_{n2}x_2 + \cdots + a_{nn}x_n = b_n.$$

Let us denote the kth iterate as above by $\mathbf{x}_k$ but use superscripts if the ith component of the vector $\mathbf{x}_k$ is referred to: $x_i^{(k)}$. Thus, $\mathbf{x}_0 = (x_1^{(0)}, x_2^{(0)}, \ldots, x_n^{(0)})^T$ is a given *initial iterate* (or *guess*).

- The **Jacobi method** (*simultaneous relaxation*). We choose $M = D$, yielding the kth iteration

$$\mathbf{x}_{k+1} = \mathbf{x}_k + D^{-1}\mathbf{r}_k.$$

Thus, assuming we have a routine that returns $A\mathbf{y}$ for any vector $\mathbf{y}$ of suitable size, at each iteration we form the residual $\mathbf{r}_k$ using one call to this routine, scale the residual by D^{-1}, and add to the current iterate.

In component form, which may be used for understanding but should not be used for computing, we have

$$x_i^{(k+1)} = \frac{1}{a_{ii}} \left[b_i - \sum_{\substack{j=1 \\ j \neq i}}^{n} a_{ij} x_j^{(k)} \right], \quad i = 1, \ldots, n.$$

The basic idea is that for each solution component i in turn, $x_i^{(k)}$ is adjusted to zero out the ith residual (or defect) with all other unknowns kept at their value in iteration k. Note that this can be done simultaneously and therefore also *in parallel* for all i.

Example 7.2. In practice we will never solve a 3×3 dense linear system using the Jacobi method (or any other iterative method, for that matter), but let us do it here anyway, just for the purpose of illustrating the mechanism of the method. Consider the linear system

$$7x_1 + 3x_2 + x_3 = 3,$$
$$-3x_1 + 10x_2 + 2x_3 = 4,$$
$$x_1 + 7x_2 - 15x_3 = 2.$$

7.2. Stationary iteration and relaxation methods

Here

$$A = \begin{pmatrix} 7 & 3 & 1 \\ -3 & 10 & 2 \\ 1 & 7 & -15 \end{pmatrix}, \text{ hence } D = \begin{pmatrix} 7 & 0 & 0 \\ 0 & 10 & 0 \\ 0 & 0 & -15 \end{pmatrix}. \text{ Also } \mathbf{b} = \begin{pmatrix} 3 \\ 4 \\ 2 \end{pmatrix}.$$

Note that the matrix A is strictly diagonally dominant.

The kth Jacobi iteration in component form reads

$$x_1^{(k+1)} = \frac{3 - 3x_2^{(k)} - x_3^{(k)}}{7},$$

$$x_2^{(k+1)} = \frac{4 + 3x_1^{(k)} - 2x_3^{(k)}}{10},$$

$$x_3^{(k+1)} = \frac{2 - x_1^{(k)} - 7x_2^{(k)}}{-15}$$

for $k = 0, 1, \ldots$. ∎

Example 7.3. For the model problem of Example 7.1 the Jacobi iteration reads

$$u_{l,m}^{(k+1)} = \frac{1}{4}\left[u_{l+1,m}^{(k)} + u_{l-1,m}^{(k)} + u_{l,m+1}^{(k)} + u_{l,m-1}^{(k)} + b_{l,m}\right],$$

where we *sweep* through the grid, $l, m = 1, \ldots, N$, in some order or in parallel. See Table 7.1 and Figure 7.5. ∎

In MATLAB we can have a function for performing matrix-vector products, say, y = mvp (A, x), which returns for any vector x the vector y = A*x calculated without necessarily having the matrix A stored in full. Assume also that a one-dimensional array d has been created which contains the diagonal elements of A. Then the Jacobi iteration can be written as

```
r = b - mvp(A,x);
x = x + r./d;
```

- The **Gauss–Seidel method**. We choose $M = E$, yielding the kth iteration

$$\mathbf{x}_{k+1} = \mathbf{x}_k + E^{-1}\mathbf{r}_k.$$

Since E is a lower triangular matrix, once the residual $\mathbf{r}_k$ has been formed a forward substitution produces the vector that gets added to the current iterate $\mathbf{x}_k$ to form the next one $\mathbf{x}_{k+1}$. In component form the iteration reads

$$x_i^{(k+1)} = \frac{1}{a_{ii}}\left[b_i - \sum_{j<i} a_{ij} x_j^{(k+1)} - \sum_{j>i} a_{ij} x_j^{(k)}\right], \quad i = 1, \ldots, n.$$

Here, newly obtained values are put to use during the same iteration, or sweep, in the hope that $|x_j^{(k+1)} - x_j| < |x_j^{(k)} - x_j|$. Note that, unlike for the Jacobi case, here the order in which the sweep is made, which corresponds to permuting rows and columns of A before the solution process begins, does matter.

It turns out that under certain conditions, Gauss–Seidel converges whenever the Jacobi iteration converges and (when they do converge) typically twice as fast. But the sweep is harder to execute in parallel.

Example 7.4. Back to Example 7.2, we have

$$E = \begin{pmatrix} 7 & 0 & 0 \\ -3 & 10 & 0 \\ 1 & 7 & -15 \end{pmatrix}.$$

In component form the Gauss–Seidel iteration reads

$$x_1^{(k+1)} = \frac{3 - 3x_2^{(k)} - x_3^{(k)}}{7},$$

$$x_2^{(k+1)} = \frac{4 + 3x_1^{(k+1)} - 2x_3^{(k)}}{10},$$

$$x_3^{(k+1)} = \frac{2 - x_1^{(k+1)} - 7x_2^{(k+1)}}{-15}$$

for $k = 0, 1, \ldots$. ∎

Example 7.5. Referring again to Example 7.1 for a grid-related case study, the resulting formula depends on the way in which the grid function **u** is ordered into a vector. If this is done by row, as in Example 7.1 (or by column; either way this is called a *lexicographic* ordering), then the resulting formula is

$$u_{l,m}^{(k+1)} = \frac{1}{4}\left[u_{l+1,m}^{(k)} + u_{l-1,m}^{(k+1)} + u_{l,m+1}^{(k)} + u_{l,m-1}^{(k+1)} + b_{l,m}\right],$$

where we *sweep* through the grid in some order, $l, m = 1, \ldots, N$.

A better ordering than lexicographic for this example can be *red-black* or *checkerboard* ordering; see Figure 7.4. Here the neighbors of a "red" grid point are all "black," and vice versa. So, a Gauss–Seidel half-sweep over all red points can be followed by a half-sweep over all black points, and each of these can be done in parallel. The general algorithm, *for this particular sparsity pattern*, can then be written as two Jacobi half-sweeps:

for $i = 1 : 2 : n$

$$x_i^{(k+1)} = \frac{1}{a_{ii}}\left[b_i - \sum_{\substack{j=1 \\ j \neq i}}^n a_{ij} x_j^{(k)}\right]$$

end

for $i = 2 : 2 : n$

$$x_i^{(k+1)} = \frac{1}{a_{ii}}\left[b_i - \sum_{\substack{j=1 \\ j \neq i}}^n a_{ij} x_j^{(k+1)}\right]$$

end ∎

7.2. Stationary iteration and relaxation methods

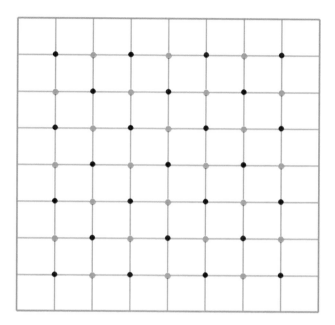

Figure 7.4. *Red-black ordering. First sweep simultaneously over the black points, then over the red.*

Successive over-relaxation

For each of the two basic methods introduced above, we can apply a simple modification leading to an ω-**Jacobi** and an ω-**Gauss–Seidel** scheme. Such a modification is defined, depending on a parameter $\omega > 0$, by replacing at the end of each sweep

$$\mathbf{x}_{k+1} \leftarrow \omega \mathbf{x}_{k+1} + (1-\omega)\mathbf{x}_k.$$

This can of course be worked directly into the definition of $\mathbf{x}_{k+1}$, i.e., the iteration does not have to be executed as a two-stage process.

It turns out that a much faster iteration method can be obtained using ω-Gauss–Seidel if ω is chosen carefully in the range $1 < \omega < 2$. This is the *successive over-relaxation (SOR)* method given by

$$x_i^{(k+1)} = (1-\omega)x_i^{(k)} + \frac{\omega}{a_{ii}}\left[b_i - \sum_{j<i} a_{ij}x_j^{(k+1)} - \sum_{j>i} a_{ij}x_j^{(k)}\right], \quad i=1,\ldots,n.$$

In matrix form, we have for SOR the lower triangular matrix $M = \frac{1-\omega}{\omega}D + E$, and the iteration reads

$$\mathbf{x}_{k+1} = \mathbf{x}_k + \omega((1-\omega)D + \omega E)^{-1}\mathbf{r}_k.$$

No corresponding improvement in speed is available when using an ω-Jacobi method.

Example 7.6. Back, yet again, to the linear system of Example 7.2, the SOR iteration reads

$$x_1^{(k+1)} = (1-\omega)x_1^{(k)} + \omega \frac{3 - 3x_2^{(k)} - x_3^{(k)}}{7},$$

$$x_2^{(k+1)} = (1-\omega)x_2^{(k)} + \omega \frac{4 + 3x_1^{(k+1)} - 2x_3^{(k)}}{10},$$

$$x_3^{(k+1)} = (1-\omega)x_3^{(k)} + \omega \frac{2 - x_1^{(k+1)} - 7x_2^{(k+1)}}{-15}$$

for $k = 0, 1, \ldots$. ∎

Example 7.7. For the model problem of Example 7.1, choosing

$$\omega = \frac{2}{1 + \sin(\pi h)}$$

turns out to yield particularly rapid convergence. (See more on the convergence of SOR in Section 7.3, and in particular in Example 7.8.) Let us demonstrate this: we set $N = 15$, $h = 1/16$, and pick the right-hand-side $b_{l,m} = h^2$, $l, m = 1, \ldots, N$. Note that $n = N^2 = 225$, i.e., our matrix A is 225×225. Then we start iterating from the grid function $\mathbf{u}_0$ given by

$$u_{l,m}^{(0)} = 0, \quad l = 1, \ldots, N, \, m = 1, \ldots, N.$$

Let us denote as before by $\mathbf{u}$ and $\mathbf{b}$ the corresponding grid functions reshaped as vectors. The maximum errors $\|\mathbf{x}_k - \mathbf{x}\|_\infty = \|\mathbf{u}_k - \mathbf{u}\|_\infty$ are recorded in Table 7.1. Clearly the errors for the SOR method with $\omega \approx 1.67$ are significantly smaller than for the other two, with Gauss–Seidel being somewhat better per iteration than Jacobi (ignoring other implementation considerations). ∎

Table 7.1. *Errors for basic relaxation methods applied to the model problem of Example 7.1 with $n = N^2 = 225$.*

Relaxation method	ω	Error after 2 iterations	Error after 20 iterations
Jacobi	1	7.15e-2	5.41e-2
Gauss–Seidel	1	6.95e-2	3.79e-2
SOR	1.67	5.63e-2	7.60e-4

Assessing SOR and other relaxation possibilities

The beauty of SOR is that it is just as simple as Gauss–Seidel, and yet a significant improvement in convergence speed is obtained. Who said there is no free lunch?!

For parameter values in the range $0 < \omega < 1$, the corresponding ω-Jacobi and ω-Gauss–Seidel iterations are called *under-relaxed* or *damped*. For instance, the damped Jacobi method reads

$$\mathbf{x}_{k+1} = \omega(\mathbf{x}_{Jacobi})_{k+1} + (1-\omega)\mathbf{x}_k = \mathbf{x}_k + \omega D^{-1}\mathbf{r}_k.$$

You will see in Exercise 11 that there is no magic whatsoever in this formula, and the fastest convergence is actually obtained for $\omega = 1$, which is nothing but standard Jacobi. But interestingly, damped Jacobi does have merit in a different context, as a *smoother*. For more on this, please see Section 7.6 and especially Figure 7.10.

Unfortunately, in more realistic situations than Example 7.7, the precise optimal value of the SOR parameter ω (i.e., the value yielding the fastest convergence) is not known in advance. A common practice then is to try a few iterations with various values of ω and choose the most promising one for the rest of the iteration process. The results typically offer a significant improvement over Gauss–Seidel, but the spectacular SOR performance depicted in Table 7.1 is rarely realized. We may then turn to other options, such as using the methods described in Section 7.4.

Before moving on let us also mention the *symmetric SOR* (SSOR) method. One iteration of SSOR consists of an SOR sweep in one direction followed by another SOR sweep in the reverse direction. The computational cost of each iteration doubles as a result, compared to SOR, but the iteration count may go down significantly, with the advantage that the iteration is less biased in terms of the sweep direction.

Specific exercises for this section: Exercises 4–5.

7.3 Convergence of stationary methods

For the purpose of analyzing the iterative methods introduced in Section 7.2 it is useful to consider their matrix representation. As before, at each iteration $\mathbf{x}_k$ the residual

$$\mathbf{r}_k = \mathbf{b} - A\mathbf{x}_k$$

is known, whereas the error

$$\mathbf{e}_k = \mathbf{x} - \mathbf{x}_k = A^{-1}\mathbf{r}_k$$

is unknown. We wish to be assured that the iteration converges, i.e., that $\mathbf{e}_k \to 0$ as $k \to \infty$.

The iteration matrix and convergence

As in the analysis for the fixed point iteration of Section 3.3, we write

$$\mathbf{x}_k = M^{-1}\mathbf{b} + (I - M^{-1}A)\mathbf{x}_{k-1},$$
$$\mathbf{x} = M^{-1}\mathbf{b} + (I - M^{-1}A)\mathbf{x}$$

and subtract. Denoting $T = I - M^{-1}A$, this yields

$$\mathbf{e}_k = T\mathbf{e}_{k-1}$$
$$= T(T\mathbf{e}_{k-2}) = \cdots = T^k \mathbf{e}_0.$$

The matrix T is the *iteration matrix*. Since $\mathbf{e}_0$ is a fixed initial error, we have convergence, i.e., $\mathbf{e}_k \to \mathbf{0}$ as $k \to \infty$, if and only if $T^k \to 0$. This is clearly the case if in any induced matrix norm

$$\|T\| < 1,$$

because

$$\|\mathbf{e}_k\| = \|T \cdot T \cdots T\mathbf{e}_0\| \leq \|T\|\|T\| \cdots \|T\|\|\mathbf{e}_0\| = \|T\|^k \|\mathbf{e}_0\|.$$

Compare this to the condition of convergence for the fixed point iteration in Section 3.3.

Proceeding more carefully, it turns out that convergence depends on the spectral radius of T (defined on page 77) rather than its norm. The resulting important theorem is given on the following page. Let us see why it holds in the case where T has n linearly independent eigenvectors $\mathbf{v}_i$. Then

> **Theorem: Stationary Method Convergence.**
> For the linear problem $A\mathbf{x} = \mathbf{b}$, consider the iterative method
> $$\mathbf{x}_{k+1} = \mathbf{x}_k + M^{-1}\mathbf{r}_k, \quad k = 0, 1, \ldots,$$
> and define the **iteration matrix** $T = I - M^{-1}A$.
> Then the method converges if and only if the spectral radius of the iteration matrix satisfies
> $$\rho(T) < 1.$$
> The smaller $\rho(T)$ the faster the convergence.

it is possible to write $\mathbf{e}_0 = \sum_{i=1}^n \gamma_i \mathbf{v}_i$ for some coefficients $\gamma_1, \ldots, \gamma_n$. So, if the corresponding eigenvalues τ_i of T satisfy

$$T\mathbf{v}_i = \tau_i \mathbf{v}_i, \quad |\tau_i| < 1, \quad i = 1, 2, \ldots, n,$$

then

$$T^k \mathbf{e}_0 = \sum_{i=1}^n \gamma_i \tau_i^k \mathbf{v}_i \to \mathbf{0}.$$

On the other hand, if there is an eigenvalue of T, say, τ_1, satisfying $|\tau_1| \geq 1$, then for the unlucky initial guess $\mathbf{x}_0 = \mathbf{x} - \mathbf{v}_1$ we get $\mathbf{e}_k = T^k \mathbf{v}_1 = \tau_1^k \mathbf{v}_1$, so there is no convergence. ♦

Of course, if $\|T\| < 1$, then also $\rho(T) < 1$. But the condition on the spectral radius is more telling than the condition on the norm, being necessary and not only sufficient for convergence of the iteration.

Convergence rate of a stationary method

Since T is independent of the iteration counter k, convergence of this general, stationary fixed point iteration is *linear* and not better (in contrast to the quadratic convergence of Newton's method in Section 3.4, for instance). How many iterations are then needed to reduce the error norm by a fixed factor, say, 10, thus reducing the error by an order of magnitude? In Section 3.3 we defined the **rate of convergence** in order to quantify this. Here, writing

$$0.1\|\mathbf{e}_0\| \approx \|\mathbf{e}_k\| \approx \rho(T)^k \|\mathbf{e}_0\|$$

and taking $\log_{10}$ of these approximate relations yields that $-1 \approx k \log_{10} \rho(T)$, so

$$k \approx -\frac{1}{\log_{10} \rho(T)}$$

iterations are needed. Thus, define the rate of convergence by

$$rate = -\log_{10} \rho(T).$$

Then $k \approx 1/rate$. The smaller the spectral radius, the larger the rate, and the fewer iterations are needed to achieve the same level of error reduction.

7.3. Convergence of stationary methods

Convergence of the relaxation methods

What can generally be said about the convergence of the basic relaxation methods defined in Section 7.2? They certainly *do not* converge for just any nonsingular matrix A. In particular, they are not even well-defined if $a_{ii} = 0$ for some i. But they converge if A is strictly diagonally dominant.

Example 7.8. Consider again the model problem presented in Example 7.1. The matrix A is symmetric positive definite, but it is not strictly diagonally dominant. Let us consider further the Jacobi iteration. Since $M = D$ we get for the model problem the iteration matrix

$$T = I - D^{-1}A = I - \frac{1}{4}A.$$

Therefore, with $N = \sqrt{n}$ and $h(N+1) = 1$, the eigenvalues are

$$\mu_{l,m} = 1 - \frac{1}{4}\lambda_{l,m} = \frac{1}{2}(\cos(l\pi h) + \cos(m\pi h)), \quad 1 \leq l, m \leq N.$$

The spectral radius is

$$\rho(T) = \mu_{1,1} = \cos(\pi h) \leq 1 - \frac{c}{N^2}$$

for some positive constant c.[28] But this convergence is very slow, because $rate = -\log \rho(T) \sim N^{-2}$. Thus, $\mathcal{O}(N^2) = \mathcal{O}(n)$ iterations are required to reduce the iteration error by a constant factor. Since each sweep (iteration) costs $\mathcal{O}(n)$ flops, the total cost is $\mathcal{O}(n^2)$, comparable to that of banded Gaussian elimination.

For the Gauss–Seidel relaxation it can be shown, for this simple problem, that the spectral radius of T is precisely squared that of T of the Jacobi relaxation. So, the rate is twice as large. But still, $\mathcal{O}(n)$ iterations, therefore $\mathcal{O}(n^2)$ flops, are required for a fixed error reduction factor.

On the other hand, the SOR method demonstrated in Example 7.7 can be shown theoretically to require only $\mathcal{O}(N)$ iterations to reduce the iteration error by a constant factor when the optimal parameter ω is used. As it turns out, for matrices like the discretized Laplacian there is a formula for the optimal parameter. It is given by

$$\omega_{opt} = \frac{2}{1 + \sqrt{1 - \rho_J^2}},$$

where ρ_J stands for the spectral radius of the Jacobi iteration matrix. For this problem we have in fact already specified the optimal SOR parameter; see Example 7.7. The spectral radius of the iteration matrix for SOR with ω_{opt} is $\rho = \omega_{opt} - 1$. Taking into account the expense per iteration, the total SOR cost is therefore $\mathcal{O}(n^{3/2})$ flops. ∎

Terminating the iteration

How should we terminate the iteration, for any of these methods, given a convergence error tolerance?

One criterion, often used in practice, is to require that the relative residual be small enough, written as

$$\|\mathbf{r}_k\| \leq \texttt{tol}\|\mathbf{b}\|.$$

---

[28] Note that $\rho(T) = -\mu_{N,N}$ as well.

Recall from Equation (5.1), however, that this implies

$$\frac{\|\mathbf{x}-\mathbf{x}_k\|}{\|\mathbf{x}\|} \leq \kappa(A)\frac{\|\mathbf{r}_k\|}{\|\mathbf{b}\|} \leq \kappa(A)\,\mathtt{tol}.$$

Thus, the tolerance must be taken suitably small if the condition number of A is large.

If we wish to control the algorithm by measuring the difference between two consecutive iterations, $\|\mathbf{x}_k - \mathbf{x}_{k-1}\|$, then care must be taken when the convergence is slow. For any vector norm and its induced matrix norm we can write

$$\|\mathbf{x}-\mathbf{x}_k\| \leq \|T\|\|\mathbf{x}-\mathbf{x}_{k-1}\| = \|T\|\|\mathbf{x}-\mathbf{x}_k + \mathbf{x}_k - \mathbf{x}_{k-1}\|$$
$$\leq \|T\|\left[\|\mathbf{x}-\mathbf{x}_k\| + \|\mathbf{x}_k - \mathbf{x}_{k-1}\|\right].$$

Thus we only have for the error that

$$\|\mathbf{e}_k\| = \|\mathbf{x}-\mathbf{x}_k\| \leq \frac{\|T\|}{1-\|T\|}\|\mathbf{x}_k - \mathbf{x}_{k-1}\|.$$

So, the tolerance on $\|\mathbf{x}_k - \mathbf{x}_{k-1}\|$ must be adjusted taking $\frac{\|T\|}{1-\|T\|}$ into account. For slowly convergent iterations this number can be quite large, as can be seen for Jacobi and Gauss–Seidel (and even SOR) applied to Example 7.1. Thus, care must be exercised when selecting a stopping criterion for the iterative method.

Specific exercises for this section: Exercises 6–12.

7.4 Conjugate gradient method

A weakness of stationary methods is that information gathered throughout the iteration is not fully utilized. Indeed, the approximate inverse M^{-1} is fixed. We might be able to do better, for example, by setting a varying splitting in each iteration, so $M = M_k$, and requiring a certain optimality property to hold. Such methods are explored in this section and the next one.

> **Note:** We assume throughout this section that A is a symmetric positive definite matrix. Methods for general nonsymmetric matrices are presented in the more advanced Section 7.5.

The methods considered in this section can all be written as

$$\mathbf{x}_{k+1} = \mathbf{x}_k + \alpha_k \mathbf{p}_k,$$

where the vector $\mathbf{p}_k$ is the **search direction** and the scalar α_k is the **step size**. Note that this includes basic stationary methods, since such methods with an associated splitting $A = M - N$ can be written as above with $\alpha_k \equiv 1$ for all k and $\mathbf{p}_k = M^{-1}\mathbf{r}_k$.

Our eventual goal is to introduce the celebrated *conjugate gradient (CG) method*. To get there, though, it is natural to start with the general family of gradient descent methods and, in particular, the method of steepest descent.

Gradient descent methods

The simplest nonstationary scheme based on setting a varying search direction and step size is obtained by setting $\mathbf{p}_k = \mathbf{r}_k$, i.e., $M_k^{-1} = \alpha_k I$, with I the identity matrix. The resulting family of methods is called *gradient descent*.

7.4. Conjugate gradient method

A popular choice for the step size α_k for gradient descent methods is available by observing that our problem $A\mathbf{x} = \mathbf{b}$ is equivalent to the problem of finding a vector $\mathbf{x}$ that minimizes

$$\phi(\mathbf{x}) = \frac{1}{2}\mathbf{x}^T A\mathbf{x} - \mathbf{b}^T\mathbf{x}.$$

Let us explain why this is so. In the scalar case of Section 3.4 we obtained critical points by setting the first derivative of the function to be optimized to zero, knowing that such a critical point is a minimizer if the second derivative is positive there. Here, similarly, a critical point is obtained by setting the *gradient*, i.e., the vector of first derivatives of ϕ, to zero. But, reminiscent to the process described in Section 6.1, the gradient of this particular function is $\nabla \phi = A\mathbf{x} - \mathbf{b} = -\mathbf{r}$, so setting this to $\mathbf{0}$ yields our problem

$$A\mathbf{x} = \mathbf{b}.$$

Moreover, the matrix of second derivatives of ϕ is A, and its assumed positive definiteness guarantees that the solution of our problem is the unique minimizer of ϕ.

To summarize our progress so far, we are looking at the gradient descent iteration

$$\mathbf{x}_{k+1} = \mathbf{x}_k + \alpha_k \mathbf{r}_k$$

and wish to determine the scalar α_k such that $\phi(\mathbf{x}_k + \alpha \mathbf{r}_k)$ is minimized over all values of α. But this is straightforward: we want to minimize over α the expression

$$\frac{1}{2}(\mathbf{x}_k + \alpha \mathbf{r}_k)^T A(\mathbf{x}_k + \alpha \mathbf{r}_k) - \mathbf{b}^T(\mathbf{x}_k + \alpha \mathbf{r}_k),$$

and hence all we need to do is set its derivative with respect to α to 0. This translates into

$$\alpha \mathbf{r}_k^T A \mathbf{r}_k + \mathbf{r}_k^T A \mathbf{x}_k - \mathbf{r}_k^T \mathbf{b} = 0.$$

Since $A\mathbf{x}_k - \mathbf{b} = -\mathbf{r}_k$, it readily follows that the desired minimizer $\alpha = \alpha_k$ is given by

$$\alpha_k = \frac{\mathbf{r}_k^T \mathbf{r}_k}{\mathbf{r}_k^T A \mathbf{r}_k} = \frac{\langle \mathbf{r}_k, \mathbf{r}_k \rangle}{\langle \mathbf{r}_k, A\mathbf{r}_k \rangle},$$

where $\langle \cdot, \cdot \rangle$ stands for the standard inner product: for any two vectors $\mathbf{p}$ and $\mathbf{q}$ of the same length, $\langle \mathbf{p}, \mathbf{q} \rangle = \mathbf{p}^T \mathbf{q}$.

This completes the description of an important basic method called **steepest descent** for the iterative solution of symmetric positive definite linear systems. In practice, when we set up the iteration, one point to keep in mind is computational efficiency. Here the basic cost of the iteration is just one matrix-vector multiplication: at the kth iteration, knowing $\mathbf{x}_k$ and $\mathbf{r}_k$, we compute $\mathbf{s}_k = A\mathbf{r}_k$, set $\alpha_k = \langle \mathbf{r}_k, \mathbf{r}_k \rangle / \langle \mathbf{r}_k, \mathbf{s}_k \rangle$, and calculate $\mathbf{x}_{k+1} = \mathbf{x}_k + \alpha_k \mathbf{r}_k$ and $\mathbf{r}_{k+1} = \mathbf{r}_k - \alpha_k \mathbf{s}_k$.

Slowness of steepest descent

Unfortunately, the steepest descent method tends to be slow: the iteration count that it takes to reduce the norm of the error by a constant factor grows *linearly* with the condition number. In our prototype Example 7.1, $\kappa(A)$ is proportional to n, which puts the steepest descent method in the same efficiency class as the basic relaxation methods of Section 7.2. This pronouncement is a bit unfair, because steepest descent is still typically faster in terms of the overall iteration count and can be further improved by preconditioning, discussed later on. Also, there are choices of step sizes which for certain problems have been experimentally observed to converge faster; see Exercise 14 for an example. However, better alternatives than any gradient descent method do exist, so we turn to them without further ado.

The CG method

The immensely popular CG method for solving large linear systems $A\mathbf{x} = \mathbf{b}$, where A is symmetric positive definite, is given on the current page.

Algorithm: Conjugate Gradient.
Given an initial guess $\mathbf{x}_0$ and a tolerance `tol`, set at first $\mathbf{r}_0 = \mathbf{b} - A\mathbf{x}_0$, $\delta_0 = \langle \mathbf{r}_0, \mathbf{r}_0 \rangle$, $b_\delta = \langle \mathbf{b}, \mathbf{b} \rangle$, $k = 0$ and $\mathbf{p}_0 = \mathbf{r}_0$. Then:

$$\text{while } \delta_k > \texttt{tol}^2 \, b_\delta$$
$$\mathbf{s}_k = A\mathbf{p}_k$$
$$\alpha_k = \frac{\delta_k}{\langle \mathbf{p}_k, \mathbf{s}_k \rangle}$$
$$\mathbf{x}_{k+1} = \mathbf{x}_k + \alpha_k \mathbf{p}_k$$
$$\mathbf{r}_{k+1} = \mathbf{r}_k - \alpha_k \mathbf{s}_k$$
$$\delta_{k+1} = \langle \mathbf{r}_{k+1}, \mathbf{r}_{k+1} \rangle$$
$$\mathbf{p}_{k+1} = \mathbf{r}_{k+1} + \frac{\delta_{k+1}}{\delta_k} \mathbf{p}_k$$
$$k = k + 1$$
end

Here the search direction $\mathbf{p}_k$ no longer equals the residual vector $\mathbf{r}_k$, except at the first iteration. Similarly to the simpler looking gradient descent method, only one matrix-vector multiplication involving A is needed per iteration. The matrix A need not be explicitly available: only a procedure to (rapidly) evaluate the matrix-vector product $A\mathbf{v}$ for any given real vector $\mathbf{v}$ of length n is required. The iteration is stopped as soon as the relative residual reaches the value of input tolerance `tol` or goes below it; see the discussion on page 181.

Example 7.9. Much as we would like to return to Example 7.2 yet again, we cannot, because the matrix A there is not symmetric positive definite. So consider another 3×3 problem $A\mathbf{x} = \mathbf{b}$, with

$$A = \begin{pmatrix} 7 & 3 & 1 \\ 3 & 10 & 2 \\ 1 & 2 & 15 \end{pmatrix}, \quad \mathbf{b} = \begin{pmatrix} 28 \\ 31 \\ 22 \end{pmatrix}.$$

The exact solution is $\mathbf{x} = A^{-1}\mathbf{b} = (3, 2, 1)^T$. The point here, as in any 3×3 example, is to demonstrate the algorithm—not to promote it as a suitable method for a trivial problem. We next follow the execution of the CG algorithm. Below we do the calculations to rounding unit but display only a few leading digits.

At first set $\mathbf{x}_0 = (0,0,0)^T$, hence $\mathbf{p}_0 = \mathbf{r}_0 = \mathbf{b}$ and $b_\delta = \delta_0 = \|\mathbf{r}_0\|^2 = 2229$. Next, $\mathbf{s}_0 = A\mathbf{p}_0 = (311, 438, 420)^T$, $\alpha_0 = 2229/(\mathbf{p}_0^T \mathbf{s}_0) = .0707$, $\mathbf{x}_1 = \mathbf{0} + \alpha_0 \mathbf{p}_0 = (1.9797, 2.1918, 1.555)^T$, $\mathbf{r}_1 = \mathbf{b} - \alpha_0 \mathbf{s}_0 = (6.0112, .031847, -7.6955)^T$, $\delta_1 = \|\mathbf{r}_1\|^2 = 95.356$, $\mathbf{p}_1 = (7.209, 1.358, -6.7543)^T$.

On to the second iteration, $\mathbf{s}_1 = (47.783, 21.699, -91.39)^T$, $\alpha_1 = .0962$, $\mathbf{x}_2 = (2.6732, 2.3225, .9057)^T$, $\mathbf{r}_2 = (1.4144, -2.0556, 1.0963)^T$, $\delta_2 = 7.428$, $\mathbf{p}_2 = (1.976, -1.9498, .5702)^T$. Note the decrease in δ from one iteration to the next.

On to the third iteration, $\mathbf{s}_2 = (8.5527, -12.43, 6.6293)^T$, $\alpha_2 = .1654$, $\mathbf{x}_3 = (3, 2, 1)^T$, $\mathbf{r}_3 = \mathbf{0}$. We have found the exact solution in $n = 3$ CG iterations! This is actually not a coincidence; it is rooted in a well-understood convergence property of CG which will be described soon. ■

7.4. Conjugate gradient method

Before developing further insight into the CG algorithm, let us consider a larger numerical example where the potential of this iterative method is demonstrated.

Example 7.10. For our prototype Example 7.1, with $N = 31$ (i.e., a linear system of size $n = N^2 = 961$), we compute the solution using Jacobi, Gauss–Seidel, SOR, and CG. The iteration is stopped when a relative residual norm $\frac{\|\mathbf{r}_k\|}{\|\mathbf{b}\|}$ smaller than 10^{-6} is reached. The convergence history is depicted in two separate plots (see Figure 7.5), where SOR appears in both. This is done because it is difficult to visualize everything on one plot, due to the very different scale of convergence between the slowly convergent Jacobi and Gauss–Seidel schemes, and the rest. Indeed, these two simple relaxation techniques entail an iteration count proportional to N^2 and are clearly not competitive here.

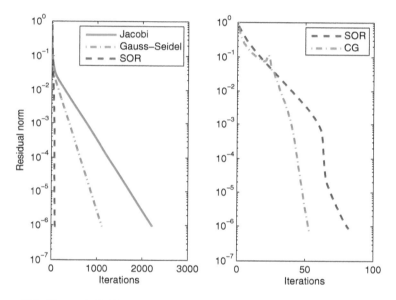

Figure 7.5. *Example* 7.10, *with* $N = 31$: *convergence behavior of various iterative schemes for the discretized Poisson equation.*

The right-hand plot shows that in terms of iteration count CG is better than, yet comparable to, SOR with its parameter set at optimal value. Theory implies that they should both require $\mathcal{O}(N)$ iterations to converge to within a fixed tolerance. But in practice the CG iteration, although more mysterious and cumbersome looking, is cheaper overall than the SOR iteration. CG has another advantage over SOR in that it does not require a special parameter. No wonder it is popular.

We emphasize that only about 50 CG iterations, far fewer than the dimension of the matrix A, are required here to achieve decent accuracy. We also note that in general, the relative residual norm does not necessarily decrease monotonically for CG. ∎

CG magic

The basic idea behind the CG method is to minimize the *energy norm* of the error, defined on the next page, with respect to $B = A$ (or, equivalently, the energy norm of the residual with respect to $B = A^{-1}$) over a *Krylov subspace*, defined on the following page. This subspace grows with each additional iteration k.

> **Energy Norm.**
> Given a symmetric positive definite matrix B, its associated energy norm is
> $$\|\mathbf{x}\|_B = \sqrt{\mathbf{x}^T B \mathbf{x}} \equiv \sqrt{\langle \mathbf{x}, B\mathbf{x} \rangle}.$$

> **Krylov Subspace.**
> For any real, $n \times n$ matrix C and vector $\mathbf{y}$ of the same length, the Krylov Subspace of C with respect to $\mathbf{y}$ is defined by
> $$\mathcal{K}_k(C; \mathbf{y}) = \text{span}\{\mathbf{y}, C\mathbf{y}, C^2\mathbf{y}, \ldots, C^{k-1}\mathbf{y}\}.$$

The CG iteration starts by setting the search direction as the direction of gradient descent, $\mathbf{r}_0$, and its first iteration coincides with steepest descent. We then proceed in the kth iteration to minimize the energy norm of the error, seeking a parameter α such that

$$\|\mathbf{x}_k + \alpha \mathbf{p}_k - \mathbf{x}\|_A$$

is minimized. But the real magic lies in the choice of the search directions. In each iteration the method minimizes the energy norm not only for the current iterate but rather for the space spanned by the current search direction and all the ones preceding it. This is achieved by making the search directions A-conjugate, meaning

$$\langle \mathbf{p}_l, A\mathbf{p}_j \rangle = 0, \quad l \neq j.$$

Imposing this requirement does not seem easy at first, but it turns out that such a search direction can be expressed quite simply as a linear combination of the previous one and the current residual. The search directions satisfy

$$\mathbf{p}_{k+1} = \mathbf{r}_{k+1} + \frac{\langle \mathbf{r}_{k+1}, \mathbf{r}_{k+1} \rangle}{\langle \mathbf{r}_k, \mathbf{r}_k \rangle} \mathbf{p}_k,$$

and the minimizing parameter turns out to be

$$\alpha_k = \frac{\langle \mathbf{r}_k, \mathbf{r}_k \rangle}{\langle \mathbf{p}_k, A\mathbf{p}_k \rangle}.$$

These formulas both appear in the algorithm definition on page 184. It can be verified that the residual vectors are all orthogonal to one another, i.e., $\langle \mathbf{r}_l, \mathbf{r}_j \rangle = 0$, $l \neq j$.

It can be easily shown (simply trace the recursion back from $\mathbf{x}_k$ to $\mathbf{x}_0$) that the CG iterates satisfy

$$\mathbf{x}_k - \mathbf{x}_0 \in \mathcal{K}_k(A; \mathbf{r}_0),$$

where $\mathbf{r}_0 = \mathbf{b} - A\mathbf{x}_0$ is the initial residual.

The minimization property in the energy norm leads to the conclusion that the iteration will terminate at most after n iterations, in the absence of roundoff errors. This explains why in Example 7.9 only $n = 3$ iterations were needed to obtain the exact solution using CG. We note though that in practice the presence of roundoff errors may make things much worse than what this theory predicts.

But waiting n iterations is out of the question, since n is typically large. The practically more important feature of CG is that the first "few" iterations already achieve much progress, as we have seen in Example 7.10 and Figure 7.5. If the eigenvalues of A are located in only a few narrow clusters, then the CG method requires only a few iterations to converge. Convergence is slower, though, if the eigenvalues are widely spread.

A bound on the error can be expressed in terms of the condition number $\kappa(A)$ of the matrix, as the theorem on the current page shows. Hence, for large κ the number of iterations k required to reduce the initial error by a fixed factor is bounded by $\mathcal{O}(\sqrt{\kappa})$, see Exercise 17. For Example 7.1 we obtain that $\mathcal{O}(\sqrt{n}) = \mathcal{O}(N)$ iterations are required, a rather significant improvement over steepest descent. See also Example 9.7 and Figure 9.6. We note that the error bound given in the CG Convergence Theorem is often quite pessimistic; in practice CG often behaves better than predicted by that bound.

> **Theorem: CG Convergence.**
> For the linear problem $A\mathbf{x} = \mathbf{b}$ with A symmetric positive definite, let S_k consist of all n-vectors that can be written as $\mathbf{x}_0 + \mathbf{w}$ with $\mathbf{w} \in \mathcal{K}_k(A; \mathbf{r}_0)$. Then the following hold:
>
> - The kth iterate of the CG method, $\mathbf{x}_k$, minimizes $\|\mathbf{e}_k\|_A$ over S_k.
>
> - In exact arithmetic the solution $\mathbf{x}$ is obtained after at most n iterations. Furthermore, if A has just m distinct eigenvalues then the number of iterations is at most m.
>
> - Only $\mathcal{O}(\sqrt{\kappa(A)})$ iterations are required to reduce the error norm $\|\mathbf{e}_k\|_A$ by a fixed amount, with the error bound
>
> $$\|\mathbf{e}_k\|_A \leq 2 \left(\frac{\sqrt{\kappa(A)} - 1}{\sqrt{\kappa(A)} + 1} \right)^k \|\mathbf{e}_0\|_A.$$

While there are several operations in every iteration, it is really the number of matrix-vector multiplications that is typically the dominant factor in the computation. Thus, it is important to observe, yet again, that *one* such multiplication is required per CG iteration.

Preconditioning

If we take $N = \sqrt{n}$ larger in Example 7.10 above, say, $N = 1023$ (corresponding to a realistic screen resolution for instance), then even the CG method requires over 1000 iterations to achieve a decently small convergence error. This is because the number of iterations depends on $\sqrt{\kappa(A)}$, which is $\mathcal{O}(N)$ in that example. In general, if the condition number of A is very large, then CG loses its effectiveness. In such a case it is worthwhile to attempt to use CG to solve a related problem with a matrix that is better conditioned or whose eigenvalues are more tightly clustered. This gives rise to a **preconditioned conjugate gradient** (PCG) method. Let P^{-1} be a symmetric positive definite approximate inverse of A. The CG iteration is then applied to the system

$$P^{-1} A \mathbf{x} = P^{-1} \mathbf{b}.$$

Note that $P^{-1}A$ is not symmetric in general even if P and A are. Strictly speaking, we should reformulate $P^{-1}A\mathbf{x} = P^{-1}\mathbf{b}$ as

$$(P^{-1/2} A P^{-1/2})(P^{1/2}\mathbf{x}) = P^{-1/2}\mathbf{b}$$

to get a linear system with a symmetric positive definite matrix. But this turns out not to be really necessary. The resulting method is given in an algorithm form on the following page. The stopping criterion is again based on the relative residual, as per the discussion on page 181, except that this time the linear system for which the relative residual is computed is $P^{-1}A\mathbf{x} = P^{-1}\mathbf{b}$ rather than $A\mathbf{x} = \mathbf{b}$.

Algorithm: Preconditioned Conjugate Gradient.
Given an initial guess $\mathbf{x}_0$ and a tolerance `tol`, set at first $\mathbf{r}_0 = \mathbf{b} - A\mathbf{x}_0$, $\mathbf{h}_0 = P^{-1}\mathbf{r}_0$, $\delta_0 = \mathbf{r}_0^T \mathbf{h}_0$, $b_\delta = \mathbf{b}^T P^{-1} \mathbf{b}$, $k = 0$ and $\mathbf{p}_0 = \mathbf{h}_0$. Then:

$$\text{while } \delta_k > \texttt{tol}^2 \, b_\delta$$
$$\mathbf{s}_k = A\mathbf{p}_k$$
$$\alpha_k = \frac{\delta_k}{\mathbf{p}_k^T \mathbf{s}_k}$$
$$\mathbf{x}_{k+1} = \mathbf{x}_k + \alpha_k \mathbf{p}_k$$
$$\mathbf{r}_{k+1} = \mathbf{r}_k - \alpha_k \mathbf{s}_k$$
$$\mathbf{h}_{k+1} = P^{-1}\mathbf{r}_{k+1}$$
$$\delta_{k+1} = \mathbf{r}_{k+1}^T \mathbf{h}_{k+1}$$
$$\mathbf{p}_{k+1} = \mathbf{h}_{k+1} + \frac{\delta_{k+1}}{\delta_k}\mathbf{p}_k$$
$$k = k + 1$$
end

To produce an effective method the preconditioner matrix P must be easily invertible. At the same breath it is desirable to have at least one of the following properties hold: $\kappa(P^{-1}A) \ll \kappa(A)$ and/or the eigenvalues of $P^{-1}A$ are much better clustered compared to those of A.

Simple preconditioners can be constructed from the relaxation methods we have seen in Section 7.2. Thus, we can set P as defined in Section 7.2 for different methods. Jacobi's $P = D$, and the preconditioner obtained by applying SSOR relaxation (page 179), are two fairly popular choices among those. But there are much more sophisticated preconditioners, and a member of the particularly popular family of incomplete factorizations is discussed next.

Incomplete Cholesky factorization

Let A be a large, sparse, symmetric positive definite matrix. Among modern classes of preconditioners for such matrices, a very popular family is that of *incomplete Cholesky* (IC) factorizations. The straightforward yet powerful idea behind them conjures memories of Sections 5.5 and 5.7, and we recall in particular that a Cholesky factorization produces factors that are much denser than A in general. The simplest IC factorization, denoted IC(0), constructs a Cholesky decomposition that follows precisely the same steps as the usual decomposition algorithms from Sections 5.2 or 5.5, except a nonzero entry of a factor is generated *only if* the matching entry of A is nonzero! Figure 7.6 illustrates this idea for our good old matrix from Example 7.1. The top left subfigure shows the sparsity pattern of the IC factor F. The top right is $P = FF^T$, effectively the preconditioner. The bottom left subfigure shows the sparsity pattern of the full Cholesky factor G such that $A = GG^T$, which we assume is too expensive to produce and work with in the present context. It is evident that the product FF^T does not restore A precisely, as there are now two additional nonzero diagonals, but the gain in terms of sparsity is significant: the full Cholesky decomposition of A results in much denser factors.

7.4. Conjugate gradient method

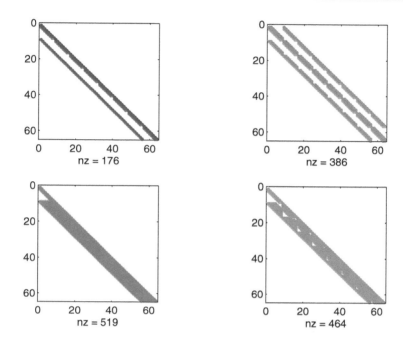

Figure 7.6. *Sparsity patterns for the matrix of Example 7.1 with $N = 8$: top left, the IC factor F with no fill-in (IC(0)); top right, the product FF^T; bottom left, the full Cholesky factor G; bottom right, the IC factor with drop tolerance .001. See Figure 7.3 for the sparsity pattern of the original matrix.*

As it turns out, using IC(0) is not always good enough. It may make more sense to construct sparse Cholesky factors in a *dynamic* fashion. That is, drop a constructed nonzero entry based not on where it is in the matrix but rather on whether the generated value is smaller than a threshold value, aptly called *drop tolerance*. (We are actually omitting some subtle details here: this value is also affected by the scaling of the matrix row in question. But let us not get into technicalities of this sort.) Of course, the smaller the drop tolerance, the closer we get to the full Cholesky factor. In addition to the drop tolerance, there is typically another parameter that limits the number of nonzeros per row of the incomplete factors. For a small drop tolerance we expect fewer PCG iterations at a higher cost per iteration.

The bottom right subfigure of Figure 7.6 shows the sparsity pattern that results using IC with a drop tolerance of .001. These figures are a bit misleading because the depicted system is so small: for N large, the number of nonzero entries in IC with the same tolerance is much smaller than in the full factor.

In MATLAB the command `pcg` applies the PCG method, underlining not the complexity of the algorithm but its importance. For example, the application of IC(0) as a preconditioner, with `tol` as the relative residual tolerance for convergence and `maxit` as the maximal allowed number of iterations, can be done by

```
R=cholinc(A,'0');
x=pcg(A,b,tol,maxit,R',R);
```

Example 7.11. Returning to Example 7.10 we now consider comparable experiments with IC preconditioners. Figure 7.7 displays plots of the relative residual norm as a function of the number of iterations using IC(0) and IC with drop tolerance 0.01. Compared to the CG performance, the iteration count using the IC(0) preconditioner is reduced by a factor of more than two, and using IC with drop tolerance 0.01 further reduces the iteration count by another factor of roughly two.

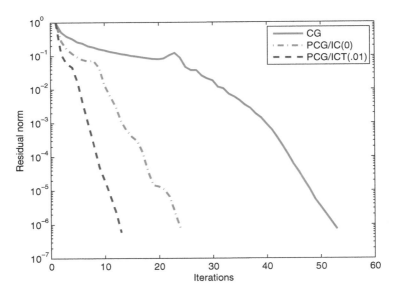

Figure 7.7. *Iteration progress for CG, PCG with the IC(0) preconditioner and PCG with the IC preconditioner using drop tolerance* `tol`= 0.01.

Note that a CG iteration is cheaper than a PCG iteration with IC(0), which is in turn cheaper than a PCG iteration using IC with a drop tolerance. But for tough problems it is worth paying the extra cost per iteration. ∎

The necessity of preconditioners

The art of preconditioning has been the topic of much recent research. If done properly it may yield great efficiency coupled with flexibility. The latter is due to the fact that unlike the given matrix A, the preconditioner P is ours to choose. However, it is important to also keep an eye on the cost of iteration and to realize that not all preconditioners are effective. While in the plain vanilla nonpreconditioned CG iterations we had to mainly concern ourselves with the cost of vector multiplication by A, here we have to perform *in addition* a vector multiplication by P^{-1}. Of course, we would not form P^{-1} explicitly, and this latter operation amounts to solving a linear system with P.

Despite the higher computational cost of preconditioned iterations compared to an iterative scheme that involves no preconditioning, it is widely agreed today that preconditioning is an inseparable and often crucial part of iterative solvers for many problems. It seems that no general-purpose preconditioner is uniformly effective, and preconditioners that are tailored to a specific class of problems are often the way to go. The construction of very good preconditioners for problems that arise from a discretization of a differential equation often involves aspects of the underlying continuous (differential) problem. There are examples of impressive preconditioning performance in the next two sections.

Specific exercises for this section: Exercises 13–18.

7.5 *Krylov subspace methods

The CG method beefed up by preconditioning as described in Section 7.4 offers significant improvement over the stationary methods considered in the previous sections of this chapter. However, strictly speaking it is applicable only to symmetric positive definite systems. In this section we extend the scope and discuss efficient iterative methods for the general case, where A is a real nonsingular matrix. We introduce the family of Krylov subspace methods, of which CG is a special member.

> **Note:** Let us say it up front: of the practical methods described in this section for general square matrices, none comes close to having the combination of efficiency, robustness, theoretical backing, and elegance that CG has. Indeed, positive definiteness is a big asset, and when it is lost the going gets tough. Nevertheless, these methods can be very efficient in practice, and developing some expertise about them beyond the scope of this section is recommended.

A general recipe with limited applicability

Given a general linear system $A\mathbf{x} = \mathbf{b}$ with A nonsingular, it is actually quite easy to turn it into a symmetric positive definite one: simply multiply both sides by A^T. Thus we consider the linear system

$$B\mathbf{x} = \mathbf{y}, \quad \text{where } B = A^T A, \ \mathbf{y} = A^T \mathbf{b},$$

and this can be solved using the CG method as in Section 7.4 because B is symmetric positive definite. Recalling Section 6.1, the system of equations $B\mathbf{x} = \mathbf{y}$ describes the *normal equations* for the least squares problem

$$\min_{\mathbf{x}} \|\mathbf{b} - A\mathbf{x}\|.$$

Of course, in Section 6.1 we were concerned with overdetermined systems, whose matrix A was of size $m \times n$ with $m > n$. Here we are interested in square matrices ($m = n$), but the principle stays the same. The computation can easily be arranged so that at each iteration only one matrix-vector product involving A and one involving A^T are required. Methods based on applying CG to $A^T A\mathbf{x} = A^T \mathbf{b}$ are known by a few names, for example, *conjugate gradient for least squares* (CGLS).

The technique can certainly be applied to quickly solve toy problems such as Example 7.2, and more generally it is practical when $\kappa(A)$ is not overly large. However, the fact already lamented in Chapter 6, namely, that the condition number of $A^T A$ is squared that of A, has potentially severe repercussions here, because the number of iterations required for a fixed convergence error typically grows linearly with $\sqrt{\kappa(B)} = \kappa(A)$, a rate which we have already found in Section 7.4 to be often unacceptable. See also Exercise 23. The quest for extensions of the CG method therefore has to go deeper than methods based on the normal equations. We next consider Krylov space methods based directly on the given general matrix A, not $B = A^T A$.

Building blocks for Krylov subspace methods

Let us assume no preconditioning for the moment. All the methods described in Section 7.4 can be characterized by

$$\mathbf{r}_{k+1} = \mathbf{r}_k - \alpha_k A\mathbf{p}_k,$$

where $\mathbf{r}_k = \mathbf{b} - A\mathbf{x}_k$ is the residual and $\mathbf{p}_k$ is the search direction in the kth iteration. This includes CG, whereby $\mathbf{p}_k$ depends on the previous search direction $\mathbf{p}_{k-1}$ and $\mathbf{r}_k$. Continuing to write $\mathbf{r}_k$ in terms of $\mathbf{r}_{k-1}$ and $\mathbf{p}_{k-1}$, and so on, we quickly arrive at the conclusion that there are coefficients $c_1,\ldots,c_k$ such that

$$\mathbf{r}_k = \mathbf{r}_0 + \sum_{j=1}^{k} c_j A^j \mathbf{r}_0.$$

We can write this as

$$\mathbf{r}_k = p_k(A)\mathbf{r}_0,$$

where $p_k(A)$ is a polynomial of degree k in A that satisfies $p_k(0) = 1$.
Moreover, $\sum_{j=1}^{k} c_j A^j \mathbf{r}_0 = \mathbf{r}_k - \mathbf{r}_0 = -A(\mathbf{x}_k - \mathbf{x}_0)$. We get

$$\mathbf{x}_k = \mathbf{x}_0 - \sum_{j=0}^{k-1} c_{j+1} A^j \mathbf{r}_0,$$

or $\mathbf{x}_k - \mathbf{x}_0 \in \mathcal{K}_k(A;\mathbf{r}_0)$, the Krylov subspace whose definition was given on page 186.

This is the essence that we retain when extending to the case where A is no longer symmetric positive definite. We thus compute approximate solutions within the shifted Krylov subspace $\mathbf{x}_0 + \mathcal{K}_k(A;\mathbf{r}_0)$, where

$$\mathcal{K}_k(A;\mathbf{r}_0) = \text{span}\{\mathbf{r}_0, A\mathbf{r}_0, A^2\mathbf{r}_0, \ldots, A^{k-1}\mathbf{r}_0\}.$$

The family of *Krylov subspace solvers* is based on finding a solution within the subspace, which satisfies a certain optimality criterion. These solvers are based on three important building blocks:

1. constructing an orthogonal basis for the Krylov subspace;

2. defining an optimality property;

3. using an effective preconditioner.

Computing an orthogonal basis for the Krylov subspace

Unfortunately, the most obvious choice of vectors to span the Krylov subspace $\mathcal{K}_k(A;\mathbf{r}_0)$, namely, $\{\mathbf{r}_0, A\mathbf{r}_0, A^2\mathbf{r}_0, \ldots, A^{k-1}\mathbf{r}_0\}$, is poorly conditioned for exactly the same reason the *power method*, described in Section 8.1, often works well: as j grows larger, vectors of the form $A^j\mathbf{r}_0$ approach the dominant eigenvector of A and therefore also one another.

Instead, let us construct an *orthonormal* basis for the Krylov subspace.

The Arnoldi process

The first vector in the Krylov subspace is $\mathbf{r}_0$, and hence to get the first basis vector all we need is to take this vector and normalize it, yielding

$$\mathbf{q}_1 = \mathbf{r}_0 / \|\mathbf{r}_0\|.$$

Next we want to find a vector $\mathbf{q}_2$ orthogonal to $\mathbf{q}_1$ such that the pair $\{\mathbf{q}_1, \mathbf{q}_2\}$ spans the same subspace that $\{\mathbf{r}_0, A\mathbf{r}_0\}$ spans. Since $A\mathbf{q}_1$ is in the same direction as $A\mathbf{r}_0$, we require

$$A\mathbf{q}_1 = h_{11}\mathbf{q}_1 + h_{21}\mathbf{q}_2.$$

For convenience, the coefficients h_{ij} are endowed with a double subscript: the first subscript is a running index signifying the coefficient of a basis vector, and the second signifies the current step. Multiplying both sides of the above equation by $\mathbf{q}_1^T$ and observing that by orthogonality $\langle \mathbf{q}_1, \mathbf{q}_2 \rangle = 0$,

7.5. *Krylov subspace methods

we obtain $h_{11} = \langle \mathbf{q}_1, A\mathbf{q}_1 \rangle$. Next, taking norms and using the fact that $\mathbf{q}_2$ is a unit vector gives $h_{21} = \|A\mathbf{q}_1 - h_{11}\mathbf{q}_1\|$, and then

$$\mathbf{q}_2 = \frac{A\mathbf{q}_1 - h_{11}\mathbf{q}_1}{h_{21}}.$$

Generalizing the above derivation, at the jth step we have

$$A\mathbf{q}_j = h_{1j}\mathbf{q}_1 + h_{2j}\mathbf{q}_2 + \cdots + h_{j+1,j}\mathbf{q}_{j+1} = \sum_{i=1}^{j+1} h_{ij}\mathbf{q}_i.$$

By the orthonormality achieved thus far, for $1 \leq m \leq j$, $\langle \mathbf{q}_m, \mathbf{q}_j \rangle = 0$ if $m \neq j$ and $\langle \mathbf{q}_m, \mathbf{q}_j \rangle = 1$ if $m = j$. This gives

$$\langle \mathbf{q}_m, A\mathbf{q}_j \rangle = \sum_{i=1}^{j+1} h_{ij} \langle \mathbf{q}_m, \mathbf{q}_i \rangle = h_{mj}.$$

For the current unknowns $h_{j+1,j}$ and $\mathbf{q}_{j+1}$, we get $h_{j+1,j}\mathbf{q}_{j+1} = A\mathbf{q}_j - \sum_{i=1}^{j} h_{ij}\mathbf{q}_i$. Taking norms and using the fact that $\|\mathbf{q}_{j+1}\| = 1$ yields

$$h_{j+1,j} = \left\| A\mathbf{q}_j - \sum_{i=1}^{j} h_{ij}\mathbf{q}_i \right\|,$$

and then

$$\mathbf{q}_{j+1} = \frac{A\mathbf{q}_j - \sum_{i=1}^{j} h_{ij}\mathbf{q}_i}{h_{j+1,j}}.$$

This incremental process can be concisely encapsulated in terms of the following matrix decomposition: for any given $k \geq 1$ there is a relation of the form

$$A[\mathbf{q}_1, \mathbf{q}_2, \ldots, \mathbf{q}_k] = [\mathbf{q}_1, \mathbf{q}_2, \ldots, \mathbf{q}_k, \mathbf{q}_{k+1}] \cdot \begin{pmatrix} h_{11} & \cdots & \cdots & \cdots & h_{1k} \\ h_{21} & \ddots & \ddots & \ddots & \vdots \\ 0 & \ddots & \ddots & \ddots & \vdots \\ \vdots & \ddots & \ddots & \ddots & \vdots \\ 0 & \cdots & 0 & h_{k,k-1} & h_{kk} \\ 0 & \cdots & 0 & 0 & h_{k+1,k} \end{pmatrix},$$

which can be written as

$$AQ_k = Q_{k+1} H_{k+1,k},$$

where Q_{k+1} is the matrix containing the $k+1$ vectors of the orthogonal basis for the Krylov subspace, Q_k is the same matrix but with the first k columns only, and $H_{k+1,k}$ is a matrix of size $(k+1) \times k$. It is easy to show that

$$Q_k^T A Q_k = H_{k,k},$$

where $H_{k,k}$ is the $k \times k$ square matrix containing the first k rows of $H_{k+1,k}$. This matrix is in upper Hessenberg form (see page 139). All these relations will see use soon when we derive solution methods.

The Arnoldi algorithm is given on the next page. It is not necessary to provide A explicitly; only a routine that returns $\mathbf{w} = A\mathbf{v}$, given a vector $\mathbf{v}$ as input, is required.

> **Algorithm: Arnoldi Process.**
> Input: initial unit vector q_1 (possibly $r_0/\|r_0\|$), matrix A, and number of steps k.
>
> $$\begin{aligned}
> &\text{for } j = 1 \text{ to } k \\
> &\quad z = A q_j \\
> &\quad \text{for } i = 1 \text{ to } j \\
> &\quad\quad h_{i,j} = \langle q_i, z \rangle \\
> &\quad\quad z \leftarrow z - h_{i,j} q_i \\
> &\quad \text{end} \\
> &\quad h_{j+1,j} = \|z\| \\
> &\quad \text{if } h_{j+1,j} = 0, \text{quit} \\
> &\quad q_{j+1} = z / h_{j+1,j} \\
> &\text{end}
> \end{aligned}$$

Example 7.12. Consider the toy system introduced in Example 7.2, given by

$$7x_1 + 3x_2 + x_3 = 3,$$
$$-3x_1 + 10x_2 + 2x_3 = 4,$$
$$x_1 + 7x_2 - 15x_3 = 2.$$

We use it to demonstrate how the Arnoldi algorithm is carried out. Suppose the initial guess is $x_0 = 0$. Then, $r_0 = b - Ax_0 = b$, and we choose $q_1 = r_0/\|r_0\| = b/\|b\| = \frac{1}{\sqrt{29}}(3,4,2)^T = (0.55709, 0.74278, 0.37139)^T$.

Next we compute q_2. The procedure outlined above yields $h_{11} = q_1^T A q_1 = 8.5172$. We can now compute $h_{21} = \|Aq_1 - h_{11}q_1\| = 3.4603$ and $q_2 = (Aq_1 - h_{11}q_1)/h_{21} = (0.50703, 0.049963, -0.86048)^T$. Proceeding in the same fashion to compute the second column of the upper Hessenberg matrix leads to $h_{12} = q_1^T A q_2 = 4.6561$, $h_{22} = q_2^T A q_2 = -10.541$, and $h_{32} = \|Aq_2 - h_{12}q_1 - h_{22}q_2\| = 8.4986$.

The next basis vector turns out to be $q_3 = (0.6577, -0.66767, 0.34878)^T$. Thus, we have constructed a decomposition of the form

$$A Q_2 = Q_3 H_{3,2},$$

where

$$Q_3 = \begin{pmatrix} 0.55709 & 0.50703 & 0.6577 \\ 0.74278 & 0.049963 & -0.66767 \\ 0.37139 & -0.86048 & 0.34878 \end{pmatrix},$$

Q_2 is the 3×2 matrix comprised of the first two columns of Q_3, and

$$H_{3,2} = \begin{pmatrix} 8.5172 & 4.6561 \\ 3.4603 & -10.541 \\ 0 & 8.4986 \end{pmatrix}.$$

It is easy to confirm that $Q_2^T A Q_2 = H_{2,2}$, where $H_{2,2}$ is the 2×2 matrix composed of the upper two rows of $H_{3,2}$, and that Q_3 is an orthogonal matrix. ∎

7.5. *Krylov subspace methods

Of course, for large problems we will never proceed with Arnoldi for a number of steps equal or nearly equal to the dimensions of the matrix. Therefore a square Q_k like Q_3 in Example 7.12 never happens in practice. Typically the $n \times k$ matrix Q_k (or Q_{k+1}, which is "thicker" by one column) is a tall and skinny rectangular matrix, since $k \ll n$.

The Lanczos method

When A is symmetric, $A = A^T$ implies that $H_{k,k}$ must be symmetric.[29] Therefore, it must be tridiagonal; let us denote it by $T_{k,k}$. In this case the Arnoldi process reduces to the well-known *Lanczos method*. The fact that we have a tridiagonal matrix simplifies the calculations and yields significant computational savings. Specifically, when dealing with the jth column we have to consider three terms only. Let us write

$$A[\mathbf{q}_1, \mathbf{q}_2, \ldots, \mathbf{q}_k] = [\mathbf{q}_1, \mathbf{q}_2, \ldots, \mathbf{q}_k, \mathbf{q}_{k+1}] \cdot \begin{pmatrix} \gamma_1 & \beta_1 & 0 & \cdots & 0 \\ \beta_1 & \gamma_2 & \beta_2 & \ddots & \vdots \\ 0 & \ddots & \ddots & \ddots & 0 \\ \vdots & \ddots & \ddots & \ddots & \beta_{k-1} \\ 0 & \cdots & 0 & \beta_{k-1} & \gamma_k \\ 0 & \cdots & 0 & 0 & \beta_k \end{pmatrix}.$$

From this it follows, for a given j, that

$$A\mathbf{q}_j = \beta_{j-1}\mathbf{q}_{j-1} + \gamma_j \mathbf{q}_j + \beta_j \mathbf{q}_{j+1}.$$

Just like for the Arnoldi algorithm, we now use orthogonality and obtain

$$\gamma_j = \langle \mathbf{q}_j, A\mathbf{q}_j \rangle.$$

When dealing with the jth column, β_{j-1} is already known, so for computing β_j we can use

$$\beta_j \mathbf{q}_{j+1} = A\mathbf{q}_j - \beta_{j-1}\mathbf{q}_{j-1} - \gamma_j \mathbf{q}_j,$$

with the value of γ_j just computed. Taking norms yields

$$\beta_j = \|A\mathbf{q}_j - \beta_{j-1}\mathbf{q}_{j-1} - \gamma_j \mathbf{q}_j\|.$$

Once β_j has been computed, $\mathbf{q}_{j+1}$ is readily available.

This outlines the *Lanczos algorithm* for computing the orthogonal basis for $\mathcal{K}_k(A; \mathbf{r}_0)$ in cases when A is symmetric. The matrix $T_{k+1,k}$ associated with the construction

$$AQ_k = Q_{k+1}T_{k+1,k}$$

---

[29] We are not assuming here that A is also positive definite, so we are generally not in CG territory. There are of course many real matrices that are symmetric and not definite, for instance, $\begin{pmatrix} 1 & \mu \\ \mu & 0 \end{pmatrix}$ for any real scalar μ.

is indeed tridiagonal. A MATLAB script for the Lanczos algorithm follows. We offer here the modified Gram–Schmidt version for Lanczos, which is preferred over classical Gram–Schmidt, see Section 6.3.

```
function [Q,T] = lanczos(A,Q,k)

% preallocate for speed
alpha=zeros(k,1);
beta=zeros(k,1);

Q(:,1) = Q(:,1)/norm(Q(:,1));
beta(1,1)=0;

for j=1:k
   w=A*Q(:,j);
   if j>1
      w=A*Q(:,j)-beta(j,1)*Q(:,j-1);
   end
   alpha(j,1)=Q(:,j)'*w;
   w=w-alpha(j,1)*Q(:,j);
   beta(j+1,1)=norm(w);

   if abs(beta(j+1,1))<1e-10
      disp('Zero beta --- returning.');
      T=spdiags([beta(2:j+1) alpha(1:j) beta(1:j)],-1:1,j+1,j);
      return
   end
   Q(:,j+1)=w/beta(j+1,1);
end
T=spdiags([beta(2:end) alpha beta(1:end-1)],-1:1,k+1,k);
```

Arnoldi, Lanczos, and eigenvalue computations

A remarkable property of the Arnoldi and Lanczos methods is that $H_{k,k}$ or $T_{k,k}$ typically does very well in approximating the eigenvalues of the original matrix. The eigenvectors can also be approximated, using the orthogonal basis for the Krylov subspace, Q_k. Of course, given that $k \ll n$, we cannot expect *all* the eigenvalues of A to be approximated, since $H_{k,k}$ or $T_{k,k}$ is much smaller. *Extremal* eigenvalues of the original matrix are well approximated, and typically it is the largest ones that are approximated in the best fashion. This fact comes in handy in advanced iterative methods for computing eigenvalues of large and sparse matrices. The eigenvalues of $H_{k,k}$ or $T_{k,k}$ are known as **Ritz values**.

Optimality criteria: CG and FOM, MINRES and GMRES

Now that the orthogonal basis construction is taken care of, we turn to discuss the second building block mentioned on page 192: optimality criteria. There are various alternatives for deciding on the type of solution we are looking for within the Krylov subspace at iteration k. Two particularly popular criteria are the following:

- force the residual $\mathbf{r}_k$ to be orthogonal to the Krylov subspace $\mathcal{K}_k(A; \mathbf{r}_0)$;

- seek the residual with minimum ℓ_2-norm within the Krylov subspace.

7.5. *Krylov subspace methods

Galerkin orthogonalization

The first alternative above is often referred to as a Galerkin orthogonalization approach, and it leads to the *full orthogonalization method* (FOM) for nonsymmetric matrices and to no other than CG (after doing some further fancy manipulations) for symmetric positive definite matrices.

Since the columns of Q_k are the orthonormal basis vectors of the Krylov subspace, this optimality criterion amounts to requiring

$$Q_k^T(\mathbf{b} - A\mathbf{x}_k) = 0.$$

But since $\mathbf{x}_k - \mathbf{x}_0 \in \mathcal{K}_k(A; \mathbf{r}_0)$ we can write $\mathbf{x}_k = \mathbf{x}_0 + Q_k \mathbf{y}$, and the criterion simplifies to

$$Q_k^T A Q_k \mathbf{y} = Q_k^T \mathbf{r}_0.$$

This is good news, because $Q_k^T A Q_k$ is nothing but $H_{k,k}$ if A is nonsymmetric or $T_{k,k}$ if A is symmetric! Furthermore, the right-hand side of this system can also be simplified. Since $\mathbf{q}_1$ is just the normalized $\mathbf{r}_0$ and all other columns in Q_k are orthogonal to it we have

$$Q_k^T \mathbf{r}_0 = \|\mathbf{r}_0\| \mathbf{e}_1,$$

where $\mathbf{e}_1 = (1, 0, \ldots, 0)^T$.

From this it follows that a method based on orthogonalization amounts to solving

$$H_{k,k} \mathbf{y} = \|\mathbf{r}_0\| \mathbf{e}_1$$

(with $T_{k,k}$ replacing $H_{k,k}$ if A is symmetric) for $\mathbf{y}$ and then setting $\mathbf{x}_k = \mathbf{x}_0 + Q_k \mathbf{y}$.

When A is symmetric positive definite, some elegant manipulations can be performed in the process of inverting $T_{k,k}$. The $k \times k$ linear system can be solved by exploiting certain algebraic connections to the linear system in the previous, $(k-1)$st step, and the resulting algorithm can be written using short recurrences. Indeed, the CG method can be derived in many ways, not only from an optimization point of view that utilizes downhill search directions, but also from an algebraic point of view based on the decomposition of a symmetric positive definite tridiagonal matrix.

GMRES for general matrices

The second optimality criterion that we have mentioned, namely, that of minimizing $\|\mathbf{r}_k\|$ within $\mathcal{K}_k(A; \mathbf{r}_0)$ in the ℓ_2-norm, leads to the popular methods *generalized minimum residual* (GMRES) for nonsymmetric matrices and *minimum residual* (MINRES) for symmetric but not necessarily positive definite matrices.

By the Arnoldi algorithm, we have that $AQ_k = Q_{k+1} H_{k+1,k}$. We can write

$$\mathbf{x}_k = \mathbf{x}_0 + Q_k \mathbf{z}$$

and aim to find $\mathbf{x}_k$ which minimizes $\|\mathbf{b} - A\mathbf{x}_k\|$. Recall that using CG we are minimizing $\|\mathbf{r}_k\|_{A^{-1}}$ instead, but that would make sense only if A is symmetric positive definite as otherwise we have no energy norm. Hence we look at minimizing $\|\mathbf{r}_k\|$ here.

We have

$$\|\mathbf{b} - A\mathbf{x}_k\| = \|\mathbf{b} - A\mathbf{x}_0 - AQ_k\mathbf{z}\| = \|\mathbf{r}_0 - Q_{k+1} H_{k+1,k} \mathbf{z}\| = \|Q_{k+1}^T \mathbf{r}_0 - H_{k+1,k} \mathbf{z}\|.$$

Now the expression on the right involves a small $(k+1) \times k$ matrix! Moreover, $Q_{k+1}^T \mathbf{r}_0 = \|\mathbf{r}_0\| \mathbf{e}_1$. So, the minimization problem $\min_{\mathbf{x}_k} \|\mathbf{b} - A\mathbf{x}_k\|$ is equivalent to $\min_{\mathbf{z}} \|\rho \mathbf{e}_1 - H_{k+1,k} \mathbf{z}\|$, where $\rho = \|\mathbf{r}_0\|$.

Recall from Section 6.2 that a linear least squares problem can be solved using the economy size QR factorization. Writing
$$H_{k+1,k} = U_{k+1,k} R_{k,k},$$
where $R_{k,k}$ is an upper triangular matrix and $U_{k+1,k}$ consists of k orthonormal vectors, yields
$$\|\rho \mathbf{e}_1 - H_{k+1,k}\mathbf{z}\| = \|\rho \mathbf{e}_1 - U_{k+1,k} R_{k,k}\mathbf{z}\| = \|\rho U_{k+1,k}^T \mathbf{e}_1 - R_{k,k}\mathbf{z}\|.$$
We therefore have
$$\mathbf{z} = R_{k,k}^{-1} U_{k+1,k}^T \|\mathbf{r}_0\| \mathbf{e}_1.$$
Once $\mathbf{z}$ is recovered, the kth iterate is simply obtained by setting $\mathbf{x}_k = \mathbf{x}_0 + Q_k \mathbf{z}$.

The procedure outlined above is known as the GMRES method. There are a few more details in its specification that we have omitted, since they are quite technical, but the essence is here. (If you will not rest until we hint at what those omitted details are, let us just say that they are related to an efficient utilization of Givens rotations or other means for solving the least squares problem involving the upper Hessenberg matrix.) The GMRES algorithm is more involved than CG, and while it is not really very long, the various details make it less concise than the algorithms we have seen so far in this chapter. Its main steps are recaptured on the current page.

GMRES Steps.
The main components of a single iteration of GMRES are

1. perform a step of the Arnoldi process;

2. update the QR factorization of the updated upper Hessenberg matrix;

3. solve the resulting least squares problem.

MINRES for symmetric matrices

For symmetric matrices A, Arnoldi is replaced by Lanczos and the upper Hessenberg matrix is really just tridiagonal. The same mechanism can be applied but the resulting iterative method is simpler and, like CG, requires short three-term recurrence relations. It is called MINRES.

In each iteration the residual for GMRES or MINRES is minimized over the current Krylov subspace, and since a $(k+1)$-dimensional Krylov subspace contains the preceding k-dimensional subspace, we are assured (at least in the absence of roundoff errors) that the residual decreases monotonically. Moreover, like CG these two schemes converge to the exact solution in n iterations. In the special case of nonsingular symmetric matrices that are indefinite (i.e., their eigenvalues are neither all positive nor all negative), MINRES is a very popular method.

CG and MINRES both work on symmetric matrices, but MINRES minimizes $\|\mathbf{r}_k\|$, whereas CG minimizes the energy norm $\|\mathbf{r}_k\|_{A^{-1}}$ on the same subspace. MINRES is applicable to a wider class of problems. However, for positive definite matrices, for which both methods work, CG is more economical and is better understood theoretically, and hence it is the preferred method.

Limited memory GMRES

For nonsymmetric matrices, where Arnoldi cannot be replaced by Lanczos, there is a significant price to pay when using GMRES. As we proceed with iterations, we accumulate more and more basis vectors of the Krylov subspace. We need to have those around since the solution is given as their linear combination. As a result, the storage requirements keep creeping up as we iterate. This

7.5. *Krylov subspace methods

may become a burden if the matrix is very large and if convergence is not very fast, which often occurs in realistic scenarios.

One way of dealing with this difficulty is by using **restarted GMRES**. We run GMRES as described above, but once a certain maximum number of vectors, say, m, is reached, we take the current iterate as our initial guess, compute the residual, and start again. The resulting method is denoted by GMRES(m). Thus, m is the number of vectors that are stored, and we may have $m \ll n$. Of course, restarted GMRES is almost certain to converge more slowly than the full GMRES. Moreover, we do not really have an optimality criterion anymore, although we still have monotonicity in residual norm reduction. But despite these drawbacks of restarted GMRES, in practice it is usually preferred over full GMRES, since the memory requirement issue is critical in large scale settings. The full GMRES is just "too good to be practical" for really large problems.

MATLAB provides commands for several Krylov subspace solvers. Check out gmres, pcg, and minres. A typical command looks like this:

[x,flag,relres,iter,resvec]=gmres(A,b,m,tol,maxit,M1,M2);

The input parameters here are set so that the iteration is terminated when either $\|\mathbf{r}_k\|/\|\mathbf{b}\| <$ tol or a maximum iteration count of maxit has been reached. The input value m is the restart parameter of GMRES(m); the other commands such as pcg or minres have a similar calling sequence except for this parameter which is not required by any other method. The (optional) input parameters M1 and M2 are factors of a preconditioner, to be discussed further below.

Example 7.13. There are many situations in which mathematical models lead to linear systems where the matrix is nonsymmetric. One such example is the numerical discretization of the steady-state **convection-diffusion equation**. This equation describes physical phenomena that involve interaction among particles of a certain material (diffusion), as well as a form of motion (convection). An example here would be the manner in which a pollutant, say, spreads as it moves in a stream of water. It moves along with the water and other objects, and at the same time it changes its shape and chemical structure and spreads as its molecules interact with each other in a certain manner that can be described mathematically.

The description that follows bears several similarities to Example 7.1, and you are urged to review that example (page 168) before proceeding. Here we highlight mainly what is *different* in the current example. In its simplest form, the convection-diffusion equation is defined on the open *unit square*, $0 < x, y < 1$, and reads

$$-\left(\frac{\partial^2 u}{\partial x^2} + \frac{\partial^2 u}{\partial y^2}\right) + \sigma \frac{\partial u}{\partial x} + \tau \frac{\partial u}{\partial y} = g(x,y).$$

The parameters σ and τ are associated with the *convection*: when they both vanish, we are back to the Poisson equation of Example 7.1. Assume also the same homogeneous Dirichlet boundary conditions as before.

Discretizing using centered differences for both the first and the second partial derivatives (see Section 14.1 or trust us), and using the notation of Example 7.1 and in particular Figure 7.2, we obtain a linear system of equations given by

$$4u_{i,j} - \tilde{\beta} u_{i+1,j} - \hat{\beta} u_{i-1,j} - \tilde{\gamma} u_{i,j+1} - \hat{\gamma} u_{i,j-1} = b_{i,j}, \quad 1 \le i, j \le N,$$
$$u_{i,j} = 0 \quad \text{otherwise},$$

where $\beta = \frac{\sigma h}{2}$, $\hat{\beta} = 1 + \beta$, $\tilde{\beta} = 1 - \beta$, and $\gamma = \frac{\tau h}{2}$, $\hat{\gamma} = 1 + \gamma$, $\tilde{\gamma} = 1 - \gamma$. The associated linear system of equations is written as

$$A\mathbf{u} = \mathbf{b},$$

where **u** consists of the $n = N^2$ unknowns $\{u_{i,j}\}$ organized as a vector, and **b** is composed likewise from the values $\{b_{i,j}\}$. The specific nonzero structure of the matrix is the same as that discussed in Example 7.1; see Figure 7.3. However, the resulting blocks are more lively. We have

$$A = \begin{pmatrix} J & L & & & \\ K & J & L & & \\ & \ddots & \ddots & \ddots & \\ & & K & J & L \\ & & & K & J \end{pmatrix}, \quad J = \begin{pmatrix} 4 & -\tilde{\beta} & & & \\ -\hat{\beta} & 4 & -\tilde{\beta} & & \\ & \ddots & \ddots & \ddots & \\ & & -\hat{\beta} & 4 & -\tilde{\beta} \\ & & & -\hat{\beta} & 4 \end{pmatrix},$$

and also $K = -\hat{\gamma} I_N$, $L = -\tilde{\gamma} I_N$, where I_N denotes the identity matrix of size N. We can see that the matrix A is diagonally dominant if $|\beta| < 1$ and $|\gamma| < 1$. Let us stay within this range of values; there is more than one good reason for this.

To generate the matrix A in sparse form with MATLAB, we can use a nice feature that comes in handy when forming highly structured matrices: the *Kronecker product*. Given two matrices of appropriate sizes, C and B, their Kronecker product is defined as

$$C \otimes B = \begin{pmatrix} c_{11}B & c_{12}B & \cdots & c_{1N}B \\ c_{21}B & c_{22}B & \cdots & c_{2N}B \\ \vdots & \vdots & \cdots & \vdots \\ c_{N1}B & c_{N2}B & \cdots & c_{NN}B \end{pmatrix}.$$

For notational convenience, let us denote an $N \times N$ tridiagonal matrix with the first subdiagonal, main diagonal, and first superdiagonal having constant values a, b, and c, respectively, by $\text{tri}_N(a,b,c)$. Then our two-dimensional convection-diffusion matrix can be written as

$$A = I_N \otimes T_1 + T_2 \otimes I_N,$$

where $T_1 = \text{tri}_N(-\hat{\beta}, 4, -\tilde{\beta})$ and $T_2 = \text{tri}_N(-\hat{\gamma}, 0, -\tilde{\gamma})$. We can thus define our A using the following MATLAB script:

```
function A=kron_conv_diff(beta,gamma,N);
ee=ones(N,1);
a=4; b=-1-gamma; c=-1-beta; d=-1+beta; e=-1+gamma;
t1=spdiags([c*ee,a*ee,d*ee],-1:1,N,N);
t2=spdiags([b*ee,zeros(N,1),e*ee],-1:1,N,N);
A=kron(speye(N),t1)+kron(t2,speye(N));
```

Note the utilization of two useful commands: `spdiags` generates a matrix in sparse form by putting the input values along the diagonals, and `kron` performs the Kronecker product.

To demonstrate the performance of a solver for this problem, we set $N = 100$, $\beta = \gamma = 0.1$, and construct the convection-diffusion matrix. For the right-hand side we used an artificially generated vector, so that the solution is a vector of all 1's. After generating the matrix and right-hand side, the linear system was solved using restarted GMRES. Here is the MATLAB command that we ran:

```
[x,flag,relres,iter,resvec]=gmres(A,b,20,1e-8,1000);
```

7.5. *Krylov subspace methods

The input parameters were set to stop the iteration when either $\|\mathbf{r}_k\|/\|\mathbf{b}\| < 10^{-8}$ or the iteration count exceeds 1000. In this example the former happened first, and the program terminated with `flag=0`.

Figure 7.8 shows the relative residual history for two runs: the preconditioned one will be explained later, so let us concentrate for now on the solid blue curve. Within fewer than 400 iterations, indeed far fewer than $n = 10,000$, the relative residual norm is about 10^{-7}. Also, the residual norm goes down monotonically, as expected. ∎

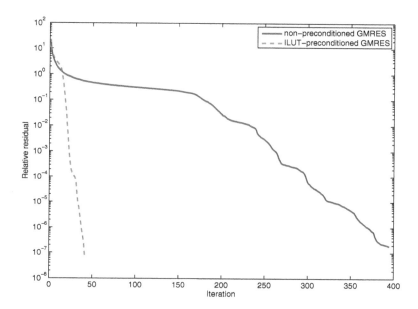

Figure 7.8. *Convergence behavior of restarted GMRES with $m = 20$, for a $10,000 \times 10,000$ matrix that corresponds to the convection-diffusion equation on a 100×100 uniform mesh.*

Krylov subspace solvers and min-max polynomials

Suppose that a given matrix A (not necessarily symmetric positive definite) has a complete set of n eigenpairs $\{\lambda_i, \mathbf{v}_i\}$, and suppose the initial residual satisfies

$$\mathbf{r}_0 = \sum_{i=1}^{n} \alpha_i \mathbf{v}_i.$$

Then

$$\mathbf{r}_k = p_k(A)\mathbf{r}_0 = \sum_{i=1}^{n} \alpha_i p_k(A)\mathbf{v}_i = \sum_{i=1}^{n} \alpha_i p_k(\lambda_i)\mathbf{v}_i.$$

This shows that the residual reduction depends on how well the polynomial dampens the eigenvalues. For the exact solution $\mathbf{x}$, denote the convergence error as before by $\mathbf{e}_k = \mathbf{x} - \mathbf{x}_k$. Since $A\mathbf{e}_k = \mathbf{r}_k$, we have

$$A\mathbf{e}_k = \mathbf{r}_k = p_k(A)\mathbf{r}_0 = p_k(A)A\mathbf{e}_0,$$

and since A commutes with powers of A, multiplying both sides by A^{-1} gives $\mathbf{e}_k = p_k(A)\mathbf{e}_0$.

The basic idea of Krylov subspace methods is to construct a "good" polynomial in this sense, and this may be posed as a problem in approximation theory: seek polynomials that satisfy the min-max property

$$\min_{\substack{p_k \in \pi_k \\ p_k(0)=1}} \max_{\lambda \in \sigma(A)} |p_k(\lambda)|,$$

where π_k denotes the space of all polynomials of degree up to k and $\sigma(A)$ signifies the spectrum of A. All this is done *implicitly*: in practice our iterative solvers never really directly solve a min-max problem.

Example 7.14. Back to the 3×3 linear system of Example 7.9, we look at the polynomial $p_k(A)$ that is obtained throughout the iteration process. Constructing the polynomial can be done, quite simply, by unrolling the CG iteration and expressing the residuals $\mathbf{r}_k, k = 1, 2, 3$, in terms of the initial residual $\mathbf{r}_0$. Figure 7.9 shows graphs of the polynomials p_1, p_2 and p_3. As evident, the values of the polynomials at the eigenvalues of A are dampened as we iterate, until they vanish for the cubic polynomial p_3. ∎

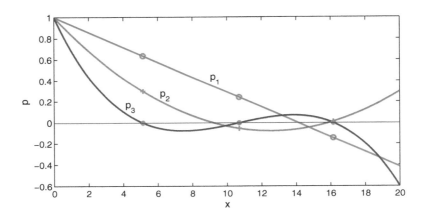

Figure 7.9. *The polynomials that are constructed in the course of three CG iterations for the small linear system of Examples 7.9 and 7.14. The values of the polynomials at the eigenvalues of the matrix are marked on the linear, quadratic, and cubic curves.*

Other Krylov solvers

There are other practical Krylov subspace methods whose detailed description belongs in a more advanced text. An important class of methods are ones that try to find short recurrence relations for nonsymmetric systems. This can be accomplished only if desirable optimality and orthogonality properties are given up. The notion of *bi-orthogonalization* is important here. This approach requires the availability of A^T for some of the schemes.

Let us just mention two favorite methods: BiCGSTAB and QMR. It is not difficult to find examples for which these methods, as well as restarted GMRES, are each better than the other ones, and other examples where they each fail. But they are worthy of your consideration nonetheless, because for many large problems that arise in practice, each of these methods can provide a powerful solution engine, especially when the system is properly preconditioned. Practical choice among these methods requires some experience or experimentation.

7.5. *Krylov subspace methods

The importance and popularity of BiCGSTAB and QMR is reflected by the fact that, just like for CG, MINRES, and GMRES, there are built-in MATLAB commands for them: `bicgstab` and `qmr`.

Preconditioning

As mentioned in Section 7.4, preconditioning is often a necessity in the application of CG, and this is in fact true for the whole family of Krylov subspace solution methods. The purpose of the preconditioner is to cluster the eigenvalues more tightly, or reduce the condition number of the matrix $P^{-1}A$. Unlike the situation for CG, and to a lesser extent MINRES or full GMRES, for which these goals are backed up by a reasonably sound theoretical justification, practical methods such as BiCGSTAB or restarted GMRES have a less complete theory to back them up. However, there is an overwhelming practical evidence that the above-stated goals in the design of a preconditioner is the right thing to do in many generic situations.

There are several basic approaches for preconditioning. Some are based on general algebraic considerations; others make use of properties of the problem underlying the linear system.

There are a few ways to perform a preconditioning operation. *Left preconditioning* is simply the operation of multiplying the system $A\mathbf{x} = \mathbf{b}$ by P^{-1} on the left, to obtain $P^{-1}A\mathbf{x} = P^{-1}\mathbf{b}$. *Right preconditioning* is based on solving $AP^{-1}\tilde{\mathbf{x}} = \mathbf{b}$, then recovering $\mathbf{x} = P^{-1}\tilde{\mathbf{x}}$. *Split preconditioning* is a combination of sorts of left and right preconditioning: it is based on solving $P_1^{-1}AP_2^{-1}\tilde{\mathbf{x}} = P_1^{-1}\mathbf{b}$, then computing $\mathbf{x} = P_2^{-1}\tilde{\mathbf{x}}$. There are a few differences between these approaches. For example, for the same preconditioner P, the Krylov subspace associated with left preconditioning is different from the Krylov subspace associated with right preconditioning in a meaningful way. (Why?) If the matrix A is symmetric, then split preconditioning with $P_1 = P_2$ is the approach that most transparently preserves symmetry, but in practice just left or right preconditioning can be used, too.

Constructing preconditioned iterates amounts in principle to replacing the original matrix for the Arnoldi or Lanczos process, as well as the other components of the solver, by the preconditioned one. But there are a few details that need to be ironed out to guarantee that there are no redundant preconditioner inversions and other basic linear algebra operations in the process. Therefore, just as we saw for PCG vs. CG in Section 7.4, iterates are computed in a slightly different fashion compared to a scheme we would obtain by simply replacing the matrix by its preconditioned counterpart.

ILU preconditioner

In Section 7.4 we described incomplete Cholesky (IC) methods. The analogous factorization for general matrices is called *incomplete LU* (ILU). The factorization ILU(0) is based on the same principle as IC(0), i.e., a complete avoidance of fill-in.

For ILU, too, we can use instead a drop tolerance `tol` to decide whether or not to implement a step in the decomposition, just like IC with a drop tolerance in the symmetric positive definite case. The resulting method is called ILUT. Pivoting issues inevitably arise here, not shared by the IC case, but a detailed description is beyond the scope of this text.

Example 7.15. Let us return to Example 7.13 and add a preconditioner to the GMRES(20) run, using MATLAB's incomplete LU built-in command with a drop tolerance `tol = 0.01`:

```
[L,U]=luinc(A,0.01);
[x,flag,relres,iter,resvec]=gmres(A,b,20,1e-8,1000,L,U);
```

The red dotted curve in Figure 7.8 shows the relative residual norms obtained using preconditioned GMRES(20) with the ILUT setting described above. As you can see, preconditioning makes a significant difference in this case: for a residual value of 10^{-6} the number of iterations required is roughly 40, about one-tenth that of the nonpreconditioned case. We have to be cautious here, since the cost of each iteration is higher when preconditioning is used. But it turns out that the overall computational expense is indeed reduced. The cost of a matrix-vector product can be roughly estimated by the number of nonzeros in the matrix. Here the number of nonzeros of A is 49,600, and the number of nonzeros of L plus the number of nonzeros of U, with the main diagonal counted only once, is 97,227, and hence approximately twice as large. But the saving in iterations is impressive, and it indicates that the preconditioner cost is a price well worth paying, despite all the forward and backward substitutions that are now required. Note also that the residual decreases monotonically, as indeed expected. ∎

For symmetric matrices that are not positive definite, there is a hitch. The most obvious method to use, MINRES, unfortunately requires the preconditioner to be symmetric positive definite, strictly speaking. Thus, a straightforward symmetric version of the ILU approach for constructing a preconditioner may be prohibited by MATLAB in this case.

Specific exercises for this section: Exercises 19–24.

7.6 *Multigrid methods

To motivate the methods briefly introduced in this section, let us consider again the Poisson problem and its discretization (7.1) (page 169) as described in Example 7.1. The condition number of the resulting matrix is $\kappa(A) = \mathcal{O}(n = N^2)$, and thus the number of CG iterations required to reduce the error by a constant amount is $\mathcal{O}(N)$. IC preconditioning helps, as we have seen, but it turns out that the number of iterations required still grows with N (see Figure 7.12(a), coming up soon). Is there an effective method that requires a small, *fixed* number of iterations independent of N?

The answer is affirmative, and to describe such a method we have to dive into the particulars of the problem that is being solved. Since there are many practical problems for which this example may serve as a simple prototype, this excursion turns out to be worthwhile.

Error components with different frequencies

Consider the same Poisson problem discretized twice: once on a finer grid with step size h such that $(N+1)h = 1$, and once on a coarser grid with the larger step size $h_c = 2h$ for which the number of unknowns per spatial variable is $N_c = 1/h_c - 1 \approx N/2$. The coarse grid matrix, A_c, has the eigenvalues

$$\lambda_{l,m}^c = 4 - 2(\cos(l\pi h_c) + \cos(m\pi h_c))$$
$$= 4 - 2(\cos(2l\pi h) + \cos(2m\pi h)), \quad 1 \leq l, m \leq N_c,$$

forming a subset of the eigenvalues of the fine grid matrix. We require here that h_c be such that N_c is an integer number. For simplicity, we may consider $N = 2^j - 1$, with j a positive integer.

To the eigenvalues that appear only in the fine set and not in the coarse set correspond eigenvectors that are, at least in one of the grid directions, *highly oscillatory*. Thus, if there is an effective way to deal with the error or residual components corresponding to just the highly oscillatory eigenvectors on the fine grid,[30] then the other components of the error or residual may be dealt with on the coarser grid, where the same operations are roughly 4 times cheaper. This rationale can obviously be repeated, leading to a recursive, multigrid method.

---

[30]These are commonly referred to as *high-frequency* components of the error or residual.

7.6. *Multigrid methods

Smoothing by relaxation

The essential observation that has led to multigrid methods is that there exist simple and cheap *relaxation* schemes such as *Gauss–Seidel, damped Jacobi*, or the ILU family of methods that reduce the highly oscillatory components of the residual (or the iteration error) in a given iteration much faster than they reduce the low frequency ones. What this means is that the error, or the residual $\mathbf{r} = \mathbf{b} - A\mathbf{x}$, becomes smoother upon applying such a relaxation, when considered as a grid function, well before it decreases significantly in magnitude.

Example 7.16. Figure 7.10 illustrates the smoothing effect of a simple relaxation scheme. At the top panel, the one-dimensional counterpart of the discretized Poisson equation of Example 7.1[31] with $N = 99$ has the highly oscillatory residual $\mathbf{r} = \sin(5\mathbf{u}) + .1\sin(100\mathbf{u})$, where $\mathbf{u}$ is the vector of grid point values. The bottom panel depicts the same residual after applying four damped Jacobi iterations with $\omega = 0.8$. It has only a slightly smaller magnitude but is much smoother. ■

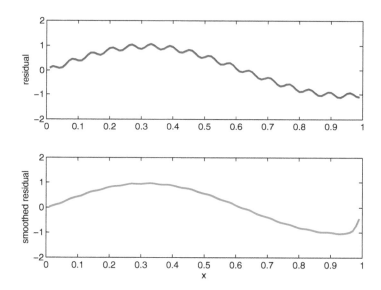

Figure 7.10. *An illustration of the smoothing effect, using damped Jacobi with $\omega = 0.8$ applied to the Poisson equation in one dimension.*

The multigrid cycle

The relaxation operator is thus a **smoother**. Now, it is possible to represent the smoothed residual error well on a coarser grid, thereby obtaining a smaller problem to solve, $A_c \mathbf{v}_c = \mathbf{r}_c$. The obtained correction $\mathbf{v}_c$ is then prolongated (interpolated) back to the finer grid and added to the current solution $\mathbf{x}$, and this is followed by additional relaxation. The process is repeated recursively: for the coarse grid problem the same idea is applied utilizing an even coarser grid. At the coarsest level the problem size is so small that it can be rapidly solved "exactly," say, using a direct method.

The multigrid algorithm given on the following page applies first ν_1 relaxations at the current finer level, then calculates the residual, coarsens it as well as the operator A, solves the coarse grid correction problem approximately γ times starting from the zero correction, prolongates (interpolates) the correction back to the finer grid, and applies ν_2 more relaxations. The whole thing

---

[31] Specifically, the ODE $-\frac{d^2 u}{dx^2} = g(x)$, $0 < x < 1$, $u(0) = u(1) = 0$, is discretized on a uniform grid with $h = .01$.

Algorithm: Multigrid Method.
Given three positive integer parameters ν_1, ν_2, and γ, start at *level = finest*:

> function $\mathbf{x}$ = multigrid($A, \mathbf{b}, \mathbf{x}, level$)
> if level = coarsest then
>   solve exactly $\mathbf{x} = A^{-1}\mathbf{b}$
> else
>   for $j = 1 : \nu_1$
> $\mathbf{x}$ = relax($A, \mathbf{b}, \mathbf{x}$)
>   end
> $\mathbf{r} = \mathbf{b} - A\mathbf{x}$
> $[A_c, \mathbf{r}_c]$ = restrict($A, \mathbf{r}$)
> $\mathbf{v}_c = 0$
>   for $l = 1 : \gamma$
> $\mathbf{v}_c$ = multigrid($A_c, \mathbf{r}_c, \mathbf{v}_c, level - 1$)
>   end
> $\mathbf{x} = \mathbf{x}$ + prolongate($\mathbf{v}_c$)
>   for $j = 1 : \nu_2$
> $\mathbf{x}$ = relax($A, \mathbf{b}, \mathbf{x}$)
>   end
> end

constitutes one multigrid iteration, or a *cycle*, costing $\mathcal{O}(n)$ operations. If $\gamma = 1$, then we have a V-cycle, while if $\gamma = 2$ we have a W-cycle. Typical relaxation values for ν_1 and ν_2 are 1 or 2. Although the W-cycle is better theoretically, in practice the cheaper V-cycle is more popular.

Example 7.17. This is the last time we consider the Poisson problem of Example 7.1: promise. But as befits a grand finale, let's do it in style.

At first, here is our script for carrying out one multigrid V-cycle for this problem. We use damped Jacobi with $\omega = .8$ for relaxation, because it is the simplest effective smoother. The prolongation from coarse to fine grid uses bilinear interpolation. (This is a "broken line" interpolation, as MATLAB does automatically when plotting a curve, in each dimension x and y; for more on this see Sections 11.2 and 11.6.) The restriction is the transpose of the prolongation, and the coarse grid operator is the same five-point formula (7.1) on the coarser grid.

```
function [x,res] = poismg(A,b,x,level);
% multigrid V-cycle to solve simplest Poisson on a square
% The uniform grid is N by N, N = 2^l-1 some l > 2,
% b is the right hand side; homogeneous Dirichlet
% A has been created by   A = delsq(numgrid('S',N+2));

coarsest = 3;              % coarsest grid
nu1 = 2;                   % relaxations before coarsening grid
nu2 = 2;                   % relaxations after return to finer level
omeg = .8;                 % relaxation damping parameter
```

7.6. *Multigrid methods

```
if level == coarsest
  x = A\b;                    % solve exactly on coarsest level
  r = b - A*x;

else % begin multigrid cycle

  % relax using damped Jacobi
  Dv = diag(A);               % diagonal part of A as vector
  for i=1:nu1
    r = b - A*x;
    x = x + omeg*r./Dv;
  end

  % restrict residual from r to rc on coarser grid
  r = b - A*x;
  N = sqrt(length(b));
  r = reshape(r,N,N);
  Nc = (N+1)/2 - 1; nc = Nc^2;   % coarser grid dimensions
  Ac = delsq(numgrid('S',Nc+2));  % coarser grid operator
  rc = r(2:2:N-1,2:2:N-1) + .5*(r(3:2:N,2:2:N-1)+...
       r(1:2:N-2,2:2:N-1) + r(2:2:N-1,3:2:N)+...
       r(2:2:N-1,1:2:N-2)) + .25*(r(3:2:N,3:2:N)+...
       r(3:2:N,1:2:N-2) + r(1:2:N-2,3:2:N) + r(1:2:N-2,1:2:N-2));
  rc = reshape(rc,nc,1);

  % descend level. Use V-cycle
  vc = zeros(size(rc));           % initialize correction to 0
  [vc,r] = poismg(Ac,rc,vc,level-1); % same on coarser grid

  % prolongate correction from vc to v on finer grid
  v = zeros(N,N);
  vc = reshape(vc,Nc,Nc);
  v(2:2:N-1,2:2:N-1) = vc;
  vz = [zeros(1,N);v;zeros(1,N)];   % embed v with a ring of 0s
  vz = [zeros(N+2,1),vz,zeros(N+2,1)];
  v(1:2:N,2:2:N-1) = .5*(vz(1:2:N,3:2:N)+vz(3:2:N+2,3:2:N));
  v(2:2:N-1,1:2:N) = .5*(vz(3:2:N,1:2:N)+vz(3:2:N,3:2:N+2));
  v(1:2:N,1:2:N) = .25*(vz(1:2:N,1:2:N)+...
      vz(1:2:N,3:2:N+2)+...
      vz(3:2:N+2,3:2:N+2)+vz(3:2:N+2,1:2:N));

  % add to current solution
  n = N^2;
  x = x + reshape(v,n,1);

  % relax using damped Jacobi
  for i=1:nu2
    r = b - A*x;
    x = x + omeg*r./Dv;
  end

end
res = norm(b - A*x);
```

We invoke this *recursive* function for a relative residual tolerance of `tol` = 1.e-6 using something like this:

```
A = delsq(numgrid('S',N+2));
b = A*ones(size(A,1),1);
for itermg = 1:100
   [xmg,rmg(itermg)] = poismg(A,b,xmg,flevel);
   if rmg(itermg)/bb < tol , break, end
end
```

For both a 255×255 grid ($n = 65,025$, $flevel = 8$) and a 511×511 grid ($n = 261,121$, $flevel = 9$), convergence is obtained after just 12 iterations!

We have run the same example also using the CG method described in Section 7.4, as well as CG with two preconditioners: IC(.01) as before, and the multigrid V-cycle as a preconditioner for CG. The progress of the iterations for a given finest grid is depicted in Figure 7.11. For better visualization, we exclude the graph that describes the convergence history of nonpreconditioned CG, which took a bit more than 400 iterations to converge.

This figure is in the spirit of Figures 7.5 and 7.7, but for a much larger matrix dimension n.

Next, let us vary N, i.e., the discretization parameter $h = 1/(N+1)$ in Example 7.1, and observe the convergence behavior for different resolutions. The results are depicted in Figure 7.12. The superiority of the multigrid-based methods for this particular example, especially for large N, is very clear. In particular, the number of iterations required to achieve convergence to within a fixed tolerance is independent of N. ∎

Multigrid method assessment

The stunning performance of the multigrid method demonstrated in Example 7.17 can be extended to many other PDE problems, but not without effort. More generally, the matrix A need not be symmetric positive definite, in fact not even symmetric, although some structure is mandatory if an effective smoother is to be found. See Exercise 27. Also, the coarse grid operator and the prolongation operator may be defined differently for more challenging problems.

There are several other variants that apply also for *nonlinear* PDE problems. The big disadvantage of the multigrid approach is that the resulting methods are fussy, and much fine-tuning is required to work with them. Essentially for this reason the option of using a multigrid iteration as a preconditioner for a Krylov subspace solver is attractive, as the resulting solver is often more robust. The fact that, at least for problems close enough to our model problem, the number of iterations is independent of the matrix size means that as a preconditioner one V-cycle moves the eigenvalues in a way that makes the interval that contains all the eigenvalues virtually independent of the grid size. Hence, also the preconditioned method requires only a fixed number of iterations. This is evident in Figure 7.12(a).

It is interesting to note that even if a "pure" multigrid method fails to converge, one cycle may still be an effective preconditioner. This is because what a Krylov subspace method requires is that (the vast majority of) the eigenvalues be nicely clustered, not that their magnitude be below some stability bound.

Specific exercises for this section: Exercises 25–27.

7.6. *Multigrid methods

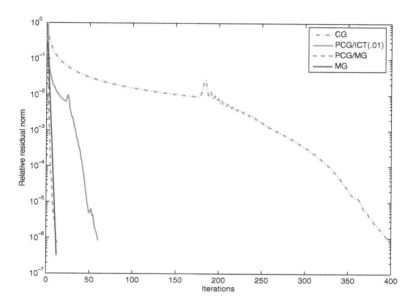

Figure 7.11. *Convergence behavior of various iterative schemes for the Poisson equation (see Example 7.17) with $n = 255^2$.*

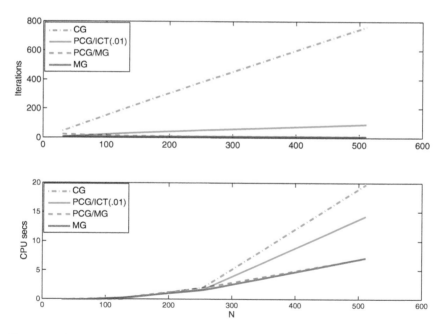

Figure 7.12. *Example 7.17: number of iterations (top panel) and CPU times (bottom panel) required to achieve convergence to* `tol=` *1.e-6 for the Poisson problem of Example 7.1 (page 168) with $N = 2^l - 1, l = 5, 6, 7, 8, 9$.*

7.7 Exercises

0. **Review questions**

    (a) State three disadvantages of direct solution methods that are well addressed by an iterative approach.

    (b) What is a splitting?

    (c) What properties should the matrix that characterizes a splitting have?

    (d) Suppose $A = M - N$. Show that the following schemes are equivalent:

    $$M\mathbf{x}_{k+1} = N\mathbf{x}_k + \mathbf{b};$$

    $$\mathbf{x}_{k+1} = (I - M^{-1}A)\mathbf{x}_k + M^{-1}\mathbf{b};$$

    $$\mathbf{x}_{k+1} = \mathbf{x}_k + M^{-1}\mathbf{r}_k, \text{ where } \mathbf{r}_k = \mathbf{b} - A\mathbf{x}_k.$$

    (e) What is an iteration matrix?

    (f) What is a necessary and sufficient condition for convergence of a stationary scheme for any initial guess?

    (g) State one advantage and one disadvantage of the Jacobi relaxation scheme over Gauss–Seidel.

    (h) Write down the iteration matrix corresponding to SOR.

    (i) What special case of the SOR scheme corresponds to Gauss–Seidel?

    (j) What are nonstationary methods?

    (k) How are search direction and step size (or step length) related to the methods of Section 7.4?

    (l) For what type of matrices is the CG method suitable?

    (m) Show that the first CG iteration coincides with that of the steepest descent method.

    (n) What is an energy norm and what is its connection to CG?

    (o) What is a preconditioner?

    (p) What makes incomplete factorizations potentially good preconditioners?

    (q) What disadvantages do iterative methods potentially have that may lead one to occasionally prefer the direct methods of Chapter 5?

1. Let A be a symmetric positive definite $n \times n$ matrix with entries a_{ij} that are nonzero only if one of the following holds: $i = 1$, or $i = n$, or $j = 1$, or $j = n$, or $i = j$. Otherwise, $a_{ij} = 0$.

    (a) Show that only $5n - 6$ of the n^2 elements of A are possibly nonzero.

    (b) Plot the zero-structure of A. (In MATLAB you can invent such a matrix for $n = 20$, say, and use spy(A).)

    (c) Explain why for $n = 100,000$ using chol (see Section 5.5) to solve $A\mathbf{x} = \mathbf{b}$ for a given right-hand-side vector would be problematic.

2. Consider the problem described in Example 7.1, where the boundary condition $u(0, y) = 0$ is replaced by

    $$\frac{\partial u}{\partial x}(0, y) = 0.$$

7.7. Exercises

(Example 4.17 considers such a change in one variable, but here life is harder.) Correspondingly, we change the conditions $u_{0,j} = 0$, $j = 1, \ldots, N$, into

$$4u_{0,j} - 2u_{1,j} - u_{0,j+1} - u_{0,j-1} = b_{0,j}, \quad 1 \leq j \leq N,$$

where still $u_{0,0} = u_{0,N+1} = 0$.

You don't need to know for the purposes of this exercise why these new linear relations make sense, only that N new unknowns and N new conditions have been introduced.

(a) What is the vector of unknowns **u** now? What is **b**? What is the dimension of the system?

(b) What does A look like?

[Hint: This exercise may not be easy; do it for the case $N = 3$ first!]

3. The linear system

$$10x_1 + x_2 + x_3 = 12,$$
$$x_1 + 10x_2 + x_3 = 12,$$
$$x_1 + x_2 + 10x_3 = 12$$

has the unique solution $x_1 = x_2 = x_3 = 1$. Starting from $\mathbf{x}_0 = (0,0,0)^T$ perform

- two iterations using Jacobi;
- one iteration using Gauss–Seidel.

Calculate error norms for the three iterations in ℓ_1. Which method seems to converge faster?

4. Consider the 2×2 matrix

$$A = \begin{pmatrix} 2 & -1 \\ -1 & 2 \end{pmatrix},$$

and suppose we are required to solve $A\mathbf{x} = \mathbf{b}$.

(a) Write down explicitly the iteration matrices corresponding to the Jacobi, Gauss–Seidel, and SOR schemes.

(b) Find the spectral radius of the Jacobi and Gauss–Seidel iteration matrices and the asymptotic rates of convergence for these two schemes.

(c) Plot a graph of the spectral radius of the SOR iteration matrix vs. the relaxation parameter ω for $0 \leq \omega \leq 2$.

(d) Find the optimal SOR parameter, ω^*. What is the spectral radius of the corresponding iteration matrix? Approximately how much faster would SOR with ω^* converge compared to Jacobi?

5. Consider the class of $n \times n$ matrices A defined in Example 7.1. Denote $N = \sqrt{n}$ and $h = 1/(N+1)$.

(a) Derive a formula for computing the condition number of A in the 2-norm without actually forming this matrix.

How does $\kappa(A)$ depend on n?

(b) Write a program that solves the system of equations $Ax = b$ for a given N, where the right-hand-side b is defined to have the same value h^2 in all its n locations. Your program should apply Jacobi iterations without ever storing A or any other $n \times n$ matrix, terminating when the residual $r = b - Ax$ satisfies

$$\|r\| \leq \texttt{tol}\|b\|.$$

In addition, your program should compute the condition number.

Test the program on a small example, say, $N = 3$, before proceeding further.

(c) Apply your program for the problem instances $N = 2^l - 1$, $l = 2, 3, 4, 5$, using $\texttt{tol} = 10^{-5}$. For each l, record n, the number of Jacobi iterations required to reduce the relative residual to below the tolerance, and the condition number. What relationship do you observe between n and the iteration count? Confirm the theoretical relationship between n and $\kappa(A)$ obtained in Part (a). Explain your observations. (You may wish to plot $\kappa(A)$ vs. n, for instance, to help you focus this observation.)

6. Continuing Exercise 3:

(a) Show that Jacobi's method will converge for this matrix regardless of the starting vector x_0.

(b) Now apply two Jacobi iterations for the problem

$$2x_1 + 5x_2 + 5x_3 = 12,$$
$$5x_1 + 2x_2 + 5x_3 = 12,$$
$$5x_1 + 5x_2 + 2x_3 = 12,$$

starting from $x_0 = (0,0,0)^T$. Does the method appear to converge? Explain why.

7. (a) Show that if the square matrix A is strictly diagonally dominant, then the Jacobi relaxation yields an iteration matrix that satisfies

$$\|T\|_\infty < 1.$$

(b) Show that if A is a 2×2 symmetric positive definite matrix, the Jacobi scheme converges for any initial guess.

8. Consider the matrix

$$A = \begin{pmatrix} 1 & a & a \\ a & 1 & a \\ a & a & 1 \end{pmatrix}.$$

Find the values of a for which A is symmetric positive definite but the Jacobi iteration does not converge.

9. Let α be a scalar, and consider the iterative scheme

$$x_{k+1} = x_k + \alpha(b - Ax_k).$$

This is the gradient descent method with a fixed step size α.

(a) If $A = M - N$ is the splitting associated with this method, state what M and the iteration matrix T are.

7.7. Exercises

(b) Suppose A is symmetric positive definite and its eigenvalues are $\lambda_1 > \lambda_2 > \cdots > \lambda_n > 0$.

   i. Derive a condition on α that guarantees convergence of the scheme to the solution $\mathbf{x}$ for any initial guess.

   ii. Show that the best value for the step size in terms of maximizing the speed of convergence is
$$\alpha = \frac{2}{\lambda_1 + \lambda_n}.$$
   Find the spectral radius of the iteration matrix in this case, and express it in terms of the condition number of A.

(c) Determine whether the following statement is true or false. Justify your answer.

   "If A is strictly diagonally dominant and $\alpha = 1$, then the iterative scheme converges to the solution for any initial guess $\mathbf{x}_0$."

10. Consider the two-dimensional partial differential equation
$$-\Delta u + \omega^2 u = f,$$
where $\Delta u = (\frac{\partial^2 u}{\partial x^2} + \frac{\partial^2 u}{\partial y^2})$ and ω is a given real scalar, on the unit square $\Omega = (0,1) \times (0,1)$, subject to homogeneous Dirichlet boundary conditions: $u = 0$ on $\partial \Omega$. The matrix that corresponds to a straightforward extension of (7.1) for this differential problem is
$$A(\omega) = A + (\omega h)^2 I,$$
where A is the block tridiagonal matrix from Example 7.1, I is the identity matrix, and h is the grid width, given by $h = \frac{1}{N+1}$. Recall that the eigenvalues of $A(0)$ are
$$\lambda_{\ell,m} = 4 - 2(\cos(\ell \pi h) + \cos(m \pi h)), \quad 1 \leq \ell, m \leq N.$$

   (a) For a fixed ω, find $\kappa(A) = \kappa_2(A)$ and show that $\kappa(A) = \mathcal{O}(N^2)$.
   (b) Explain why for any real ω the matrix $A(\omega)$ is symmetric positive definite.
   (c) For $\omega \neq 0$, which of $A(\omega)$ and $A(0)$ is better conditioned?
   (d) What is the spectral radius of the Jacobi iteration matrix associated with $A(\omega)$?

11. The *damped Jacobi* (or *under-relaxed Jacobi*) method is briefly described in Sections 7.2 and 7.6. Consider the system $A\mathbf{x} = \mathbf{b}$, and let D be the diagonal part of A.

   (a) Write down the corresponding splitting in terms of D, A, and ω.
   (b) Suppose A is again the same block tridiagonal matrix arising from discretization of the two-dimensional Poisson equation.

      i. Find the eigenvalues of the iteration matrix of the damped Jacobi scheme.
      ii. Find the values of $\omega > 0$ for which the scheme converges.
      iii. Determine whether there are values $\omega \neq 1$ for which the performance is better than the performance of standard Jacobi, i.e., with $\omega = 1$.

12. Consider the linear system $A\mathbf{x} = \mathbf{b}$, where A is a symmetric matrix. Suppose that $M - N$ is a splitting of A, where M is symmetric positive definite and N is symmetric. Show that if $\lambda_{min}(M) > \rho(N)$, then the iterative scheme $M\mathbf{x}_{k+1} = N\mathbf{x}_k + \mathbf{b}$ converges to $\mathbf{x}$ for any initial guess $\mathbf{x}_0$.

13. Repeat parts (b) and (c) of Exercise 5 with the CG method (without preconditioning) replacing the Jacobi method, and running for the problem instances $N = 2^l - 1$, $l = 2, 3, 4, 5, 6$. All other details and requirements remain the same. Explain your observations.

14. Repeat parts (b) and (c) of Exercise 5 with the gradient descent method (without preconditioning) replacing the Jacobi method, and running for the problem instances $N = 2^l - 1$, $l = 2, 3, 4, 5, 6$. All other details and requirements remain the same. Try two choices for the step size:

    (a) Steepest descent, given by
    $$\alpha_k = \frac{\mathbf{r}_k^T \mathbf{r}_k}{\mathbf{r}_k^T A \mathbf{r}_k} = \frac{\langle \mathbf{r}_k, \mathbf{r}_k \rangle}{\langle \mathbf{r}_k, A\mathbf{r}_k \rangle}.$$

    (b) The same formula applied only every second step, i.e., for k even use the steepest descent formula and for k odd use the already calculated α_{k-1}.

    What are your observations?

15. Show that the energy norm is indeed a norm when the associated matrix is symmetric positive definite.

16. Let A be symmetric positive definite and consider the CG method. Show that for $\mathbf{r}_k$ the residual in the kth iteration and $\mathbf{e}_k$ the error in the kth iteration, the following energy norm identities hold:

    (a) $\|\mathbf{r}_k\|_{A^{-1}} = \|\mathbf{e}_k\|_A$.

    (b) If $\mathbf{x}_k$ minimizes the quadratic function $\phi(\mathbf{x}) = \frac{1}{2}\mathbf{x}^T A \mathbf{x} - \mathbf{x}^T \mathbf{b}$ (note that $\mathbf{x}$ here is an argument vector, not the exact solution) over a subspace S, then the same $\mathbf{x}_k$ minimizes the error $\|\mathbf{e}_k\|_A$ over S.

17. Show that the error bound on $\|\mathbf{e}_k\|_A$ given in the CG Convergence Theorem on page 187 implies that the number of iterations k required to achieve an error reduction by a constant amount is bounded by $\sqrt{\kappa(A)}$ times a modest constant.

    [Hint: Recall the discussion of convergence rate from Section 7.3. Also, by Taylor's expansion we can write
    $$\ln(1 \pm \varepsilon) \approx \pm \varepsilon,$$
    for $0 \leq \varepsilon \ll 1$.]

18. Write a program that solves the problem described in Exercise 10, with a right-hand-side vector defined so that the solution is a vector of all 1s. For example, for $\omega = 0$ the matrix and right-hand side can be generated by

    ```
    A = delsq(numgrid('S',N+2));
    b = A*ones(N^2,1);
    ```

    Your program will find the numerical solution using the Jacobi, Gauss–Seidel, SOR, and CG methods. For CG use the MATLAB command `pcg` and run it once without preconditioning and once preconditioned with incomplete Cholesky IC(0). For SOR, use the formula for finding the optimal ω^* and apply the scheme only for this value. As a stopping criterion use $\|\mathbf{r}_k\|/\|\mathbf{r}_0\| < 10^{-6}$. Also, impose an upper bound of 2000 iterations. That is, if a scheme fails to satisfy the accuracy requirement on the relative norm of the residual after 2000 iterations, it should be stopped. For each of the methods, start with a zero initial guess. Your program should print out the following:

7.7. Exercises

- Iteration counts for the five cases (Jacobi, Gauss–Seidel, SOR, CG, PCG).
- Plots of the relative residual norms $\frac{\|r_k\|}{\|b\|}$ vs. iterations. Use the MATLAB command `semilogy` for plotting.

Use two grids with $N = 31$ and $N = 63$, and repeat your experiments for three values of ω: $\omega = 0$, $\omega^2 = 10$, and $\omega^2 = 1000$. Use your conclusions from Exercise 16 to explain the observed differences in speed of convergence.

19. Suppose CG is applied to a symmetric positive definite linear system $Ax = b$ where the right-hand-side vector b happens to be an eigenvector of the matrix A. How many iterations will it take to converge to the solution? Does your answer change if A is not SPD and instead of CG we apply GMRES?

20. (a) Write a program for solving the linear least squares problems that arise throughout the iterations of the GMRES method, using Givens rotations, where the matrix is a nonsquare $(k+1) \times k$ upper Hessenberg matrix. Specifically, solve
$$\min_z \|\rho e_1 - H_{k+1,k} z\|.$$
Provide a detailed explanation of how this is done in your program, and state what Q and R in the associated QR factorization are.

(b) Given $H_{k+1,k}$, suppose we perform a step of the Arnoldi process. As a result, we now have a new upper Hessenberg matrix $H_{k+2,k+1}$. Describe the relationship between the old and the new upper Hessenberg matrices and explain how this relationship can be used to solve the new least squares problem in an economical fashion.

(c) The least squares problems throughout the iterations can be solved using a QR decomposition approach. Show that the upper triangular factor cannot be singular unless $x_k = x$, the exact solution.

21. Consider the saddle point linear system
$$\underbrace{\begin{pmatrix} A & B^T \\ B & 0 \end{pmatrix}}_{\mathcal{K}} \begin{pmatrix} x \\ y \end{pmatrix} = \begin{pmatrix} d \\ b \end{pmatrix},$$
where A is $n \times n$, symmetric positive definite and B is $m \times n$ with $m < n$. Consider the preconditioner
$$\mathcal{M} = \begin{pmatrix} A & 0 \\ 0 & BA^{-1}B^T \end{pmatrix}.$$

(a) Show that if B has full row rank, the matrix $\mathcal{K}$ is nonsingular.

(b) Show that $\mathcal{K}$ is symmetric indefinite.

(c) How many iterations does it take for a preconditioned minimum residual scheme to converge in this case, if roundoff errors are ignored? To answer this question, it is recommended to find the eigenvalues of $\mathcal{M}^{-1}\mathcal{K}$.

(d) Practical preconditioners based on this methodology approximate the matrix $BA^{-1}B^T$ (or its inverse), rather than use it as is. Give two good reasons for this.

22. A *skew-symmetric* matrix is a matrix S that satisfies
$$S^T = -S.$$

   (a) Show that for any general matrix A, the matrix $(A - A^T)/2$ is skew-symmetric. (This matrix is in fact referred to as the "skew-symmetric part" of A.)
   (b) Show that the diagonal of a skew-symmetric matrix S must be zero component-wise.
   (c) Show that the eigenvalues of S must be purely imaginary.
   (d) If S is $n \times n$ with n an odd number, then it is necessarily singular. Why?
   (e) Suppose that the skew-symmetric S is nonsingular and sparse. In the process of solving a linear system associated with S, a procedure equivalent to Arnoldi or Lanczos is applied to form an orthogonal basis for the corresponding Krylov subspace. Suppose the resulting matrices satisfy the relation
   $$SQ_k = Q_{k+1} U_{k+1,k},$$
   where Q_k is an $n \times k$ matrix whose orthonormal columns form the basis for the Krylov subspace, Q_{k+1} is the matrix of basis vectors containing also the $(k+1)$st basis vector, and $U_{k+1,k}$ is a $(k+1) \times k$ matrix.

   i. Determine the nonzero structure of $U_{k+1,k}$. Specifically, state whether it is tridiagonal or upper Hessenberg, and explain what can be said about the values along the main diagonal.
   ii. Preconditioners for systems with a dominant skew-symmetric part often deal with the possibility of singularity by solving a *shifted* skew-symmetric system, where instead of solving for S one solves for $S + \beta_k I$ with β_k a scalar. Suppose we have the same right-hand-side, but we need to solve the system for several values of β_k. Can the Arnoldi or Lanczos type procedure outlined above be applied once and for all and then be easily adapted?
   iii. Describe the main steps of a MINRES-like method for solving a skew-symmetric linear system.

23. Define a linear problem with $n = 500$ using the script

    ```
    A = randn(500,500); xt = randn(500,1); b = A * xt;
    ```

    Now we save **xt** away and solve $A\mathbf{x} = \mathbf{b}$. Set `tol` = 1.e-6 and maximum iteration limit of 2000. Run three solvers for this problem:

    (a) CG on the normal equations: $A^T A\mathbf{x} = A^T \mathbf{b}$.
    (b) GMRES(500).
    (c) GMRES(100), i.e., restarted GMRES with $m = 100$.

    Record residual norm and solution error norm for each run.

    What are your conclusions?

24. Let A be a general nonsymmetric nonsingular square matrix, and consider the following two alternatives. The first is applying GMRES to solve the linear system $A\mathbf{x} = \mathbf{b}$; the second is applying CG to the normal equations
    $$A^T A\mathbf{x} = A^T \mathbf{b}.$$
    We briefly discussed this in Section 7.5; the method we mentioned in that context was CGLS.

(a) Suppose your matrix A is nearly orthogonal. Which of the two solvers is expected to converge faster?

(b) Suppose your matrix is block diagonal relative to 2×2 blocks, where the jth block is given by
$$\begin{pmatrix} 1 & j-1 \\ 0 & 1 \end{pmatrix}$$
with $j = 1, \ldots, n/2$. Which of the two solvers is expected to converge faster?

[Hint: Consider the eigenvalues and the singular values of the matrices.]

25. Consider the Helmholtz equation
$$-\left(\frac{\partial^2 u}{\partial x^2} + \frac{\partial^2 u}{\partial y^2}\right) - \omega^2 u = g(x, y),$$
defined in the unit square with homogeneous Dirichlet boundary conditions.

(a) Suppose this problem is discretized on a uniform grid with step size $h = 1/(N+1)$ using a five-point scheme as in Example 7.1 plus an additional term. Write down the resulting difference method.

(b) Call the resulting matrix A. Find a value ω_c such that for $\omega^2 < \omega_c^2$ and h arbitrarily small, A is still positive definite, but for $\omega^2 > \omega_c^2$ the positive definiteness is lost.

(c) Solve the problem for $\omega = 1$ and $\omega = 10$ for $N = 2^7 - 1$ using an appropriate preconditioned Krylov subspace method of your choice, or a multigrid method. Use `tol = 1.e-6`. Verify that the last residual norm is below `tol` and tell us how many iterations it took to get there.

26. The *smoothing factor* μ^* for a discrete operator is defined as the worst (i.e., smallest) factor by which high frequency components are reduced in a single relaxation step. For the two-dimensional Laplacian we have discussed throughout this chapter and a basic relaxation scheme, this can be stated as follows. Suppose $\mathbf{e}_0$ is the error before a relaxation step associated with a stationary iteration matrix T and $\mathbf{e}_1$ the error after that step, and write
$$\mathbf{e}_0 = \sum_{l,m=1}^{N} \alpha_{l,m} \mathbf{v}_{l,m},$$
where $\{\mathbf{v}_{l,m}\}_{l,m=1}^{N}$ are the eigenvectors of the iteration matrix. Then
$$\mathbf{e}_1 = \sum_{l,m=1}^{N} \alpha_{l,m} \mu_{l,m} \mathbf{v}_{l,m},$$
where $\{\mu_{l,m}\}_{l,m=1}^{N}$ are eigenvalues of the iteration matrix. The smoothing factor is thus given by
$$\mu^* = \max\left\{|\mu_{l,m}| : \frac{N+1}{2} \leq l \leq N, 1 \leq m \leq N\right\}.$$

(a) Denote the discrete Laplacian by A and the iteration matrix for damped Jacobi by T_ω. Confirm that the eigenvectors of A are the same as the eigenvectors of T_ω for this scheme. (If you have already worked on Exercise 11, this should be old news.)

(b) Show that the optimal ω that gives the smallest smoothing factor over $0 \leq \omega \leq 1$ for the two-dimensional Laplacian is $\omega^* = \frac{4}{5}$, and find the smoothing factor $\mu^* = \mu^*(\omega^*)$ in this case. Note: μ^* should not depend on the mesh size.

(c) Show that Jacobi (i.e., the case $\omega = 1$) is not an effective smoother.

27. Write a program that solves the problem of Example 7.13 for $N = 127$ and $N = 255$ using a multigrid method. The script in that example and the code for a V-cycle in Section 7.6 should prove useful with appropriate modification. Set your coarsest level (where the problem is solved exactly, say, by a direct solver) to $N = 31$.

You should be able to obtain considerably faster convergence than in Example 7.13 at the price of a considerable headache.

7.8 Additional notes

The basic relaxation methods of Section 7.2 are described by just about any text that includes a section on iterative methods for linear systems. They are analyzed in more detail, it almost seems, the older the text is.

The CG method forms the basis of modern day solution techniques for large, sparse, symmetric positive definite systems. See LeVeque [50], Saad [62], and Greenbaum [32] for nice presentations of this method from different angles.

As described in Section 7.5, various Krylov subspace methods have been developed in recent years which extend the CG method to general nonsingular matrices. The book of Saad [62] provides a thorough description, along with a nice framework in the form of projection methods. We have briefly described only a few of those methods. A popular family of methods that we have not discussed are bi-orthogonalization techniques, of which BiCGSTAB is a member; see van der Vorst [72].

While CG is the clear "winner" for symmetric positive definite systems, declaring a winner for general systems is not easy, since different solvers rely on different optimality and spectral properties. A paper by Nachtigal, Reddy, and Trefethen [55] makes this point by showing that for a small collection of methods, there are different problems for which a given method is the winner and another is the loser.

Modern methods for large, sparse systems of equations are rarely as simple as the relaxation methods we have described in Sections 7.2 and 7.3. However, the simple relaxation methods are occasionally used as building blocks for more complex methods. We saw this in Section 7.6: both Gauss–Seidel and damped Jacobi turn out to be effective *smoothers*. Much more on this and other aspects of multigrid methods (including the important class of *algebraic multigrid methods*) can be found in Trottenberg, Oosterlee, and Schuller [71].

For a given practical problem, the challenge usually boils down to finding a good preconditioner. Indeed, these days the focus has shifted from attempting to find new iterative solvers to attempting to find effective preconditioners. Incomplete factorizations are very popular as preconditioners but are by no means the only alternative. If the linear system arises from partial differential equations, then often very effective preconditioners are based on properties of the underlying differential operators. Similarly, if the matrix has a special block structure, preconditioners that exploit that structure are sought. The book by Elman, Silvester, and Wathen [23] illustrates these concepts very well for fluid flow problems.

Chapter 8

Eigenvalues and Singular Values

In this chapter we consider algorithms for solving the *eigenvalue problem*

$$A\mathbf{x} = \lambda \mathbf{x}$$

and for computing the *singular value decomposition* (SVD)

$$A = U\Sigma V^T.$$

We also discuss several relevant applications.

Recall from Section 4.1 (page 71) that eigenvalues, and therefore also singular values, generally cannot be computed precisely in a finite number of steps, even in the absence of floating point error. All algorithms for computing eigenvalues and singular values are therefore necessarily iterative, unlike those in Chapters 5 and 6.

Methods for finding eigenvalues can be split into two distinct categories. The first is generally based on decompositions involving similarity transformations for finding several or all eigenvalues. The other category mainly deals with very large and typically sparse matrices for which just a few eigenvalues and/or eigenvectors are sought; for that there are algorithms that are largely based on matrix-vector products.

We start in Section 8.1 by discussing the effect of repeated multiplication of a vector by the given matrix, which may bring back memories of Sections 7.4 and 7.5. This leads to the fundamental *power method*, which finds application in searching algorithms for large networks. The important concept of shift and invert is introduced next, leading to the *inverse iteration*. We will see how it can accelerate convergence, albeit at a price.

In Section 8.2 we turn to the SVD, introduced in Section 4.4, and explain how it can be used in practice. One such application brings us back to solving least squares problems as in Chapter 6, but with the important difference that the columns of the given matrix may be linearly dependent.

Complete descriptions of robust algorithms for finding all eigenvalues or singular values of a given general matrix involve several nontrivial aspects and belong in a more advanced text. However, we discuss some details of such algorithms in Section 8.3.

8.1 The power method and variants

The power method is a simple technique for computing the dominant eigenvalue and eigenvector of a matrix. It is based on the idea that repeated multiplication of a random vector (almost any vector) by the given matrix yields vectors that eventually tend towards the direction of the dominant eigenvector.

> **Note:** Compared to the sequence of Chapters 5–7, there appears to be an inverted order here: methods based on matrix-vector products and aiming at possibly finding only a few eigenvalues are discussed first, while the general methods for finding all eigenvalues for matrices without a special structure are delayed until Section 8.3. This is because those latter methods rely on the ones introduced in Section 8.1 as building blocks.

But why should we ever want to compute the dominant eigenvector of a matrix? Here is a motivating case study.

Example 8.1. Google's search engine is the dominant Internet search technology these days: it is a mechanism for ranking pages and displaying top hits—webpages that are, in the search engine's judgment, the most relevant sites to the user's query.

The company will not reveal the exact details of its current algorithms—indeed these are rumored to be continually modified—but the basic algorithm that has produced an Internet search giant is called **PageRank** and was published in 1998 by Google's founders, Sergey Brin and Larry Page. This algorithm amounts to a computation of the dominant eigenvector of a large and very sparse matrix.

Before diving into the gory details, though, let us give some necessary context. The Web is an enormous creature of billions of webpages, and it dynamically changes: pages are updated, added, and deleted constantly. Search engines are continuously busy using their vast computing power to "crawl" the Web and update their records. When a user enters a search query, the search engine is ready with billions of records, and according to the specifics of the query, the relevant webpages can be instantly retrieved. A tricky part, though, is to rank those relevant pages and determine the order in which they are displayed to the user; indeed, the typical surfer rarely flips through more than a few top hits. The idea that formed the basis for the search engine was that beyond questions of contents (and rest assured that these are not overlooked) it is also crucial to take the link structure of the Web into account.

Given a network linkage graph with n nodes (webpages), the importance of a webpage is given by the number and the importance of pages that link to it. Mathematically, suppose the importance, or *rank*,[32] of page i is given by x_i. To determine the value of x_i we first record all the pages that link to it. Suppose the locations, or indices, of these pages are given by the set $\{B_i\}$. If a webpage whose index is $j \in B_i$ points to N_j pages including page i, and its own rank is x_j, then we say that it contributes a share of $\frac{x_j}{N_j}$ to the rank x_i of page i. Thus, in one formula, we set

$$x_i = \sum_{j \in B_i} \frac{1}{N_j} x_j, \quad i = 1, \ldots, n.$$

Looking carefully at this expression, we can see that in fact it is nothing but an eigenvalue problem! We seek a vector **x** such that $\mathbf{x} = A\mathbf{x}$, where the nonzero values a_{ij} of the matrix A are the elements $1/N_j$ associated with page i. In other words, we are seeking to compute an eigenvector of A that corresponds to an eigenvalue equal to 1. Since the number of links in and out of a given webpage is smaller by many orders of magnitude than the overall number of webpages, the matrix A is extremely sparse.

There are a few unanswered questions at this point. For example, how do we at all know that the matrix A has an eigenvalue equal to 1? If it does, is that eigenvalue large or small in magnitude compared to the other eigenvalues of the matrix? If there is a solution **x** to the problem, is it unique up to magnitude? Even if it is unique, is it guaranteed to be real? And even if it is real, would all

---

[32]This term relates better to an army rank than to a matrix rank, in the current context.

8.1. The power method and variants

its entries be necessarily positive, as philosophically befits the term "rank"? Some of the answers to these questions are simple and some are more involved. We will discuss them in Example 8.3. For now, please take our word that with a few adjustments to this basic model (see Example 8.3), A indeed has an eigenvalue 1, which happens to be the *dominant* eigenvalue, with algebraic multiplicity 1, and the problem has a unique real positive solution $\mathbf{x}$. This is exactly the vector of ranking that we are looking for, and the power method discussed next can be used to find it. ∎

Developing the power method

Suppose that the eigenvalues of a matrix A are given by $\{\lambda_j, \mathbf{x}_j\}$ for $j = 1, \ldots, n$. Note that here $\mathbf{x}_j$ are eigenvectors as in Section 4.1, not iterates as in Chapter 7. Let $\mathbf{v}_0$, $\|\mathbf{v}_0\| = 1$, be an arbitrary initial guess, and consider the following algorithm:

for $k = 1, 2, \ldots$ until termination do
    $\tilde{\mathbf{v}} = A\mathbf{v}_{k-1}$
    $\mathbf{v}_k = \tilde{\mathbf{v}}/\|\tilde{\mathbf{v}}\|$
end

The output of this algorithm at the kth step is a vector $\mathbf{v}_k = \gamma_k A^k \mathbf{v}_0$, where γ_k is a scalar that guarantees that $\|\mathbf{v}_k\| = 1$. The reason for the repeated normalization is that there is nothing in the definition of an eigenvector to prevent our iterates from growing in magnitude, which would indeed happen for large eigenvalues, accelerating roundoff error growth. Hence we keep the iterate magnitude in check.

We will assume throughout that the matrix A has n linearly independent eigenvectors. Hence, it is possible to express $\mathbf{v}_0$ as a linear combination of the eigenvectors $\{\mathbf{x}_j\}$: there are coefficients β_j such that

$$\mathbf{v}_0 = \sum_{j=1}^{n} \beta_j \mathbf{x}_j.$$

Upon multiplying $\mathbf{v}_0$ by A we obtain

$$A\mathbf{v}_0 = A\left(\sum_{j=1}^{n} \beta_j \mathbf{x}_j\right) = \sum_{j=1}^{n} \beta_j A\mathbf{x}_j = \sum_{j=1}^{n} \beta_j \lambda_j \mathbf{x}_j.$$

The eigenvectors that correspond to the larger eigenvalues are therefore more pronounced in the new linear combination. Continuing, for any positive integer k we have

$$A^k \mathbf{v}_0 = \sum_{j=1}^{n} \beta_j \lambda_j^k \mathbf{x}_j.$$

Now, assume that the eigenvalues $\lambda_1, \lambda_2, \ldots, \lambda_n$ are sorted in decreasing order in terms of their magnitude, and the magnitude of the second eigenvalue is smaller than that of the first, so

$$|\lambda_1| > |\lambda_j|, \quad j = 2, \ldots, n.$$

(That is, the possibility of equality of magnitudes is excluded.) Suppose also that $\mathbf{v}_0$ has a component in the direction of $\mathbf{x}_1$, that is, $\beta_1 \neq 0$. Then

$$\mathbf{v}_k = \gamma_k \lambda_1^k \sum_{j=1}^{n} \beta_j \left(\frac{\lambda_j}{\lambda_1}\right)^k \mathbf{x}_j = \gamma_k \lambda_1^k \beta_1 \mathbf{x}_1 + \gamma_k \lambda_1^k \sum_{j=2}^{n} \beta_j \left(\frac{\lambda_j}{\lambda_1}\right)^k \mathbf{x}_j,$$

where γ_k is a normalization factor, there to ensure that $\|\mathbf{v}_k\| = 1$. Now, by our assumption about the dominance of the first eigenvalue it follows that for $j > 1$ we have $|\frac{\lambda_j}{\lambda_1}|^k \to 0$ as $k \to \infty$. Thus, the larger k is, the more dominant $\mathbf{x}_1$ is in $\mathbf{v}_k$, and in the limit we obtain a unit vector in the direction of $\mathbf{x}_1$.

We thus have a simple way of approximately computing the dominant eigenvector. What about the dominant eigenvalue? In the case of Examples 8.1 and 8.3 this eigenvalue is known, but in most cases it is not. Here we can use the **Rayleigh quotient**, defined for any given vector by

$$\mu(\mathbf{v}) = \frac{\mathbf{v}^T A \mathbf{v}}{\mathbf{v}^T \mathbf{v}}.$$

If $\mathbf{v}$ were an eigenvector,[33] then $\mu(\mathbf{v})$ would simply give the associated eigenvalue. If $\mathbf{v}$ is not an eigenvector, then the Rayleigh quotient of $\mathbf{v}$ gives the closest possible approximation to the eigenvalue in the least squares sense. (We ask you to show this in Exercise 2. It is not difficult!)

Note that by the normalization of $\mathbf{v}_k$, we have

$$\mu(\mathbf{v}_k) = \mathbf{v}_k^T A \mathbf{v}_k.$$

This then is our estimate for the eigenvalue λ_1 in the kth iteration. The power method for computing the dominant eigenpair of a matrix, under the conditions we have stated so far, is given on the current page.

Algorithm: Power Method.
Input: matrix A and initial guess $\mathbf{v}_0$.

$\quad$ for $k = 1, 2, \ldots$ until termination
$\quad\quad \tilde{\mathbf{v}} = A \mathbf{v}_{k-1}$
$\quad\quad \mathbf{v}_k = \tilde{\mathbf{v}} / \|\tilde{\mathbf{v}}\|$
$\quad\quad \lambda_1^{(k)} = \mathbf{v}_k^T A \mathbf{v}_k$
$\quad$ end

Selecting a stopping criterion for this method depends somewhat on the purpose of the computation.

Example 8.2. It is surprising how much a diagonal matrix can tell us about the qualities of an eigenvalue computation method for a more general symmetric matrix A. This is mainly because such matrices have a spectral decomposition $A = QDQ^T$ with Q orthogonal and D diagonal, and so the diagonal matrix here is representative of the "nonorthogonal" part, loosely speaking. Thus, we may be excused for presenting in Figure 8.1 two experiments applying the power method to diagonal matrices.

In MATLAB syntax the matrices are

```
u = [1:32]; v = [1:30,30,32];
A = diag(u); B = diag(v);
```

---

[33] Let us assume for notational simplicity that $\mathbf{v}$ has only real components. Otherwise we replace $\mathbf{v}^T$ by $\mathbf{v}^H$ throughout the above expression, where $\mathbf{v}^H$ stands for "transposed and conjugated $\mathbf{v}$."

8.1. The power method and variants

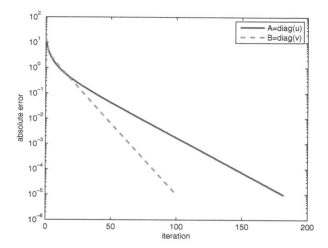

Figure 8.1. *Convergence behavior of the power method for two diagonal matrices, Example 8.2.*

Thus, in both A and B the largest eigenvalue is $\lambda_1 = 32$, but while in A the second eigenvalue is $\lambda_2 = 31$, in B it is $\lambda_2 = 30$.

The plot depicts the absolute error, given by

$$|\lambda_1^{(k)} - \lambda_1| \equiv |\lambda_1^{(k)} - 32|.$$

The results in the graph show the significant improvement in convergence for the second matrix B. The accelerated convergence can best be understood by comparing $\log(\frac{31}{32})$ to $\log(\frac{30}{32})$. The ratio between these two values is approximately 2.03, which indicates a roughly similar factor in speed of convergence.

Exercise 3 considers a similar experiment for nondiagonal matrices. ∎

> **Note:** Example 8.3 below continues the fascinating PageRank saga. The eigenvector computation itself, however, is not what gives it special flavor. If you are curious like most people, then do read it, but if not, then skipping it smoothly is possible.

Example 8.3. As promised, let us return to the problem described in Example 8.1 and illustrate how the power method works to yield the PageRank vector defined there. Let $\mathbf{e}$ be a vector of length n with all elements equal to 1, and define $\mathbf{v}_0 = \frac{1}{n}\mathbf{e}$. Then, for $k = 0, 1, \ldots$, the iteration is defined by

$$v_i^{(k+1)} = \sum_{j \in B_i} \frac{1}{N_j} v_j^{(k)}, \quad i = 1, \ldots, n.$$

This is equivalent to applying the power method without vector normalization and eigenvalue estimation—operations that are not required for the present simple example.

To illustrate this, consider the link structure of the toy network depicted in Figure 8.2. Here the webpages are nodes in the graph, numbered 1 through 6. Notice that the graph is *directed* or,

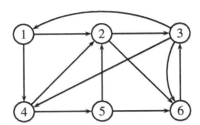

Figure 8.2. *A toy network for the PageRank Example 8.3.*

equivalently, the matrix associated with it is nonsymmetric. Indeed, a page pointing to another page does not necessarily imply that the latter also points to the former.

The task of constructing the link matrix becomes straightforward now. The jth column represents the outlinks from page j. For instance, the third column of the matrix indicates that node 3 has outlinks to the three nodes numbered 1, 4, and 6. Each of them is thus assigned the value 1/3 in their corresponding location in the column, and the rest of the column entries are set to zero. We obtain

$$A = \begin{pmatrix} 0 & 0 & \frac{1}{3} & 0 & 0 & 0 \\ \frac{1}{2} & 0 & 0 & \frac{1}{2} & \frac{1}{2} & 0 \\ 0 & \frac{1}{2} & 0 & 0 & 0 & 1 \\ \frac{1}{2} & 0 & \frac{1}{3} & 0 & 0 & 0 \\ 0 & 0 & 0 & \frac{1}{2} & 0 & 0 \\ 0 & \frac{1}{2} & \frac{1}{3} & 0 & \frac{1}{2} & 0 \end{pmatrix}.$$

In the matrix A the *rows* indicate the inlinks. For example, row 2 indicates that the nodes linking to node 2 are numbered 1, 4, and 5. Note that only the columns sum up to 1; we have no control over the sum of elements in any particular row. Such a matrix is called *column stochastic*.[34]

The matrix A is elementwise nonnegative. There is a very elegant, century-old theory which ensures for this matrix that a unique, simple eigenvalue 1 exists.[35] It follows that all other eigenvalues of A are smaller in magnitude and the associated eigenvector of the dominant eigenvalue has real entries.

To compute PageRank, we start with the vector $\mathbf{v}_0 = (1/6, 1/6, 1/6, 1/6, 1/6, 1/6)^T$ and repeatedly multiply it by A. (This choice of an initial guess is based more on a "democratic" state of mind than any mathematical or computational hunch. In the absence of any knowledge on the solution, why not start from a vector that indicates equal ranking to all?!) Eventually, the method

---

[34] In the literature the matrix A is often denoted by P^T, indicating that its transpose, P, is *row stochastic* and contains the information on outlinks per node in its rows. To avoid notational confusion, we do not use a transpose throughout this example.

[35] The theory we are referring to here is called Perron–Frobenius, and it can be found within a very vast body of literature on nonnegative matrices.

8.1. The power method and variants

converges to

$$\mathbf{x} = \begin{pmatrix} 0.0994 \\ 0.1615 \\ 0.2981 \\ 0.1491 \\ 0.0745 \\ 0.2174 \end{pmatrix},$$

which is the desired PageRank vector. This shows that node 3 is the top-ranked entry. In fact, the output of a query will not typically include the values of $\mathbf{x}$; rather, the ranking is given, which according to the values of $\mathbf{x}$ is $(3, 6, 2, 4, 1, 5)$.

Note that in this special case, since $\|\mathbf{v}_0\|_1 = 1$, all subsequent iterates satisfy $\|\mathbf{v}_k\|_1 = 1$ (why?), and hence we are justified in skipping the normalization step in the power iteration.

There is also a probabilistic interpretation to this model. A random surfer follows links in a random walk fashion, with the probability of following a particular link from webpage j given by the value $1/N_j$ in the corresponding edge. We ask ourselves what is the probability of a surfer being at a given webpage after a long ("infinite") time, regardless of where they started their journey.

Can we expect this basic model to always work for a general Internet network? Not without making adjustments. Figure 8.3 illustrates two potential difficulties. A *dangling node* is depicted in the left diagram. Such a situation occurs when there are no outlinks from a certain webpage. Note that this implies nothing about the webpage's rank! It can indeed be a very important webpage that simply has no links; think, for example, about a country's constitution. The difficulty that a dangling node generates is that the corresponding column in A is zero. In our toy example, imagine that node 6 did not have a link to node 3. In that case the matrix would have a zero 6th column. This practically means that once we hit node 6, we are "stuck" and cannot continue to follow links.

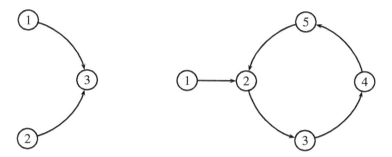

Figure 8.3. *Things that can go wrong with the basic model: depicted are a dangling node (left) and a terminal strong component featuring a cyclic path (right).*

One simple way this can be fixed in the PageRank model is by changing the whole column from zeros to entries all equal to $1/n$ (in our example, $1/6$). In terms of the probabilistic interpretation, this amounts to assuming that if surfers follow links and arrive at a webpage without outlinks, they can jump out to any webpage with equal probability. This is called a dangling node correction.

Another potential difficulty with the basic model is the possibility of entering a "dead end," as is illustrated in the right diagram of Figure 8.3. In graph-speak, this is called a *terminal strong component.* In the diagram, node 1 leads to a cyclic path, and once a surfer arrives at it there is no way out. The situation could be more grave if node 1 did not exist. There is no reason to think that "closed loops" that are isolated from the outer world do not exist in abundance on the Web. There are many instances where this general situation (of course, not necessarily with a cyclic path) can happen; think, for example, of a large body of documentation for a programming language like Java. Webpages point to one another, and it is likely that there are external links to this component of the graph, but it may be the case that there are only interlinks, and no links from this body of documentation back to the outside world.

In the matrix A, this is translated to a diagonal block that barely "connects," or does not at all, to any rows or columns in the matrix which do not belong to that block. For one, this may eliminate the uniqueness of the PageRank vector. The way this is fixed in the PageRank model is by forming a convex combination of the matrix A (*after* the dangling node correction) with a rank-1 matrix. A popular strategy is to replace A by

$$\alpha A + (1-\alpha)\mathbf{u}\mathbf{e}^T,$$

where α is a *damping factor*, $\mathbf{e}$ is a vector of all 1s, and $\mathbf{u}$ is a *personalization vector*. We then compute the dominant eigenvector of this new matrix. For example, if $\alpha = 0.85$ and $\mathbf{u} = \frac{1}{n}\mathbf{e}$, then the fix can be interpreted as a model where with probability 0.85 a surfer follows links, but there is a 0.15 probability that the surfer will jump randomly to anywhere on the Web. If $\mathbf{u}$ contains zeros, then the random jump is done selectively, only to those portions of the graph that correspond to nonzero entries of $\mathbf{u}$; this may clarify the term "personalization."

The matrix $\alpha A + (1-\alpha)\mathbf{u}\mathbf{e}^T$ has some very interesting spectral properties. In particular, while its dominant eigenvalue is 1, the rest of its eigenvalues (which are generally complex) are bounded in magnitude by α. This readily leads to the conclusion that the smaller α is, the faster the power method converges. The problem is that a small α means that we give less weight to the link graph and the notion of following links, and more weight to the option of jumping randomly to webpages, without following links. Whether or not this is "good" is purely a modeling question; indeed, the choice of the "optimal" damping parameter, if such a thing at all exists, has been subject to much debate in recent years. ■

Assessing limitations of the power method

Let us briefly contemplate further on a few issues related to the power method. We have made several assumptions along the way, some more restrictive than others.

Requiring $\beta_1 \neq 0$ in our initial guess looks arbitrary at first but is not very restrictive in practice. Interestingly, even if $\beta_1 = 0$, roundoff errors usually save the day! This is because a roundoff error in the computation would typically introduce a small component in all eigenvector directions.

More importantly, the situation where the matrix has several dominant eigenvalues with the same magnitude has not been considered. This excludes from the discussion some very important instances, including a dominant complex eigenvalue (for which its conjugate is also an eigenvalue and has the same magnitude) and dominant eigenvalues of opposite signs. It even excludes the identity matrix! (Fortunately we know the eigenvalues and eigenvectors of that one.)

Furthermore, we have assumed that all the eigenvectors of the matrix are linearly independent. This is not always the case. Recall from Section 4.1 that matrices whose eigenvectors do not span $\mathcal{R}^n$ are termed *defective*. Here the notion of *geometric* multiplicity of eigenvalues plays a role. The power method for such matrices can still be applied, but convergence may be painfully slow, and definitely much slower than anticipated by ratios of dominant eigenvalues. There is a fairly complete theory for such cases, but it belongs in a more advanced text.

The inverse and Rayleigh quotient iterations

The power method converges linearly and its asymptotic error constant is $\left|\frac{\lambda_2}{\lambda_1}\right|$. If λ_2 is very close in magnitude to λ_1 (a situation that often occurs in applications), convergence may be extremely slow. Just to illustrate this, if $\lambda_1 = 1$ and $\lambda_2 = 0.99$, then for $k = 100$ we have $\left(\frac{\lambda_2}{\lambda_1}\right)^k = 0.99^{100} \approx 0.36$, which means that after 100 iterations we are not even close to gaining a single decimal digit!

The *inverse iteration* overcomes this difficulty by what is known as a shift and invert technique, which gives rise to a substantially faster convergence at the considerable price of having to solve a linear system in each iteration.

Shift and invert technique

The idea is as follows. If the eigenvalues of A are λ_j, the eigenvalues of $A - \alpha I$ are $\lambda_j - \alpha$, and the eigenvalues of $B = (A - \alpha I)^{-1}$ are

$$\mu_j = \frac{1}{\lambda_j - \alpha}.$$

Now, the closer α is to λ_1, the more dominant the largest eigenvalue of B is. Indeed, in the limit if $\alpha \to \lambda_1$, then the first eigenvalue of B tends to ∞ while the other eigenvalues tend to finite values, namely, $\frac{1}{\lambda_j - \lambda_1}$. So, suppose we were to apply the power method to $B = (A - \alpha I)^{-1}$ rather than to A, and suppose that λ_2 is the eigenvalue of A that is closest to λ_1. Then such an iteration, though still converging linearly, does so at the improved rate

$$\left|\frac{\mu_2}{\mu_1}\right| = \left|\frac{\frac{1}{\lambda_2 - \alpha}}{\frac{1}{\lambda_1 - \alpha}}\right| = \left|\frac{\lambda_1 - \alpha}{\lambda_2 - \alpha}\right|.$$

The closer this number is to zero, the faster the convergence. Indeed, if α is very close to λ_1, convergence is expected to be very fast, and likely much faster than the convergence of the power method applied to A.

The quick reader may have already noted at this point that the same technique is possible using an α near *any* eigenvalue λ_j, not necessarily only the dominant one, λ_1. This is indeed true, and the approach described here works in general for computing any eigenvalue, as long as one knows roughly what it is. In particular, using shifts we can easily overcome the difficulty of the power method where more than one simple eigenvalue is dominant.

There are two issues to address here. The first is regarding the choice of the parameter α. How do we select an α that is (i) easy to come up with and, at the same time, (ii) sufficiently close to λ_1, thus guaranteeing fast convergence? Fortunately, effective and computationally cheap estimates, in particular for the dominant eigenvalue, are available and can be used. For example, it is known that for a given matrix A we have $\rho(A) \leq \|A\|$, where ρ is the spectral radius, $\rho(A) = \max_i |\lambda_i(A)|$. Certain matrix norms, such as the 1-norm or the ∞-norm, are easy to compute, and so taking $\alpha = \|A\|_1$, say, may in many cases be a reasonable choice of a shift for computing the dominant eigenvalue of A.

Another issue, and a major one at that, is computational cost. In the power method each iteration essentially involves a matrix-vector product, whereas the inverse iteration requires solving a linear system. This brings to the fore the considerations of Chapters 5 and 7. Recalling iterative methods for linear systems in particular, we could be entertaining methods that require hundreds and thousands of matrix-vector multiplications for one iteration of the inverse iteration. Thus, convergence of the inverse iteration must be very fast for it to be effective, and for huge examples such as Internet searching, using this method is out of the question. For smaller problems, or problems with special structure that enables fast direct methods, the inverse iteration is more attractive. Note that since α is fixed it would make sense to factor the shifted and inverted matrix B once and for all before the iteration starts, and then the cost during the iteration is that of forward/backward solves.

The inverse iteration algorithm is given below.

Algorithm: Inverse Iteration.
Input: matrix A, initial guess $\mathbf{v}_0$, and shift α.

$$\begin{aligned}
&\text{for } k = 1, 2, \ldots \text{ until termination} \\
&\quad \text{solve } (A - \alpha I)\tilde{\mathbf{v}} = \mathbf{v}_{k-1} \\
&\quad \mathbf{v}_k = \tilde{\mathbf{v}}/\|\tilde{\mathbf{v}}\| \\
&\quad \lambda^{(k)} = \mathbf{v}_k^T A \mathbf{v}_k \\
&\text{end}
\end{aligned}$$

Let us discuss for another moment the choice of α. We have already established that the Rayleigh quotient is a good approximation to an eigenvalue for a given vector. So, we may choose the shift α *dynamically*, i.e., $\alpha = \alpha_k$, setting it to be the Rayleigh quotient. With this approach the convergence speed increases as we get closer to the sought eigenvalue; thus the convergence order is better than linear. In fact, in most cases it is *cubic*! In this case it may be worth paying the price of having to refactor the matrix in every iteration.

The Rayleigh quotient iteration algorithm is given below.

Algorithm: Rayleigh Quotient Iteration.
Input: matrix A and normalized initial guess $\mathbf{v}_0$; set $\lambda^{(0)} = \mathbf{v}_0^T A \mathbf{v}_0$.

$$\begin{aligned}
&\text{for } k = 1, 2, \ldots \text{ until termination} \\
&\quad \text{solve } (A - \lambda^{(k-1)} I)\tilde{\mathbf{v}} = \mathbf{v}_{k-1} \\
&\quad \mathbf{v}_k = \tilde{\mathbf{v}}/\|\tilde{\mathbf{v}}\| \\
&\quad \lambda^{(k)} = \mathbf{v}_k^T A \mathbf{v}_k \\
&\text{end}
\end{aligned}$$

Example 8.4. For the matrix A in Example 8.2 we run the inverse iteration with two fixed parameters: $\alpha = 33$ and $\alpha = 35$. Again we take absolute errors, as done in Example 8.2. The results are recorded in Figure 8.4. Observe that a shift closer to the dominant eigenvalue of 32 yields much faster convergence. Also, the iteration counts are substantially smaller in Figure 8.4 than the corresponding values depicted in Figure 8.1 for the power method, in the previous example.

If we now run the Rayleigh quotient iteration, things are even faster. For this example and a random initial guess, a typical sequence of absolute errors that we have obtained in one of our runs is 3.71e-1, 9.46e-2, 2.34e-4, 2.16e-11. Convergence is thus extremely fast, and the number of digits approximately triples in every iteration. ∎

When shifts are employed, typically fewer than ten iterations are needed to obtain the same level of accuracy that may take hundreds (if not more) of iterations when the power method is applied. Of course, each iteration of the inverse iteration is typically significantly more expensive than an iteration of the power method, and Rayleigh quotient iterations are even more expensive, as they require refactoring the matrix in every iteration. It is not difficult to imagine several scenarios

8.2. Singular value decomposition

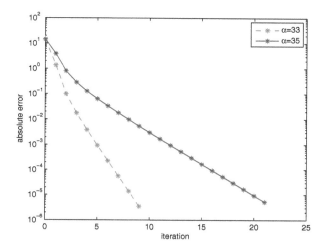

Figure 8.4. *Convergence behavior of the inverse iteration in Example 8.4.*

in which each of these methods is superior to the others, but in most settings the additional overhead may be worthwhile.

Specific exercises for this section: Exercises 1–6.

8.2 Singular value decomposition

When a given square matrix A has a very large condition number, i.e., when it is close to being singular, various computations can go wrong because of the large magnification that small errors (such as roundoff) may be subjected to. Many algorithms which otherwise work well become less reliable, including those for estimating the condition number itself.

This extends directly to overdetermined systems. Recall from (6.1) (page 153) that if A is $m \times n$ with rank$(A) = n$, then the condition number of A is given by $\kappa(A) = \kappa_2(A) = \frac{\sigma_1}{\sigma_n}$, the ratio of largest to smallest singular values. Now, if this condition number is very large, corresponding to the columns of A being almost linearly dependent, then solving a corresponding least squares problem is prone to large errors.

The SVD, which was described in Section 4.4, can come in handy here. Below we demonstrate use of the SVD in several challenging or otherwise interesting situations. Recall also Example 4.18.

Solving almost singular linear systems

Let A be $n \times n$ and real. If $\kappa(A)$ is very large, then solving a linear system $A\mathbf{x} = \mathbf{b}$ can be an ill-conditioned problem.[36]

Of course, given the SVD of A it is easy to find that $\mathbf{x} = V\Sigma^{-1}U^T\mathbf{b}$. But this does not make the resulting numerical solution more meaningful, as we have discussed in Section 5.8. Basically, for an ill-conditioned problem the smallest singular values, although positive, are very small ("almost zero" in some sense). If the problem is too ill-conditioned to be of use, then people often seek to

---

[36]However, a problem $A\mathbf{x} = \mathbf{b}$ with $\kappa(A)$ large *does not have* to be ill-conditioned in the sense described here; in particular, the problem of Example 7.1 can be safely solved for any large n.

regularize it. This means, replace the given problem intelligently by a nearby problem which is better conditioned.

> **Note:** Note the departure made here from everything else done thus far. Instead of seeking numerical algorithms for solving the given problem, we change the problem first and solve a nearby one in a hopefully more meaningful way.

Using SVD this can be done by setting the singular values below a cutoff tolerance to 0, and minimizing the ℓ_2-norm of the solution to the resulting underdetermined problem. We proceed as follows:

1. Starting from n go backward until r is found such that $\frac{\sigma_1}{\sigma_r}$ is tolerable in size. This is the condition number of the problem that we actually solve.

2. Calculate $\mathbf{z} = U^T \mathbf{b}$; in fact just the first r components of $\mathbf{z}$ are needed. In other words, if $\mathbf{u}_i$ is the ith column vector of U, then $z_i = \mathbf{u}_i^T \mathbf{b}$, $i = 1, \ldots, r$.

3. Calculate $y_i = \sigma_i^{-1} z_i$, $i = 1, 2, \ldots, r$, and set $y_i = 0$, $i = r+1, \ldots, n$.

4. Calculate $\mathbf{x} = V\mathbf{y}$. This really involves only the first r columns of V and the first r components of $\mathbf{y}$. In other words, if $\mathbf{v}_i$ is the ith column vector of V, then $\mathbf{x} = \sum_{i=1}^{r} y_i \mathbf{v}_i$.

Of course the resulting $\mathbf{x}$ may not satisfy $A\mathbf{x} = \mathbf{b}$ in general (although it does in Example 8.5 below). But it's the best one can do under certain circumstances, and it produces a solution $\mathbf{x}$ of the smallest norm for a sufficiently well-conditioned approximate problem.

Example 8.5. Recall Example 4.3 and consider

$$A = \begin{pmatrix} 1 & 1 \\ 3 & 3 \end{pmatrix}, \quad \mathbf{b} = \begin{pmatrix} 2 \\ 6 \end{pmatrix}.$$

Here, A is singular, but $\mathbf{b}$ is in its range. So, there are many solutions to the equations $A\mathbf{x} = \mathbf{b}$. A straightforward Gaussian elimination, however, results in a division by 0.

MATLAB yields the SVD

$$U = \begin{pmatrix} -0.316227766016838 & -0.948683298050514 \\ -0.948683298050514 & 0.316227766016838 \end{pmatrix},$$

$$V = \begin{pmatrix} -0.707106781186547 & 0.707106781186548 \\ -0.707106781186548 & -0.707106781186547 \end{pmatrix},$$

$$\Sigma = \begin{pmatrix} 4.47213595499958 & 0 \\ 0 & 4.01876204512712e-16 \end{pmatrix}.$$

We decide (wisely) that the smaller singular value is too small, and thus that the effective rank of this matrix is $r = 1$. We then calculate

$$z_1 = -6.32455532033676 \rightarrow y_1 = -1.4142135623731 \rightarrow \mathbf{x} = \begin{pmatrix} 1 \\ 1 \end{pmatrix}.$$

8.2. Singular value decomposition

Recall that all solutions for this problem have the form $\tilde{\mathbf{x}} = (1+\alpha, 1-\alpha)^T$. Since $\|\tilde{\mathbf{x}}\|_2^2 = (1+\alpha)^2 + (1-\alpha)^2 = 2(1+\alpha^2)$, we see that our SVD procedure has stably found the solution of minimum ℓ_2-norm for this singular problem. ∎

Let us reemphasize the subtle point that our SVD procedure solves a different problem from the original $A\mathbf{x} = \mathbf{b}$, namely, a problem having precisely one solution which generally satisfies the given equations only approximately. The procedure outlined above for solving highly ill-conditioned systems makes use of the Best Lower Rank Approximation Theorem given on this page. The approximation is known as **truncated SVD**.

> **Theorem: Best Lower Rank Approximation.**
> The best rank-r approximation A_r of a matrix $A = U\Sigma V^T$, in the sense that $\|A - A_r\|_2 = \sigma_{r+1}$ is at a minimum, is the matrix
> $$A_r = \sum_{i=1}^{r} \sigma_i \mathbf{u}_i \mathbf{v}_i^T,$$
> where $\mathbf{u}_i$ and $\mathbf{v}_i$ are the ith column vectors of U and V, respectively.

Compressing image information

The fact that the best lower rank approximation can be so directly obtained by the SVD makes it possible to devise a *compression scheme*: by storing the first r columns of U and V, as well as the first r singular values, we obtain an approximation of the matrix A using only $r(m+n+1)$ locations in place of the original mn.

Example 8.6. Consider the MATLAB commands

```
colormap('gray')
load clown.mat;
figure(1)
image(X);

[U,S,V] = svd(X);
figure(2)
r = 20;
colormap('gray')
image(U(:,1:r)*S(1:r,1:r)*V(:,1:r)');
```

These instructions load a clown image from MATLAB's cellars into a 200×320 array X, display the image in one figure, find the SVD of A, and display the image obtained from a rank-20 SVD approximation of A in another figure. The original image is displayed in Figure 8.5 and the compression result is in Figure 8.6.

The original storage requirements for A are $200 \cdot 320 = 64,000$, whereas the compressed representation requires $20 \cdot (200 + 320 + 1) \approx 10,000$ storage locations. This result is certainly better than some random approximation, though it is not exactly jaw-dropping. See further comments in Example 4.18 in this regard. ∎

Latent semantic analysis

A central task in the field of **information retrieval** is to rapidly identify documents that are relevant to a user's query. The relevance here is not necessarily in the sense of search engines, discussed

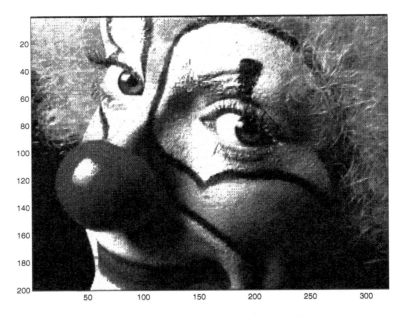

Figure 8.5. *Original* 200×320 *pixel image of a clown.*

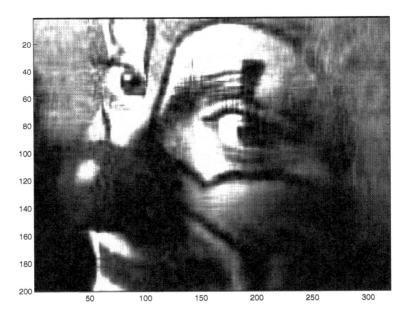

Figure 8.6. *A rank-20 SVD approximation of the image of a clown.*

in considerable length in Section 8.1. Rather, the question is whether a query and a document share a common *theme*. This task becomes complicated in part by the large scale of databases and by various linguistic issues such as the existence of many synonyms. Large databases are not only harder to deal with because of their scale, they also inevitably contain "noise" in the form of

8.2. Singular value decomposition

irrelevant documents, rarely used words, and so on. It is impossible in practice to rely on human experts, not only because humans may not have the time or the stamina to sort through millions of documents, but also because there could be large differences in interpretation of the same data among different experts. Therefore, mathematical models are necessary here. Several such models are based on projecting the data sets, as well as the queries, into a smaller space that can deal with the above mentioned issues in a better fashion. This, in a nutshell, is the idea underlying *latent semantic analysis* (or *latent semantic indexing*).

An important object for storing data is a *term-document matrix*. This is an $n \times m$ matrix, where n is the number of terms (words) and m is the number of documents. Entry i in column j of the matrix represents a function of the frequency of word i in document j. The simplest function is a simple count. But there are more sophisticated measures, which dampen the importance of very frequently used terms and give a higher weight to rarely used ones. The latter are often more effective for discriminating among documents. Let us show what this matrix is all about by presenting a tiny example.

Example 8.7. Consider a toy data set that contains the following two one-sentence documents:

"Numerical computations are fun."

"Numerical algorithms and numerical methods are interesting."

We have two documents and eight different words here. However, in a realistic information retrieval application words like "are" and "and" are too frequent and too generic to possibly provide any useful information and are typically excluded. We thus define a 6×2 term-document matrix, for the six (alphabetically ordered) words "algorithms," "computations," "fun," "interesting," "methods," and "numerical," and for the two documents.

Storing in the matrix entries the frequency of the words used, the term-document matrix in this instance is

$$A = \begin{pmatrix} 0 & 1 \\ 1 & 0 \\ 1 & 0 \\ 0 & 1 \\ 0 & 1 \\ 1 & 2 \end{pmatrix}.$$

We also define 6-long vectors for representing queries. For example, the query "numerical methods" is represented as $\mathbf{q} = (0,0,0,0,1,1)^T$.

It is likely that a reasonable person who sees the two documents in our example will agree that they deal with a similar issue. And yet, the overlap between the two columns in the matrix A is relatively small. A similar issue may apply to queries that use similar but not identical words. ∎

A general *vector space model* seeks to answer the question whether a query and a document have much in common. One popular measure relates to inner products or the cosine of the angle between a vector representation of a query and the vectors representing the documents. Suppose the term-document matrix, A, is $n \times m$. For a given query $\mathbf{q}$, we set

$$\cos(\theta_j) = \frac{(A\mathbf{e}_j)^T \mathbf{q}}{\|A\mathbf{e}_j\| \, \|\mathbf{q}\|}, \qquad j = 1,\ldots,m,$$

where $\mathbf{e}_j$ is the jth column of the $m \times m$ identity matrix.

As already mentioned, latent semantic analysis aims to project the data onto a smaller space, which is easier to deal with computationally and where similar documents and queries can be identified in a relatively easy fashion. An obvious candidate for such information compression is the best lower rank approximation, defined on page 231 and obtained by the truncated SVD. If $A_r = U_r \Sigma_r V_r^T$ with r small, then the above-defined angles are now approximated in this reduced space by

$$\cos(\theta_j) = \frac{\mathbf{e}_j^T V_r \Sigma_r (U_r^T \mathbf{q})}{\|\Sigma_r V_r^T \mathbf{e}_j\| \|\mathbf{q}\|}.$$

So, in the reduced space, the n-long query $\mathbf{q}$ is transformed into the r-long vector

$$\tilde{\mathbf{q}} = U_r^T \mathbf{q}.$$

One is now ready to work in the reduced space, made possible by the truncated SVD.

Of course, to fully appreciate the effect of this technique, you would have to look at a large term-document matrix, certainly one that has many more than the two documents that Example 8.7 features.

Note that while there is a similarity in the techniques used in Examples 8.6 and 8.7, the problems lead to different computational challenges: term-document matrices are almost always large and sparse, and while images could indeed be large too, they are often smaller and typically dense. Therefore, algorithms for obtaining the truncated SVD would also be different, using techniques similar in spirit to those presented in Section 7.5. A relevant MATLAB command for the present application is `svds`.

There are many issues that we have not resolved in this short discussion. One of them is, what should r be? In fact, is there any r that is both small enough and effective? The answer to this cannot typically be stated analytically and is often dependent on the data. See also Example 4.18 on page 87.

Rank deficient linear least squares

The SVD can in some way be thought of as a generalization of the spectral decomposition to nonsquare matrices. An important class of problems leading to nonsquare matrices involves overdetermined systems of equations.

The problem and its usual least squares solution methods are defined and discussed in Chapter 6. But when the $m \times n$ matrix A ($n \leq m$) has a deficient or almost deficient *column* rank the following alternative is worthwhile. In fact, it includes the case for regularizing square systems considered above as a special case.

Let us consider minimizing

$$\|\mathbf{b} - A\mathbf{x}\| = \|\mathbf{b} - U\Sigma V^T \mathbf{x}\|$$

in the ℓ_2-norm, where the $m \times n$ matrix Σ has only r nonzero singular values $\sigma_1, \ldots, \sigma_r$ on its main diagonal, $r \leq n \leq m$. Similarly to the reasoning in Section 6.2, we can further write

$$\|\mathbf{b} - A\mathbf{x}\| = \|\mathbf{z} - \Sigma \mathbf{y}\|, \quad \mathbf{z} = U^T \mathbf{b}, \quad \mathbf{y} = V^T \mathbf{x}.$$

If $r = n$, i.e., A has full column rank, then the unique solution is given by $\mathbf{x} = V\mathbf{y}$, where

$$y_i = \frac{z_i}{\sigma_i}, \quad i = 1, \ldots, r.$$

8.2. Singular value decomposition

Minimum norm solution

If $r < n$, then still the best we can do to minimize $\|\mathbf{z} - \Sigma \mathbf{y}\|$ is to set $y_1, \ldots, y_r$ as above. However, this is not sufficient to define a unique $\mathbf{x}$. In fact, for any choice $y_{r+1}, \ldots, y_n$ forming the n-vector $\mathbf{y}$ together with the first fixed r components, there corresponds an $\mathbf{x}$ which minimizes $\|\mathbf{b} - A\mathbf{x}\|$. We then choose, as before, the solution $\mathbf{x}$ with *minimum norm*. Thus, we solve the problem of minimizing $\|\mathbf{x}\|$ over all solutions of the given linear least squares problem.

The (now unique!) solution of this double-decker optimization problem is mercifully easy. Obviously, $\|\mathbf{y}\|$ is minimized by the choice

$$y_i = 0, \quad i = r+1, r+2, \ldots, n.$$

But then $\mathbf{x} = V\mathbf{y}$ also has minimum norm because V is orthogonal.

In terms of the columns $\mathbf{u}_i$ of U and $\mathbf{v}_i$ of V, we can concisely write

$$\mathbf{x} = \sum_{i=1}^{r} \frac{\mathbf{u}_i^T \mathbf{b}}{\sigma_i} \mathbf{v}_i = A^\dagger \mathbf{b}$$

with the *pseudo-inverse* defined by $A^\dagger = V \Sigma^\dagger U^T$, where

$$\Sigma^\dagger = \begin{cases} 0, & \sigma_i = 0, \\ \frac{1}{\sigma_i}, & \sigma_i \neq 0. \end{cases}$$

As is the case for almost singular linear systems, here the matrix A may have almost linearly dependent columns, in the sense that $\kappa(A) = \sigma_1/\sigma_n$ is finite but intolerably large. A cutoff procedure similar to the case $m = n$ described earlier is then applied. Thus, solving the linear least squares problem via the SVD includes the five steps specified in the algorithm defined on this page.

Algorithm: Least Squares via the SVD.

1. Form $A = U \Sigma V^T$.

2. Decide on cutoff r.

3. Compute $\mathbf{z} = U^T \mathbf{b}$.

4. Set $y_i = \begin{cases} z_i/\sigma_i, & 1 \leq i \leq r, \\ 0, & i > r. \end{cases}$

5. Compute $\mathbf{x} = V\mathbf{y}$.

Example 8.8. Let us take the data of Example 6.8 for A and $\mathbf{b}$, and maliciously add another column to A which is the sum of the existing three. In MATLAB this is achieved by B = [A, sum(A,2)]. Thus we form a 5×4 matrix B which, in the absence of roundoff error, would have rank 3. In particular, in exact arithmetic, $r < n < m$. We next check how the methods that are based on orthogonal transformations fare.

The instruction x = A \ b produces for the well-conditioned A the solution given in Example 6.8, which satisfies $\|\mathbf{x}\| \approx .9473$, $\|\mathbf{b} - A\mathbf{x}\| \approx 5.025$.

The instruction x = B \ b, which employs a method based on QR decomposition, produces a warning regarding rank deficiency and a solution which satisfies $\|\mathbf{x}\| \approx 1.818$, $\|\mathbf{b} - B\mathbf{x}\| \approx 5.025$. Note the growth in $\|\mathbf{x}\|$, due to roundoff magnification near singularity, although the norm of the optimal residual remains the same in exact arithmetic and about the same also in the reported calculation.

Next, we solve the problem $\min_{\mathbf{x}} \{\|\mathbf{x}\| \text{ s.t. } \mathbf{x} \text{ minimizes } \|B\mathbf{x} - \mathbf{b}\|\}$ in MATLAB using svd as described in the algorithm on the previous page. This produces

$$\mathbf{x} \approx (.3571, .4089, -.7760, -.9922 \times 10^{-2})^T, \quad \|\mathbf{x}\| \approx .9471, \quad \|\mathbf{b} - B\mathbf{x}\| \approx 5.025.$$

Here, the solution $\mathbf{x}$ with minimal norm also yields the same residual norm, and no major effect of ill-conditioning is detected. In fact, this solution is a perturbation of an augmented version of the solution for the well-conditioned problem of Example 6.8. ■

Efficiency of the SVD-based least squares algorithm

The fact that in this application always $n \leq m$ may be used to devise an *economy-size SVD*, as for the QR decomposition in Sections 6.2 and 6.3, where U has only n columns and Σ is square $n \times n$. In fact, note that in the least squares algorithm only the first r columns of U and V are utilized. However, it may be argued that in general we don't know r in advance, only that $r \leq n$, so all n singular values must be calculated.

The cost of the algorithm is dominated by forming the SVD, which turns out to be approximately $2mn^2 + 11n^3$ flops. For $m \gg n$ this is approximately the same cost as the QR-based approach, but for $m \approx n$ the SVD approach is substantially more expensive.

Specific exercises for this section: Exercises 7–10.

8.3 General methods for computing eigenvalues and singular values

Section 8.1 describes in detail how to compute the dominant eigenpair of a matrix, using the power method. The shift and invert approach leading to the inverse iteration and the Rayleigh quotient iteration allows for computing an eigenpair that is not necessarily dominant, and very rapidly at that, but still, it is a single pair. A natural question is, how to compute several eigenvalues or *all* the eigenvalues of a given matrix? The answer to this question touches upon some very elegant numerical algorithms. We provide a brief description in this section and also explain how the SVD is computed. Only algorithms for matrices that are potentially dense but small enough to comfortably fit into core are discussed.

Orthogonal similarity transformation

Recall from the discussion on page 71 that if S is a nonsingular matrix of the same size as a given matrix A, then $B = S^{-1}AS$ has the same eigenvalues as those of A. Furthermore, if $\mathbf{x}$ is an eigenvector of A, then $S^{-1}\mathbf{x}$ is an eigenvector of B. If $Q = S$ is orthogonal, i.e., $Q^T Q = I$, then $B = S^{-1}AS = Q^T AQ$. The use of an orthogonal similarity transformation is computationally attractive because the inversion in this case is trivial and because orthogonal transformations preserve the 2-norm and hence are less prone than alternatives to roundoff error accumulation.

The basic idea of advanced algorithms for computing all eigenvalues or singular values of a given matrix is to separate the computation into two stages. The first stage, which involves a fixed number of steps, is aimed at orthogonally transforming the matrix into a "simple" one. In general this

8.3. General methods for computing eigenvalues and singular values

simpler matrix would be in upper Hessenberg form (see page 139). For the symmetric eigenvalue problem this form reduces to a tridiagonal matrix; see Figure 8.7. For the SVD the process is slightly different, and we seek to transform the matrix into a bidiagonal one, as in Figure 8.8.

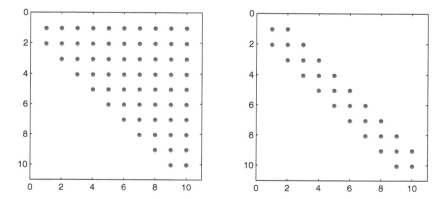

Figure 8.7. *The result of the first stage of a typical eigensolver: a general upper Hessenberg matrix for nonsymmetric matrices (left) and a tridiagonal form for symmetric matrices (right).*

The second stage cannot in principle yield the exact eigenvalues in finitely many operations except for very special cases, and it typically involves a sequence of orthogonal similarity transformations whose goal is to bring the transformed matrix as close as possible to upper triangular form. Then the approximate eigenvalues can be extracted from the diagonal. This upper triangular form is in fact diagonal if A is symmetric. The process for singular values is similar.

You may ask yourself what the point is of separating the computation into two stages, with the first being exact (in the absence of roundoff errors) and the second being iterative. The answer is that without transforming the matrix in the first stage into an upper Hessenberg form, the iterations in the second stage could be unacceptably costly. This will become clearer in the discussion that follows.

The QR algorithm for eigenvalues

Let A be real, square, and not necessarily symmetric. We will assume for simplicity that A has only real eigenvalues and eigenvectors. This simplifying assumption is not necessary for the first stage of the algorithm described below, but it greatly simplifies the description of the second stage. Recall from our discussion in Section 4.1 that in general, nonsymmetric matrices have complex eigenvalues and eigenvectors, so the code qreig to be developed below is not really as general as MATLAB's built-in function eig, but on the other hand its form and function will hopefully be more lucid.

The first stage

Recall from Section 6.2 the compact Householder reflections. We used them for forming the QR decomposition, and we can use them successfully here, too, to form an orthogonal similarity transformation of the form $Q^T A Q$, which is upper Hessenberg. Explaining formally how this can be done is a bit tedious, so let us instead show it for a small example, followed by a general program.

Example 8.9. Consider the 4×4 matrix

$$A = \begin{pmatrix} .5 & -.1 & -.5 & .4 \\ -.1 & .3 & -.2 & -.3 \\ -.3 & -.2 & .6 & .3 \\ .1 & -.3 & .3 & 1 \end{pmatrix}.$$

The reduction into Hessenberg form is done in two steps, for the first and the second rows and columns. Let us denote the elementary transformation matrices for these two steps by Q_1 and Q_2. First, for $k = 1$, we apply a Householder reflection that zeros out the last two elements in the first column. In other words, we look for a reflector $\mathbf{u}_1$ such that the vector $(-.1, -.3, .1)^T$ turns into a vector of the form $(\alpha, 0, 0)^T$ with the same ℓ_2-norm. Such a vector is given to four decimal digits by $\mathbf{u}_1 = (-.8067, -.5606, .1869)^T$, and we have an orthogonal 3×3 matrix of the form $P^{(1)} = I_3 - 2\mathbf{u}_1\mathbf{u}_1^T$. We then define

$$Q_1^T = \begin{pmatrix} 1 & \\ & P^{(1)} \end{pmatrix},$$

where the first row and the first column have zeros except in the $(1, 1)$ position. The matrix $Q_1^T A$ now has two zeros in the $(3, 1)$ and $(4, 1)$ positions.

The significant point here is that in the same way that multiplying by Q_1^T on the left does not touch the first row of A, multiplying by Q_1 on the right does not touch the first column of $Q_1^T A$. Note that had we wanted to obtain an upper triangular form, we would have easily been able to find a matrix Q_1^T that zeros out the three entries below $a_{1,1} = .5$, but then upon multiplying by Q_1 on the right all that hard work would have gone down the drain, since all four columns would have been affected.

Thus, the first similarity transformation gives

$$Q_1^T A Q_1 = \begin{pmatrix} .5 & .6030 & -.0114 & .2371 \\ .3317 & .3909 & -.1203 & .0300 \\ 0 & -.1203 & .6669 & .5255 \\ 0 & .03 & .5255 & .8422 \end{pmatrix},$$

and by construction the eigenvalues of A are preserved. The process is completed by following a similar step for $k = 2$. This time the reflector is $\mathbf{u}_2 = (-.9925, .1221)^T$, and the matrix Q_2^T is defined as a concatenation of I_2 with the 2×2 orthogonal matrix $P^{(2)} = I_2 - 2\mathbf{u}_2\mathbf{u}_2^T$. We get an upper Hessenberg form, as desired, given by

$$Q_2^T Q_1^T A Q_1 Q_2 = \begin{pmatrix} .5 & .6030 & .0685 & .2273 \\ .3317 & .3909 & .1240 & 0 \\ 0 & .1240 & .4301 & -.4226 \\ 0 & 0 & -.4226 & 1.0790 \end{pmatrix}.$$

8.3. General methods for computing eigenvalues and singular values

The real effect of this stage is fully realized only for larger matrices, where only $n-1$ nonzeros out of $\frac{(n-1)n}{2}$ elements are left in the entire strictly lower left triangle. ∎

The numbers in Example 8.9 were obtained using the following function. Note in particular the last two lines, where we multiply the current A by the constructed (symmetric) orthogonal transformation from the left and then from the right.

```
function A = houseeig (A)
%
% function A = houseeig (A)
%
% reduce A to upper Hessenberg form using Householder reflections
n = size(A,1);
for k = 1:n-2
  z=A(k+1:n,k);
  e1=[1; zeros(n-k-1,1)];
  u=z+sign(z(1))*norm(z)*e1;
  u = u/norm(u);
  % multiply from left and from right by Q = eye(n-k)-2*u*u';
  A(k+1:n,k:n) = A(k+1:n,k:n) - 2*u*(u'*A(k+1:n,k:n));
  A(1:n,k+1:n) = A(1:n,k+1:n) - 2*(A(1:n,k+1:n)*u)*u';
end
```

The second stage: Subspace iteration

At this point we may assume that the given matrix A is in upper Hessenberg form. The question is how to proceed iteratively. To the rescue comes an extension of the power method. Suppose we have an initial orthogonal matrix V_0 of size $n \times m$, $1 \leq m \leq n$, and consider the iteration

for $k = 1, 2, \ldots$
  Set $\widehat{V} = AV_{k-1}$
  Compute QR decomposition of $\widehat{V}$: $\widehat{V} = V_k R_k$

Since the columns of V_k span the same space as the columns of $\widehat{V}$ for a given k, and since $\widehat{V} = AV_{k-1}$, we can proceed recursively and conclude that this space is spanned by $A^k V_0$. The orthogonalization step (i.e., the QR decomposition) is crucial: without it, all we would produce are several approximations of the dominant eigenvector (see Section 8.1), which is of course useless.

The above iteration is known by a few names: *subspace iteration*, *orthogonal iteration*, or *simultaneous iteration*. Note that the number of columns of the initial guess V_0, and hence of the subsequent iterates V_k has been left open. If we wish to find *all* the eigenvalues of A, then set $m = n$. It is possible to show that if the eigenvalues of A are all distinct and all the principal submatrices of A have full rank, then starting with $V_0 = I$ the iterative process yields, as $k \to \infty$, an upper triangular matrix whose diagonal elements are the eigenvalues of A.

The second stage: Inefficient QR eigenvalue algorithm

Let us proceed with the assumption that $m = n$. From $AV_{k-1} = V_k R_k$ it follows that $V_k^T A V_{k-1} = R_k$. On the other hand, we also have that

$$V_{k-1}^T A V_{k-1} = V_{k-1}^T V_k R_k.$$

The last equation is nothing but a QR factorization: indeed $V_{k-1}^T A V_{k-1} = QR$, with $Q = V_{k-1}^T V_k$ and $R = R_k$. But from this it follows that

$$V_k^T A V_k = (V_k^T A V_{k-1})(V_{k-1}^T V_k) = RQ.$$

We have just defined the celebrated QR eigenvalue algorithm in its simplest form:

Set $A_0 = A$;
for $k = 0, 1, 2, \ldots$ until termination
  Decompose $A_k = Q_k R_k$
  Construct $A_{k+1} = R_k Q_k$

This is amazingly palindromic! Notice that the QR eigenvalue algorithm is not the same as the QR decomposition algorithm; rather, the former uses the latter. Clearly, A_{k+1} and A_k are orthogonally similar, and because we can interpret this process in terms of the above-mentioned subspace iteration, A_k will eventually converge to an upper triangular matrix whose diagonal elements approximate the eigenvalues of A_0. Since we are dealing with iterates that each involves computing the QR decomposition, it is important to make those calculations as inexpensive as possible. Computing the QR decomposition of an upper Hessenberg matrix requires $\mathcal{O}(n^2)$ operations compared to the $\mathcal{O}(n^3)$ count required by the same decomposition for a full matrix. Furthermore, forming the matrix product RQ preserves the nonzero structure! You are asked to verify these facts in Exercises 12 and 13.

Example 8.10. Continuing with Example 8.9, let us move on to applying the QR eigenvalue algorithm in the form just defined. The eigenvalues of the original matrix, rounded to the number of digits shown, are $-0.0107, 0.2061, 0.9189$, and 1.2857, and the upper Hessenberg matrix obtained at the end of that example preserves them.

Using the upper Hessenberg matrix as an initial guess A_0, and applying three iterations of this algorithm, the elements on the main diagonal of the matrix become $(-0.0106, 0.2111, 0.9100, 1.2896)$. The subdiagonal entries, meanwhile, gradually decrease to $(0.0613, 0.0545, -0.0001)^T$.

Thus, by the third iteration we have a rough approximation to the eigenvalues and entries in the subdiagonal that are much smaller than at the beginning, but admittedly they are not very small. ■

The second stage: Efficient QR eigenvalue algorithm

The results of Example 8.10 show that it is possible to obtain estimates of the eigenvalues from the main diagonal after a few iterations. But convergence, which to a large degree is related to the rate of decay of the subdiagonal entries of the upper Hessenberg matrix, is slow even for this case where the matrix is extremely small and the eigenvalues are reasonably well separated.

Indeed, the QR algorithm in the form described above is not particularly efficient. We run here into the same problems observed before for the power method, which motivated us to pursue the better alternative of a shift and invert approach. Luckily, shifts are easy to incorporate into the QR iteration and often yield spectacular results.

If α_k is a shift near an eigenvalue, then we can define the *QR algorithm with shifts*. Here, too, it is straightforward to show that A_k and A_{k+1} are orthogonally similar. The algorithm is given on the next page.

But how should we select these shifts? In practical applications there is no way to avoid the use of shifts to accelerate convergence, and choosing these is in fact a sophisticated art. Notice that we can change α_k dynamically throughout the iteration. Fortunately, it often suffices to take it as a value along the diagonal, and we proceed here to do so. It can be shown that such a choice is

8.3. General methods for computing eigenvalues and singular values

similar to taking a Rayleigh quotient. Indeed, this is typically done in the so-called *QR algorithm with explicit single shifts*.

> **Algorithm: QR Iteration.**
> Let A_0 be the given matrix, transformed into upper Hessenberg form;
> for $k = 0, 1, 2, \ldots$ until termination
> $$A_k - \alpha_k I = Q_k R_k$$
> $$A_{k+1} = R_k Q_k + \alpha_k I$$
> end

A possible implementation works as follows. We use the last diagonal entry of the matrix as a shift and apply the QR algorithm given on this page. In each iteration we check to see whether all the elements in the last row except the diagonal element are sufficiently small: for an upper Hessenberg matrix there is only one such element to check. If affirmative, then we can declare to have converged to a single eigenvalue, which is no other than the corresponding diagonal entry. We can then get rid of the row and column that correspond to this diagonal entry and start all over again, applying the same iteration to a matrix whose dimension is reduced by one. Here is our program:

```
function [lambda,itn] = qreig (A,tol)
%
% function [lambda,itn] = qreig (A,Tol)
%
% Find all real eigenvalues lambda of A
% Return also iteration counters in itn

% First stage, bring to upper Hessenberg form
A = houseeig(A);

% second stage: deflation loop
n = size(A,1); lambda = []; itn = [];
for j = n:-1:1
  % find jth eigenvalue
  [lambda(j),itn(j),A] = qrshift (A(1:j,1:j),tol);
end

function [lam,iter,A] = qrshift (A,tol)
%
% function [lam,iter,A] = qrshift (A,tol)
%
% Find one eigenvalue lam of A in upper Hessenberg form,
% return iteration count, too. Also improve A for future

m = size(A,1); lam = A(m,m); iter=0; I = eye(m);
if m == 1, return, end
while (iter < 100) % max number of iterations
  if (abs(A(m,m-1)) < tol), return, end    % check convergence
  iter=iter+1;
  [Q,R]=qr(A-lam*I);          % compute the QR decomposition
  A=R*Q+lam*I;                % find the next iterate
  lam = A(m,m);               % next shift
end
```

Example 8.11. We run `qreig` for the toy 4×4 matrix from Example 8.9 with a tight tolerance value of 1e-12. This yields the output

$$\texttt{lambda} = [-0.010679, 0.20608, 0.91887, 1.2857]$$
$$\texttt{itn} = [0, 3, 3, 5].$$

Now, this is more like it! Note that the eigenvalues are retrieved in reverse order, so the iteration counts are decreasing through the deflation loop. Thus we see that the algorithm, while retrieving the jth eigenvalue, works also to improve the lot of the remaining $j-1$ eigenvalues. ∎

Can our program solve more serious eigenvalue problems, too? Let us consider a more challenging case study next.

Example 8.12. Eigenvalue problems arise naturally also in differential equations. But you need not really understand the differential equation aspects detailed below to appreciate the results.

Consider the problem of finding eigenvalues λ and corresponding *eigenfunctions* $u(t)$ that satisfy the differential equation

$$u''(t) - u'(t) = \lambda u(t), \quad 0 < t < L,$$

as well as the boundary conditions $u(0) = u(L) = 0$. It would be useful to recall the notation in Example 4.17 on page 87 at this point. We regard the length of the interval, L, as a parameter and seek nontrivial solutions, meaning that $u(t) \neq 0$ for some value of t.

As it turns out, the (real and countably many) eigenvalues for this differential problem are given by

$$\lambda_j^{de} = -\frac{1}{4} - \left(\frac{j\pi}{L}\right)^2, \quad j = 1, 2, \ldots.$$

For each of these values there is a corresponding eigenfunction that, just like an eigenvector in the algebraic case, is a nontrivial solution for a singular linear system.

Next, to obtain a numerical approximation to the first n values λ_j^{de} we discretize this problem in a way that directly extends the derivation in Example 4.17. For a chosen small value h, we look for an eigenvalue λ and an eigenvector $\mathbf{u} = (u_1, u_2, \ldots, u_{N-1})^T$ satisfying

$$\frac{u_{i+1} - 2u_i + u_{i-1}}{h^2} - \frac{u_{i+1} - u_{i-1}}{2h} = \lambda u_i, \quad i = 1, 2, \ldots, N-1,$$

setting $u_0 = u_N = 0$, for $N = L/h$. Writing this as an algebraic eigenvalue problem $A\mathbf{u} = \lambda \mathbf{u}$, we have a nonsymmetric, potentially large tridiagonal matrix A of size $n = N-1$ and hopefully real eigenvalues.

We have applied `qreig` to this problem for $L = 10$ using $h = .1$, i.e., the matrix size is $n = 99$, with tolerance 1.e-4. The results are reassuring, with an average number of 2.7 iterations per eigenvalue. Sorting the eigenvalues in descending order, the maximum absolute difference between the first six eigenvalues λ_j and their corresponding continuous comrades λ_j^{de} is .015. The discrepancy arises because of the discretization error that depends in turn on the value of h and is of no concern to us here: it does not change meaningfully if we decrease the tolerance in our eigenvalue routine.

Next we have tried to solve for $L = 80$ with $h = .1$, i.e., the matrix size is $n = 799$. Unfortunately, our algorithm did not converge.

To get an idea on what happened we applied the MATLAB workhorse `eig` to solve the same problems. For $L = 10$ the results are comparable to those of our `qreig`. But for $L = 80$ this results in complex eigenvalues, i.e., eigenvalues with nonzero imaginary parts, even though λ_j^{de} stay real.

8.3. General methods for computing eigenvalues and singular values

The reason for the sudden appearance of complex eigenvalues has to do with ill-conditioning of the differential problem and is not our focus here. The lesson, however, is that the use of `qreig` as a general investigation tool for nonsymmetric matrices can be dangerous, as the assumption that the eigenvalues and eigenvectors are real may become violated, possibly without much advance notice. ∎

Indeed, when considering the QR algorithm given on page 241 and the way we chose the shifts, there is simply no mechanism to lift us off the real line and into the complex plane (i.e., the bug is in the specification rather than the implementation of the corresponding program).

State-of-the-art algorithms rely on what is known as *implicit shifts*. In general, for nonsymmetric matrices the eigenvalues can be complex, and wishing to stick with real arithmetic, this leads to the notion of double shifts. Another important component, which guarantees that the QR decompositions throughout the QR iteration are computed in $\mathcal{O}(n^2)$ operations, is known as the *implicit Q theorem*. Unfortunately, we do not have enough space to delve into these details.

We end with a note about symmetric matrices. Not only are their associated eigenvalues and eigenvectors real (provided A is), but also the corresponding upper Hessenberg form is a tridiagonal symmetric matrix. The computations are significantly cheaper and more stable than in the general case. And so, if you know how to solve nonsymmetric eigenvalue problems, applying the same algorithms we discuss in this section to symmetric problems may be painless. Of course, by their nature, symmetric eigenvalue problems invite specialized algorithms and theory, but again we defer to a more specialized text.

Computing the SVD

We end our survey with a brief description of how the SVD is computed. Computing the SVD yields flexibility that eigenvalue computations do not have. For the latter we were forced to apply the same orthogonal transformation on the left and on the right (transposed), since it was necessary to preserve the eigenvalues by employing a similarity transformation. For the SVD, however, there is no need to perform the same operations, since we have U on the left and V^T on the right. Note also that the eigenvalues of the symmetric positive semidefinite matrix $A^T A$ are the squares of the singular values of A, and for the former we have a technique of reducing the matrix into tridiagonal form; see Figure 8.8.

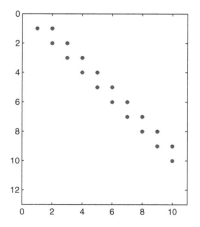
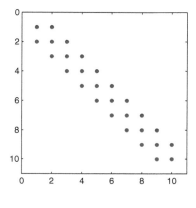

Figure 8.8. *The result of the first stage of the computation of the SVD is a bi-diagonal matrix C (left). The corresponding tridiagonal matrix $C^T C$ is given on the right.*

These facts lead to searching for a procedure that would reduce A into *bidiagonal form*, using different orthogonal transformations on the left and on the right. We proceed in a similar fashion to the Hessenberg reduction for eigenvalue computations. To illustrate the idea, suppose the nonzero structure of a 5×4 matrix A is given by

$$A = \begin{pmatrix} \times & \times & \times & \times \\ \times & \times & \times & \times \\ \times & \times & \times & \times \\ \times & \times & \times & \times \\ \times & \times & \times & \times \end{pmatrix}.$$

Applying U_1^T on the left using Householder transformations, we have

$$U_1^T A = \begin{pmatrix} \times & \times & \times & \times \\ 0 & \times & \times & \times \\ 0 & \times & \times & \times \\ 0 & \times & \times & \times \\ 0 & \times & \times & \times \end{pmatrix}.$$

Now, we can apply a different orthogonal transformation on the right, but since we do not want to touch the first column, we settle for zeroing out the entries to the right of the $(1,2)$ element, namely,

$$U_1^T A V_1 = \begin{pmatrix} \times & \times & 0 & 0 \\ 0 & \times & \times & \times \\ 0 & \times & \times & \times \\ 0 & \times & \times & \times \\ 0 & \times & \times & \times \end{pmatrix}.$$

Another step gives

$$U_2^T U_1^T A V_1 V_2 = \begin{pmatrix} \times & \times & 0 & 0 \\ 0 & \times & \times & 0 \\ 0 & 0 & \times & \times \\ 0 & 0 & \times & \times \\ 0 & 0 & \times & \times \end{pmatrix},$$

and this continues until we get a bidiagonal form. Of course, the number of orthogonal transformations on the left is not necessarily equal to that on the right.

Once a bidiagonal form has been obtained, there are various ways to proceed. Traditional methods involve an adaptation of the QR eigenvalue algorithm from this point on. Recent, faster methods employ a divide-and-conquer approach. We leave it at this.

Specific exercises for this section: Exercises 11–16.

8.4 Exercises

0. **Review questions**

    (a) What is an orthogonal similarity transformation?

    (b) Show that the order of convergence of the power method is linear, and state what the asymptotic error constant is.

    (c) What assumptions should be made to guarantee convergence of the power method?

    (d) What is a shift and invert approach?

    (e) Show that the order of convergence of the inverse iteration with constant shift is linear, and state what the asymptotic error constant is.

    (f) What is the difference in cost of a single iteration of the power method, compared to the inverse iteration?

    (g) What is a Rayleigh quotient and how can it be used for eigenvalue computations?

    (h) What is the singular value decomposition (SVD)?

    (i) How is the SVD defined for a rectangular matrix whose number of rows is smaller than its number of columns?

    (j) When should the SVD be used for solving a least squares problem? When should it not be used?

    (k) What is the connection between singular values and the 2-norm of a matrix?

    (l) What is the connection between singular values and the spectral condition number, $\kappa_2(A)$?

    (m) What are the two main stages of eigensolvers, and what is the main purpose of the first stage?

    (n) What is the sparsity pattern of the matrix that is obtained when applying an upper Hessenberg reduction procedure to a *symmetric* matrix? Why?

    (o) What is the QR iteration, and how is it connected to the QR decomposition?

    (p) Why is it useful to introduce shifts in the QR iteration?

    (q) What is bidiagonalization and how is it related to the computation of SVD?

1. A projection matrix (or *a projector*) is a matrix P for which $P^2 = P$.

    (a) Find the eigenvalues of a projector.

    (b) Show that if P is a projector, then so is $I - P$.

2. Show that the Rayleigh quotient of a real matrix A and vector $\mathbf{v}$, $\mu(\mathbf{v}) = \frac{\mathbf{v}^T A \mathbf{v}}{\mathbf{v}^T \mathbf{v}}$, is the least squares solution of the problem
$$\min_{\mu} \| A\mathbf{v} - \mu\mathbf{v} \|,$$
where $\mathbf{v}$ is given.

3. The MATLAB script

    ```
    u = [1:32]; v = [1:30,30,32];
    M = randn(32,32);
    [Q,R] = qr(M);
    A = Q*diag(u)*Q';
    B = Q*diag(v)*Q';
    ```

    generates two full, mysterious-looking matrices A and B.

Repeat the calculations and considerations of Example 8.2 for these two matrices. Make observations and explain them.

4. Use the setup of Exercise 3 to repeat the calculations and considerations of Example 8.4 for the matrix A with shifts $\alpha = 33$ and $\alpha = 35$. Make observations and explain them.

5. Let
$$A = \begin{pmatrix} \lambda_1 & 1 \\ 0 & \lambda_2 \end{pmatrix},$$
with $\lambda_2 = \lambda_1$.

    How fast will the power method converge in this case to the lone eigenvalue and its eigenvector? How is this different from the observations made in the analysis given in Section 8.1, and why?

6. A column-stochastic matrix P is a matrix whose entries are nonnegative and whose column sums are all equal to 1. In practice such matrices are often large and sparse.

    Let E be a matrix of the same size as P, say, $n \times n$, all of whose entries are equal to $1/n$, and let α be a scalar, $0 < \alpha < 1$.

    (a) Show that $A(\alpha) = \alpha P + (1-\alpha)E$ is also a column-stochastic matrix.

    (b) What is the largest eigenvalue of $A(\alpha)$?

    (c) Show that the second largest eigenvalue of $A(\alpha)$ is bounded (in absolute value) by α.

    (d) Suppose the dominant eigenvector of $A(\alpha)$ is to be computed using the power method. This vector, if normalized so that its ℓ_1-norm is equal to 1, is called the stationary distribution vector.

        i. Show how matrix-vector products with $P(\alpha)$ can be performed in an efficient manner in terms of storage. (Assume n is very large, and recall that E is dense.)

        ii. Show that if the power method is applied and the initial guess $\mathbf{v}_0$ satisfies $\|\mathbf{v}_0\|_1 = 1$, then in the absence of roundoff errors all subsequent iterates $\mathbf{v}_k$ also have a unit ℓ_1-norm.

    [Warning: Item (d) and even more so item (c) above are significantly tougher nuts to crack than items (a) and (b).]

7. Use the definition of the pseudo-inverse of a matrix A in terms of its singular values and singular vectors, as given in the discussion on solving linear least squares problems via the SVD, to show that the following relations hold:

    (a) $AA^\dagger A = A$.

    (b) $A^\dagger AA^\dagger = A^\dagger$.

    (c) $(AA^\dagger)^T = AA^\dagger$.

    (d) $(A^\dagger A)^T = A^\dagger A$.

8. Consider the linear least squares problem of minimizing $\|\mathbf{b} - A\mathbf{x}\|_2$, where A is an $m \times n$ ($m > n$) matrix of rank n.

    (a) Use the SVD to show that $A^T A$ is nonsingular.

    (b) Given an $m \times n$ matrix A that has full column rank, show that $A(A^T A)^{-1} A^T$ is a projector which is also symmetric. Such operators are known as *orthogonal projectors*.

(c) Show that the solution of the linear least squares problem satisfies

$$\mathbf{r} = \mathbf{b} - A\mathbf{x} = P\mathbf{b},$$

where P is an orthogonal projector. Express the projector P in terms of A.

(d) Let Q and R be the matrices associated with the QR decomposition of A. Express the matrix P in terms of Q and R. Simplify your result as much as possible.

(e) With $\mathbf{r}$ defined as usual as the residual, consider replacing $\mathbf{b}$ by $\hat{\mathbf{b}} = \mathbf{b} + \alpha \mathbf{r}$ for some scalar α. Show that we will get the same least squares solution to $\min_\mathbf{x} \| A\mathbf{x} - \hat{\mathbf{b}} \|_2$ regardless of the value of α.

9. Consider the least squares problem

$$\min_\mathbf{x} \| \mathbf{b} - A\mathbf{x} \|_2,$$

where we know that A is ill-conditioned. Consider the *regularization* approach that replaces the normal equations by the modified, better-conditioned system

$$(A^T A + \gamma I)\mathbf{x}_\gamma = A^T \mathbf{b},$$

where $\gamma > 0$ is a parameter.

(a) Show that $\kappa_2^2(A) \geq \kappa_2(A^T A + \gamma I)$.

(b) Reformulate the equations for $\mathbf{x}_\gamma$ as a linear least squares problem.

(c) Show that $\| \mathbf{x}_\gamma \|_2 \leq \| \mathbf{x} \|_2$.

(d) Find a bound for the relative error $\frac{\| \mathbf{x} - \mathbf{x}_\gamma \|_2}{\| \mathbf{x} \|_2}$ in terms of either the largest or the smallest singular value of the matrix A.

State a sufficient condition on the value of γ that would guarantee that the relative error is bounded below a given value ε.

(e) Write a short program to solve the 5×4 problem of Example 8.8 regularized as above, using MATLAB's backslash command. Try $\gamma = 10^{-j}$ for $j = 0, 3, 6$, and 12. For each γ, calculate the ℓ_2-norms of the residual, $\| B\mathbf{x}_\gamma - \mathbf{b} \|$, and the solution, $\| \mathbf{x}_\gamma \|$. Compare to the results for $\gamma = 0$ and to those using SVD as reported in Example 8.8. What are your conclusions?

(f) For large ill-conditioned least squares problems, what is a potential advantage of the regularization with γ presented here over minimum norm truncated SVD?

10. In this question we will play with two pictures (see Figure 8.9) that can be found in MATLAB's repository of images and can be retrieved by entering `load mandrill` and `load durer`. After loading these files, enter for each the command `colormap(gray)` and then `image(X)`. As you can see, these pictures feature the handsome mandrill and a drawing by the artist Albrecht Dürer, who lived in the 1500s.

(a) Write a short MATLAB script for computing the truncated SVD of these images. For both pictures, start with rank $r = 2$ and go up by powers of 2, to $r = 64$. For a compact presentation of your figures, use the command `subplot` for each of the pictures, with 3 and 2 as the first two arguments. (Check out `help subplot` for more information.)

(b) Comment on the performance of the truncated SVD for each of the pictures. State how much storage is required as a function of r and how much storage is required for the original pictures. Explain the difference in the effectiveness of the technique for the two images for small r.

[The mandrill picture file is in fact in color, and you may see it at its full glory by avoiding entering `colormap(gray)`, or simply by entering `colormap(map)` at any time. However, for your calculations please use grayscale.]

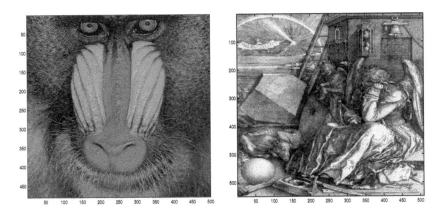

Figure 8.9. *Mandrill image and a drawing by Albrecht Dürer; see Exercise* 11.

11. Show that two matrices in adjacent iterations of the QR eigenvalue algorithm with a single explicit shift, A_k and A_{k+1}, are orthogonally similar.

12. Suppose A is a symmetric tridiagonal $n \times n$ square matrix.

    (a) Describe the nonzero structure of the factors of the QR factorization of A.

    (b) Explain how Givens rotations can be used in the computation of the QR factorization of A, and show briefly that the operation count is far below what would be required for a full matrix.

    (c) What is the nonzero structure of RQ, and how is this useful for applying the QR iteration for computing eigenvalues?

13. Repeat Exercise 12 for a general upper Hessenberg matrix A.

14. Recall from Exercise 4.3 that a real matrix A is said to be skew-symmetric if $A^T = -A$.

    Write a program for computing the eigenvalues and eigenvectors of a skew-symmetric matrix. Do the following:

    (a) Reduce A to tridiagonal form $A = QJQ^T$ using Householder transformations. Show that the diagonal elements of the reduced matrix J are all zero.

    (b) Develop a QR iteration program for the tridiagonal matrix J.

    (c) Apply your program to the skew-symmetric part of the discrete convection-diffusion operator described in Example 7.13.

15. Apply the QR iteration with shifts to the matrix of Exercise 3. Run your program with various tolerance parameters, and comment on the speed of convergence and the overall computational work that is required.

16. Suggest an efficient way of computing the eigenvalues of

$$M = \begin{pmatrix} A & C \\ B & D \end{pmatrix},$$

where $A \in \mathbb{R}^{k \times k}$, $B \in \mathbb{R}^{j \times k}$, $C \in \mathbb{R}^{k \times j}$, and $D \in \mathbb{R}^{j \times j}$ are given real, diagonal matrices.

[Notice that the sizes of the matrices appearing in M are generally different and they are not all square.]

8.5 Additional notes

The topics covered in this chapter are treated much more thoroughly in several specialized numerical linear algebra texts. We mention here Demmel [21], Golub and van Loan [30], and Trefethen and Bau [70]. An encyclopedic text that covers extensive ground is Stewart [63]. An attractively accessible text is presented by Watkins [74]. A classic that stays relevant more than 40 years after its publication is Wilkinson [75].

There are numerous applications where some or all of the eigenvalues or singular values of a matrix are required, and we have presented two data mining examples in this chapter. See Langville and Meyer [47] for much more on PageRank, and Berry and Browne [7] for a description of various information retrieval methodologies, including latent semantic analysis.

As we have already mentioned, although all methods for computing eigenvalues are iterative, they are divided into two classes, one more reminiscent of the direct methods of Chapter 5 and the other more like the iterative methods of Chapter 7. The first class of methods are based on decompositions and do not really take into consideration the sparsity pattern of the matrix. An example is the QR iteration. This algorithm, like direct methods, is based on (repeated) decompositions and is quite robust, though not without failures. Interestingly, the person who originally derived it, John Francis, disappeared from the world of numerical analysis shortly after publishing his seminal paper in 1961 and was made aware only a few years ago of the huge impact that his algorithm has made!

The second class of methods are based mainly of matrix-vector products, and as such, they accommodate sparsity. Indeed, they are typically applied to large and sparse matrices. Only a few eigenvalues and eigenvectors are sought. The power method is a basic such method. The Lanczos and Arnoldi methods, described in Section 7.5, are the workhorses for such eigensolvers.

In MATLAB, a way to distinguish between the above mentioned two classes of methods is by understanding the difference between eig and svd, which are "direct," and eigs and svds, which are "iterative."

There are several reliable software packages for eigenvalue computations. The mathematical software repository Netlib contains many dependable routines for computing all eigenvalues of non-huge matrices. Among the packages for computing a few eigenvalues of large and sparse matrices we mention in particular the state-of-the-art code ARPACK [48].

Chapter 9
Nonlinear Systems and Optimization

Optimization seems to be almost a primal urge among scientists and engineers, and there is a vast number of applications giving rise to the mathematical problems and the numerical methods described in this chapter.

There are several types of optimization problems. Our prototype here is the minimization of a scalar function ϕ in n variables $\mathbf{x} = (x_1, x_2, \ldots, x_n)^T$. We write this as

$$\min_{\mathbf{x}} \phi(\mathbf{x})$$

and require that $\mathbf{x}$ be in $\mathcal{R}^n$ or a subset of it that is characterized by one or more constraints.

A necessary condition for the function $\phi(\mathbf{x})$ to have an unconstrained minimum at a certain point is that all its first derivatives, i.e., the gradient vector, vanish there. This generalizes the setting and results discussed in Section 3.5 and is further justified in Section 9.2 below. We must require $\frac{\partial \phi}{\partial x_1} = 0$, $\frac{\partial \phi}{\partial x_2} = 0$, $\ldots$, $\frac{\partial \phi}{\partial x_n} = 0$. In general, these are n nonlinear equations in n unknowns. Thus, we consider in Section 9.1 the problem of solving *systems of nonlinear equations*. In fact, we do it in a more general context, without a minimization problem necessarily lurking in the background.

The problem of minimizing a scalar, sufficiently smooth function in several variables without constraints is considered in Section 9.2. We derive several useful methods, including a special method for *nonlinear least squares* problems, assuming that the gradient can be evaluated.

Finally, in the more advanced Section 9.3 we briefly consider *constrained optimization*, where a function is to be minimized subject to constraints, so $\mathbf{x}$ is restricted to be in some set Ω strictly contained in $\mathcal{R}^n$. This usually makes the problem harder to solve. The important special case of *linear programming*, where the objective function and the constraints are all linear, will be discussed in slightly more detail than general constrained nonlinear programming.

9.1 Newton's method for nonlinear systems

Consider a system of n nonlinear equations in n unknowns, written as

$$\mathbf{f}(\mathbf{x}) = \mathbf{0},$$

where

$$\mathbf{x} = \begin{pmatrix} x_1 \\ x_2 \\ \vdots \\ x_n \end{pmatrix}, \quad \mathbf{f}(\mathbf{x}) = \begin{pmatrix} f_1(\mathbf{x}) \\ f_2(\mathbf{x}) \\ \vdots \\ f_n(\mathbf{x}) \end{pmatrix}.$$

In component form the system reads

$$f_i(x_1, x_2, \ldots, x_n) = 0, \quad i = 1, 2, \ldots, n.$$

This is a generalization of the case $n = 1$, considered in Chapter 3. We will not assume in this section that the problem necessarily arises from the minimization of any scalar function. Our purpose is to present the basic Newton method in a general context.

Example 9.1. Let $f_1(x_1, x_2) = x_1^2 - 2x_1 - x_2 + 1$ and $f_2(x_1, x_2) = x_1^2 + x_2^2 - 1$. The equation $f_1(x_1, x_2) = 0$ describes a parabola, while $f_2(x_1, x_2) = 0$ describes a circle; see Figure 9.1.

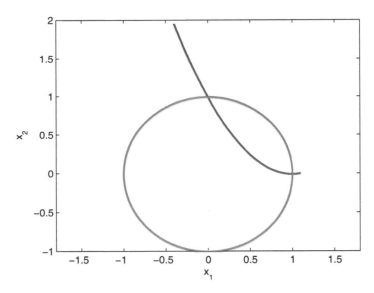

Figure 9.1. *A parabola meets a circle.*

There are two roots to the system of equations

$$x_1^2 - 2x_1 - x_2 + 1 = 0,$$
$$x_1^2 + x_2^2 - 1 = 0.$$

These are the two intersection points, $\mathbf{x}^{*(1)} = (1, 0)^T$ and $\mathbf{x}^{*(2)} = (0, 1)^T$. ∎

The more general picture

We will assume throughout that $\mathbf{f}$ has bounded derivatives at least up to order two.

9.1. Newton's method for nonlinear systems

Similarly to the case of one nonlinear equation considered in Chapter 3, a system of nonlinear equations may have any number of solutions in general; indeed the number of solutions is not directly related to n. Also as before, we will seek an *iterative* method for solving nonlinear equations: starting from an initial guess $\mathbf{x}_0$, such a method generates a sequence of iterates $\mathbf{x}_1, \mathbf{x}_2, \ldots, \mathbf{x}_k, \ldots$, hopefully converging to a solution $\mathbf{x}^*$ satisfying $\mathbf{f}(\mathbf{x}^*) = \mathbf{0}$.

Systems of nonlinear equations can be much more complicated to solve than one nonlinear equation in one unknown, although some of the techniques developed in Chapter 3 do extend to the present family of problems. Unfortunately, no method such as bisection is available here. We will soon see an extension of Newton's method and variants, which provide a powerful tool for *local convergence*. That is, if we start with $\mathbf{x}_0$ already close in some sense to $\mathbf{x}^*$, then under reasonable conditions the method will converge rapidly. However, obtaining convergence for cases where the starting point $\mathbf{x}_0$ is remote can be really problematic, due to the lack of a tool like the bisection method that would allow getting closer to a solution safely.

Theorem: Taylor Series for Vector Functions.
Let $\mathbf{x} = (x_1, x_2, \ldots, x_n)^T$, $\mathbf{f} = (f_1, f_2, \ldots, f_m)^T$, and assume that $\mathbf{f}(\mathbf{x})$ has bounded derivatives up to order at least two. Then for a direction vector $\mathbf{p} = (p_1, p_2, \ldots, p_n)^T$, the Taylor expansion for each function f_i in each coordinate x_j yields

$$\mathbf{f}(\mathbf{x}+\mathbf{p}) = \mathbf{f}(\mathbf{x}) + J(\mathbf{x})\mathbf{p} + \mathcal{O}(\|\mathbf{p}\|^2),$$

where $J(\mathbf{x})$ is the **Jacobian** matrix of first derivatives of $\mathbf{f}$ at $\mathbf{x}$, given by

$$J(\mathbf{x}) = \begin{pmatrix} \frac{\partial f_1}{\partial x_1} & \frac{\partial f_1}{\partial x_2} & \cdots & \frac{\partial f_1}{\partial x_n} \\ \frac{\partial f_2}{\partial x_1} & \frac{\partial f_2}{\partial x_2} & \cdots & \frac{\partial f_2}{\partial x_n} \\ \vdots & \vdots & \ddots & \vdots \\ \frac{\partial f_m}{\partial x_1} & \frac{\partial f_m}{\partial x_2} & \cdots & \frac{\partial f_m}{\partial x_n} \end{pmatrix}.$$

Thus we have

$$f_i(\mathbf{x}+\mathbf{p}) = f_i(\mathbf{x}) + \sum_{j=1}^{n} \frac{\partial f_i}{\partial x_j} p_j + \mathcal{O}(\|\mathbf{p}\|^2), \quad i = 1, \ldots, m.$$

To start, we need an extension of the scalar Taylor Series Theorem to systems. The corresponding theorem is given on the current page. Note that for the present purposes, $n = m$. Observe that there is no evaluation of the next, remainder term, at some intermediate ξ for a system. Moreover, the precise form of the term involving the *second* derivatives of $\mathbf{f}$ is a monster that we luckily need not look at too often in our applications.

It is helpful to think of $\mathbf{x}$ as a *point* in $\mathcal{R}^n$ and of $\mathbf{p} = (p_1, p_2, \ldots, p_n)^T$ as a *direction* vector. We then move from a point $\mathbf{x}$, in the direction $\mathbf{p}$, to the point $\mathbf{x} + \mathbf{p}$. See Figure 9.2.

Deriving Newton's method

Let us return to the system of nonlinear equations

$$\mathbf{f}(\mathbf{x}) = \mathbf{0}.$$

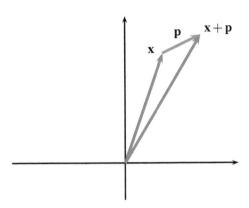

Figure 9.2. *The point* **x**, *the direction* **p**, *and the point* **x** + **p**.

Newton's method for solving such a system is derived in a fashion similar to the scalar case. Starting from an initial guess $\mathbf{x}_0$, a sequence of iterates $\mathbf{x}_1, \mathbf{x}_2, \ldots, \mathbf{x}_k, \ldots$ is generated where $\mathbf{x}_{k+1}$ is obtained upon approximating the Taylor expansion of **f** about $\mathbf{x}_k$ by its linear terms.

Thus, at $\mathbf{x} = \mathbf{x}_k$, if we knew $\mathbf{p} = \mathbf{p}^* = \mathbf{x}^* - \mathbf{x}_k$, then for small $\mathbf{p}^*$

$$0 = \mathbf{f}(\mathbf{x}_k + \mathbf{p}^*) \approx \mathbf{f}(\mathbf{x}_k) + J(\mathbf{x}_k)\mathbf{p}^*.$$

Of course, we do not know $\mathbf{p}^*$, because we do not know $\mathbf{x}^*$, but we use the above to *define* $\mathbf{p} = \mathbf{p}_k$ by requiring

$$\mathbf{f}(\mathbf{x}_k) + J(\mathbf{x}_k)\mathbf{p}_k = \mathbf{0}.$$

This yields the algorithm defined below.

Algorithm: Newton's Method for Systems.

for $k = 0, 1, \ldots$, until convergence

$\qquad\qquad$ solve $J(\mathbf{x}_k)\mathbf{p}_k = -\mathbf{f}(\mathbf{x}_k)$ for $\mathbf{p}_k$

$\qquad\qquad$ set $\mathbf{x}_{k+1} = \mathbf{x}_k + \mathbf{p}_k$.

end

The following MATLAB function implements this algorithm:

```
function [x,k] = newtons(f,x,tol,nmax)
%
% function [x,k] = newtons(f,x,tol,nmax)
%
% This function returns in x a column vector x_k such that
%       || x_k - x_{k-1} || < tol (1 + ||x_k||)
% and in k the number of iterations (Jacobian evaluations) required.
% On entry, x contains an initial guess.
% If k equals nmax then no convergence has been reached.
%
```

9.1. Newton's method for nonlinear systems

```
% The iterates ||f(x_k)|| are recorded. This option
% can be easily turned off

%Initialize
x = x(:); % ensure x is a column vector
fprintf ('k          ||f(x_k)||  \n')
format long g

%Newton
for k=1:nmax
  [fx,Jx] = feval(f,x);
  fprintf ('%d      %e     \n',k-1,norm(fx) )
  p = -Jx \ fx;
  x = x + p;
  if norm(p) < tol*(1+norm(x))
    fx = feval(f,x);
    fprintf ('%d      %e     \n',k,norm(fx) )
    return
  end
end
k = nmax;
```

The stopping criterion in newtons can be easily modified to include a test on $\|\mathbf{f}(\mathbf{x}_k)\|$ as well.

Example 9.2. For the problem defined in Example 9.1 there are two solutions (roots), $\mathbf{x}^{*(1)} = (1,0)^T$ and $\mathbf{x}^{*(2)} = (0,1)^T$. The Jacobian matrix for this problem is

$$J(\mathbf{x}) = \begin{pmatrix} 2x_1 - 2 & -1 \\ 2x_1 & 2x_2 \end{pmatrix}.$$

Note that the Jacobian matrix is not necessarily nonsingular for just *any* choice of x_1 and x_2. For example, starting the iteration from $\mathbf{x}_0 = \mathbf{0}$ is troublesome because $J(\mathbf{0})$ is singular. But the Jacobian is nonsingular at the two roots and at a neighborhood of these roots.

Starting Newton's method from the two initial guesses $\mathbf{x}_0^{(1)} = (1,1)^T$ and $\mathbf{x}_0^{(2)} = (-1,1)^T$ yields quick convergence to the two roots, respectively. The progress of $\|\mathbf{f}(\mathbf{x}_k)\|$ is recorded in Table 9.1. Note how rapidly the iteration converges once the residual $\|\mathbf{f}(\mathbf{x}_k)\|$ becomes small enough. ∎

It can be shown that if at a neighborhood of an isolated root $\mathbf{x}^*$ the Jacobian matrix $J(\mathbf{x})$ has a bounded inverse and continuous derivatives, then Newton's method converges locally **quadratically**, i.e., there is a constant M such that

$$\|\mathbf{x}^* - \mathbf{x}_{k+1}\| \leq M \|\mathbf{x}^* - \mathbf{x}_k\|^2,$$

provided $\|\mathbf{x}^* - \mathbf{x}_k\|$ is already small enough. This quadratic convergence order is nicely demonstrated in Table 9.1.

Example 9.3. Nonlinear systems of equations are of course not restricted to two components. In fact, n can easily become large in applications. Let us show a simple example that gives rise to a larger set of nonlinear equations.

Table 9.1. *Convergence of Newton's method to the two roots of Example 9.1.*

k	$\|\mathbf{f}(\mathbf{x}_k)\|$ first guess	$\|\mathbf{f}(\mathbf{x}_k)\|$ second guess
0	1.414214e+00	3.162278e+00
1	1.274755e+00	7.218033e-01
2	2.658915e-01	1.072159e-01
3	3.129973e-02	4.561589e-03
4	3.402956e-04	9.556657e-06
5	7.094460e-08	4.157885e-11
6	1.884111e-15	

Extending Example 4.17 (see page 87), suppose we are now required to find a function $v(t)$ that satisfies the nonlinear boundary value ODE

$$v''(t) + e^{v(t)} = 0, \quad 0 < t < 1,$$
$$v(0) = v(1) = 0,$$

and we do this approximately by applying a *finite difference* discretization. Such methods and problems are in general the subject of the more advanced Section 16.7, but all we do here is extend the method in Example 4.17 directly and concentrate on the resulting nonlinear system of algebraic equations.

Thus we subdivide the interval $[0, 1]$ into $n + 1$ equal subintervals and set $t_i = ih$, $i = 0, 1, \ldots, n+1$, where $(n+1)h = 1$. We then look for an approximate solution $v_i \approx v(t_i)$, $i = 1, \ldots, n$, using the boundary conditions to set $v_0 = v_{n+1} = 0$.

The discretization of the differential equation is given by

$$\frac{v_{i+1} - 2v_i + v_{i-1}}{h^2} + e^{v_i} = 0, \quad i = 1, 2, \ldots, n.$$

This is a system of nonlinear equations, $\mathbf{f}(\mathbf{x}) = \mathbf{0}$. Here $\mathbf{x} \leftarrow \mathbf{v}$ and $f_i(\mathbf{v}) = \frac{v_{i+1} - 2v_i + v_{i-1}}{h^2} + e^{v_i}$.

The Jacobian matrix has in its (i, j)th element $\frac{\partial f_i}{\partial v_j}$. We obtain

$$J = \frac{1}{h^2} \begin{pmatrix} -2 + h^2 e^{v_1} & 1 & & & \\ 1 & -2 + h^2 e^{v_2} & 1 & & \\ & \ddots & \ddots & \ddots & \\ & & 1 & -2 + h^2 e^{v_{n-1}} & 1 \\ & & & 1 & -2 + h^2 e^{v_n} \end{pmatrix}.$$

All we need to fire up Newton's method is an initial guess $\mathbf{v}_0$. Let us choose the corresponding mesh values of $\alpha t(1-t)$, so that

$$\mathbf{v}_0 = \alpha \left(t_1(1-t_1), \ldots, t_n(1-t_n)\right)^T,$$

and try different values for the parameter α.

9.1. Newton's method for nonlinear systems

Setting `tol` = 1.e-8 and $h = .04$ (i.e., $n = 24$), the results are as follows. For a zero initial guess $\alpha = 0$ convergence is reached within 4 iterations to a solution denoted $\mathbf{v}^{*1}$ in Figure 9.3. For $\alpha = 10$ the same solution is reached in 6 iterations. For $\alpha = 20$ another solution, $\mathbf{v}^{*2}$, is reached within 6 iterations. Figure 9.3 depicts this solution, too.

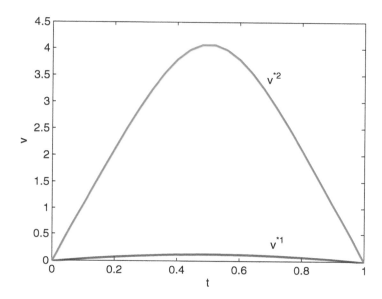

Figure 9.3. *Two solutions for a boundary value ODE.*

For $\alpha = 50$ the algorithm diverges. Unfortunately, divergence is not a rare occasion when using Newton's method without due care. ∎

Modifying the Newton method

In Chapter 3 we have discussed ways to deal with various deficiencies of Newton's method, including the fact that it may diverge when the initial guess is not sufficiently close to the solution and the requirement that the user specify what corresponds to the Jacobian matrix. Here, in addition, there is the need to solve a potentially large system of equations at each iteration k, so techniques from Chapters 5 and 7 are directly relevant.

We discuss such issues and more in this chapter, but let's do it in the context of unconstrained optimization, considered next.

Before moving on, note that the general problem considered in this section can be artificially cast as that of minimizing $\|\mathbf{f}(\mathbf{x})\|$ or $\|\hat{J}^{-1}\mathbf{f}(\mathbf{x})\|$ for some constant matrix $\hat{J}$ somehow representing the Jacobian matrix. We do not recommend *solving* nonlinear equations through such a setting, but it is possible to use them as criteria for improvement of a sequence of iterates. For instance, the next iterate $\mathbf{x}_{k+1}$ of the same Newton method described on page 254 may be considered "better" than the current $\mathbf{x}_k$ if $\|\mathbf{f}(\mathbf{x}_{k+1})\| < \|\mathbf{f}(\mathbf{x}_k)\|$.

Specific exercises for this section: Exercises 1–9.

9.2 Unconstrained optimization

In this section we consider numerical methods for the *unconstrained minimization* of a function in n variables. Thus, the prototype problem reads

$$\min_{\mathbf{x} \in \mathcal{R}^n} \phi(\mathbf{x}).$$

Here $\phi : \mathcal{R}^n \to \mathcal{R}$. That is to say, the argument $\mathbf{x} = (x_1, x_2, \ldots, x_n)^T$ is a vector as in Section 9.1, but ϕ takes on scalar values.

Example 9.4. Here is a very simple example. For $n = 2$, $\mathbf{x} = (x_1, x_2)^T$, we specify the function

$$\phi(\mathbf{x}) = x_1^2 + x_2^4 + 1.$$

This function obviously has the minimum value of 1 at $\mathbf{x}^* = (0,0)^T = \mathbf{0}$. See Figure 9.4.

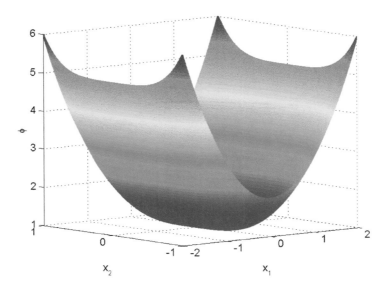

Figure 9.4. *The function $x_1^2 + x_2^4 + 1$ has a unique minimum at the origin $(0,0)$ and no maximum. Upon flipping it, the resulting function $-(x_1^2 + x_2^4 + 1)$ would have a unique maximum at the origin $(0,0)$ and no minimum.*

In general, finding a maximum for $\phi(\mathbf{x})$ is the same as finding a minimum for $-\phi(\mathbf{x})$. Note that there is no finite argument at which the function in this example attains a maximum. Therefore, the function defined by

$$\phi(\mathbf{x}) = -x_1^2 - x_2^4 - 1$$

does not attain a minimum value anywhere, even though it has a maximum at $\mathbf{x} = \mathbf{0}$. ∎

For realistic, nontrivial problems we typically cannot find minimum points by inspection. Indeed, it is often unclear in practice how many local minima a function $\phi(\mathbf{x})$ has and, if it has more than one local minimum, how to efficiently find the *global* minimum, which has the overall smallest value for ϕ.

9.2. Unconstrained optimization

Unconstrained optimization problems provide one very rich source for systems of nonlinear equations. But there is more at stake here. As we have seen in Section 7.4, the special case of function minimization where the objective function is quadratic yields a linear system of equations, which then need to be solved as a necessary stage. Minimization algorithms then translate into methods for solving linear systems of equations. We can recover the methods of Section 7.4 in this way—see Example 9.7 below—which is important in the case where the matrix is very large yet a matrix-vector multiplication is cheap. More generally, for large problems it is possible and useful to think of some variant of Newton's method for the nonlinear equations coupled with a method from Sections 7.4–7.6 for the resulting linear system in each nonlinear iteration.

Conditions for a minimum

Theorem: Taylor Series in Several Variables.
Let $\mathbf{x} = (x_1, x_2, \ldots, x_n)^T$ and assume that $\phi(\mathbf{x})$ has bounded derivatives up to order at least 3. Then for a direction vector $\mathbf{p} = (p_1, p_2, \ldots, p_n)^T$, the Taylor expansion in each coordinate yields

$$\phi(\mathbf{x} + \mathbf{p}) = \phi(\mathbf{x}) + \nabla\phi(\mathbf{x})^T \mathbf{p} + \frac{1}{2}\mathbf{p}^T \nabla^2 \phi(\mathbf{x}) \mathbf{p} + \mathcal{O}(\|\mathbf{p}\|^3).$$

Here, $\nabla\phi(\mathbf{x})$ is the **gradient** vector of first derivatives of ϕ at $\mathbf{x}$, and $\nabla^2 \phi(\mathbf{x})$ is the **Hessian** matrix of second derivatives of ϕ at $\mathbf{x}$. They are given by

$$\nabla\phi(\mathbf{x}) = \begin{pmatrix} \frac{\partial \phi}{\partial x_1} \\ \frac{\partial \phi}{\partial x_2} \\ \vdots \\ \frac{\partial \phi}{\partial x_n} \end{pmatrix}, \quad \nabla^2\phi(\mathbf{x}) = \begin{pmatrix} \frac{\partial^2 \phi}{\partial x_1^2} & \frac{\partial^2 \phi}{\partial x_1 \partial x_2} & \cdots & \frac{\partial^2 \phi}{\partial x_1 \partial x_n} \\ \frac{\partial^2 \phi}{\partial x_2 \partial x_1} & \frac{\partial^2 \phi}{\partial x_2^2} & \cdots & \frac{\partial^2 \phi}{\partial x_2 \partial x_n} \\ \vdots & \vdots & \ddots & \vdots \\ \frac{\partial^2 \phi}{\partial x_n \partial x_1} & \frac{\partial^2 \phi}{\partial x_n \partial x_2} & \cdots & \frac{\partial^2 \phi}{\partial x_n^2} \end{pmatrix}.$$

The remainder $\mathcal{O}(\|\mathbf{p}\|^3)$ term depends on derivatives of ϕ of order 3 and higher.

Assume that $\phi(\mathbf{x})$ has continuous derivatives up to order at least 2, which we denote by $\phi \in C^2$. It can then be expanded in a Taylor series as given on this page.

The scalar function

$$\nabla\phi(\mathbf{x})^T \mathbf{p} = \sum_{i=1}^{n} p_i \frac{\partial \phi}{\partial x_i}$$

is the *directional derivative* of ϕ at $\mathbf{x}$ in the direction $\mathbf{p}$. The next term in the expansion is the *quadratic form*

$$\frac{1}{2}\mathbf{p}^T \nabla^2 \phi(\mathbf{x}) \mathbf{p} = \frac{1}{2} \sum_{i,j=1}^{n} \frac{\partial^2 \phi}{\partial x_i \partial x_j} p_i p_j.$$

Now, if $\mathbf{x}^*$ is a local minimum, i.e., where the value of ϕ is less than or equal to its value at all neighboring points, then for any direction $\mathbf{p}$ we have

$$\phi(\mathbf{x}^* + \mathbf{p}) = \phi(\mathbf{x}^*) + \nabla\phi(\mathbf{x}^*)^T \mathbf{p} + \frac{1}{2}\mathbf{p}^T \nabla^2 \phi(\mathbf{x}^*) \mathbf{p} + \mathcal{O}(\|\mathbf{p}\|^3) \geq \phi(\mathbf{x}^*).$$

From this it transpires that a *necessary condition* for a minimum is

$$\nabla \phi(\mathbf{x}^*) = \mathbf{0}.$$

Such a point where the gradient vanishes is called a *critical point*. For Example 9.4 we obtain $2x_1 = 0$ and $4x_2^3 = 0$, trivially yielding $x_1 = x_2 = 0$.

Furthermore, a *sufficient condition* that a critical point be actually a minimum is that the Hessian matrix $\nabla^2 \phi(\mathbf{x}^*)$ be symmetric positive definite.[37]

How do we see all this? Think of a very small $\|\mathbf{p}\|$, such that $\|\mathbf{p}\|^2 \ll \|\mathbf{p}\|$. Now, if $\nabla \phi(\mathbf{x}^*) \neq \mathbf{0}$, then it is always possible to find a direction $\mathbf{p}$ such that $\nabla \phi(\mathbf{x}^*)^T \mathbf{p} < 0$. So, $\phi(\mathbf{x}^* + \mathbf{p}) < \phi(\mathbf{x}^*)$ and there is no minimum at $\mathbf{x}^*$. Thus, we must have $\nabla \phi(\mathbf{x}^*) = \mathbf{0}$. A similar argument also works to show that the gradient must vanish at points where $\phi(\mathbf{x})$ attains a local maximum. The gradient must also vanish at *saddle points* where a maximum is reached with respect to some variables and a minimum is reached with respect to other components. A saddle point is neither a minimizer nor a maximizer. For a simple instance, the function $\phi(\mathbf{x}) = x_1^2 - x_2^4 + 1$ has a saddle point at the origin; see Exercise 10.

Next, at a critical point which is a strict minimum we must also have for all directions $\mathbf{p}$ satisfying $0 < \|\mathbf{p}\| \ll 1$ that

$$\phi(\mathbf{x}^* + \mathbf{p}) = \phi(\mathbf{x}^*) + \frac{1}{2} \mathbf{p}^T \nabla^2 \phi(\mathbf{x}^*) \mathbf{p} + \mathcal{O}(\|\mathbf{p}\|^3) > \phi(\mathbf{x}^*).$$

This will happen if the Hessian matrix $\nabla^2 \phi(\mathbf{x}^*)$ is positive definite, since this guarantees that $\mathbf{p}^T \nabla^2 \phi(\mathbf{x}^*) \mathbf{p} > 0$. ♦

Theorem: Unconstrained Minimization Conditions.
Assume that $\phi(\mathbf{x})$ is smooth enough, e.g., suppose it has all derivatives up to third order, bounded. Then:

- A necessary condition for having a local minimum at a point $\mathbf{x}^*$ is that $\mathbf{x}^*$ be a critical point, i.e.,

$$\nabla \phi(\mathbf{x}^*) = \mathbf{0},$$

  and that the symmetric Hessian matrix $\nabla^2 \phi(\mathbf{x}^*)$ be positive semidefinite.

- A sufficient condition for having a local minimum at a point $\mathbf{x}^*$ is that $\mathbf{x}^*$ be a critical point and that $\nabla^2 \phi(\mathbf{x}^*)$ be positive definite.

Basic methods for unconstrained optimization

Generally speaking, the condition for a critical point yields a system of nonlinear equations

$$\mathbf{f}(\mathbf{x}) \equiv \nabla \phi(\mathbf{x}) = \mathbf{0}.$$

Thus, solving for a critical point is a special case of solving a system of nonlinear equations, and methods such as Newton's may be directly applied. Note that what corresponds to the Jacobian matrix $J(\mathbf{x})$ is the Hessian matrix of ϕ, $J(\mathbf{x}) = \nabla^2 \phi(\mathbf{x})$. Since the Hessian is usually symmetric positive definite near a miniumum point, this is indeed an important *special case* of systems of nonlinear equations.

---

[37] To recall, a symmetric matrix A is *positive definite* if $\mathbf{x}^T A \mathbf{x} > 0$ for all $\mathbf{x} \neq \mathbf{0}$. The eigenvalues of such a matrix are all positive, and an immediate corollary is that A is necessarily nonsingular. A symmetric matrix A is *positive semidefinite* if $\mathbf{x}^T A \mathbf{x} \geq 0$ for all $\mathbf{x}$. In this case the eigenvalues are all nonnegative, and here the possibility for having a singular matrix is not excluded.

9.2. Unconstrained optimization

Newton's method for unconstrained minimization

Newton's algorithm for the unconstrained minimization problem appears below.

> **Algorithm: Newton's Method for Unconstrained Minimization.**
> Consider the problem of minimizing $\phi(\mathbf{x})$ over $\mathcal{R}^n$, and let $\mathbf{x}_0$ be a given initial guess.
>
> for $k = 0, 1, \ldots$, until convergence
>
> $$\text{solve } \nabla^2 \phi(\mathbf{x}_k) \mathbf{p}_k = -\nabla \phi(\mathbf{x}_k) \text{ for } \mathbf{p}_k$$
> $$\text{set } \mathbf{x}_{k+1} = \mathbf{x}_k + \mathbf{p}_k.$$
>
> end

Example 9.5. Consider minimizing the function

$$\phi(\mathbf{x}) = \frac{1}{2} \left([1.5 - x_1(1 - x_2)]^2 + [2.25 - x_1(1 - x_2^2)]^2 + [2.625 - x_1(1 - x_2^3)]^2 \right).$$

A critical point is then defined by the equations

$$\nabla \phi(\mathbf{x}) = \mathbf{f}(\mathbf{x}) = \mathbf{0},$$

where

$$f_1(x_1, x_2) = -(1.5 - x_1(1 - x_2))(1 - x_2) - (2.25 - x_1(1 - x_2^2))(1 - x_2^2)$$
$$-(2.625 - x_1(1 - x_2^3))(1 - x_2^3),$$
$$f_2(x_1, x_2) = x_1(1.5 - x_1(1 - x_2)) + 2x_1 x_2(2.25 - x_1(1 - x_2^2))$$
$$+ 3x_1 x_2^2(2.625 - x_1(1 - x_2^3)).$$

As it turns out, there is a unique minimum for this problem at $\mathbf{x}^* = (3, .5)^T$, $\phi(\mathbf{x}^*) = 0$. There is also a saddle point, though: at $\hat{\mathbf{x}} = (0, 1)^T$ the gradient vanishes, so $\mathbf{f}(\hat{\mathbf{x}}) = \mathbf{0}$, but the Hessian of ϕ (which is the Jacobian of $\mathbf{f}$) equals $J = \begin{pmatrix} 0 & 13.875 \\ 13.875 & 0 \end{pmatrix}$. The eigenvalues are ± 13.875, hence the Hessian is not positive semidefinite at $\hat{\mathbf{x}}$ and the necessary conditions for a minimum are not satisfied there. For a visual illustration, see Figure 9.5.

Starting Newton's method from $\mathbf{x}_0 = (8, .2)^T$ we obtain the iterations recorded in the left part of Table 9.2. Rapid convergence to the minimum is observed. Note that the values of ϕ decrease monotonically here, even though the Hessian at the starting iterate, $\nabla^2 \phi(\mathbf{x}_0)$, is not positive definite: luck is on our side with this one.

Starting from $\mathbf{x}_0 = (8, .8)^T$ we obtain the iterations recorded in the right part of Table 9.2. Rapid convergence is again observed but, alas, to the saddle point $\hat{\mathbf{x}} = (1, 0)^T$ rather than to the minimum point! Note that the directions encountered are all descent directions: for a sufficiently small $\alpha > 0$, $\phi(\mathbf{x}_k + \alpha \mathbf{p}_k) < \phi(\mathbf{x}_k)$. This can happen if the critical point is a saddle point (and not a maximum), although the Hessian matrices encountered are not all positive definite, of course. ∎

A class of methods

The lesson we learn from Example 9.5 is that there is more to the unconstrained minimization problem than simply being a special case of nonlinear equation solving, although the latter is a useful point of view. There is more structure here. In particular, for any direction $\mathbf{p}$ at a point $\mathbf{x}$

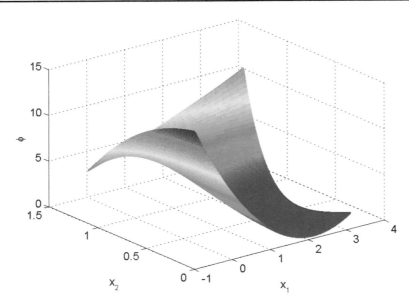

Figure 9.5. *The function of Example 9.5 has a unique minimum at* $\mathbf{x}^* = (3, .5)^T$ *as well as a saddle point at* $\hat{\mathbf{x}} = (0, 1)^T$.

Table 9.2. *Example 9.5. The first three columns to the right of the iteration counter track convergence starting from* $\mathbf{x}_0 = (8, .2)^T$*: Newton's method yields the minimum quickly. The following (rightmost) three columns track convergence starting from* $\mathbf{x}_0 = (8, .8)^T$*: Newton's method finds a critical point, but it's not the minimizer.*

k	$\|\mathbf{x}_k - \mathbf{x}^*\|$	$\phi_k - \phi(\mathbf{x}^*)$	$\mathbf{f}_k^T \mathbf{p}_k$	$\|\mathbf{x}_k - \mathbf{x}^*\|$	$\phi_k - \phi(\mathbf{x}^*)$	$\mathbf{f}_k^T \mathbf{p}_k$
0	5.01e+00	4.09e+01	−7.27e+01	5.01e+00	1.02e+00	−1.65e+00
1	8.66e-01	1.21e+00	−2.26e+00	3.96e+00	1.28e-01	−2.94e-02
2	6.49e-02	1.20e-02	−2.32e-02	4.22e+00	1.16e-01	−3.21e-01
3	1.39e-01	1.72e-03	−3.16e-03	1.66e+01	2.87e+02	−4.65e+02
4	2.10e-02	6.91e-05	−1.35e-04	6.14e+00	2.61e+01	−3.16e+01
5	1.38e-03	1.43e-07	−2.86e-07	3.43e+00	7.98e+00	−1.85e+00
6	3.03e-06	1.09e-12	−2.19e-12	2.97e+00	7.10e+00	−4.22e-03
7	2.84e-11	6.16e-23	−1.23e-22	3.04e+00	7.10e+00	−2.84e-08
8				3.04e+00	7.10e+00	−5.42e-19

where $\nabla \phi(\mathbf{x}) \neq \mathbf{0}$, if $\|\mathbf{p}\|$ is very small, then to have $\phi(\mathbf{x} + \mathbf{p}) < \phi(\mathbf{x})$ the directional derivative must be negative, written as

$$\nabla \phi(\mathbf{x})^T \mathbf{p} < 0.$$

This follows directly from the Taylor expansion given on page 259. A vector $\mathbf{p}$ for which this inequality holds is called a **descent direction** at the point $\mathbf{x}$.

9.2. Unconstrained optimization

Most methods for unconstrained minimization utilize iterations of the form

$$\mathbf{x}_{k+1} = \mathbf{x}_k + \alpha_k \mathbf{p}_k, \quad \text{where}$$
$$\mathbf{p}_k = -B_k^{-1} \nabla \phi(\mathbf{x}_k).$$

Thus, if B_k is symmetric positive definite, then $\mathbf{p}_k$ is a descent direction because $\nabla \phi(\mathbf{x}_k)^T \mathbf{p}_k = -\nabla \phi(\mathbf{x}_k)^T B_k^{-1} \nabla \phi(\mathbf{x}_k) < 0$. We next consider several possibilities for choosing the matrix B_k.

To motivate and focus such a method search, it is worthwhile to elaborate on the advantages and disadvantages of the basic Newton method. *Advantages* include (in addition to the fact that we have already seen it before) a quadratic local convergence—see Exercise 3—and the fact that for large problems, if the Hessian is sparse, then B_k is trivially also sparse. *Disadvantages* include the following:

- the method requires existence of the Hessian matrix;
- worse, it requires evaluation of the Hessian;
- the method requires solving a linear system of equations at each iteration;
- the matrix $B_k = \nabla^2 \phi(\mathbf{x}_k)$ may not be symmetric positive definite away from the minimum point;
- there is no control over convergence (e.g., does it converge at all? if yes, is it to a minimum point?)

The methods described below can all be considered as attempts to alleviate some of the difficulties listed above.

Gradient descent

The simplest choice for B_k is the identity matrix I. This yields the *gradient descent* direction

$$\mathbf{p}_k = -\nabla \phi(\mathbf{x}_k).$$

The gradient descent direction is guaranteed to be a direction of descent. Moreover, solving a linear system of equations at each iteration, as required by Newton's method, is avoided!

On the other hand, we must always determine the step size α_k. Worse, as demonstrated in Example 9.7 (page 266), coming up soon, this method may converge painfully slowly.

One use of the gradient descent direction is to fix a disadvantage of Newton's method by using a combination of the two methods, choosing

$$B_k = \nabla^2 \phi(\mathbf{x}_k) + \mu_k I.$$

The mix parameter $\mu_k \geq 0$ is selected to ensure that B_k is symmetric positive definite and hence that $\mathbf{p}_k$ is a descent direction. Sophisticated **trust region** methods exist to determine this parameter at each iteration k. These are beyond the scope of our text.

Line search

Techniques of *line search* address the selection of the scalar step length α_k for the iteration update

$$\mathbf{x}_{k+1} = \mathbf{x}_k + \alpha_k \mathbf{p}_k,$$

where $\mathbf{p}_k$ is the already-determined search direction. For the ("pure") gradient descent method this is a must.

For the unmodified Newton method, by the rationale of its derivation, the default choice is $\alpha_k = 1$. This is effective when $\|\mathbf{p}_k\|$ is small but not necessarily when $\|\mathbf{p}_k\|$ is large. The lack of guarantee of convergence from remote starting points $\mathbf{x}_0$, demonstrated, for instance, in Example 9.3 with $\alpha = 50$, is often a major cause for concern in practice. The step length α_k, $0 < \alpha_k \leq 1$, is therefore selected so as to guarantee a decrease in the objective function ϕ.

Thus, given $\mathbf{x}_k$ and a descent direction $\mathbf{p}_k$ we search along the *line* $\mathbf{x}_k + \alpha \mathbf{p}_k$ for a value $\alpha = \alpha_k$ such that

$$\phi(\mathbf{x}_{k+1}) \equiv \phi(\mathbf{x}_k + \alpha_k \mathbf{p}_k) \leq \phi(\mathbf{x}_k) + \sigma \alpha_k \nabla \phi(\mathbf{x}_k)^T \mathbf{p}_k,$$

where σ is a guard constant, e.g., $\sigma = 10^{-4}$. Typically, we start with a value $\alpha = \alpha_{\max}$ and decrease it if necessary.

A simple **backtracking** algorithm is to test the improvement inequality for geometrically decreasing values of α, which can be chosen by

$$\alpha/\alpha_{\max} = 1,\ 1/2,\ 1/4,\ \ldots,\ (1/2)^j,\ \ldots,$$

stopping as soon as a suitable value for α_k satisfying the above criterion is found. This is a particular **weak line search** strategy.

Another weak line search strategy is the following. Note that for the function

$$\psi(\alpha) = \phi(\mathbf{x}_k + \alpha \mathbf{p}_k),$$

we know $\psi(0) = \phi(\mathbf{x}_k)$, $\psi'(0) = \mathbf{p}^T \nabla \phi(\mathbf{x}_k)$, and $\psi(\tilde{\alpha}_k) = \phi(\mathbf{x}_k + \tilde{\alpha}_k \mathbf{p}_k)$, where $\tilde{\alpha}_k$ is the current, unsatisfactory value for α_k. Thus, we pass a quadratic polynomial through these three data points (see Section 10.7 and in particular Exercise 10.25) and minimize that quadratic interpolant, obtaining

$$\alpha_k = \frac{-\psi'(0)\tilde{\alpha}_k^2}{2(\psi(\tilde{\alpha}_k) - \psi(0) - \tilde{\alpha}_k \psi'(0))}.$$

Note that this may work only if $\psi'(0) < 0$. Here is a script mixing the above two strategies, with the search direction p, the function value phix, and the gradient gphix given at the current $\mathbf{x} = \mathbf{x}_k$.

```
pgphi = p' * gphix;
alpha = alphamax;
xn = x + alpha * p; phixn = feval(phi,xn);
while (phixn > phix + sigma * alpha * pgphi) * (alpha > alphamin)
    mu = -0.5 * pgphi * alpha / (phixn - phix - alpha * pgphi );
    if mu < .1 || pgphi >= 0
        mu = .5; % don't trust quadratic interpolation from far away
    end
    alpha = mu * alpha;
    xn = x + alpha * p;
    phixn = feval(phi,xn);
end
```

For Newton's method, once $\mathbf{x}_k$ gets close to $\mathbf{x}^*$ the best value for α_k is expected to be $\alpha_k = 1$, recovering the fast convergence of the pure method. If we are using the gradient descent method, however, then there is no such natural value for α_k in general.

Example 9.6. Consider

$$\phi(\mathbf{x}) = x_1^4 + x_1 x_2 + (1 + x_2)^2.$$

There is a unique minimum at $\mathbf{x}^* \approx (.695884386, -1.34794219)^T$, where $\phi(\mathbf{x}^*) \approx -.582445174$.

9.2. Unconstrained optimization

Starting from $\mathbf{x}_0 = (.75, -1.25)^T$ and using the pure Newton method we obtain very rapid convergence.

However, starting the pure Newton iteration from $\mathbf{x}_0 = (0, 0.3)^T$ yields no convergence. To understand why this happens, note that $\nabla^2 \phi(\mathbf{x}) = \begin{pmatrix} 12x_1^2 & 1 \\ 1 & 2 \end{pmatrix}$ is positive definite only when $24x_1^2 > 1$. In particular, the Hessian matrix is singular for $x_1 = \pm \frac{1}{\sqrt{24}}$. The initial iterate $\mathbf{x}_0 = (0, 0.3)^T$ and subsequent Newton iterates are trapped in the "bad region" of the Hessian.

Turning on weak line search, the "good" starting point yields the same sequence of pure Newton iterates, i.e., $\alpha_k = 1$ for all k encountered. On the other hand, starting from $\mathbf{x}_0 = (0, 0.3)^T$ and using the line search strategy we obtain convergence in 6 iterations. Here caution does pay off! The first three step lengths are less than 0.5 each, whereas in the last three, pure Newton iterations yield rapid convergence. ∎

There is theory to show that, under appropriate conditions, convergence to a local minimum starting from any initial guess $\mathbf{x}_0$ in a no longer small domain is guaranteed. But there are also concerns. One is that if Newton's method actually requires line search, then it may well be a sign that the search direction $\mathbf{p}_k$ itself is no good. Modifying that direction as in a trust region method may then be more fortuitous. Another concern for all these approaches is that they are **greedy**, trying to do what is best locally, rather than globally.

Example 9.7 (Convex quadratic minimization). Let us consider the simplest nonlinear optimization problem, namely, that of unconstrained minimization of quadratic functions of the form

$$\phi(\mathbf{x}) = \frac{1}{2} \mathbf{x}^T A \mathbf{x} - \mathbf{b}^T \mathbf{x},$$

where A is a given $n \times n$ symmetric positive definite matrix and $\mathbf{b}$ is a given n-vector.

Note that $\nabla \phi(\mathbf{x}) = A\mathbf{x} - \mathbf{b}$ and $\nabla^2 \phi(\mathbf{x}) = A$ for all $\mathbf{x}$. Hence, the necessary condition for a minimum, $\nabla \phi(\mathbf{x}) = \mathbf{0}$, becomes

$$A\mathbf{x} = \mathbf{b},$$

and this is sufficient for a unique minimum as well.

Newton's method becomes trivial here, reading

$$\mathbf{x}^* = \mathbf{x}_1 = \mathbf{x}_0 + A^{-1}(\mathbf{b} - A\mathbf{x}_0) = A^{-1} \mathbf{b}.$$

Thus, the solution is found in one iteration for any starting guess. Life becomes interesting again only when we think of a case where the direct solution of $A\mathbf{x} = \mathbf{b}$ is not advisable and iterative methods as described in Chapter 7 are required. Now simpler optimization methods act as iterative solvers for the linear system of equations.

Section 7.4 becomes particularly relevant here because we have used the same objective function $\phi(\mathbf{x})$ to derive the CG and gradient descent methods. Both steepest descent (SD) and CG methods can be written as

$$\mathbf{x}_{k+1} = \mathbf{x}_k + \alpha_k \mathbf{p}_k, \quad \alpha_k = \frac{\langle \mathbf{r}_k, \mathbf{r}_k \rangle}{\langle \mathbf{p}_k, A\mathbf{p}_k \rangle}.$$

This is the result of an *exact line search*. The SD method is a particular case of gradient descent, i.e., $\mathbf{p}_k = \mathbf{r}_k$, where $\mathbf{r}_k = -\nabla \phi(\mathbf{x}_k) = \mathbf{b} - A\mathbf{x}_k$, whereas for the CG method we have a more complex formula for $\mathbf{p}_k$.

But for a gradient descent method we can choose the step size α_k in different ways, not just SD. For instance, consider lagging the SD step size, using at the kth iteration

$$\alpha_k = \frac{\langle \mathbf{r}_{k-1}, \mathbf{r}_{k-1} \rangle}{\langle \mathbf{r}_{k-1}, A\mathbf{r}_{k-1} \rangle}.$$

We call this the lagged steepest descent (LSD) method; see also Exercise 7.14.

Figure 9.6 is directly comparable to Figure 7.5. It shows the convergence behavior for the same Poisson problem as in Examples 7.1 and 7.10 with $n = 31^2 = 961$ and $\kappa(A) \sim n$. The SD method is slow, requiring $\mathcal{O}(n)$ iterations. The CG method converges in an orderly fashion, requiring $\mathcal{O}(\sqrt{n})$ iterations. Interestingly, the gradient descent variant LSD converges much faster than SD, although the convergence is both slightly slower and significantly more erratic than CG.

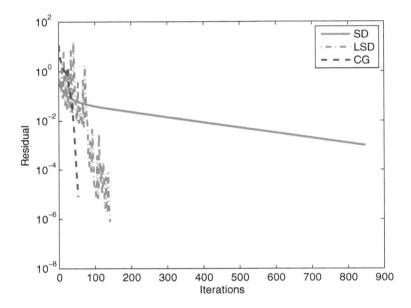

Figure 9.6. *Convergence behavior of gradient descent and conjugate gradient iterative schemes for the Poisson equation of Example* 7.1.

The first few SD iterations (say, up to 10) can actually be rather effective in applications. It is only when many such steps are taken that efficiency drops significantly, with the method falling into a pattern where the additional iterations do not add much. ∎

Inexact Newton

For large problems with sparse Hessian matrices, an iterative method for finding the direction $\mathbf{p}_k$ at each iteration may be called for. Incorporating this yields a method with an inner loop within an outer loop.

The outer loop is for a Newton-like iteration: a favorite choice is $B_k = \nabla^2 \phi(\mathbf{x}_k) + \mu_k I$ as mentioned earlier, with $\mu_k \geq 0$ at least guaranteeing positive semidefiniteness and B_k retaining sparsity.

The inner loop is for a linear system solution method such as those considered in Section 7.4 and Example 9.7. The PCG method is a great favorite here. The combination typically produces an

9.2. Unconstrained optimization

inexact Newton method, where the iterative method for the linear problem (the inner iteration) is stopped early. However, the accuracy of the linear iterative solver is increased, i.e., the convergence tolerance is tightened, if the outer iteration solution is already very close to a root $\mathbf{x}^*$, so as to realize the fast local convergence of the outer iteration. Some very popular methods in applications are included in this class.

Quasi-Newton methods

Of course, the difficulties in evaluating derivatives, arising already in Section 9.1 because we have to evaluate a Jacobian matrix there, are even more highlighted here because we have both first and second derivatives to evaluate. Moreover, extending the secant method discussed in Section 3.4 becomes more involved and more interesting because here we have the opportunity to insist on maintaining positive definiteness in B_k, thus ensuring descent directions throughout the iteration process.

So, at the kth iteration the approximation B_k of the Hessian is updated to form the next approximation B_{k+1}. These methods not only circumvent the need for evaluating $\nabla^2 \phi(\mathbf{x})$ explicitly, but are also often (though not always) much cheaper to use than other variants of Newton's method. Moreover, under appropriate conditions they guarantee convergence to a point $\mathbf{x}^*$ satisfying, to within an error tolerance, $\nabla \phi(\mathbf{x}^*) = \mathbf{0}$.

Note that by Taylor's expansion for the gradient we have

$$\nabla \phi(\mathbf{x}_k) \approx \nabla \phi(\mathbf{x}_{k+1}) - \nabla^2 \phi(\mathbf{x}_{k+1})(\mathbf{x}_{k+1} - \mathbf{x}_k).$$

The essential action of the Hessian is therefore deemed to be in the direction of $\mathbf{w}_k = \mathbf{x}_{k+1} - \mathbf{x}_k$. We require

$$B_{k+1}\mathbf{w}_k = \mathbf{y}_k, \quad \mathbf{y}_k = \nabla \phi(\mathbf{x}_{k+1}) - \nabla \phi(\mathbf{x}_k),$$

so that this action is reproduced by the approximation of the Hessian even though otherwise B_{k+1} and $\nabla^2 \phi(\mathbf{x}_{k+1})$ may not be particularly close.

BFGS method

The most popular variant of such methods, the Broyden–Fletcher–Goldfarb–Shanno update, commonly known as the BFGS method, is given on the next page. It forms the backbone of several general-purpose optimization software packages. Rather than updating B_k it updates directly its inverse $G_k = B_k^{-1}$.

The local convergence speed of this method is only superlinear, not quadratic, but that is not a big concern. Unfortunately, for ill-conditioned Hessian matrices and for large, sparse Hessian matrices the BFGS method may lose some of its attraction. For the sparse Hessian case there are popular *limited memory* versions of BFGS called L-BFGS. Describing these is beyond the scope of this text. In any case, other variants of Newton's method reappear on the scene.

Nonlinear least squares and Gauss–Newton

Let us recall the fundamental data fitting problem described in the opening paragraphs of Chapter 6 and demonstrated in Examples 8.1 and 8.2. Given observed data $\mathbf{b}$ and a model function $\mathbf{g}(\mathbf{x}) = A\mathbf{x}$ which predicts the data for each $\mathbf{x}$, the problem was to find the best $\mathbf{x}$ in the sense that the predicted data most closely agree with the observed data.

There are many instances in practice, however, where the model function depends *nonlinearly* on the argument vector $\mathbf{x}$. Then we are led to the unconstrained nonlinear optimization problem

$$\min_{\mathbf{x}} \|\mathbf{g}(\mathbf{x}) - \mathbf{b}\|,$$

Algorithm: The BFGS Iteration.
Choose $\mathbf{x}_0$ and G_0 (e.g., $G_0 = I$)
for $k = 0, 1, \ldots,$ until convergence

$$\mathbf{p}_k = -G_k \nabla \phi(\mathbf{x}_k)$$

find a suitable step size α_k

$$\mathbf{x}_{k+1} = \mathbf{x}_k + \alpha_k \mathbf{p}_k$$

$$\mathbf{w}_k = \alpha_k \mathbf{p}_k$$

$$\mathbf{y}_k = \nabla \phi(\mathbf{x}_{k+1}) - \nabla \phi(\mathbf{x}_k)$$

$$G_{k+1} = \left(I - \frac{\mathbf{w}_k \mathbf{y}_k^T}{\mathbf{y}_k^T \mathbf{w}_k}\right) G_k \left(I - \frac{\mathbf{y}_k \mathbf{w}_k^T}{\mathbf{y}_k^T \mathbf{w}_k}\right) + \frac{\mathbf{w}_k \mathbf{w}_k^T}{\mathbf{y}_k^T \mathbf{w}_k}.$$

end

where $\mathbf{g} \in C^2$ has m components, $\mathbf{g} : \mathcal{R}^n \to \mathcal{R}^m$, $m \geq n$, and the norm is by default the 2-norm. Obviously the canonical problem considered in Chapter 6 is a special case of the present, *nonlinear least squares* problem.

As in Section 6.1, it is more convenient to consider the problem in the form

$$\min_{\mathbf{x}} \phi(\mathbf{x}) = \frac{1}{2} \|\mathbf{g}(\mathbf{x}) - \mathbf{b}\|^2.$$

With the Jacobian matrix of $\mathbf{g}$ defined as

$$A(\mathbf{x}) = \begin{pmatrix} \frac{\partial g_1}{\partial x_1} & \frac{\partial g_1}{\partial x_2} & \cdots & \frac{\partial g_1}{\partial x_n} \\ \frac{\partial g_2}{\partial x_1} & \frac{\partial g_2}{\partial x_2} & \cdots & \frac{\partial g_2}{\partial x_n} \\ \vdots & \vdots & \ddots & \vdots \\ \frac{\partial g_{m-1}}{\partial x_1} & \frac{\partial g_{m-1}}{\partial x_2} & \cdots & \frac{\partial g_{m-1}}{\partial x_n} \\ \frac{\partial g_m}{\partial x_1} & \frac{\partial g_m}{\partial x_2} & \cdots & \frac{\partial g_m}{\partial x_n} \end{pmatrix},$$

the necessary conditions for a minimum are given by

$$\nabla \phi(\mathbf{x}^*) = A(\mathbf{x}^*)^T (\mathbf{g}(\mathbf{x}^*) - \mathbf{b}) = \mathbf{0}.$$

(Exercise 14 verifies that this is indeed true.) The obtained expression is an obvious generalization of the normal equations, and indeed we also assume as in Chapter 6 that the Jacobian matrix has full column rank, at least in a neighborhood of $\mathbf{x}^*$, so that $A^T A$ is symmetric positive definite. Here, however, we still have to solve a nonlinear system of n equations.

Another useful calculus manipulation gives the Hessian matrix

$$\nabla^2 \phi(\mathbf{x}) = A(\mathbf{x})^T A(\mathbf{x}) + L(\mathbf{x}),$$

where L is an $n \times n$ matrix with elements

$$L_{i,j} = \sum_{l=1}^{m} \frac{\partial^2 g_k}{\partial x_i \partial x_j} (g_l - b_l).$$

9.2. Unconstrained optimization

Gauss–Newton method

Turning to numerical methods, a classical one can be derived as follows. As usual we start with an initial guess $\mathbf{x}_0$ and consider at the kth iteration the question of obtaining the next iterate, $\mathbf{x}_{k+1}$, given the current one, $\mathbf{x}_k$. It is natural to approximate $\mathbf{g}(\mathbf{x}_{k+1})$ by $\mathbf{g}(\mathbf{x}_k) + A(\mathbf{x}_k)\mathbf{p}_k$, as in Newton's method. Thus, we solve for the correction vector the linear least squares problem

$$\min_{\mathbf{p}} \| A(\mathbf{x}_k)\mathbf{p} - (\mathbf{b} - \mathbf{g}(\mathbf{x}_k)) \|,$$

calling the minimizer $\mathbf{p}_k$, and set

$$\mathbf{x}_{k+1} = \mathbf{x}_k + \mathbf{p}_k.$$

This is the *Gauss–Newton* iteration. Observing that $\mathbf{b} - \mathbf{g}(\mathbf{x}_k)$ is the residual vector for the current iterate, the next direction vector is then obtained by fitting this residual from the range space of the current Jacobian matrix $A(\mathbf{x}_k)$. The normal equations that are mathematically equivalent to the linear least squares formulation of the kth iteration are

$$A(\mathbf{x}_k)^T A(\mathbf{x}_k)\mathbf{p}_k = A(\mathbf{x}_k)^T (\mathbf{b} - \mathbf{g}(\mathbf{x}_k)).$$

Algorithm: Gauss–Newton Method for Least Squares.
for $k = 0, 1, \ldots$, until convergence

1. Solve the linear least squares problem

$$\min_{\mathbf{p}} \| A(\mathbf{x}_k)\mathbf{p} - (\mathbf{b} - \mathbf{g}(\mathbf{x}_k)) \|,$$

   calling the minimizer $\mathbf{p} = \mathbf{p}_k$.

2. Set

$$\mathbf{x}_{k+1} = \mathbf{x}_k + \mathbf{p}_k.$$

end

Example 9.8. We generate data using the function $u(t) = e^{-2t} \cos(20t)$, depicted in Figure 9.7 as the solid blue curve. At the 51 points $t_i = .02(i-1)$, $i = 1, \ldots, 51$, we add 20% random noise to $u(t_i)$ to generate data b_i. The data obtained this way are depicted as green circles in Figure 9.7, and we now pretend that we have never seen the blue curve and attempt to fit to this $\mathbf{b}$ a function of the form

$$v(t) = x_1 e^{x_2 t} \cos(x_3 t).$$

In the above notation we have $m = 51$, $n = 3$, and the ith rows of $\mathbf{g}$ and A are subsequently defined by

$$g_i = x_1 e^{x_2 t_i} \cos(x_3 t_i), \quad a_{i,1} = e^{x_2 t_i} \cos(x_3 t_i),$$
$$a_{i,2} = t_i g_i \quad a_{i,3} = -t_i x_1 e^{x_2 t_i} \sin(x_3 t_i), \quad 1 \leq i \leq m.$$

Firing the Gauss–Newton method starting with $\mathbf{x}_0 = (1.2, -1.9, 18)^T$, we obtain after 7 iterations $\|\mathbf{p}_k\| < 10^{-7}$, $\mathbf{x} \approx (1.0471, -2.1068, 19.9588)^T$, and $\|\mathbf{g} - \mathbf{b}\| \approx .42$. The predicted data are displayed as red diamonds in Figure 9.7. This is a very satisfactory result.

With the noise level at 100% we obtain after 15 iterations $\|\mathbf{g} - \mathbf{b}\| \approx 1.92$ for $\mathbf{x} \approx (1.0814, -3.0291, 19.8583)^T$. Note the peculiar property that when the model approximates the data less well, more iterations are required.

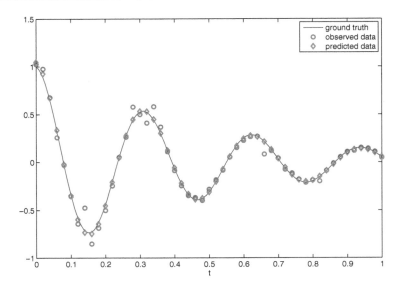

Figure 9.7. *Nonlinear data fitting; see Example 9.8.*

However, using the rougher initial guess $\mathbf{x}_0 = (1.2, -1.9, 10)^T$ quickly leads to another solution for this sensitive problem which does not approximate the data well. The world of nonlinearities is full of surprises, a fact that should not be underestimated: a careful assessment of any computed result is required at all times! ∎

Gauss–Newton vs. Newton

Note that the Gauss–Newton method differs from Newton's method for minimizing $\phi(\mathbf{x})$. Specifically, Newton's method reads

$$\nabla^2 \phi(\mathbf{x}_k)\, \mathbf{p}_k = -\nabla \phi(\mathbf{x}_k) = A(\mathbf{x}_k)^T (\mathbf{b} - \mathbf{g}(\mathbf{x}_k)),$$

so the right-hand side is the same as for the normal equations of the Gauss–Newton iteration, but the matrix $\nabla^2 \phi(\mathbf{x}_k)$ contains the extra term L, which is dropped in the Gauss–Newton iteration. This leads to several interesting conclusions:

- The Gauss–Newton direction, unlike Newton's, is guaranteed to be a descent direction with respect to ϕ. This is because $A^T A$ is symmetric positive definite even when $A^T A + L$ is not.

- The Gauss–Newton iteration is cheaper and can be better conditioned (more stable) than Newton's iteration.

- The convergence order of Gauss–Newton is not guaranteed in general to be quadratic, as the difference between it and Newton's iteration does not vanish in the limit.

- Gauss–Newton converges faster for problems where the model fits the data well! This is because then $\|\mathbf{g}(\mathbf{x}) - \mathbf{b}\|$ is "small" near the solution, hence $\|L\|$ is small and Gauss–Newton is closer to Newton.

Specific exercises for this section: Exercises 10–18.

9.3 *Constrained optimization

The general constrained optimization problem that we consider here consists as before of minimizing a scalar function in n variables, $\phi(\mathbf{x})$. The difference is that now there are *equality* and *inequality constraints* that any eligible $\mathbf{x}$ must satisfy. Many optimization problems in applications arise in this form and we will survey approaches and some methods for their solution.

Constraints

The general problem is written as

$$\min_{\mathbf{x} \in \Omega} \phi(\mathbf{x}), \quad \text{where}$$
$$\Omega = \{\mathbf{x} \in \mathcal{R}^n \mid c_i(\mathbf{x}) = 0, \, i \in \mathcal{E}, \, c_i(\mathbf{x}) \geq 0, \, i \in \mathcal{I}\}.$$

Thus, $\mathcal{E}$ is the set of equality constraints and $\mathcal{I}$ is the set of inequality constraints. We shall assume that $c_i(\mathbf{x}) \in C^1$, for all i. Any point $\mathbf{x} \in \Omega$ is called a *feasible solution* (as distinct from optimal solution). Let us see this notation in action for a simple example.

Example 9.9.

1. Consider the set depicted in Figure 9.8. Here $\mathcal{E} = \{1\}$, $\mathcal{I} = \{2, 3\}$. The set Ω consists of the straight line segment of $c_1 = 0$ between the curve $c_2 = 0$ and the x_1-axis.

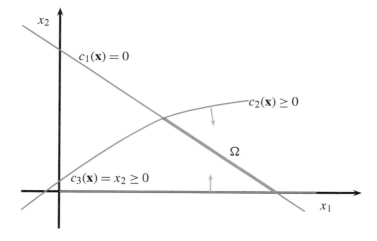

Figure 9.8. *Equality and inequality constraints. The feasible set Ω consists of the points on the thick red line.*

2. Often $\mathcal{E}$ is empty. Then Ω can have a nonempty interior as in Figure 9.9.

The basic difference above in the forms of Ω may well find its way into corresponding algorithms. ∎

A very common example of constraints is where some variables in $\mathbf{x}$ are required to be nonnegative, because they correspond to a physical quantity such as conductivity or a commodity such as the amount of flour left in the bakery's storage room, which cannot be negative.

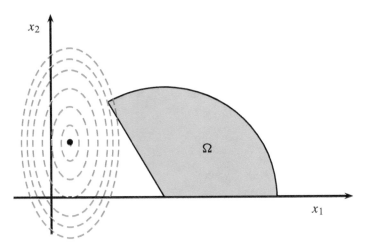

Figure 9.9. *A feasible set Ω with a nonempty interior, and level sets of ϕ; larger ellipses signify larger values of ϕ.*

If the unconstrained minimum of ϕ is inside Ω, then the problem is essentially that of unconstrained minimization. Thus, to make things interesting and different, assume that the unconstrained minimizer of ϕ is not inside Ω, as in Figure 9.9. In this figure and elsewhere it is convenient to plot level sets of the objective function ϕ. A **level set** of ϕ consists of all points $\mathbf{x}$ at which $\phi(\mathbf{x})$ attains the same value. The concentric ellipses in Figure 9.9 correspond to level sets for different values of ϕ. The gradient of ϕ at some point $\mathbf{x}$ is orthogonal to the level set that passes through that point.

Define at each point $\mathbf{x} \in \mathcal{R}^n$ the *active set*

$$\mathcal{A}(\mathbf{x}) = \mathcal{E} \cup \{i \in \mathcal{I} \mid c_i(\mathbf{x}) = 0\}.$$

We are looking, then, at problems where $\mathcal{A}(\mathbf{x}^*)$ is nonempty.

Our brief exposition starts with the necessary conditions for an optimum. We then survey various algorithm families in practical use for constrained optimization. Finally, we discuss in more detail a very useful algorithm for the linear programming problem.

Conditions for a constrained minimum

As in the unconstrained case, there are first order necessary conditions for a critical point, and there are second order necessary and sufficient conditions for a local minimum. However, they are all significantly more complicated than in the unconstrained case. We start with first order necessary conditions, extending the unconstrained requirement that $\nabla \phi(\mathbf{x}^*)$ vanish. Let us consider a motivating example first.

Example 9.10. We start with one equality constraint in $\mathcal{R}^2$: $c(x_1, x_2) = 0$. See Figure 9.10. At any point $\mathbf{x}$ the gradient $\nabla \phi(\mathbf{x})$ is orthogonal to the tangent of the level set at that point. At the minimum point $\mathbf{x}^*$ the constraint and level set for $\phi(\mathbf{x}^*)$ have the same tangent direction, hence $\nabla \phi(\mathbf{x}^*)$ is parallel to $\nabla c(\mathbf{x}^*)$. This means that there is a constant of proportionality, λ^*, such that

$$\nabla \phi(\mathbf{x}^*) = \lambda^* \nabla c(\mathbf{x}^*).$$

9.3. *Constrained optimization

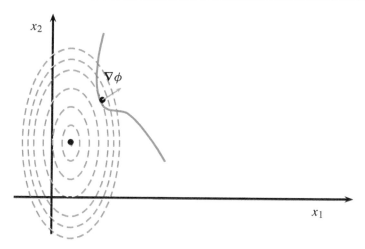

Figure 9.10. *One equality constraint and the level sets of ϕ. At $\mathbf{x}^*$ the gradient is orthogonal to the tangent of the constraint.*

Next, suppose that there is only one inequality constraint, $c(x_1, x_2) \geq 0$. The feasible region Ω is the domain to the right of the solid blue curve in Figure 9.10. Thus, not only is $\nabla \phi(\mathbf{x}^*)$ parallel to $\nabla c(\mathbf{x}^*)$, it must also point into the interior of the feasible set, which means that the constant of proportionality λ^* cannot be negative. We therefore have

$$\nabla \phi(\mathbf{x}^*) = \lambda^* \nabla c(\mathbf{x}^*), \quad \lambda^* \geq 0.$$

This λ^* is called a Lagrange multiplier. ∎

The general case is not an immediate extension of Example 9.10, but we omit a complete derivation and just state it. First we must make the *constraint qualification* assumption, specified on this page.

Constraint qualification.
Let $\mathbf{x}^*$ be a local critical point and denote by $\nabla c_i(\mathbf{x})$ the gradients of the constraints. Further, let A_*^T be the matrix whose columns are the gradients $\nabla c_i(\mathbf{x}^*)$ of all active constraints, that is, those belonging to $\mathcal{A}(\mathbf{x}^*)$.
The *constraint qualification* assumption is that the matrix A_*^T has full column rank.

Next, define the **Lagrangian**

$$\mathcal{L}(\mathbf{x}, \lambda) = \phi(\mathbf{x}) - \sum_{i \in \mathcal{E} \cup \mathcal{I}} \lambda_i c_i(\mathbf{x}).$$

Then the celebrated Karush–Kuhn–Tucker (KKT) conditions, given on the next page, are necessary first order conditions for a minimum.

With our constraint qualification holding there exists a locally unique λ^* that together with $\mathbf{x}^*$ satisfies the KKT conditions. Further, it can be shown that without constraint qualification these conditions are not quite necessary.

> **Theorem: Constrained Minimization Conditions.**
> Assume that $\phi(\mathbf{x})$ and $c_i(\mathbf{x})$ are smooth enough near a critical point $\mathbf{x}^*$ and that constraint qualification holds. Then there is a vector of *Lagrange multipliers* $\boldsymbol{\lambda}^*$ such that
> $$\nabla_{\mathbf{x}}\mathcal{L}(\mathbf{x}^*, \boldsymbol{\lambda}^*) = \mathbf{0},$$
> $$c_i(\mathbf{x}^*) = 0 \quad \forall i \in \mathcal{E},$$
> $$c_i(\mathbf{x}^*) \geq 0 \quad \forall i \in \mathcal{I},$$
> $$\lambda_i^* \geq 0 \quad \forall i \in \mathcal{I},$$
> $$\lambda_i^* c_i(\mathbf{x}^*) = 0 \quad \forall i \in \mathcal{E} \cup \mathcal{I}.$$

Example 9.11. Consider minimizing a quadratic function under linear equality constraints. The problem is specified as

$$\min_{\mathbf{x}} \phi(\mathbf{x}) = \frac{1}{2}\mathbf{x}^T H \mathbf{x} - \mathbf{d}^T \mathbf{x}$$
$$\text{s.t. } A\mathbf{x} = \mathbf{b},$$

where A is $m \times n$, $m \leq n$. The constraint qualification condition translates to requiring that A has full row rank.

This is a **quadratic programming** problem with *equality constraints*. We have

$$\mathcal{L}(\mathbf{x}, \boldsymbol{\lambda}) = \frac{1}{2}\mathbf{x}^T H \mathbf{x} - \mathbf{d}^T \mathbf{x} - \boldsymbol{\lambda}^T(\mathbf{b} - A\mathbf{x}),$$

and the KKT conditions read

$$H\mathbf{x} - \mathbf{d} + A^T \boldsymbol{\lambda} = \mathbf{0},$$
$$\mathbf{b} - A\mathbf{x} = \mathbf{0}.$$

The resulting linear system of equations, also known as a **saddle point system**, can be arranged as

$$\begin{pmatrix} H & A^T \\ A & 0 \end{pmatrix} \begin{pmatrix} \mathbf{x} \\ \boldsymbol{\lambda} \end{pmatrix} = \begin{pmatrix} \mathbf{d} \\ \mathbf{b} \end{pmatrix}.$$

The KKT, or saddle point, matrix $K = \begin{pmatrix} H & A^T \\ A & 0 \end{pmatrix}$, is symmetric, but it is *indefinite* even if H is positive definite. Since A has full row rank, K is nonsingular if $\mathbf{y}^T H \mathbf{y} \neq 0$ for all $\mathbf{y} \in \text{null}(A)$, $\mathbf{y} \neq \mathbf{0}$; see Exercise 19.

If K is nonsingular, then obviously there is exactly one critical point $\mathbf{x}^*$ satisfying the necessary conditions. ∎

The situation is more complicated, even for quadratic programming problems, when there are inequality constraints.

Given what has happened to the very simple unconstrained first order condition defining a critical point, you can imagine that the second order conditions for the constrained case (involving the Hessian matrix of the Lagrangian $\mathcal{L}$ with respect to $\mathbf{x}$) would not be simple. Indeed they are not, and are thus excluded from this brief presentation.

9.3. *Constrained optimization

An overview of algorithms

Generally speaking, equality constraints have a more algebraic flavor (searching in a reduced space, on a constraint manifold, etc.), whereas inequality constraints may introduce an additional combinatorial flavor to the problem (namely, which constraints are active at the solution and which are not). It is possible to distinguish between two general approaches for constrained optimization:

- **Active set**

    Assuming that the unconstrained minimum of ϕ is not in the interior of the feasibility set Ω, our solution must be on the boundary $\partial\Omega$. Active set methods thus search for the optimum along the boundary. For inequality constraints there are active set methods where we keep track of $\mathcal{A}(\mathbf{x}_k)$, shuffling constraints in and out of the active set as we go downhill along the boundary.

- **Other approaches**

    Here the optimal solution is approached in an iterative fashion, either from within the feasible region Ω (these are *interior point* methods) or, more generally, by a method that may use infeasible points as well but does not move along the boundary.

    Methods of the second type include those where the objective function is modified sequentially: for each such modification the corresponding unconstrained minimization problem is solved.

- **Penalty methods**

    These methods, like gradient descent, are a favorite among common folks due to their simplicity. Consider, for instance, the unconstrained minimization of the penalized objective function

    $$\min_{\mathbf{x}} \psi(\mathbf{x}, \mu) = \phi(\mathbf{x}) + \frac{1}{2\mu} \sum_{i \in \mathcal{E}} c_i^2(\mathbf{x}),$$

    where $\mu > 0$ is a parameter. This makes sense for problems with only equality constraints: notice that $c_i(\mathbf{x}^*) = 0$. One then solves a sequence of such problems for decreasing values of nonnegative μ, $\mu \downarrow 0$, using $\mathbf{x}(\mu_{k-1})$ to construct the initial iterate for the unconstrained iteration for $\mathbf{x}(\mu_k)$.

- **Barrier methods**

    These are interior point methods. Starting from a point within the feasible set, a sequence of unconstrained problems is solved such that at each stage the objective function is modified to ensure that the solution on the boundary is approached from within Ω. For instance, consider

    $$\min_{\mathbf{x}} \psi(\mathbf{x}, \mu) = \phi(\mathbf{x}) - \mu \sum_{i \in \mathcal{I}} \log c_i(\mathbf{x}),$$

    where $\mu \downarrow 0$.

- **Augmented Lagrangian**

    For a problem with only equality constraints, consider a variation of the penalty method given by the unconstrained problem

    $$\min_{\mathbf{x}} \psi(\mathbf{x}, \boldsymbol{\lambda}, \mu) = \phi(\mathbf{x}) - \sum_{i \in \mathcal{E}} \lambda_i c_i(\mathbf{x}) + \frac{1}{2\mu} \sum_{i \in \mathcal{E}} c_i^2(\mathbf{x}).$$

    Given estimates λ_k, μ_k, we solve the unconstrained minimization problem for $\mathbf{x} = \mathbf{x}_{k+1}$, then update the multipliers to λ_{k+1}, μ_{k+1}.

The most popular method in practice for general-purpose codes is **sequential quadratic programming** (SQP), although there are some augmented Lagrangian variants breathing down its neck. At each iteration one solves a quadratic program (QP) given by

$$\min_{\mathbf{p}} \frac{1}{2}\mathbf{p}^T W_k \mathbf{p} + \nabla \phi(\mathbf{x}_k)^T \mathbf{p},$$

$$c_i(\mathbf{x}_k) + \nabla \mathbf{c}_i(\mathbf{x}_k)^T \mathbf{p} = 0, \ i \in \mathcal{E}, \quad c_i(\mathbf{x}_k) + \nabla \mathbf{c}_i(\mathbf{x}_k)^T \mathbf{p} \geq 0, \ i \in \mathcal{I},$$

yielding a direction $\mathbf{p}_k$ at iterate $(\mathbf{x}_k, \lambda_k)$. The objective function here approximates the Lagrangian $\mathcal{L}$ near $(\mathbf{x}_k, \lambda_k)$ and the linear constraints form the linearization of the original constraints at the current iterate. An active set method is used to solve the QP with inequality constraints.

Linear programming

The linear programming (LP) problem has a huge number of applications. It has been a focal point in the discipline of **operations research** for decades. In fact, it is a special case of the general constrained optimization problem, where both objective function and constraints are linear, but as befits such an important class of problems it has its own well-established folklore and typical notation.

Primal and dual forms

In **primal form** the LP problem is written as

$$\min \phi(\mathbf{x}) = \mathbf{c}^T \mathbf{x}$$
$$\text{s.t. } A\mathbf{x} = \mathbf{b},$$
$$\mathbf{x} \geq \mathbf{0}.$$

Note that this notation does not fit exactly into our previous conventions: $\mathbf{c}$ is just a constant ("cost") vector and the inequality constraints are separated from the matrix notation. We assume that $A \in \mathcal{R}^{m \times n}$ has full row rank m. Also, let $n > m$ to avoid trivial cases, and assume that there exists an optimal solution.

A fundamental concept throughout optimization is that of *duality*. For our LP the **dual form** is given by

$$\max_{\mathbf{y}} \mathbf{b}^T \mathbf{y}$$
$$\text{s.t. } A^T \mathbf{y} \leq \mathbf{c}.$$

The dual of the dual is the primal. If $\mathbf{x}$ is feasible for the primal and $\mathbf{y}$ is feasible for the dual (meaning they each satisfy the corresponding constraints), then $\mathbf{b}^T \mathbf{y} \leq \mathbf{c}^T \mathbf{x}$, with equality holding at optimum. This is an essential model in commerce, where the seller's goal of maximum profit and the buyer's goal of minimum expenditure are related in this way.

Example 9.12. A company manufactures two agricultural products, high grade and low grade, that have selling prices of $150 and $100 per ton, respectively. The products require three raw materials of which there are 75, 60, and 25 tons available, respectively, per hour. If x_1 and x_2 are the corresponding production levels in tons, then the raw material requirements are given by

$$2x_1 + x_2 \leq 75,$$
$$x_1 + x_2 \leq 60,$$
$$x_1 \leq 25.$$

9.3. *Constrained optimization

The challenge is to determine these production levels so as to maximize the profit $150x_1 + 100x_2$. Of course, the production levels cannot be negative, so $x_1 \geq 0$ and $x_2 \geq 0$.

To cast this problem in primal form, we first turn the inequality constraints into equalities by introducing additional nonnegative unknowns $x_3 = 75 - 2x_1 - x_2$, $x_4 = 60 - x_1 - x_2$ and $x_5 = 25 - x_1$. These extra variables are also nonnegative but do not have a cost associated with them. Furthermore, we can write the objective as that of minimizing $-150x_1 - 100x_2$. Hence we have the LP in primal form as given on the facing page defined by

$$\mathbf{c}^T = \begin{pmatrix} -150 & -100 & 0 & 0 & 0 \end{pmatrix},$$

$$A = \begin{pmatrix} 2 & 1 & 1 & 0 & 0 \\ 1 & 1 & 0 & 1 & 0 \\ 1 & 0 & 0 & 0 & 1 \end{pmatrix}, \quad \mathbf{b} = \begin{pmatrix} 75 \\ 60 \\ 25 \end{pmatrix},$$

where the unknown vector $\mathbf{x} = (x_1, x_2, x_3, x_4, x_5)^T$ is nonnegative componentwise. Please stay tuned: Example 9.13 reveals the optimal dollar figure.

Now, try to interpret the commercial significance that the dual form has. ∎

A **basic feasible solution** $\hat{\mathbf{x}}$ for this problem satisfies the equality and inequality constraints and in addition has at most only m nonzero components. In fact, the boundary of the constraint set Ω is polygonal for LP, and a basic feasible solution corresponds to a vertex of this polygon. Moreover, there is always an optimal solution for the LP problem that is a basic feasible solution!

The Lagrangian can be written as

$$\mathcal{L} = \mathbf{c}^T \mathbf{x} - \sum_{i=1}^{m} y_i \left(\sum_{j=1}^{n} a_{ij} x_j - b_i \right) - \sum_{j=1}^{n} s_j x_j,$$

i.e., we use both y_i and s_j (s is for "slack") to denote the Lagrange multipliers λ. The KKT conditions therefore read

$$\mathbf{c} - A^T \mathbf{y} - \mathbf{s} = \mathbf{0},$$
$$A\mathbf{x} = \mathbf{b},$$
$$s_i x_i = 0, \quad i = 1, \ldots, n,$$
$$\mathbf{x} \geq \mathbf{0}, \mathbf{s} \geq \mathbf{0}.$$

These conditions are necessary as well as sufficient for a minimum. They are called **complementarity slackness** conditions, and they express the LP problem without any objective function to minimize or maximize. Note that in view of the nonnegativity constraints the equalities $x_i s_i = 0$ can be replaced by the more compact (if enigmatic) condition

$$\mathbf{x}^T \mathbf{s} = 0.$$

The KKT equations are also called the *primal-dual form*.

LP algorithms

Turning to algorithms, both general classes of methods mentioned above have popular instances in the LP context. The famous, veteran, **simplex** method moves along the boundary of Ω from one basic feasible solution to the next in search for an optimal one. It had been the mainstay of the entire field of optimization for several decades ending in the 1980s. Sophisticated variants of the simplex method remain very competitive today.

Here, however, we concentrate on a **primal-dual** iterative method that, in its pure form, does not actually touch the boundary of the constraint set: all iterates satisfy $\mathbf{x}_k > \mathbf{0}$, $\mathbf{s}_k > \mathbf{0}$. It is both simple to implement and performs very well in most situations. A detailed computational assessment would largely depend on the efficiency of the linear solver that is used, but this is beyond the scope of this brief introduction to the topic.

Let us denote by X the diagonal matrix with elements $x_1, \ldots, x_n$ on the main diagonal. Likewise, denote by S the diagonal matrix with elements $s_1, \ldots, s_n$ on the main diagonal, and in addition set $\mathbf{e} = (1, \ldots, 1)^T$. Next, define the **center path**, depending on a parameter $\tau \geq 0$, by

$$A^T \mathbf{y} + \mathbf{s} = \mathbf{c},$$
$$A\mathbf{x} = \mathbf{b},$$
$$XS\mathbf{e} = \tau \mathbf{e},$$
$$\mathbf{x} > \mathbf{0}, \mathbf{s} > \mathbf{0}.$$

The solution $\mathbf{x}(\tau)$, $\mathbf{y}(\tau)$, $\mathbf{s}(\tau)$ is feasible for $\tau \geq 0$, and methods can be constructed which follow this path to optimality as $\tau \downarrow 0$. In the limit the KKT conditions are active; see Figure 9.11.

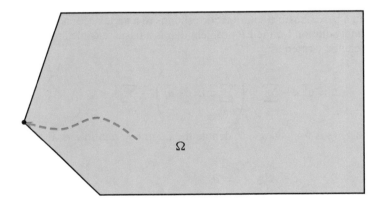

Figure 9.11. *Center path in the LP primal feasibility region.*

We need one more definition. For a current point $\mathbf{z} = (\mathbf{x}, \mathbf{y}, \mathbf{s})$ satisfying $\mathbf{x} > \mathbf{0}$, $\mathbf{s} > \mathbf{0}$, but not necessarily anything else, the *duality measure* or **duality gap** is defined by

$$\mu = \frac{1}{n} \mathbf{x}^T \mathbf{s}.$$

On the center path, obviously $\mu = \tau$. More generally, $\mu > 0$ and we want $\mu \downarrow 0$.

Now, a Newton step for the mildly nonlinear subsystem consisting of the equalities in both the KKT system and the system defining the center path reads

$$\begin{pmatrix} 0 & A^T & I \\ A & 0 & 0 \\ S & 0 & X \end{pmatrix} \begin{pmatrix} \delta \mathbf{x} \\ \delta \mathbf{y} \\ \delta \mathbf{s} \end{pmatrix} = \begin{pmatrix} \mathbf{c} - A^T \mathbf{y} - \mathbf{s} \\ \mathbf{b} - A\mathbf{x} \\ \sigma \mu \mathbf{e} - XS\mathbf{e} \end{pmatrix},$$

9.3. *Constrained optimization

where $\sigma \in [0, 1]$ is a *centering parameter*. The value $\sigma = 1$ yields a (cautious) step towards the center path with $\tau = \mu$, while $\sigma = 0$ gives a (euphoric) Newton step for the KKT system. Once this linear system is solved we update the current solution by adding suitable multiples of $\delta\mathbf{x}$, $\delta\mathbf{y}$, and $\delta\mathbf{s}$ to $\mathbf{x}$, $\mathbf{y}$, and $\mathbf{s}$, respectively, such that the positivity conditions still hold.

A predictor-corrector step

A practical algorithm of the above sort is obtained using a predictor-corrector approach. The idea can be described as follows. Assuming for the moment that a full step can be taken, we ideally want at the end of the step to obtain

$$A^T(\mathbf{y} + \delta\mathbf{y}) + (\mathbf{s} + \delta\mathbf{s}) = \mathbf{c},$$
$$A(\mathbf{x} + \delta\mathbf{x}) = \mathbf{b},$$
$$(X + \delta X)(S + \delta S)\mathbf{e} = \tau\mathbf{e},$$
$$\mathbf{x} > \mathbf{0}, \mathbf{s} > \mathbf{0}.$$

Here we write $\tau = \sigma\mu$, with μ the (computable) duality gap and σ the centering parameter. Two questions then arise:

1. How to set σ such that a large step can be taken?
2. How to approximate the curvature term $\Delta = \delta X \delta S$? This is the only nonlinear term here; the rest indeed is precisely a linearizing Newton iteration.

Both of these questions are settled by a *predictor* step. We set $\sigma = \sigma_p = 0$, $\Delta_p = 0$, and solve the linear equations arising from the Newton step. This direction is meant to achieve maximum progress towards the optimum and would be successful if positivity constraints do not prove too restrictive. Denote the result $\delta\mathbf{x}^p, \delta\mathbf{s}^p$. For one thing, this gives an approximation

$$\Delta_c = \delta X^p \delta S^p$$

for the curvature term. We next figure out how far we can go in the predicted direction while still retaining positivity, by evaluating

$$\alpha_p = \min\left\{1, \min_{\delta x_i < 0} \frac{x_i}{-\delta x_i^p}\right\}, \quad \beta_p = \min\left\{1, \min_{\delta s_i < 0} \frac{s_i}{-\delta s_i^p}\right\}.$$

If we were to take the maximum allowed step, then the new duality gap would be

$$\mu_p = \frac{1}{n}(\mathbf{x} + \alpha_p \delta\mathbf{x}^p)^T(\mathbf{s} + \beta_p \delta\mathbf{s}^p).$$

This quantity can be evaluated. Now, if μ_p is much smaller than μ, then great progress can be made by the predictor. If not, then a significant centering has to be added to the search direction. This is captured by choosing the hack

$$\sigma = \left(\frac{\mu_p}{\mu}\right)^3.$$

With the value of the centering parameter thus fixed and with the curvature approximated by Δ_c, we next solve the system

$$\begin{pmatrix} 0 & A^T & I \\ A & 0 & 0 \\ S & 0 & X \end{pmatrix} \begin{pmatrix} \delta\mathbf{x} \\ \delta\mathbf{y} \\ \delta\mathbf{s} \end{pmatrix} = \begin{pmatrix} \mathbf{c} - A^T\mathbf{y} - \mathbf{s} \\ \mathbf{b} - A\mathbf{x} \\ \sigma\mu\mathbf{e} - (XS + \Delta_c)\mathbf{e} \end{pmatrix}$$

for the corrected direction. Note that the same matrix and most of the same right-hand side are used for both the predictor and the corrector.

Further details

Since X and S are diagonal, it is easy (and common!) to perform block Gaussian elimination and reduce the above matrix to the saddle point matrix

$$\begin{pmatrix} -X^{-1}S & A^T \\ A & 0 \end{pmatrix},$$

and even further to the symmetric positive definite matrix

$$AS^{-1}XA^T.$$

The code that follows on the next page demonstrates the latter formulation (specifically, in function newtlp).

Finally, we want the next solution point to be positive, too, so the update reads

$$\mathbf{x} = \mathbf{x} + \alpha\, \delta\mathbf{x}, \quad \alpha = \min\left\{1, \min_{\delta x_i < 0} \frac{(1-\texttt{tolf})x_i}{-\delta x_i}\right\},$$

$$\mathbf{s} = \mathbf{s} + \beta\, \delta\mathbf{s}, \quad \beta = \min\left\{1, \min_{\delta s_i < 0} \frac{(1-\texttt{tolf})s_i}{-\delta s_i}\right\},$$

$$\mathbf{y} = \mathbf{y} + \beta\, \delta\mathbf{y}.$$

Set, e.g., $\texttt{tolf} = 0.01$.

Note that our algorithm is not strictly an interior point method, because the iterates do not necessarily satisfy the equality constraints of the KKT equations. The combined scaled norm of these residuals, which we refer to as the "infeasibility," is expected of course to shrink to 0 as the optimal solution is approached.

This ends the description of the primal-dual algorithm, but there is another item to consider below, before we see an implementation.

Hopping to a nearby feasible solution

Recall that the exact solution does have (potentially many) zeros. Thus, as the current iterate gets closer to the optimum there will be very large elements in S^{-1} and X^{-1}, giving the methods of Sections 7.4 and 7.5 the creeps. On the other hand, the main difficulty with active set methods, which is to search among an exponentially large number of feasible solutions, may be greatly reduced by considering only nearby ones, this being justified because we are near the optimum.

When the duality gap μ gets small below some tolerance $\texttt{tolb}$, say, $\texttt{tolb} = 0.0001$, it can be worthwhile to hop to a nearby vertex of Ω (which corresponds to a basic solution) and check optimality. This can be done as follows:

1. Let $\mathcal{B} \subset \{1, 2, \ldots, n\}$ be the set of m indices corresponding to largest components of $\mathbf{x}$. This is the "suspected optimal basis."

2. Set

$$\hat{x}_j = 0 \quad \text{if } j \notin \mathcal{B}, \quad \hat{s}_j = 0 \quad \text{if } j \in \mathcal{B}.$$

Compose the $m \times m$ matrix B out of the m basic columns of A and likewise the vector $\mathbf{c}_B$ out of $\mathbf{c}$.

9.3. *Constrained optimization

3. Solve
$$B\hat{x}_B = b$$
and insert the values of $\hat{x}_B$ as the basic values of $\hat{x}$ for $j \in \mathcal{B}$.

4. Set
$$\hat{y} = B^{-T} c_B, \quad \hat{s} = c - A^T \hat{y}.$$

5. If $\hat{x} > -\epsilon e$, $\hat{s} > -\epsilon e$, and $\frac{1}{n}\hat{s}^T \hat{x} < \epsilon$ for a very small positive tolerance ϵ, then an optimal solution has been found in $(\hat{x}, \hat{y}, \hat{s})$. If not, then this information is discarded and the algorithm continues from (x, y, s).

Here is our program. It does not do fancy checking for special situations such as degenerate cases, empty feasibility sets, or the potential for an unbounded solution. Moreover, as usual we have preferred readability over performance optimization. Still, it is amazing how simple an efficient program for such a central, nontrivial problem can finally become.

```
function [x,gap,nbas] = lpm (A,b,c)
%
% function [x,gap,nbas] = lpm (A,b,c)
%
% solve the linear programming problem
% min c^T x   s.t. Ax = b, x >= 0 .
%
% A is l x m,  b is l x 1, c is m x 1.
% return solution x and duality gap
% (should be close to 0 if all well).
% Also, nbas is the number of hops to check basic solutions
% before optimality is reached.

[l,m] = size(A);
scaleb = norm(b) + norm(A,inf) + 1;
scalec = norm(c) + norm(A,inf) + 1;
tolf = 0.01; otol = 1-tolf;
toln = 1.e-9; tolp = 1.e-10; tolb = 1.e-4;
nbas = 0;

% Initial guess
x = ones(m,1); s = ones(m,1); y = zeros(l,1);

fprintf('itn          gap            infeas             mu\n')
for it = 1:2*m+10 % iterate, counting pred-cor as one

  % duality measure
  mu = (x'*s)/m;
  % predict correction
  [dx,dy,ds] = newtlp(A,b,c,x,y,s,0);

  % incorporate positivity into constraints
  alfa = 1; beta = 1;
  for i=1:m
    if dx(i) < 0, alfa = min(alfa, -x(i)/dx(i)); end
```

282   Chapter 9. Nonlinear Systems and Optimization

```
      if ds(i) < 0, beta = min(beta, -s(i)/ds(i)); end
   end

   % the would-be duality measure
   muaff = ( (x+alfa*dx)' * (s+beta*ds) ) / m;
   % centering parameter
   sigma = (muaff/mu)^3;

   % correct towards center path
   smu = sigma * mu;
   [dx,dy,ds] = newtlp(A,b,c,x,y,s,smu,dx,ds);

   % incorporate positivity into constraints
   alfa = 1; beta = 1;
   for i=1:m
     if dx(i) < 0, alfa = min(alfa, -otol*x(i)/dx(i)); end
     if ds(i) < 0, beta = min(beta, -otol*s(i)/ds(i)); end
   end

   % update solution
   x = x + alfa*dx;
   s = s + beta*ds;
   y = y + beta*dy;

   % check progress
   infeas = norm(b - A*x)/scaleb + norm(c - A'*y - s)/scalec;
   gap = (c'*x - b'*y) / m;
   if (infeas > 1.e+12)+(gap < -toln)
     fprintf('no convergence: perhaps no solution')
     return
   end
   fprintf('%d    %e    %e    %e\n',it,gap,infeas,mu)

   if (abs(infeas) < toln)*(abs(gap) < toln), return, end

   % hop to next basic solution
   if gap < tolb
     nbas = nbas + 1;
     [xx,sortof] = sort(-x);
     [xx,yy,ss] = basln(A,b,c,sortof);
     gap = (c'*xx - b'*yy) / m;
     if (sum(xx+tolp >= 0) > m-1)*(sum(ss+tolp >= 0) > m-1)...
           *(abs(gap) < toln)
       x = xx;
       return
     end
   end

end

function [dx,dy,ds] = newtlp(A,b,c,x,y,s,mu,dx,ds)
%
% function [dx,dy,ds] = newtlp(A,b,c,x,y,s,mu,dx,ds)
```

9.3. *Constrained optimization

```
%
% A Newton step for lp

[l,m] = size(A);
rc = A'*y + s - c;
rb = A*x - b;
rt = x.*s - mu;
if nargin == 9, rt = rt + dx.*ds; end

rhs = [-rb + A * ((rt - x.*rc)./s)];
Mat = A * diag(x./s) * A';
dy = Mat \ rhs;
ds = -rc - A'*dy;
dx = -(x.*ds + rt)./s;

function [x,y,s] = basln(A,b,c,sort)
%
% function [x,y,s] = basln(A,b,c,sort)
%
% given a vector of indices, the first l indicate a basis
% out of A. Find corresponding basic solution.

[l,m] = size(A);
B = zeros(l,l); cb = zeros(l,1);

% construct basis
for j=1:l
  B(:,j) = A(:,sort(j));
  cb(j)  = c(sort(j));
end

xb = B \ b;
x = zeros(m,1);
for j=1:l
  x(sort(j)) = xb(j);
end
y = B' \ cb;
s = c - A'*y;
```

Example 9.13. Running lpm for the small problem of Example 9.12, convergence was obtained in 13 iterations. One hop to a nearby basic solution found the optimum $x_1 = 15$, $x_2 = 45$. The profit is therefore $150x_1 + 100x_2 = \$6750$.

The performance of our program on this small problem, where $m = 3$, $n = 5$, is not really exceptional, but this is not very important: more interesting is what happens with larger problems. ∎

Example 9.14. Let us generate a test problem as follows. We set $m = 260$, $n = 570$, and generate a random $m \times n$ matrix A using MATLAB's randn. Then we generate two more random (nonnegative) n-vectors, $\hat{x}$ and c, using MATLAB's rand. Then set $b = A\hat{x}$ and forget about $\hat{x}$. This guarantees that the feasibility region is not empty.

Calling lpm(A,b,c) yields convergence in 9 iterations, with the convergence history depicted in Table 9.3. There were two hops to nearby boundary vertices, the first unsuccessful and the second successful, terminating the iteration with the optimal solution for which both measures recorded in Table 9.3 are zero.

Table 9.3. *Example 9.14. Tracking progress of the primal-dual LP algorithm.*

Iteration	Infeasibility norm	Duality gap
1	6.38e-02	1.78e-01
2	1.36e-02	5.60e-02
3	3.50e-03	1.78e-02
4	7.80e-04	5.89e-03
5	1.65e-04	1.77e-03
6	5.07e-05	6.79e-04
7	1.74e-05	2.15e-04
8	3.53e-06	5.78e-05
9	5.66e-07	2.01e-05

Note how much smaller than m the total number of iterations is! ∎

Example 9.15 (minimum norm solutions of underdetermined systems). This case study is longer than usual and solves a problem which is a focus of current general interest.

Consider the underdetermined linear system of equations

$$J\mathbf{y} = \mathbf{b},$$

where J is a given $m \times \hat{n}$ matrix with full row rank, and $\mathbf{b}$ is a given right-hand side. To have some instance in mind, think of J coinciding with A of the previous example, except that here there is no objective function ϕ to minimize and $\mathbf{y}$ has nothing to do with dual variables. If $\hat{n} > m$, then there can be many solutions to this system, because the nullspace of J is not empty. Which one shall we choose?!

If you have recently read Section 8.2, then the answer can be almost automatic: choose the one with the smallest ℓ_2-norm. Thus, we solve the easy, non-LP, constrained optimization problem

$$\min_{\mathbf{y}} \|\mathbf{y}\|_2$$
$$\text{s.t. } J\mathbf{y} = \mathbf{b}.$$

This can be done using the SVD of J, when J is small enough. Recalling Section 4.4, let us write the decomposition as $J = U\Sigma V^T$. Then $\Sigma(V^T\mathbf{y}) = U^T\mathbf{b}$, with Σ essentially a diagonal matrix with the singular values σ_i in decreasing order on its main diagonal. For $\mathbf{z} = V^T\mathbf{y}$ we therefore set

$$z_i = \begin{cases} \frac{(U^T\mathbf{b})_i}{\sigma_i}, & 1 \leq i \leq m, \\ 0, & m < i \leq \hat{n}, \end{cases}$$

and the minimum ℓ_2-norm solution is recovered as $\mathbf{y} = V\mathbf{z}$. These transformations back and forth are allowed, to recall, because U and V are orthogonal matrices, so the transformations using them preserve the ℓ_2-norm. See also Exercise 22 for an alternative solution method.

9.3. *Constrained optimization

The above-described solution process is carried out often in many practical applications. But it is not always what we actually want, because typically all components of the obtained solution **y** are nonzero. There are important applications in which we know a priori that the right-hand-side **b** can be written as a linear combination of a relatively small number l, $l \leq m < \hat{n}$, of the columns of J. A **sparse solution** for the given linear system can then be constructed, if we only knew which l columns of J to choose, by setting all other components of **y** to zero. This is very interesting because such a sparse solution corresponds to picking a selection of what is important in the model that the linear system of equations represents.

So how do we find such a sparse solution? As it turns out, the problem of finding a solution with the minimal number of nonzeros can be very tough indeed.

Fortunately, it also turns out that often (though not always) a sparse solution is obtained by solving the constrained optimization problem

$$\min_{\mathbf{y}} \|\mathbf{y}\|_1$$
$$\text{s.t. } J\mathbf{y} = \mathbf{b}.$$

Thus, we are seeking the solution with the smallest ℓ_1-norm!

The latter problem can be formulated as LP. Let us write

$$y_i = u_i - v_i, \quad 1 \leq i \leq \hat{n},$$

and set

$$A = [J, -J], \quad \mathbf{x} = \begin{pmatrix} \mathbf{u} \\ \mathbf{v} \end{pmatrix}, \quad \mathbf{c} = \mathbf{e}.$$

Setting $n = 2\hat{n}$ we face an LP in standard, primal form using our previous notation, to which our program `lpm` can be applied.

One additional improvement that can be made to `lpm` for this special but rather important class of problems is regarding early checks for an optimal solution. As the iteration gets well on its way to convergence the dominant components of **y** start to emerge well above the rest in magnitude. So, instead of the general "hopping to the nearest basis," here we can collect the columns of J corresponding to the first l largest current values $|y_i|$ into an $m \times l$ matrix. Solving a linear least squares overdetermined problem to which the techniques of Chapter 6 apply, trying to fit **b** with these l components, we check the resulting residual and declare victory if its norm is small enough, recovering the full solution by setting all other components of **y** to 0.

For a numerical instance we create J exactly as A in Example 9.14, so $m = 260$ and $n = 2 \cdot 570 = 1140$. Then we set $\hat{y}_i = 1 - 10/i$ for $i = 20, 40, 60, \ldots, 260$, $\hat{y}_i = 0$ otherwise, then form $\mathbf{b} = J\hat{\mathbf{y}}$ and not use $\hat{\mathbf{y}}$ any further.

The ℓ_2 minimum norm solution $\mathbf{y}^{(2)}$ yields

$$\|\mathbf{y}^{(2)}\|_2 = 2.12, \quad \|\mathbf{y}^{(2)}\|_1 = 34.27.$$

For the ℓ_1 minimum norm solution $\mathbf{y}^{(1)}$ we apply `lpm` with the search for early minimum as described above, using the estimate $l = 20$. Only 3 iterations are required before a solution is found satisfying

$$\|\mathbf{y}^{(1)}\|_2 = 3.20, \quad \|\mathbf{y}^{(1)}\|_1 = 11.41.$$

The two solutions are plotted in Figure 9.12. Clearly the ℓ_1 solution is highly sparse, reproducing $\hat{\mathbf{y}}$ in this particular case, while the ℓ_2 solution is not. Many researchers are rather excited about this ℓ_1 phenomenon nowadays. ∎

Specific exercises for this section: Exercises 19–22.

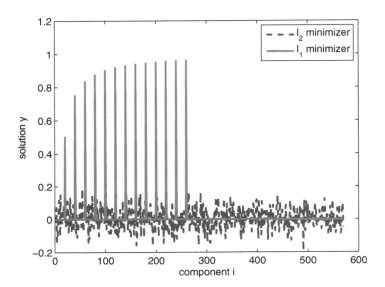

Figure 9.12. *Two minimum norm solutions for an underdetermined linear system of equations. The ℓ_1 solution is nicely sparse.*

9.4 Exercises

0. **Review questions**

    (a) What is constrained optimization and what is unconstrained optimization? State a significant difference between the two.

    (b) What is a Jacobian matrix?

    (c) How many solutions is a nonlinear system of n algebraic equations expected to have?

    (d) What are the fundamental additional difficulties in the numerical solution of problems considered in Section 9.1 as compared to those considered in Chapter 3?

    (e) State the condition for quadratic convergence for a nonlinear system and explain its importance.

    (f) Explain how minimizing a nonlinear function in several variables (Section 9.2) leads to solving systems of algebraic equations (Section 9.1).

    (g) What are the necessary and the sufficient conditions for an unconstrained minimum?

    (h) What is the difference between a local minimum and a global minimum? Which of the two types do the methods discussed in this chapter typically attempt to find?

    (i) Why is it important to keep a symmetric positive definite iteration matrix B_k while seeking a minimum of a smooth function $\phi(\mathbf{x})$? Does Newton's method automatically guarantee this?

    (j) Define descent direction and line search, and explain their relationship.

    (k) What is a gradient descent method? State two advantages and two disadvantages that it has over Newton's method for unconstrained optimization.

    (l) What is a quasi-Newton method? Name three advantages that such a method may have over Newton's method.

9.4. Exercises

(m) State the KKT conditions and explain their importance.

(n) What is an active set method? Name a famous active set method for the problem of linear programming.

(o) How does the primal-dual form for linear programming relate to the primal form of such problems?

(p) Define central path and duality gap for linear programming problems.

1. Express the Newton iteration for solving each of the following systems:

   (a)
   $$x_1^2 + x_1 x_2^3 = 9,$$
   $$3x_1^2 x_2 - x_2^3 = 4.$$

   (b)
   $$x_1 + x_2 - 2x_1 x_2 = 0,$$
   $$x_1^2 + x_2^2 - 2x_1 + 2x_2 = -1.$$

   (c)
   $$x_1^3 - x_2^2 = 0,$$
   $$x_1 + x_1^2 x_2 = 2.$$

2. Consider the system
   $$x_1 - 1 = 0,$$
   $$x_1 x_2 - 1 = 0.$$

   It is trivial to solve this system immediately, you will surely agree, but suppose we apply Newton's method anyway. For what initial guesses will the method fail? Explain.

3. This exercise is concerned with proofs.

   Let $\mathbf{f}(\mathbf{x})$ be Lipschitz continuously differentiable in an open convex set $\mathcal{D} \subset \mathcal{R}^n$, i.e., there is a constant $\gamma \geq 0$ such that
   $$\|J(\mathbf{x}) - J(\mathbf{y})\| \leq \gamma \|\mathbf{x} - \mathbf{y}\| \quad \forall \mathbf{x}, \mathbf{y} \in \mathcal{D},$$
   where J is the $n \times n$ Jacobian matrix of $\mathbf{f}$. It is possible to show that if $\mathbf{x}$ and $\mathbf{x} + \mathbf{p}$ are in $\mathcal{D}$, then
   $$\mathbf{f}(\mathbf{x} + \mathbf{p}) = \mathbf{f}(\mathbf{x}) + \int_0^1 J(\mathbf{x} + \tau \mathbf{p}) \mathbf{p} \, d\tau.$$

   (a) Assuming the above as given, show that
   $$\|\mathbf{f}(\mathbf{x} + \mathbf{p}) - \mathbf{f}(\mathbf{x}) - J(\mathbf{x})\mathbf{p}\| \leq \frac{\gamma}{2} \|\mathbf{p}\|^2.$$

   (b) Suppose further that there is a root $\mathbf{x}^* \in \mathcal{D}$ satisfying
   $$\mathbf{f}(\mathbf{x}^*) = \mathbf{0}, \quad J(\mathbf{x}^*) \text{ nonsingular}.$$
   Show that for $\mathbf{x}_0$ sufficiently close to $\mathbf{x}^*$, Newton's method converges quadratically.

4. Consider the nonlinear PDE in two dimensions given by

$$-\left(\frac{\partial^2 u}{\partial x^2} + \frac{\partial^2 u}{\partial y^2}\right) + e^u = g(x,y).$$

Here $u = u(x,y)$ is a function in the two variables x and y, defined on the unit square $(0,1) \times (0,1)$ and subject to homogeneous Dirichlet boundary conditions. We discretize it on a uniform mesh, just like in Example 7.1, using $h = 1/(N+1)$, and obtain the equations

$$4u_{i,j} - u_{i+1,j} - u_{i-1,j} - u_{i,j+1} - u_{i,j-1} + h^2 e^{u_{i,j}} = h^2 g_{i,j}, \quad 1 \leq i, j \leq N,$$
$$u_{i,j} = 0 \text{ otherwise}.$$

(a) Find the Jacobian matrix, J, and show that it is always symmetric positive definite.

(b) Write a MATLAB program that solves this system of nonlinear equations using Newton's method for $N = 8, 16, 32$. Generate a right-hand-side function $g(x,y)$ such that the exact solution of the differential problem is $u(x,y) = \sin(\pi x)\sin(\pi y)$. Start with an initial guess of all zeros, and stop the iteration when $\|\delta u^{(k)}\|_2 < 10^{-6}$. Plot the norms $\|\delta u^{(k)}\|_2$ and explain the convergence behavior you observe.

5. An $n \times n$ linear system of equations $A\mathbf{x} = \mathbf{b}$ is modified in the following manner: for each i, $i = 1, \ldots, n$, the value b_i on the right-hand side of the ith equation is replaced by $b_i - x_i^3$. Obviously, the modified system of equations (for the unknowns x_i) is now nonlinear.

(a) Find the corresponding Jacobian matrix.

(b) Given that A is strictly diagonally dominant with positive elements on its diagonal, state whether or not it is guaranteed that the Jacobian matrix at each iterate is nonsingular.

(c) Suppose that A is symmetric positive definite (not necessarily diagonally dominant) and that Newton's method is applied to solve the nonlinear system. Is it guaranteed to converge?

6. (a) Suppose Newton's method is applied to a linear system $A\mathbf{x} = \mathbf{b}$. How does the iterative formula look and how many iterations does it take to converge?

(b) Suppose the Jacobian matrix is singular at the solution of a nonlinear system of equations. Speculate what can occur in terms of convergence and rate of convergence. Specifically, is it possible to have a situation where the Newton iteration converges but convergence is not quadratic?

7. Use Newton's method to solve a discretized version of the differential equation

$$y'' = -(y')^2 - y + \ln x, \quad 1 \leq x \leq 2, \quad y(1) = 0, \quad y(2) = \ln 2.$$

The discretization on a uniform mesh, with the notation of Example 9.3, can be

$$\frac{y_{i+1} - 2y_i + y_{i-1}}{h^2} + \left(\frac{y_{i+1} - y_{i-1}}{2h}\right)^2 + y_i = \ln(1 + ih), \quad i = 1, 2, \ldots, n.$$

The actual solution of this problem is $y(x) = \ln x$. Compare your numerical results to the solution $y(x)$ for $n = 8, 16, 32$, and 64. Make observations regarding the convergence behavior of Newton's method in terms of the iterations and the mesh size, as well as the solution error.

9.4. Exercises

8. Consider the nonlinear problem

$$-\left(\frac{\partial^2 u}{\partial x^2} + \frac{\partial^2 u}{\partial y^2}\right) - e^u = 0$$

defined in the unit square $0 < x, y < 1$ with homogeneous Dirichlet boundary conditions (see Example 7.1). This problem is known to have two solutions.

Using a discretization on a uniform grid with step size $h = 1/(N+1) = 2^{-7}$, extending that described in Example 7.1, find approximations for the two solution functions. Use Newton's method with appropriate initial guesses and solve the resulting linear system directly (i.e., using MATLAB's `backslash`). Plot the two solutions and display their scaled norms $\|\mathbf{u}\|_2/\sqrt{n}$ as well as $\|\exp(\mathbf{u})\|_\infty$. How many iterations does it take Newton's method to converge?

9. Repeat the process described in Exercise 8, this time using a Krylov subspace method of your choice from among those described in Section 7.5 for solving the linear systems that arise. Instead of requiring the iterative linear solver to converge to a tight tolerance, impose a "loose" stopping criterion: stop when the relative residual reaches a value below 0.01. This is an example of an *inexact Newton's method*. Comment on the rate of convergence and the overall computational cost.

   [Warning: This question is more advanced than others nearby. Check out Exercise 7.25 before attempting to solve it.]

10. Show that the function

$$\phi(\mathbf{x}) = x_1^2 - x_2^4 + 1$$

has a saddle point at the origin, i.e., the origin is a critical point that is neither a minimum nor a maximum.

What happens if Newton's method is applied to find this saddle point?

11. Which of the systems of nonlinear equations in Exercise 1 can be expressed as $\nabla \phi(\mathbf{x}) = \mathbf{0}$ for some function ϕ to be minimized?

12. The Rosenbrock function $\phi(\mathbf{x}) = 100(x_2 - x_1^2)^2 + (1 - x_1)^2$ has a unique minimizer that can be found by inspection.

    Apply Newton's method and the BFGS method, both with weak line search, starting from $\mathbf{x}_0 = (0,0)^T$. Compare performance.

13. Consider minimizing the function $\phi(\mathbf{x}) = \mathbf{c}^T \mathbf{x} + \frac{1}{2} \mathbf{x}^T H \mathbf{x}$, where $\mathbf{c} = (5.04, -59.4, 146.4, -96.6)^T$ and

$$H = \begin{pmatrix} .16 & -1.2 & 2.4 & -1.4 \\ -1.2 & 12.0 & -27.0 & 16.8 \\ 2.4 & -27.0 & 64.8 & -42.0 \\ -1.4 & 16.8 & -42.0 & 28.0 \end{pmatrix}.$$

Try both Newton and BFGS methods, starting from $\mathbf{x}_0 = (-1, 3, 3, 0)^T$. Explain why the BFGS method requires significantly more iterations than Newton's.

14. Consider the nonlinear least square problem of minimizing

$$\phi(\mathbf{x}) = \frac{1}{2}\|\mathbf{g}(\mathbf{x}) - \mathbf{b}\|^2.$$

(a) Show that
$$\nabla \phi(\mathbf{x}) = A(\mathbf{x})^T (\mathbf{g}(\mathbf{x}) - \mathbf{b}),$$
where A is the $m \times n$ Jacobian matrix of $\mathbf{g}$.

(b) Show that
$$\nabla^2 \phi(\mathbf{x}) = A(\mathbf{x})^T A(\mathbf{x}) + L(\mathbf{x}),$$
where L is an $n \times n$ matrix with elements
$$L_{i,j} = \sum_{k=1}^{m} \frac{\partial^2 g_k}{\partial x_i \partial x_j} (g_k - b_k).$$

[You may want to check first what $\frac{\partial \phi}{\partial x_i}$ looks like for a fixed i; later on, look at $\frac{\partial^2 \phi}{\partial x_i \partial x_j}$ for a fixed j.]

15. In Exercise 6.4 you are asked to devise a variable transformation trick in order to solve a simple nonlinear data fitting problem as a linear one.

    Solve the same problem in the given variables, i.e., $\mathbf{x} = (\gamma_1, \gamma_2)^T$, without the special transformation, using the Gauss–Newton method. Your iterative process should stop when $\|\mathbf{p}_k\| < $ `tol`$(\|\mathbf{x}_k\| + 1)$. Experiment with a few initial guesses and tolerances `tol`. What are your observations?

    If you have also solved Exercise 6.4, then please use the solution obtained in that way as an initial guess for the Gauss–Newton method. Assess the relative quality of the obtained solutions by comparing the resulting residuals.

16. Solve the problem of Exercise 15 (i.e., the problem given in Exercise 6.4 without the special transformation applied there) using Newton's method. Apply the same stopping criterion as in Exercise 15, and compare the iteration counts of the two methods (i.e., compare Newton to Gauss–Newton) for tolerance values `tol` $= 10^{-6}$ and `tol` $= 10^{-10}$. Discuss your observations.

17. This exercise is concerned with recovering a function $u(t)$ on the interval $[0, 1]$ given noisy data b_i at points $t_i = ih$, $i = 0, 1, \ldots, N$, with $N = 1/h$. Because the data values are noisy, we cannot simply set $u(t_i) \equiv u_i = b_i$: knowing that $u(t)$ should be piecewise smooth, we add a regularization term to penalize excessive roughness in u. For the unknown vector $\mathbf{u} = (u_0, u_1, \ldots, u_N)^T$ we therefore solve
$$\min \phi_2(\mathbf{u}) = \frac{h}{2} \sum_{i=1}^{N} \frac{1}{2} \left[(u_i - b_i)^2 + (u_{i-1} - b_{i-1})^2 \right] + \frac{\beta h}{2} \sum_{i=1}^{N} \left(\frac{u_i - u_{i-1}}{h} \right)^2.$$

    (a) Write down the gradient and the Hessian of this objective function. To describe the regularization term use the matrix
    $$W = \frac{1}{\sqrt{h}} \begin{pmatrix} -1 & 1 & & & \\ & -1 & 1 & & \\ & & \ddots & \ddots & \\ & & & -1 & 1 \end{pmatrix} \in \mathfrak{R}^{N \times (N+1)}.$$

9.4. Exercises

(b) Solve this problem numerically for the following problem instances. To "synthesize data" for a given N, start with

$$b_p(t) = \begin{cases} 1, & 0 \le t < .25, \\ 2, & .25 \le t < .5, \\ 2 - 100(t - .5)(.7 - t), & .5 \le t < .7, \\ 4, & .7 \le t \le 1. \end{cases}$$

Evaluate this at the grid points and then add noise as follows:

```
noisev = randn(size(b_p)) * mean(abs(b_p)) * noise;
data = b_p + noisev;
```

The resulting values are the data that your program "sees." An illustration is given in Figure 9.13.

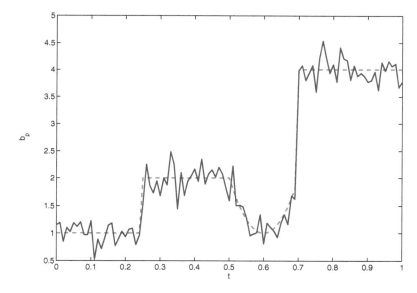

Figure 9.13. *A depiction of the noisy function (in solid blue) and the function to be recovered (in red) for Exercise* 17.

Plot this data and the recovered curve **u** for the parameter values $(\beta, noise) = (10^{-3}, .01)$, $(10^{-3}, .1)$, $(10^{-4}, .01)$ and $(10^{-4}, .1)$. Try $N = 64$ or $N = 128$. What are your observations?

18. Continuing Exercise 17, unfortunately if the data function contains jump discontinuities, then these discontinuities will be smeared by the regularization proposed there. So, consider instead the minimization problem

$$\min \phi_1(\mathbf{u}) = \frac{h}{2} \sum_{i=1}^{N} \frac{1}{2}\left[(u_i - b_i)^2 + (u_{i-1} - b_{i-1})^2\right] + \gamma h \sum_{i=1}^{N} \sqrt{\left(\frac{u_i - u_{i-1}}{h}\right)^2 + \varepsilon},$$

where $\varepsilon = 10^{-6}$, say.

(a) Let $T_{1,h}$ denote what γ multiplies in the objective function ϕ_1. Show that

$$\frac{\partial T_{1,h}}{\partial u_j} = \frac{u_j - u_{j-1}}{\sqrt{(u_j - u_{j-1})^2 + \varepsilon h^2}} + \frac{u_j - u_{j+1}}{\sqrt{(u_j - u_{j+1})^2 + \varepsilon h^2}}.$$

Moreover, letting

$$\hat{D} = \text{diag}\left\{h/\sqrt{(u_j - u_{j-1})^2 + \varepsilon h^2}\right\}, \quad \hat{B} = \sqrt{\hat{D}},$$

it is possible to write

$$\nabla T_{1,h} = W^T \hat{B}^T \hat{B} W\mathbf{u} = W^T \hat{D} W\mathbf{u}.$$

(b) A method for solving this problem then suggests itself, whereby at the beginning of each iteration we fix $\hat{D}$ based on the current iterate and apply the usual algorithms for a linear weighted least squares functional. This is called *iterated least squares*.

Use this method to solve the same problems as in the previous question (i.e., same synthesized data), for $\gamma = 10^{-2}$ and $\gamma = 10^{-3}$. Discuss your observations.

19. Consider the saddle point matrix

$$K = \begin{pmatrix} H & A^T \\ A & 0 \end{pmatrix},$$

where the matrix H is symmetric positive semidefinite and the matrix A has full row rank.

(a) Show that K is nonsingular if $\mathbf{y}^T H \mathbf{y} \neq 0$ for all $\mathbf{y} \in \text{null}(A)$, $\mathbf{y} \neq \mathbf{0}$.
(b) Show by example that K is symmetric but indefinite, i.e., it has both positive and negative eigenvalues.

20. For the saddle point matrix of Exercise 19, if we are forced to solve the corresponding linear system of equations iteratively, which of the methods described in Sections 7.4 and 7.5 would you choose?

21. Consider the problem

$$\min_{\mathbf{y}} \|\mathbf{y}\|_p$$
$$\text{s.t. } J\mathbf{y} = \mathbf{b},$$

where J is $m \times n$, $m \leq n$.
Devise an example where solving with $p = 2$ actually makes more sense than with $p = 1$.

22. With the notation of Exercise 21, for $p = 2$ the problem can be solved as in Example 9.15 using SVD. But it can also be solved using the techniques of constrained optimization:

(a) Explain why it is justified to replace the objective function by $\frac{1}{2}\|\mathbf{y}\|_2^2$.
(b) Form the KKT conditions of the resulting constrained optimization problem, obtaining a linear system of equations.
(c) Devise an example and solve it using the method just developed as well as the one from Example 9.15. Compare and discuss.

9.5 Additional notes

The topic of optimization is vast. The *Mathematical Optimization Society* is dedicated to various academic aspects of optimization, and there is much more in addition to what it usually covers. Interestingly, this society was previously named the *Mathematical Programming Society* and decided to change its name just recently, in 2010. Here we have quickly described only part of the elephant that most directly relates to other topics of our text.

There are many textbooks on numerical optimization. One that deals with topics in a manner close to our heart is Nocedal and Wright [57]. Possibly more accessible though less computationally savvy is Griva, Nash, and Sofer [34]. The books of Fletcher [25], Dennis and Schnabel [22], and Gill, Murray, and Wright [29] are favorite classics.

There are many state-of-the-art software packages that offer solutions to a variety of constrained optimization problems. An example of such a package is CPLEX: http://www-01.ibm.com/software/integration/optimization/cplex-optimizer/.

The predictor-corrector method in Section 9.3 is a variation of the description in [57] of an algorithm due to Mehrotra [53].

The veteran field of continuous constrained optimization is still seeing much action these days. One emerging area deals with very large but structured problems where the constraints are partial differential equations; see, e.g., Biegler at al. [8].

Another currently hot research area is sparse reconstruction and sparse prior techniques, which largely rely on remarkable sparse solution recovery properties that ℓ_1-norm optimization, and also ℓ_p-norms with $p < 1$, possess. A taste of this was given in Example 9.15. Basically, these provide often credible approximation methods for the exponentially hard combinatorial problem of selecting a small number of basis functions, or solution components, for describing an observed phenomenon to acceptable accuracy. Mallat [52] addresses uses of this tool in signal processing, as do many others.

Linear systems that arise from constrained optimization problems are often called saddle point systems, mainly by researchers who do not focus solely on optimization. Methods for such problems have seen a surge of interest in the last two decades. A comprehensive study of iterative solvers for saddle point systems appears in Benzi, Golub, and Liesen [6].

As we have already said, optimization is a vast field, and several optimization problem classes are not considered here at all. These include discrete optimization problems, where some variables are restricted to take on only one of several discrete values (e.g., a decision variable x_1 can only be 0 for "no" or 1 for "yes"), and stochastic optimization. Also, we have not discussed in any depth problems of *global* optimization. Finally, the methods we have considered all assume availability of the gradient of the objective function. There is a separate branch of research for methods that address problems that do not satisfy this assumption.

Chapter 10
Polynomial Interpolation

Polynomial interpolants are rarely the end product of a numerical process. Their importance is more as building blocks for other, more complex algorithms in differentiation, integration, solution of differential equations, approximation theory at large, and other areas. Hence, polynomial interpolation arises frequently; indeed, it is one of the most ubiquitous tasks, both within the design of numerical algorithms and in their analysis. Its importance and centrality help explain the considerable length of the present chapter.

Section 10.1 starts the chapter with a general description of approximation processes in one independent variable, arriving at polynomial interpolation as one such fundamental family of techniques. In Sections 10.2, 10.3, and 10.4 we shall see no less than three different forms (different bases) of interpolating polynomials. They are all of fundamental importance and are used extensively in the practical construction of numerical algorithms.

Estimates and bounds for the error in polynomial interpolation are derived in Section 10.5. If the choice of locations for the interpolation data is up to the user, then a special set of abscissae (nodes) called Chebyshev points is an advantageous choice, and this is discussed in Section 10.6. Finally, Section 10.7 considers the case where not only function values but also derivative values are available for interpolation.

10.1 General approximation and interpolation

Interpolation is a special case of approximation. In this section we consider different settings in which approximation problems arise, explain the need for finding approximating functions, describe a general form for interpolants and important special cases, and end up with polynomial interpolation.

Discrete and continuous approximation in one dimension

It is possible to distinguish between approximation techniques for two types of problems:

1. **Data fitting** (Discrete approximation problem):

   Given a set of data points $\{(x_i, y_i)\}_{i=0}^{n}$, find a *reasonable* function $v(x)$ that fits the data points.

   If the data are accurate it might make sense to require that $v(x)$ **interpolate** the data, i.e., that the curve pass through the data exactly, satisfying

   $$v(x_i) = y_i, \quad i = 0, 1, \ldots, n.$$

   See Figure 10.1.

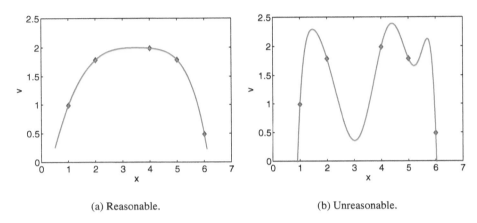

Figure 10.1. *Different interpolating curves through the same set of points.*

2. **Approximating functions**:

   For a complicated function $f(x)$ (which may be given explicitly, or only implicitly), find a simpler function $v(x)$ that approximates $f(x)$.

   For instance, suppose we need to quickly find an approximate value for sin(1.2) (that's 1.2 in radians, not degrees) with only a primitive calculator at hand. From basic trigonometry we know the values of sin(x) for $x = 0, \pi/6, \pi/4, \pi/3$, and $\pi/2$: how can we use these to estimate sin(1.2)?

   For another instance, suppose we have a complex, expensive program that calculates the final point (say, the landing location) of the trajectory of a space shuttle for each given value of a certain control parameter. We perform this calculation, i.e., invoke the program, for several parameter values. But then we may want to use this computed information to have an idea of the resulting landing location for other values of the control parameter—without resorting to the full calculation for each parameter value.

   Interpolation techniques for such function approximation are *identical* to those of data fitting once we specify the data points $\{(x_i, y_i = f(x_i))\}_{i=0}^n$. The *difference* between function interpolation and data fitting interpolation is that in the former we

   - have some freedom to choose x_i cleverly, and
   - may be able to consider the *global* interpolation error.

The need for interpolation

Why do we want to find an approximating function $v(x)$ in general?

- For *prediction*: we can use $v(x)$ to find approximate values of the underlying function at locations x other than the data abscissae, $x_0, \ldots, x_n$.

  If x is inside the smallest interval containing all the data abscissae, then this is called *interpolation*; if x is outside that interval, then we have *extrapolation*. For instance, we may have data regarding the performance of some stock at each trading week's end during the past year. Interpolating for the value of the stock during other days of last year is a relatively safe undertaking. Extrapolating for the value of the stock sometime next year is much more risky (although potentially more interesting).

10.1. General approximation and interpolation

- For *manipulation*: an instance is finding approximations for derivatives and integrals of the underlying function.

The interpolating function must be not only easy to evaluate and manipulate, but also "reasonable"; i.e., it must resemble a curve that we would actually draw through the points—see Figure 10.1. Just exactly what qualifies as reasonable is context-dependent and hard to define in precise terms for general purposes.

Interpolants and their representation

We generally assume a *linear form* for all interpolating (or, more generally, approximating) functions $v(x)$.[38] Thus we write

$$v(x) = \sum_{j=0}^{n} c_j \phi_j(x) = c_0 \phi_0(x) + \cdots + c_n \phi_n(x),$$

where $\{c_j\}_{j=0}^n$ are *unknown coefficients*, or **parameters** determined from the data, and $\{\phi_j(x)\}_{j=0}^n$ are predetermined **basis functions**. These basis functions are further assumed to be *linearly independent*, which means that if coefficients $\{c_j\}_{j=0}^n$ are found such that $v(x) = 0$ for all x, then all these coefficients themselves must vanish: $c_j = 0$, $j = 0, 1, \ldots, n$.

Notice our default assumption that the number of basis functions equals the number of data points $n + 1$. If there are fewer basis functions than data, then we cannot hope for interpolation of all data values and we typically end up resorting to a least squares approach as in Chapter 6. Other techniques of approximating functions which do not fit the data exactly are considered in Chapters 12 and 13.

In general, the interpolation conditions can be written as $n + 1$ linear relations for the $n + 1$ unknown coefficients. The resulting linear system of equations is

$$\begin{pmatrix} \phi_0(x_0) & \phi_1(x_0) & \phi_2(x_0) & \cdots & \phi_n(x_0) \\ \phi_0(x_1) & \phi_1(x_1) & \phi_2(x_1) & \cdots & \phi_n(x_1) \\ \vdots & \vdots & \vdots & & \vdots \\ \phi_0(x_n) & \phi_1(x_n) & \phi_2(x_n) & \cdots & \phi_n(x_n) \end{pmatrix} \begin{pmatrix} c_0 \\ c_1 \\ \vdots \\ c_n \end{pmatrix} = \begin{pmatrix} y_0 \\ y_1 \\ \vdots \\ y_n \end{pmatrix}. \quad (10.1)$$

We may not actually directly form and solve the system (10.1) in a given situation, but it is always there as a result of the assumption of linear representation of $v(x)$ in terms of the basis functions.

Here are some common examples of interpolants:

- In the present chapter we consider *polynomial interpolation*

$$v(x) = \sum_{j=0}^{n} c_j x^j = c_0 + c_1 x^1 + \cdots + c_n x^n.$$

This simplest and most familiar form of representing a polynomial implies the choice of a *monomial* basis

$$\phi_j(x) = x^j, \quad j = 0, 1, \ldots, n,$$

but we will see other choices as well.

---

[38] Note that the form of the interpolating function is linear in the sense that it is a linear combination of basis functions in some appropriate space, *not* that $v(x)$ itself is a linear function of x.

- In the next chapter we discuss *piecewise polynomial interpolation*, which is based on performing polynomial interpolation in "pieces," rather than on the entire given interval.

- *Trigonometric interpolation* is also extremely useful, especially in signal processing and for describing wave and other periodic phenomena. For instance, consider

$$\phi_j(x) = \cos(jx), \quad j = 0, 1, \ldots, n.$$

We elaborate on more general choices in this spirit in Chapter 13.

In general, it is important to distinguish two stages in the interpolation process:

1. **Constructing** the interpolant. These are operations that are independent of where we would then evaluate $v(x)$. An instance of this is determining the coefficients $c_0, c_1, \ldots, c_n$ for a given basis $\phi_0(x), \phi_1(x), \ldots, \phi_n(x)$.

2. **Evaluating** the interpolant at a given point x.

The interpolant construction is done once for a given set of data. After that, the evaluation may be applied many times.

Polynomial interpolation

> **Note:** The rest of this chapter is devoted exclusively to polynomial interpolation.

The main reason polynomial approximation is desirable is its simplicity. Polynomials

- are easy to construct and evaluate (recall also the nested form from Example 1.4);
- are easy to sum and multiply (and the result is also a polynomial);
- are easy to differentiate and integrate (and the result is also a polynomial); and
- have widely varying characteristics despite their simplicity.

10.2 Monomial interpolation

Let us denote a polynomial interpolant of degree at most n by

$$p(x) = p_n(x) = \sum_{j=0}^{n} c_j x^j = c_0 + c_1 x + \cdots + c_n x^n.$$

For $n+1$ data points

$$(x_0, y_0), (x_1, y_1), \ldots, (x_n, y_n),$$

we want to find $n+1$ coefficients[39] $c_0, c_1, \ldots, c_n$ such that

$$p(x_i) = y_i, \quad i = 0, 1, \ldots, n.$$

We will assume, until Section 10.7, that the abscissae of the data points are distinct, meaning

$$x_i \neq x_j \quad \text{whenever} \quad i \neq j.$$

---

[39] Remember that a polynomial of degree n has $n+1$, not n, coefficients.

10.2. Monomial interpolation

Example 10.1. Let $n = 1$ and let our two data points be $(x_0, y_0) = (1,1)$ and $(x_1, y_1) = (2,3)$. We want to fit a polynomial of degree at most 1 of the form

$$p_1(x) = c_0 + c_1 x$$

through these two points.

The interpolating conditions are

$$p_1(x_0) = c_0 + 1c_1 = 1,$$
$$p_1(x_1) = c_0 + 2c_1 = 3.$$

These are two linear equations for the two unknowns c_0 and c_1, which can be solved using high school algebra techniques. We obtain $c_0 = -1$ and $c_1 = 2$, so

$$p_1(x) = 2x - 1.$$

This linear interpolant is depicted in Figure 10.2.

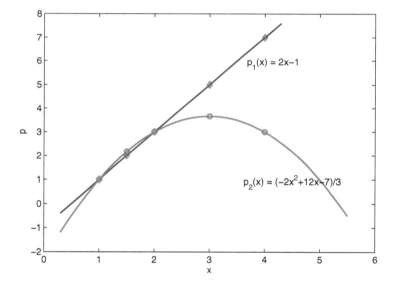

Figure 10.2. *Quadratic and linear polynomial interpolation.*

Next, let $n = 2$ and let our three data points be $(1,1)$, $(2,3)$, and $(4,3)$. The first two are the same as above, but the third data pair specifies a significantly different value at $x = 4$ than what $p_1(x)$ predicts: while $p_1(4) = 7$, here we seek a polynomial whose value at $x = 4$ equals 3. Thus, we want to fit a polynomial of degree at most 2 of the form

$$p_2(x) = c_0 + c_1 x + c_2 x^2$$

through these three points. Note that the coefficients c_0 and c_1 in $p_2(x)$ are not expected to be the same as for $p_1(x)$, in general. The interpolating conditions are

$$p_2(x_0) = c_0 + 1c_1 + 1c_2 = 1,$$
$$p_2(x_1) = c_0 + 2c_1 + 4c_2 = 3,$$
$$p_2(x_2) = c_0 + 4c_1 + 16c_2 = 3.$$

This is a 3 × 3 linear system for the unknown coefficients c_j, which in matrix form reads

$$\begin{pmatrix} 1 & 1 & 1 \\ 1 & 2 & 4 \\ 1 & 4 & 16 \end{pmatrix} \begin{pmatrix} c_0 \\ c_1 \\ c_2 \end{pmatrix} = \begin{pmatrix} 1 \\ 3 \\ 3 \end{pmatrix}.$$

The MATLAB commands

```
A = [1 1 1; 1 2 4; 1 4 16];
y = [1; 3; 3];
c = A \ y
```

yield (of course, up to a rounding error)

$$c_0 = -\frac{7}{3}, \quad c_1 = 4, \quad c_2 = -\frac{2}{3}.$$

This completes the construction of the quadratic interpolant. The desired interpolating polynomial p_2 is

$$p_2(x) = (-2x^2 + 12x - 7)/3,$$

and this can be evaluated for any given value of x. For instance, at $x = 3$ we have

$$p_2(3) = \frac{11}{3},$$

which is quite lower than the value $p_1(3) = 5$. Observe also the different values of the two polynomials at $x = 1.5$ in Figure 10.2, as the pair of close but not coinciding magenta symbols illustrates. ∎

Unique interpolating polynomial

Generalizing the example above to the case of $n+1$ data points, the interpolation conditions lead to a system of $(n+1)$ linear equations in the $(n+1)$ unknowns $c_0, c_1, \ldots, c_n$ given by

$$\begin{pmatrix} 1 & x_0^1 & x_0^2 & \cdots & x_0^n \\ 1 & x_1^1 & x_1^2 & \cdots & x_1^n \\ \vdots & \vdots & \vdots & & \vdots \\ 1 & x_n^1 & x_n^2 & \cdots & x_n^n \end{pmatrix} \begin{pmatrix} c_0 \\ c_1 \\ \vdots \\ c_n \end{pmatrix} = \begin{pmatrix} y_0 \\ y_1 \\ \vdots \\ y_n \end{pmatrix}.$$

The coefficient matrix X is called a *Vandermonde* matrix; in particular, it is known from an introductory linear algebra text that

$$\det(X) = \prod_{i=0}^{n-1} \left[\prod_{j=i+1}^{n} (x_j - x_i) \right],$$

i.e., this determinant is the product of all possible differences in the x_i. So, as long as the abscissae are distinct, $\det(X) \neq 0$ and hence X is nonsingular.[40] This argument provides a simple proof to

---

[40] It also verifies our assumption that the basis functions, in this case the monomials $\phi_j(x) = x^j$, $j = 0, 1, \ldots, n$, are linearly independent, because the matrix is nonsingular for an arbitrary choice of distinct points $x_0, x_1, \ldots, x_n$.

10.2. Monomial interpolation

Product Notation.
Let $z_1, z_2, \ldots, z_n$ be n scalar values. We are used to the notation for their sum, given by

$$\Sigma_{i=1}^{n} z_i = z_1 + z_2 + \cdots + z_n.$$

Analogously, the notation for their product is

$$\Pi_{i=1}^{n} z_i = z_1 \cdot z_2 \cdots z_n.$$

Theorem: Polynomial Interpolant Unique Existence.
For any real data points $\{(x_i, y_i)\}_{i=0}^{n}$ with distinct abscissae x_i there exists a unique polynomial $p(x)$ of degree at most n which satisfies the interpolation conditions

$$p(x_i) = y_i, \quad i = 0, 1, \ldots, n.$$

the theorem given on this page that there exists a *unique interpolating polynomial p*. The uniqueness of polynomial interpolation is particularly important because of the different forms that this polynomial can take: the same result is obtained, no matter which method or basis is used to obtain the interpolating polynomial.

Later on, in Section 10.7, we will generalize this existence and uniqueness result for the case when data abscissae are not necessarily distinct.

Using the monomial basis for constructing interpolants

Our discussion so far not only shows uniqueness but also suggests a general way for obtaining the interpolating polynomial $p(x)$: form the Vandermonde matrix and solve a linear system of equations. The big advantage of this approach is its intuitive simplicity and straightforwardness. However, if we consider a general-purpose use of polynomial interpolation, then this approach has disadvantages:

1. The calculated coefficients c_j are not directly indicative of the interpolated function, and they may completely change if we wish to slightly modify the interpolation problem; more on this in Sections 10.3 and 10.4.

2. The Vandermonde matrix X is often ill-conditioned (see Section 5.8), so the coefficients thus determined are prone to inaccuracies.

3. This approach requires about $\frac{2}{3}n^3$ operations (flops) to carry out Gaussian elimination (see Section 5.1) for the construction stage; another method exists which requires only about n^2 operations. The evaluation stage, however, is as quick as can be; using the nested form, it requires about $2n$ flops per evaluation point.

There are situations in which the last two disadvantages are not important. First, the higher computational cost, if there is any, is important only when n is "large," not when it equals 2 or 3. Also, the ill-conditioning, i.e., the claim that the basis $\phi_j(x) = x^j$, $j = 0, 1, \ldots, n$, is "not good" in the sense that roundoff error gets unreasonably magnified, occurs mostly when the interval of interpolation is wide or n is not small. If all points of concern, including all data points x_i and evaluation points x, are in a small interval near, say, some point $\hat{x}$, then the basis

$$\{1, (x - \hat{x}), (x - \hat{x})^2, (x - \hat{x})^3\}$$

is perfectly reasonable for a cubic polynomial. More on this in Section 11.2. The polynomial

$$p(x) = c_0 + c_1(x - \hat{x}) + \cdots + c_n(x - \hat{x})^n$$

is, in fact, reminiscent of a *Taylor series* expansion (see page 5)

$$f(x) = f(\hat{x}) + f'(\hat{x})(x - \hat{x}) + \cdots + \frac{f^{(n)}(\hat{x})}{n!}(x - \hat{x})^n + R_n(x),$$

where the remainder term can be written as

$$R_n(x) = \frac{f^{(n+1)}(\xi(x))}{(n+1)!}(x - \hat{x})^{n+1}$$

for some ξ between x and $\hat{x}$.

Throughout most of this chapter we consider values of n in the single decimal digit territory. Larger polynomial degrees appear only in Section 10.6, where the monomial basis is indeed inadequate.

There are several additional desirable features, often far more important than the efficiency considerations mentioned above, which the simple monomial basis $\{\phi_j(x) = x^j\}$ does not have. These are introduced in the next two sections, together with bases that do possess such features and which allow a more intuitive understanding in context of both the problems and their resolutions.

Specific exercises for this section: Exercises 1–3.

10.3 Lagrange interpolation

The coefficients c_j of the polynomials $p_1(x)$ and $p_2(x)$ in Example 10.1 do not relate directly to the given data values y_j. It would be nice if a polynomial basis is found such that $c_j = y_j$, giving

$$p(x) = p_n(x) = \sum_{j=0}^{n} y_j \phi_j(x).$$

Such a representation would be particularly easy to manipulate, for instance, when we seek formulas for differentiation or integration. This is shown in later chapters, particularly Chapters 14 through 16.

Such a polynomial basis is provided by the Lagrange interpolation process. For this we define the **Lagrange polynomials**, $L_j(x)$, which are polynomials of degree n that satisfy

$$L_j(x_i) = \begin{cases} 0, & i \neq j, \\ 1, & i = j. \end{cases}$$

Given data y_i at abscissae x_i as before, the unique polynomial interpolant of degree at most n can now be written as

$$p(x) = \sum_{j=0}^{n} y_j L_j(x).$$

Indeed, p is of degree at most n (being the linear combination of polynomials of degree n), and it satisfies the interpolation conditions because

$$p(x_i) = \sum_{j=0}^{n} y_j L_j(x_i) = 0 + \cdots + 0 + y_i L_i(x_i) + 0 + \cdots + 0 = y_i.$$

10.3. Lagrange interpolation

Example 10.2. Let us use the same three data pairs of Example 10.1, namely, $(1,1)$, $(2,3)$, and $(4,3)$, to demonstrate the construction of Lagrange polynomials. To make $L_0(x)$ vanish at $x=2$ and $x=4$ we write
$$L_0(x) = a(x-2)(x-4).$$
Requiring $L_0(1) = 1$ then determines $a(1-2)(1-4) = 1$, i.e., $a = \frac{1}{3}$, and thus
$$L_0(x) = \frac{1}{3}(x-2)(x-4).$$
Similarly we determine that
$$L_1(x) = -\frac{1}{2}(x-1)(x-4), \quad L_2(x) = \frac{1}{6}(x-1)(x-2).$$
These Lagrange polynomials are depicted in Figure 10.3. We thus obtain the interpolant
$$p_2(x) = \frac{y_0}{3}(x-2)(x-4) - \frac{y_1}{2}(x-1)(x-4) + \frac{y_2}{6}(x-1)(x-2)$$
$$= \frac{1}{3}(x-2)(x-4) - \frac{3}{2}(x-1)(x-4) + \frac{3}{6}(x-1)(x-2).$$

Despite the different form, this is precisely the same quadratic interpolant as the one in Example 10.1, so in fact, $p_2(x) = (-2x^2 + 12x - 7)/3$. It is also easy to verify that here, too, $p_2(3) = \frac{11}{3}$. All this should not come as a surprise—it's an illustration of the uniqueness of polynomial interpolation; see the theorem on page 301. ∎

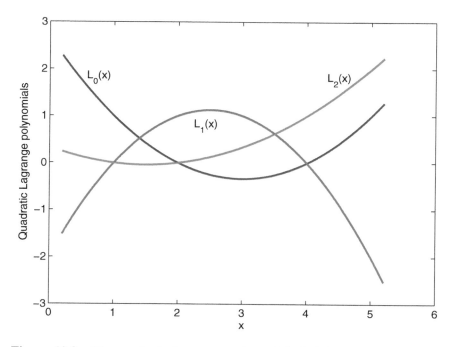

Figure 10.3. *The quadratic Lagrange polynomials $L_0(x)$, $L_1(x)$, and $L_2(x)$ based on points $x_0 = 1$, $x_1 = 2$, $x_2 = 4$, used in Example 10.2.*

Properties of Lagrange polynomials

What properties do Lagrange polynomials have? What do they look like?

In the general case where $n+1$ data abscissae x_i are specified, the Lagrange polynomials uniquely exist, because they are really nothing other than polynomial interpolants for special data.[41] This is also straightforward to verify directly, by explicitly specifying

$$L_j(x) = \frac{(x-x_0)\cdots(x-x_{j-1})(x-x_{j+1})\cdots(x-x_n)}{(x_j-x_0)\cdots(x_j-x_{j-1})(x_j-x_{j+1})\cdots(x_j-x_n)} = \prod_{\substack{i=0 \\ i\neq j}}^{n} \frac{(x-x_i)}{(x_j-x_i)}.$$

Indeed, the polynomial of degree n, written in terms of its roots as

$$(x-x_0)\cdots(x-x_{j-1})(x-x_{j+1})\cdots(x-x_n),$$

clearly interpolates the 0-values at all data abscissae other than x_j, and dividing by its value at x_j normalizes the expression to yield $L_j(x_j) = 1$. Another picture of a Lagrange polynomial is provided in Figure 10.4.

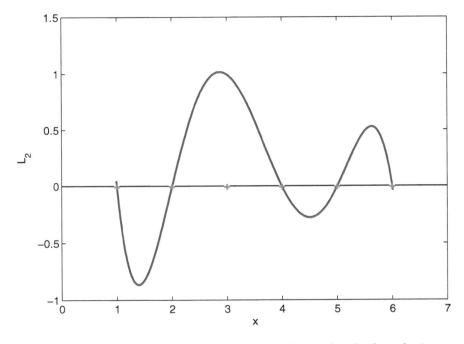

Figure 10.4. *The Lagrange polynomial $L_2(x)$ for $n = 5$. Guess what the data abscissae x_i are.*

The Lagrange polynomials form an ideally conditioned basis $\phi_j(x) = L_j(x)$ for all polynomials of degree at most n. In the case of Lagrange interpolation, the matrix elements in the system (10.1) on page 297 are $\phi_j(x_i) = L_j(x_i)$, and we see that what was a potentially problematic Vandermonde matrix for the monomial basis in Section 10.2 is now an *identity* matrix. Thus, the system (10.1) (which is never formed as such) yields the solution $c_j = y_j$, $j = 1,\ldots,n$.

Note that L_j has n zeros and thus $n-1$ extrema (alternating minima and maxima). Incidentally, $L_j(x_j) = 1$ need not be a maximum for $L_j(x)$.

---

[41] For each j, set $y_i \leftarrow 1$ if $i = j$, $y_i \leftarrow 0$ if $i \neq j$, $i = 0,1,\ldots,n$.

10.3. Lagrange interpolation

Algorithm: Lagrange Polynomial Interpolation.

1. *Construction*: Given data $\{(x_i, y_i)\}_{i=0}^n$, compute barycentric weights $w_j = 1/\prod_{i \neq j}(x_j - x_i)$, and also the quantities $w_j y_j$, for $j = 0, 1, \ldots n$.

2. *Evaluation*: Given an evaluation point x not equal to one of the data points $\{x_i\}_{i=0}^n$, compute
$$p(x) = \frac{\sum_{j=0}^n \frac{w_j y_j}{(x-x_j)}}{\sum_{j=0}^n \frac{w_j}{(x-x_j)}}.$$

Construction and evaluation

The discussion so far should make it clear what the Lagrange polynomials are and how they work for interpolation. Next, let us devise a way to carry out the construction and evaluation stages efficiently. This would be important if n is large.

The construction stage involves what does not depend on an evaluation point x. Here this means constructing the denominators of the $n+1$ Lagrange polynomials. Let us then define

$$\rho_j = \prod_{i \neq j}(x_j - x_i), \quad w_j = \frac{1}{\rho_j}, \quad j = 0, 1, \ldots, n.$$

This construction requires about n^2 flops. The quantities w_j are called **barycentric weights**.

Moving on to the evaluation stage, for a given point x different from the abscissae at which the value of $p(x)$ is required, note that all numerators of the $L_j(x)$ involve small variations of the function

$$\psi(x) = (x - x_0) \cdots (x - x_n) = \prod_{i=0}^n (x - x_i).$$

The interpolant is then given by

$$p(x) = \psi(x) \sum_{j=0}^n \frac{w_j y_j}{(x - x_j)}.$$

Clearly, for each argument value x the evaluation costs $\mathcal{O}(n)$ flops (about $5n$, to be more precise).

A slight further modification, essentially for the purpose of beauty, can be made by observing that for the particular function $f(x)$ that equals 1 everywhere we have $y_j = 1$ for all j. Moreover, the interpolant is also identically 1, so

$$1 = \psi(x) \sum_{j=0}^n \frac{w_j \cdot 1}{(x - x_j)}$$

for any x. This determines $\psi(x)$ in terms of quantities that are computed anyway. The resulting formula is called **barycentric interpolation**, leading to the algorithm given on this page. It would be worth your while to apply this algorithm to Example 10.2 and check what intermediate quantities arise.

Specific exercises for this section: Exercises 4–6.

10.4 Divided differences and Newton's form

In this section we continue to consider polynomial interpolation at $n+1$ data points with distinct abscissae. Yet another representation for polynomial interpolation is introduced (it's the last one: promise!), because our previous representations do not cover two important aspects. The first is the desirability of introducing interpolation data (x_i, y_i) one pair at a time, rather than all at once from the start. The other aspect that we have avoided thus far is estimating the error in the interpolation approximation, and the developments here help in setting up the discussion in Section 10.5.

The Newton polynomial basis

We have seen the standard monomial basis, $\{\phi_j(x) = x^j\}_{j=0}^n$. This has led to a relatively inferior procedure for constructing $p = p_n$ but an easy procedure for evaluating $p_n(x)$ at a given argument x. On the other hand, with the Lagrange polynomial basis, $\{\phi_j(x) = \prod_{i \neq j, i=0}^n \frac{(x-x_i)}{(x_j-x_i)}\}_{j=0}^n$, the construction stage is easy, but the evaluation of $p_n(x)$ is relatively involved conceptually. The Newton polynomial basis can be viewed as a useful compromise: we set

$$\phi_j(x) = \prod_{i=0}^{j-1}(x - x_i), \quad j = 0, 1, \ldots, n.$$

The discussion that follows illustrates the merits of this choice.

Example 10.3. For a quadratic interpolant we have the basis functions

$$\phi_0(x) = 1, \quad \phi_1(x) = x - x_0, \quad \phi_2(x) = (x - x_0)(x - x_1).$$

Consider again the same data set as in Examples 10.1 and 10.2, namely, $(1,1)$, $(2,3)$, and $(4,3)$. Then the interpolation condition at $x_0 = 1$ yields

$$1 = p(1) = c_0 \phi_0(1) + c_1 \phi_1(1) + c_2 \phi_2(1) = c_0 \cdot 1 + c_1 \cdot 0 + c_2 \cdot 0 = c_0.$$

Next, with $c_0 = 1$, the interpolation condition at $x_1 = 2$ reads

$$3 = p(2) = 1 + c_1(2-1) + c_2 \cdot 0,$$

yielding $c_1 = 2$.
Finally, the third interpolation condition gives

$$3 = p(4) = 1 + 2(4-1) + c_2(4-1)(4-2),$$

yielding $c_2 = -\frac{4}{6}$. Therefore, the interpolating polynomial is

$$p_2(x) = 1 + 2(x-1) - \frac{2}{3}(x-1)(x-2).$$

At $x = 3$ this evaluates, of course, to $p_2(3) = \frac{11}{3}$, which is the same value as in Examples 10.1 and 10.2, because the interpolating polynomial is unique: only its form changes with the representation. As in our previous examples, we encourage the skeptics (and a healthy dose of skepticism is a trait of a good scientist!) to open the brackets and verify that the same polynomial as in those previous examples is obtained, yet again.
  The important point to take away from this example is that to evaluate c_0 we used only the first data point, and to evaluate c_1 we needed only the first two. ∎

10.4. Divided differences and Newton's form

Existence of an adaptive interpolant

The most important feature of the Newton representation is that it is evolutionary, or recursive: having determined $p_{n-1}(x)$ that interpolates at the first n data points, we use it to cheaply construct $p_n(x)$ which interpolates all previous data as well as (x_n, y_n). Thus, we do not need to know all data points, or determine the degree n, ahead of time. Rather, we can do this *adaptively*. This feature may be of great value when working with, say, laboratory measurements where not all data become available at once. See also Exercises 9 and 10.

The present form of $p_n(x)$ is

$$p_n(x) = c_0 + c_1(x - x_0) + c_2(x - x_0)(x - x_1) + \cdots + c_n(x - x_0)(x - x_1) \cdots (x - x_{n-1}).$$

Is it always possible to determine the coefficients c_j? And if yes, is this particular representation unique?

The (affirmative) answers can be obtained upon considering the general system (10.1) on page 297. Given the form of the basis functions, we clearly have in the present case

$$\phi_j(x_i) = 0, \ i = 0, 1, \ldots, j-1, \quad \phi_j(x_j) \neq 0.$$

Hence, the matrix in (10.1) is *lower triangular* and the elements on its diagonal are nonzero. Since the determinant of a triangular matrix is the product of its diagonal elements we obtain that for arbitrary (distinct) data abscissae the matrix is nonsingular. Hence there is a unique solution for the unknown coefficients $c_0, \ldots, c_n$ in terms of the data $y_0, \ldots, y_n$.

The argument above relies on basic notions of linear algebra. In fact, an efficient *forward substitution* algorithm given on page 96 exists as well. Thus, we could form the system (10.1) and apply the algorithm of Section 5.1 to end the present discussion right here. However, we proceed instead to carry out the forward substitution algorithm symbolically, without forming (10.1) and without relying on knowledge from Chapter 5, because this results in important additional information regarding approximation processes as well as a more direct, efficient algorithm.

Representation in terms of divided differences

For ease of notation let us switch below from y_i to $f(x_i)$: we want to keep track of which data is used through the recursive construction process. Thus we proceed as follows.

- Determine c_0 using the condition $p_n(x_0) = f(x_0)$:

$$f(x_0) = p_n(x_0) = c_0 + 0 + \cdots + 0 = c_0$$
$$\Rightarrow c_0 = f(x_0).$$

- Next, determine c_1 using the condition $p_n(x_1) = f(x_1)$:

$$f(x_1) = p_n(x_1) = c_0 + c_1(x_1 - x_0) + 0 + \cdots + 0 = c_0 + c_1(x_1 - x_0)$$
$$\Rightarrow c_1 = \frac{f(x_1) - f(x_0)}{x_1 - x_0}.$$

- Next, impose also the condition $p_n(x_2) = f(x_2)$. With c_0 and c_1 already determined this yields a condition involving c_2 alone. Please verify (you may use the statement of Exercise 13 for this purpose) that the result can be arranged as

$$c_2 = \frac{\frac{f(x_2) - f(x_1)}{x_2 - x_1} - \frac{f(x_1) - f(x_0)}{x_1 - x_0}}{x_2 - x_0}.$$

- Continue this process along until all the coefficients c_j have been determined.

The coefficient c_j of the interpolating polynomial in Newton's form is called the jth *divided difference*, denoted $f[x_0, x_1, \ldots, x_j]$. So we write

$$f[x_0] = c_0, \ f[x_0, x_1] = c_1, \ \ldots, \ f[x_0, x_1, \ldots, x_n] = c_n.$$

This explicitly indicates which of the data points each coefficient c_j depends upon.

> **Note:** The notation of divided differences is rather detailed, perhaps in an unappealing way at first. But please hang on, as several important ideas soon make an appearance.

Thus, the *Newton divided difference interpolation formula* in its full glory is given by

$$\begin{aligned} p_n(x) &= f[x_0] + f[x_0, x_1](x - x_0) + f[x_0, x_1, x_2](x - x_0)(x - x_1) \\ &\quad + \cdots + f[x_0, x_1, \ldots, x_n](x - x_0)(x - x_1) \cdots (x - x_{n-1}) \\ &= \sum_{j=0}^{n} \left(f[x_0, x_1, \ldots, x_j] \prod_{i=0}^{j-1}(x - x_i) \right). \end{aligned}$$

The divided difference coefficients satisfy the recursive formula

$$f[x_0, x_1, \ldots, x_j] = \frac{f[x_1, x_2, \ldots, x_j] - f[x_0, x_1, \ldots, x_{j-1}]}{x_j - x_0}.$$

See Exercise 7. More generally, x_0 can be replaced by x_i for any $0 \leq i < j$. The resulting formula is given on this page. This recursion means that we do not have to build the coefficients from scratch each time we add a new interpolation point.

> **Divided Differences.**
> Given points $x_0, x_1, \ldots, x_n$, for arbitrary indices $0 \leq i < j \leq n$, set
>
> $$f[x_i] = f(x_i),$$
> $$f[x_i, \ldots, x_j] = \frac{f[x_{i+1}, \ldots, x_j] - f[x_i, \ldots, x_{j-1}]}{x_j - x_i}.$$

Divided difference table and interpolant evaluation

How do we use all of these recurrences and formulas in practice? Note that in order to compute $\gamma_{n,n} = f[x_0, x_1, \ldots, x_n]$ we must compute all of

$$\gamma_{j,l} = f[x_{j-l}, x_{j-l+1}, \ldots, x_j], \quad 0 \leq l \leq j \leq n.$$

Thus, we construct a *divided difference table*, which is a lower triangular array depicted next.

i	x_i	$f[x_i]$	$f[x_{i-1}, x_i]$	$f[x_{i-2}, x_{i-1}, x_i]$	$\cdots$	$f[x_{i-n}, \ldots, x_i]$
0	x_0	$f(x_0)$				
1	x_1	$f(x_1)$	$\frac{f[x_1] - f[x_0]}{x_1 - x_0}$			
2	x_2	$f(x_2)$	$\frac{f[x_2] - f[x_1]}{x_2 - x_1}$	$f[x_0, x_1, x_2]$		
$\vdots$	$\vdots$	$\vdots$	$\vdots$	$\vdots$	$\ddots$	
n	x_n	$f(x_n)$	$\frac{f[x_n] - f[x_{n-1}]}{x_n - x_{n-1}}$	$f[x_{n-2}, x_{n-1}, x_n]$	$\cdots$	$f[x_0, x_1, \ldots, x_n]$

10.4. Divided differences and Newton's form

Extracting the diagonal entries yields the coefficients $c_j = \gamma_{j,j} = f[x_0,\ldots,x_j]$ for the Newton interpolation polynomial.

Example 10.4. For the same problem as in Example 10.3 we have

$$f[x_0,x_1] = \frac{3-1}{2-1} = 2, \quad f[x_1,x_2] = \frac{3-3}{4-2} = 0, \quad f[x_0,x_1,x_2] = \frac{0-2}{4-1} = -\frac{2}{3}.$$

The corresponding divided difference table is

i	x_i	$f[\cdot]$	$f[\cdot,\cdot]$	$f[\cdot,\cdot,\cdot]$
0	1	1		
1	2	3	2	
2	4	3	0	$-\frac{2}{3}$

Thus, the interpolating polynomial $p_2(x)$ is as given in Example 10.3.

Note that the first two terms of $p_2(x)$ form the linear interpolant $p_1(x)$ obtained in Example 10.1. This demonstrates the specialty of the Newton basis. Also, in nested form we have

$$p_2(x) = 1 + (x-1)\left(2 - \frac{2}{3}(x-2)\right).$$

If we wish to add another data point, say, $(x_3, f(x_3)) = (5,4)$, we need only add another row to the divided difference table. We calculate

$$f[x_3] = 4, \quad f[x_2,x_3] = \frac{4-3}{5-4} = 1, \quad f[x_1,x_2,x_3] = \frac{1-0}{5-2} = \frac{1}{3},$$

$$f[x_0,x_1,x_2,x_3] = \frac{(1/3)-(-2/3)}{5-1} = \frac{1}{4}.$$

The corresponding table is

i	x_i	$f[\cdot]$	$f[\cdot,\cdot]$	$f[\cdot,\cdot,\cdot]$	$f[\cdot,\cdot,\cdot,\cdot]$
0	1	1			
1	2	3	2		
2	4	3	0	$-\frac{2}{3}$	
3	5	4	1	$\frac{1}{3}$	$\frac{1}{4}$

Thus, for p_3 we have the expression

$$p_3(x) = p_2(x) + f[x_0,x_1,x_2,x_3](x-x_0)(x-x_1)(x-x_2)$$
$$= 1 + (x-1)\left(2 - \frac{2}{3}(x-2)\right) + \frac{1}{4}(x-1)(x-2)(x-4).$$

In nested form, this reads

$$p_3(x) = 1 + (x-1)\left(2 + (x-2)\left(-\frac{2}{3} + \frac{1}{4}(x-4)\right)\right).$$

Obtaining a higher degree approximation is simply a matter of adding a term.

Figure 10.5 displays the two polynomials p_2 and p_3. Note that p_2 predicts a rather different value for $x = 5$ than the additional datum $f(x_3)$ later imposes, which explains the significant difference between the two interpolating curves. ∎

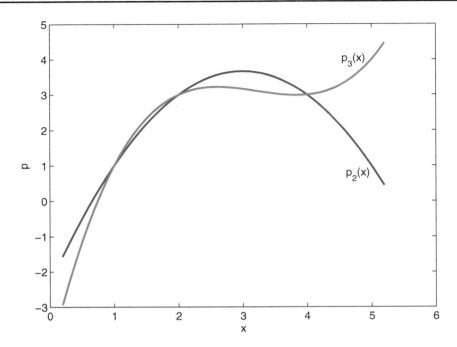

Figure 10.5. *The interpolating polynomials p_2 and p_3 for Example* 10.4.

Here are three MATLAB functions that implement the method described above. The first constructs the divided difference table and returns the coefficients of the polynomial interpolant.

```
function [coef,table] = divdif (xi, yi)
%
% function [coef,table] = divdif (xi, yi)
%
% Construct a divided difference table based on data points (xi,yi).
% Upon return, the Newton interpolation coefficients are in coef

np1 = length(xi); n = np1-1;
table = zeros(np1,np1); xi = shiftdim(xi); yi = shiftdim(yi);
% construct divided difference table one column at a time
table(1:np1,1) = yi;
for k = 2:np1
   table(k:np1,k) = (table(k:np1,k-1) - table(k-1:n,k-1)) ./ ...
                   (xi(k:np1) - xi(1:np1-k+1));
end
coef = diag(table);   % the diagonal elements of table
```

Note that the variable `table` is a two-dimensional array, and you can safely ignore the zero values in its strictly upper triangular part. In fact, you should be able to see from the code that it is not strictly necessary to store this entire table as such, but we do this anyway for illustration purposes.

Next, the following function uses nested evaluation to evaluate the polynomial interpolant in Newton's form:

10.4. Divided differences and Newton's form

```
function p = evalnewt (x, xi, coef)
%
%   function p = evalnewt (x, xi, coef)
%
%   evaluate at x the interpolating polynomial in Newton form
%   based on interpolation points xi and coefficients coef

np1 = length(xi);
p = coef(np1)*ones(size(x));
for j=np1-1:-1:1
  p = p.*(x - xi(j)) + coef(j);
end
```

An algorithm for a more general case (in which derivative values are also involved) is provided on page 321 in Section 10.7.

Example 10.4 also demonstrates the process of adding just one more data pair $(x_{n+1}, f(x_{n+1}))$ to an existing interpolant p_n of the first $n+1$ data pairs, obtaining

$$p_{n+1}(x) = p_n(x) + f[x_0, x_1, \ldots, x_n, x_{n+1}] \prod_{i=0}^{n} (x - x_i).$$

This is implemented in the following routine:

```
function [coef,table] = divdifadd (xi, yi, table)
%
% function [coef,table] = divdifadd (xi, yi, table)
%
% Construct one more row of an existing divided
% difference table, and add interpolation coefficient, based
% on one additional (last in xi and yi) data point

np1 = length(xi); n = np1 - 1;
table = [table zeros(n,1); yi(np1) zeros(1,n)];
for k=2:np1
  table(np1,k) = (table(np1,k-1) - table(n,k-1)) / ...
                 (xi(np1) - xi(np1-k+1));
end
coef = diag(table);
```

Here is the script (minus some plotting instructions) used to generate Figure 10.5:

```
x = .2:.01:5.2; % evaluation mesh
%quadratic interpolant
xi = [1,2,4]; yi = [1,3,3];
[coef2,table] = divdif(xi,yi);
%evaluate quadratic at x
y2 = evalnewt(x,xi,coef2);
%add data point
xi = [xi,5]; yi = [yi,4];
%cubic interpolant
[coef3,table] = divdifadd(xi,yi,table);
%evaluate cubic at x
y3 = evalnewt(x,xi,coef3);
plot (x,y2,'b',x,y3,'g')
```

Algorithms comparison

Although a precise operation count is not awfully important in practice (for one thing because n is not expected to be large, except perhaps in Section 10.6) we note that the construction of the divided difference table requires $n^2/2$ divisions and n^2 additions/subtractions. Moreover, nested evaluation of the Newton form requires roughly $2n$ multiplications and additions: these operation counts at each stage are at least as good as for the previous interpolants we saw.

Having described no less than three bases for polynomial interpolation, and in considerable detail at that, let us pause and summarize their respective properties in a table. The costs of construction and evaluation at one point are given to leading term in flop count.

Basis name	$\phi_j(x)$	Construction cost	Evaluation cost	Selling feature
Monomial	x^j	$\frac{2}{3}n^3$	$2n$	simple
Lagrange	$L_j(x)$	n^2	$5n$	$c_j = y_j$ most stable
Newton	$\prod_{i=0}^{j-1}(x-x_i)$	$\frac{3}{2}n^2$	$2n$	adaptive

Divided differences and derivatives

> **Note:** The important connection between divided differences and derivatives, derived below, is given in the theorem on the current page. If you are prepared to believe it, then you can skip the rather technical discussion that follows.

The divided difference function can be viewed as an extension of the concept of function derivative. Indeed, if we consider a differentiable function $f(x)$ applied at two points z_0 and z_1, and imagine z_1 approaching z_0, then

$$f[z_0, z_1] = \frac{f(z_1) - f(z_0)}{z_1 - z_0} \to f'(z_0).$$

But even if z_0 and z_1 remain distinct, the Mean Value Theorem given on page 10 asserts that there is a point ζ between them such that

$$f[z_0, z_1] = f'(\zeta).$$

This directly relates the first divided difference with the first derivative of f, in case the latter exists.

A similar connection exists between the kth divided difference of f and its kth derivative. Let us show this in detail.

> **Theorem: Divided Difference and Derivative.**
> Let the function f be defined and have k bounded derivatives in an interval $[a,b]$ and let $z_0, z_1, \ldots, z_k$ be $k+1$ distinct points in $[a,b]$. Then there is a point $\zeta \in [a,b]$ such that
>
> $$f[z_0, z_1, \ldots, z_k] = \frac{f^{(k)}(\zeta)}{k!}.$$

Thus, let $z_0, z_1, \ldots, z_k$ be $k+1$ distinct points contained in an interval on which a function f is defined and has k bounded derivatives. Note before we start that divided difference coefficients are symmetric in the arguments. That is, if $(\hat{z}_0, \hat{z}_1, \ldots, \hat{z}_k)$ is a permutation of the abscissae $(z_0, z_1, \ldots, z_k)$, then
$$f[\hat{z}_0, \hat{z}_1, \ldots, \hat{z}_k] = f[z_0, z_1, \ldots, z_k];$$
see Exercise 13. Thus, we may as well assume that the points z_i are sorted in increasing order and write
$$a = z_0 < z_1 < \cdots < z_k = b.$$

Without computing anything let p_k be the interpolating polynomial of degree at most k satisfying $p_k(z_i) = f(z_i)$, and denote the error $e_k(x) = f(x) - p_k(x)$. Then $e_k(z_i) = 0$, $i = 0, 1, \ldots, k$, i.e., e_k has $k+1$ zeros (roots).

Applying Rolle's Theorem (see page 10) we observe that between each two roots z_{i-1} and z_i of $e_k(x)$ there is a root of the derivative error, $e'_k(x)$, for a total of k roots in the interval $[a,b]$. Repeatedly applying Rolle's Theorem further we obtain that the function $e_k^{(k-l)}(x)$ has $l+1$ roots in $[a,b]$, $l = k, k-1, \ldots, 1, 0$. In particular, $e_k^{(k)}$ has one such root, denoted ζ, where
$$0 = e_k^{(k)}(\zeta) = f^{(k)}(\zeta) - p_k^{(k)}(\zeta).$$

Next, we must characterize $p_k^{(k)}$, the kth derivative of the interpolating polynomial of degree k. This task is straightforward if the Newton interpolation form is considered, because using this form we can write the interpolating polynomial as
$$p_k(x) = f[z_0, z_1, \ldots, z_k]\, x^k + q_{k-1}(x),$$
with $q_{k-1}(x)$ a polynomial of degree $< k$.

But the kth derivative of $q_{k-1}(x)$ vanishes everywhere: $q_{k-1}^{(k)} \equiv 0$. Hence, the only element that remains is the kth derivative of $f[z_0, z_1, \ldots, z_k] x^k$. The kth derivative of x^k is the constant $k!$ (why?), and hence $p_k^{(k)} = f[z_0, z_1, \ldots, z_k](k!)$. Substituting in the expression for $e_k^{(k)}(\zeta)$ we obtain the important formula
$$f[z_0, z_1, \ldots, z_k] = \frac{f^{(k)}(\zeta)}{k!}.$$

Thus we have proved the theorem stated on the facing page which methodically connects between divided differences and derivatives even when the data abscissae are not close to one another. ♦

This formula stands in line with the view that the Newton interpolation form is an extension of sorts of the Taylor series expansion. It proves useful in Section 10.7 as well as in estimating the interpolation error, a task to which we proceed without further ado.

Specific exercises for this section: Exercises 7–14.

10.5 The error in polynomial interpolation

In our discussion of divided differences and derivatives at the end of the previous section we have focused on data ordinates $f(x_i) = y_i$, $i = 0, 1, \ldots, n$, and paid no attention to any other value of $f(x)$. The function f that presumably gave rise to the interpolated data could even be undefined at other argument values. But now we assume that $f(x)$ is defined on an interval $[a,b]$ containing the interpolation points, and we wish to assess how large the difference between $f(x)$ and $p_n(x)$ may be at any point of this interval. For this we will also assume that some derivatives of $f(x)$ exist and are bounded in the interval of interest.

An expression for the error

Let us define the *error function* of our constructed interpolant as

$$e_n(x) = f(x) - p_n(x).$$

We can obtain an expression for this error using a simple trick: at any given point $x \in [a,b]$ where we want to evaluate the error in $p_n(x)$, pretend that x is another interpolation point! Then, from the formula in Section 10.4 connecting p_{n+1} to p_n, we obtain

$$f(x) = p_{n+1}(x) = p_n(x) + f[x_0, x_1, \ldots, x_n, x]\psi_n(x),$$

where

$$\psi_n(x) = \prod_{i=0}^{n}(x - x_i),$$

defined also on page 305. Therefore, the error is

$$e_n(x) = f(x) - p_n(x) = f[x_0, x_1, \ldots, x_n, x]\psi_n(x).$$

However, simple as this formula for the error is, it depends explicitly both on the data and on an individual evaluation point x. Hence we want a more general, qualitative handle on the error. So, we proceed to obtain further forms of error estimates and bounds.

Error estimates and bounds

The first thing we do is to replace the divided difference in the error formula by the corresponding derivative, assuming f is smooth enough. Thus, we note that if the approximated function f has $n+1$ bounded derivatives, then by the theorem given on page 312 there is a point $\xi = \xi(x)$, where $a \leq \xi \leq b$, such that $f[x_0, x_1, \ldots, x_n, x] = \frac{f^{(n+1)}(\xi)}{(n+1)!}$. This relationship is obtained upon identifying k with $n+1$ and $(z_0, z_1, \ldots, z_k)$ with $(x_0, \ldots, x_n, x)$ in that theorem. It yields the *error estimate*

$$f(x) - p_n(x) = \frac{f^{(n+1)}(\xi)}{(n+1)!}\psi_n(x).$$

Theorem: Polynomial Interpolation Error.
If p_n interpolates f at the $n+1$ points $x_0, \ldots, x_n$ and f has $n+1$ bounded derivatives on an interval $[a,b]$ containing these points, then for each $x \in [a,b]$ there is a point $\xi = \xi(x) \in [a,b]$ such that

$$f(x) - p_n(x) = \frac{f^{(n+1)}(\xi)}{(n+1)!}\prod_{i=0}^{n}(x - x_i).$$

Furthermore, we have the error bound

$$\max_{a \leq x \leq b}|f(x) - p_n(x)| \leq \frac{1}{(n+1)!}\max_{a \leq t \leq b}|f^{(n+1)}(t)|\max_{a \leq s \leq b}\prod_{i=0}^{n}|s - x_i|.$$

10.5. The error in polynomial interpolation

Although we do not know $\xi = \xi(x)$, we can get a bound on the error at all evaluation points x in some interval $[a,b]$ containing the data abscissae $x_0, \ldots, x_n$ if we have an upper bound on

$$\|f^{(n+1)}\| = \max_{a \leq t \leq b} |f^{(n+1)}(t)|.$$

Likewise, we can get a bound on the interpolation error if in addition we maximize $|\psi_n(x)|$ for the given abscissae over the evaluation interval. Moreover, since a similar error expression holds for each evaluation point $x \in [a,b]$, the maximum error magnitude on this interval is also bounded by the same quantity. We obtain the Polynomial Interpolation Error Theorem given on the facing page.

Note that although the error expression above was derived using Newton's form, it is independent of the basis used for the interpolating polynomial. There is only one such polynomial, regardless of method of representation!

Practical considerations

In practice, f is usually unknown, and so are its derivatives. We can sometimes do the following:

- Gauge the accuracy of such an approximation by computing a sequence of polynomials $p_k(x), p_{k+1}(x), \ldots$ using different subsets of the points and comparing results to see how well they agree. (If they agree well, then this would suggest convergence, or sufficient accuracy.) This should work when f does not vary excessively and the interval $[a,b]$ containing the abscissae of the data as well as the evaluation points is not too large.

- Hold the data points in reserve to use for construction of higher order polynomials to estimate the error. Thus, if we have more than the $n+1$ data pairs used to construct $p_n(x)$, then we can evaluate p_n at these additional abscissae and compare against the given values to obtain an underestimate of the maximum error in our interpolant. We may need many extra points to appraise realistically, though.

Example 10.5. The following maximum daily temperatures (in Celsius) were recorded every third day during one August month in a Middle Eastern city:

Day	3	6	9	12	15	18	21	24	27
Temperature (C)	31.2	32.0	35.3	34.1	35.0	35.5	34.1	35.1	36.0

Note that these data values are *not* monotonically decreasing or increasing. Suppose we wish to estimate the maximum temperature at day $x = 13$ of that month.

It is useful to practice some judgment here: just because there is a table available with lots of data does not necessarily imply that they should *all* be used to get an idea of what is happening around a specific point. What we could do, for example, is construct a cubic ($n = 3$) polynomial using four specific points near $x = 13$. Following this strategy, we choose $x_0 = 9$, $x_1 = 12$, $x_2 = 15$, $x_3 = 18$. Doing so, we get $p_3(13) = 34.29$. If on the other hand we go for linear interpolation of the two closest neighbors, at $x = 12$ and $x = 15$, we get $p_1(13) = 34.4$. This provides some confidence that our calculations are probably on the mark.

In summary, it was hot on that day, too. ∎

How should you choose the data abscissae, if possible, for lower degree approximations?

To keep $|\psi_n(x)|$ small, try to cluster data near where you want to make the approximations. Also, avoid extrapolation if there is that option. Last but not least, treat polynomial interpolation as a device for *local* approximation; for *global* interpolation of given data, use piecewise polynomials—see Chapter 11.

Specific exercises for this section: Exercises 15–16.

10.6 Chebyshev interpolation

Suppose that we are interested in a good quality interpolation of a given smooth function $f(x)$ on the entire interval $[a,b]$. We are free to choose the $n+1$ data abscissae, $x_0, x_1, \ldots, x_n$: how should we choose these points?

Consider the expression for the interpolation error from the previous section, specifically given on page 314. Let us further assume absence of additional information about ξ or even f itself, other than the assurance that it can be sampled anywhere and that its bounded $(n+1)$st derivative exists: this is a common situation in many applications. Then the best we can do to minimize the error $\max_{a \leq x \leq b} |f(x) - p_n(x)|$ is to choose the data abscissae so as to minimize the quantity $\max_{a \leq x \leq b} |\psi_n(x)|$.

The latter minimization task leads to the choice of **Chebyshev points**. These points are defined on the interval $[-1,1]$ by

$$x_i = \cos\left(\frac{2i+1}{2(n+1)}\pi\right), \quad i = 0, \ldots, n.$$

For a general interval $[a,b]$ we apply the affine transformation that maps $[-1,1]$ onto $[a,b]$ to shift and scale the Chebyshev points. So, by the transformation

$$x = a + \frac{b-a}{2}(t+1), \quad t \in [-1,1],$$

we redefine the interpolation abscissae as

$$x_i \longleftarrow a + \frac{b-a}{2}(x_i + 1), \quad i = 0, \ldots, n.$$

Interpolation error using Chebyshev points

Let us stay with the interval $[-1,1]$, then. The Chebyshev points are zeros (roots) of the *Chebyshev polynomial*, defined and discussed in detail in Section 12.4. Thus, the monic Chebyshev polynomial (i.e., the Chebyshev polynomial that is scaled so as to have its leading coefficient equal 1) is given by $\psi_n(x)$, where the x_i are the Chebyshev points. As explained in detail in Section 12.4, the maximum absolute value of this polynomial over the interpolation interval is 2^{-n}. Thus, the $n+1$ Chebyshev points defined above solve the **min-max** problem

$$\beta = \min_{x_0, x_1, \ldots, x_n} \max_{-1 \leq x \leq 1} |(x - x_0)(x - x_1) \cdots (x - x_n)|,$$

yielding the value $\beta = 2^{-n}$. This leads to the interpolation error bound

$$\max_{-1 \leq x \leq 1} |f(x) - p_n(x)| \leq \frac{1}{2^n(n+1)!} \max_{-1 \leq t \leq 1} |f^{(n+1)}(t)|.$$

Example 10.6. A long time ago, C. Runge gave the innocent-looking example

$$f(x) = \frac{1}{1 + 25x^2}, \quad -1 \leq x \leq 1.$$

10.6. Chebyshev interpolation

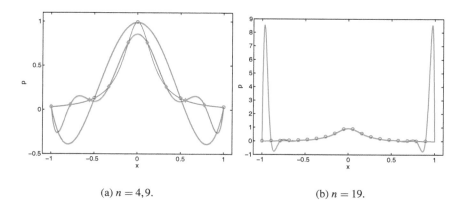

(a) $n = 4, 9$.   (b) $n = 19$.

Figure 10.6. *Global polynomial interpolation at uniformly spaced abscissae can be bad. Here the blue curve with one maximum point is the Runge function of Example 10.6, and the other curves are polynomial interpolants of degree n.*

Results of polynomial interpolation at 5, 10 and 20 *equidistant* points are plotted in Figure 10.6. Note the different scaling on the y-axis between the two graphs: the approximation gets worse for larger n!

Calculating $\frac{f^{(n+1)}}{(n+1)!}$ we can see growth in the error term near the interval ends, explaining the fact that the results do not improve as the degree of the polynomial is increased.[42]

Next, we repeat the experiment, this time using the Chebyshev points for abscissae of the data points; nothing else changes. The much better results are plotted in Figure 10.7. ∎

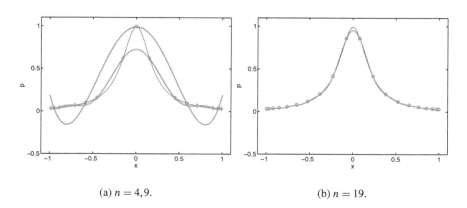

(a) $n = 4, 9$.   (b) $n = 19$.

Figure 10.7. *Polynomial interpolation at Chebyshev points, Example 10.6. Results are much improved as compared to Figure 10.6, especially for larger n.*

The improvement in Example 10.6 upon employing interpolation at Chebyshev points over polynomial interpolation at uniformly spaced points is rather remarkable. You may think that it

---

[42] Although there is a classical theorem by Weierstrass stating that it is possible to find arbitrarily close polynomial approximations to any continuous function.

is a bit "special," in that the Chebyshev points concentrate more near the interval ends, which is precisely where $\frac{f^{(n+1)}}{(n+1)!}$ gets large for this particular example, while our working assumption was that we don't know much about f. It is not difficult to construct examples where Chebyshev interpolation would not do well. Nonetheless, interpolation at Chebyshev points yields very good results in many situations, and important numerical methods for solving differential equations are based on it.

> **Note:** A fuller explanation of the remarkable behavior of Chebyshev interpolation goes much deeper into approximation theory than we will in this book. A practical outcome is that this is the only place in the present chapter where it is natural to allow n to grow large. One conclusion from the latter is that the Lagrange (barycentric) interpolation (see page 305) should generally be employed for accurate Chebyshev interpolation.

Example 10.7. Suppose we wish to obtain a really accurate polynomial approximation for the function
$$f(x) = e^{3x} \sin(200x^2)/(1+20x^2), \quad 0 \le x \le 1.$$
A plot of this function is given in the top panel of Figure 10.8. It certainly does not look like a low degree polynomial.

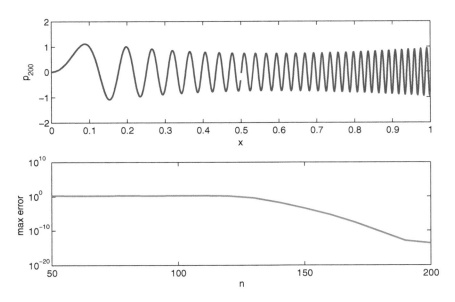

Figure 10.8. *Top panel: the function of Example* 10.7 *is indistinguishable from its polynomial interpolant at* 201 *Chebyshev points. Bottom panel: the maximum polynomial interpolation error as a function of the polynomial degree. When doubling the degree from* $n = 100$ *to* $n = 200$ *the error decreases from unacceptable* (> 1) *to almost rounding unit level.*

We have interpolated this function at $n+1$ Chebyshev points for n from 50 to 200 in increments of 10. The Lagrange representation was employed. The overall maximum error (measured over the uniform evaluation mesh `x = 0:.001:1`) is graphed in Figure 10.8 as well. Notice that for polynomial degrees as high as 100 and more the interpolation error is rather lousy. But as n is increased further the error eventually goes down, and rather fast: the error then looks like $\mathcal{O}(q^{-n})$

10.7. Interpolating also derivative values

for some $q > 1$. This is called **spectral accuracy**. For $n = 200$ the error is already at rounding unit level.

For the Runge function in Example 10.6 things look even better: spectral accuracy is observed already for moderate values of n (see Exercise 19). But then that function is somewhat special.

In many applications extreme accuracy is not required, but reasonable accuracy is, and the lack of initial improvement in the bottom panel of Figure 10.8 can be unnerving: only at about $n = 150$ results are visually acceptable and the fact that for even larger n we start getting phenomenal accuracy is not always crucial, although no one would complain about too much accuracy. ∎

Specific exercises for this section: Exercises 17–20.

10.7 Interpolating also derivative values

Often, an interpolating polynomial $p_n(x)$ is desired that interpolates not only values of a function but also values of its *derivatives* at given points.

Example 10.8. A person throws a stone at an angle of 45°, and the stone lands after flying for five meters. The trajectory of the stone, $y = f(x)$, obeys the equations of motion, taking gravity and perhaps even air-drag forces into account. But we are after a quick approximation, and we know right away about f that $f(0) = 1.5$, $f'(0) = 1$, and $f(5) = 0$. Thus, we pass a quadratic through these points.

Using a monomial basis we write

$$p_2(x) = c_0 + c_1 x + c_2 x^2, \quad \text{hence } p_2'(x) = c_1 + 2c_2 x.$$

Substituting the data gives three equations for the coefficients, namely, $1.5 = p_2(0) = c_0$, $1 = p_2'(0) = c_1$, and

$$0 = p_2(5) = c_0 + 5c_1 + 25c_2.$$

The resulting interpolant (please verify) is

$$p_2(x) = 1.5 + x - 0.26x^2,$$

and it is plotted in Figure 10.9. ∎

The general problem

The general case involves more notation, but there is not much added complication beyond that. Suppose that there are $q + 1$ distinct abscissae

$$t_0, t_1, t_2, \ldots, t_q$$

and $q + 1$ nonnegative integers

$$m_0, m_1, m_2, \ldots, m_q,$$

and consider finding the unique **osculating polynomial**[43] of lowest degree satisfying

$$p_n^{(k)}(t_i) = f^{(k)}(t_i) \quad (k = 0, \ldots, m_i), \ i = 0, 1, \ldots, q.$$

By counting how many conditions there are to satisfy, clearly the degree of our interpolating polynomial is at most

$$n = \sum_{k=0}^{q} m_k + q.$$

---

[43] The word "osculating" is derived from Latin for "kiss."

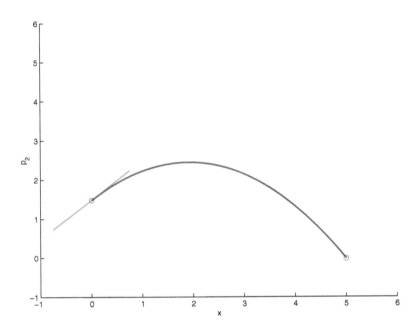

Figure 10.9. *A quadratic interpolant $p_2(x)$ satisfying $p_2(0) = 1.5$, $p'_2(0) = 1$, and $p_2(5) = 0$.*

Thus, the values of p_n and its first m_i derivatives coincide with the corresponding values of the first m_i derivatives of f at each of the interpolation points. This generalizes what we have seen before and more. Instances include the following:

1. If $q = n$, $m_i = 0$ for each i, then we have the polynomial interpolation at function values as we have seen before.

2. If $q = 0$, then we have the *Taylor polynomial* of degree m_0 at t_0.

3. If $n = 2q + 1$, $m_i = 1$ for each i, then we have **Hermite interpolation**.

This list provides the most useful instances, although the above formulation is *not* restricted to these choices.

Hermite cubic interpolation

The most popular osculating polynomial interpolant is the *Hermite cubic*, obtained by setting $q = m_0 = m_1 = 1$. Using the monomial basis we can write the interpolation conditions as

$$c_0 + c_1 t_0 + c_2 t_0^2 + c_3 t_0^3 = f(t_0), \quad c_1 + 2c_2 t_0 + 3c_3 t_0^2 = f'(t_0),$$
$$c_0 + c_1 t_1 + c_2 t_1^2 + c_3 t_1^3 = f(t_1), \quad c_1 + 2c_2 t_1 + 3c_3 t_1^2 = f'(t_1),$$

and solve these four linear equations for the coefficients c_j.

Constructing the osculating polynomial

A *general method* of constructing the osculating polynomial can be easily devised by extending Newton's form and the divided differences of Section 10.4. We define the set of abscissae by repeat-

10.7. Interpolating also derivative values

ing each of the points t_i in the sequence $m_i + 1$ times, resulting in the sequence

$$(x_0, x_1, x_2, \ldots, x_n) = \Big(\underbrace{t_0, t_0, \ldots, t_0}_{m_0+1}, \underbrace{t_1, \ldots, t_1}_{m_1+1}, \ldots, \underbrace{t_q, \ldots, t_q}_{m_q+1} \Big).$$

The corresponding data values are

$$(y_0, y_1, y_2, \ldots, y_n)$$
$$= \Big(f(t_0), f'(t_0), \ldots, f^{(m_0)}(t_0), f(t_1), \ldots, f^{(m_1)}(t_1), \ldots, f(t_q), \ldots, f^{(m_q)}(t_q) \Big).$$

Then use the Newton interpolating form

$$p_n(x) = \sum_{j=0}^{n} f[x_0, x_1, \ldots, x_j] \prod_{i=0}^{j-1} (x - x_i),$$

where

$$f[x_k, \ldots, x_j] = \begin{cases} \frac{f[x_{k+1}, \ldots, x_j] - f[x_k, \ldots, x_{j-1}]}{x_j - x_k}, & x_k \neq x_j, \\ \frac{f^{(j-k)}(x_k)}{(j-k)!}, & x_k = x_j, \end{cases}$$

for $0 < k \leq j \leq n$. Note that repeated values must be consecutive in the sequence $\{x_i\}_{i=0}^{n}$. The resulting algorithm is given on this page.

Algorithm: Polynomial Interpolation in Newton Form.

1. *Construction*: Given data $\{(x_i, y_i)\}_{i=0}^{n}$, where the abscissae are not necessarily distinct,

   for $j = 0, 1, \ldots, n$
   for $l = 0, 1, \ldots, j$

   $$\gamma_{j,l} = \begin{cases} \frac{\gamma_{j,l-1} - \gamma_{j-1,l-1}}{x_j - x_{j-l}} & \text{if } x_j \neq x_{j-l}, \\ \frac{f^{(l)}(x_j)}{l!} & \text{otherwise.} \end{cases}$$

2. *Evaluation*: Given an evaluation point x,

   $p = \gamma_{n,n}$
   for $j = n-1, n-2, \ldots, 0$,
   $p = p(x - x_j) + \gamma_{j,j}$

With this method definition, the expression for the error in polynomial interpolation derived in Section 10.5 also extends seamlessly to the case of osculating interpolation.

Example 10.9. For the function $f(x) = \ln(x)$ we have the values $f(1) = 0$, $f'(1) = 1$, $f(2) = .693147$, $f'(2) = .5$. Let us construct the corresponding Hermite cubic interpolant.

Using the simple but not general monomial basis procedure outlined above, we readily obtain the interpolant

$$p(x) = -1.53426 + 2.18223x - 0.761675x^2 + 0.113706x^3.$$

Using instead the general algorithm given on the previous page yields the alternative representation

$$p(x) = (x-1) - .30685(x-1)^2 + .113706(x-1)^2(x-2).$$

The function and its interpolant are depicted in Figure 10.10.

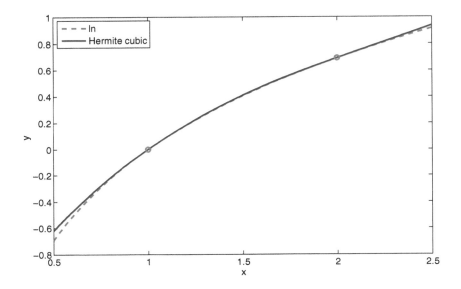

Figure 10.10. *The osculating Hermite cubic for* $\ln(x)$ *at the points 1 and 2.*

For example, $p_3(1.5) = .409074$, whereas $\ln(1.5) = .405465$.

To estimate the error at $x = 1.5$, pretending we cannot evaluate f there, we use the same expression as before: since $|f''''(x)| = |6x^{-4}| \leq 6$, we have that the error is bounded by

$$\frac{1}{4}(1.5-1)^2(1.5-2)^2 = .015625.$$

This turns out to be about 4 times larger than the actual error. ∎

Example 10.10. Suppose we are given the five data values

t_i	$f(t_i)$	$f'(t_i)$	$f''(t_i)$
8.3	17.564921	3.116256	0.120482
8.6	18.505155	3.151762	

We set up the divided difference table

$$(x_0, x_1, x_2, x_3, x_4) = \left(\underbrace{8.3, 8.3, 8.3}_{m_0=2}, \underbrace{8.6, 8.6}_{m_1=1}\right),$$

$$f[x_0, x_1] = \frac{f'(t_0)}{1!} = f'(8.3), \quad f[x_1, x_2] = \frac{f'(t_0)}{1!},$$

$$f[x_0, x_1, x_2] = \frac{f''(t_0)}{2!} = \frac{f''(8.3)}{2}, \quad f[x_3, x_4] = \frac{f'(t_1)}{1!} = f'(8.6),$$

and so on. In the table below the originally given derivative values are underlined:

x_i	$f[\cdot]$	$f[\cdot,\cdot]$	$f[\cdot,\cdot,\cdot]$	$f[\cdot,\cdot,\cdot,\cdot]$	$f[\cdot,\cdot,\cdot,\cdot,\cdot]$
8.3	17.564921				
8.3	17.564921	<u>3.116256</u>			
8.3	17.564921	3.116256	<u>0.060241</u>		
8.6	18.505155	3.134113	0.059524	-0.002389	
8.6	18.505155	<u>3.151762</u>	0.058829	-0.002319	0.000233

The resulting quartic interpolant is

$$p_4(x) = \sum_{k=0}^{4} f[x_0, \ldots, x_k] \prod_{j=0}^{k-1} (x - x_j)$$

$$= 17.564921 + 3.116256(x - 8.3) + 0.060241(x - 8.3)^2$$
$$- 0.002389(x - 8.3)^3 + 0.000233(x - 8.3)^3(x - 8.6). \quad \blacksquare$$

Specific exercises for this section: Exercises 21–25.

10.8 Exercises

0. **Review questions**

   (a) Distinguish between the terms data fitting, interpolation, and polynomial interpolation.

   (b) Distinguish between (discrete) data fitting and approximating a given function.

   (c) What are basis functions? Does an approximant $v(x)$ that is written as a linear combination of basis functions have to be linear in x?

   (d) An interpolating polynomial is unique regardless of the choice of the basis. Explain why.

   (e) State one advantage and two disadvantages of using the monomial basis for polynomial interpolation.

   (f) What are Lagrange polynomials? How are they used for polynomial interpolation?

   (g) What are barycentric weights?

(h) State the main advantages and the main disadvantage for using the Lagrange representation.

(i) What is a divided difference table and how is it constructed?

(j) Write down the formula for polynomial interpolation in Newton form.

(k) State two advantages and two disadvantages for using the Newton representation for polynomial interpolation.

(l) Describe the linear systems that are solved for the monomial basis, the Lagrange representation, and the Newton representation.

(m) Describe the connection between the kth divided difference of a function f and its kth derivative.

(n) Provide an expression for the error in polynomial interpolation as well as an error bound expression.

(o) How does the smoothness of a function and its derivatives affect the quality of polynomial interpolants that approximate it, in general?

(p) Give an example where the error bound is attained.

(q) When we interpolate a function f given only data points, i.e., we do not know f or its derivatives, how can we gauge the accuracy of our approximation?

(r) What are Chebyshev points and why are they important?

(s) Describe osculating interpolation. How is it different from the usual polynomial interpolation?

(t) What is a Hermite cubic interpolant?

1. Derive the linear interpolant through the two data points $(1.0, 2.0)$ and $(1.1, 2.5)$. Derive also the quadratic interpolant through these two pairs as well as $(1.2, 1.5)$. Show that the situation can be depicted as in Figure 10.11.

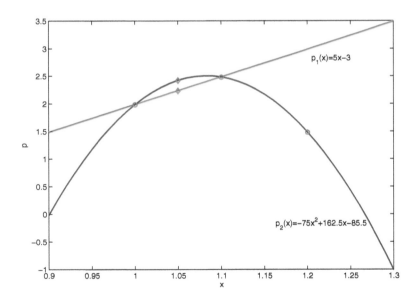

Figure 10.11. *Quadratic and linear polynomial interpolation.*

10.8. Exercises

2. Some modeling considerations have mandated a search for a function

$$u(x) = \gamma_0 e^{\gamma_1 x + \gamma_2 x^2},$$

where the unknown coefficients γ_1 and γ_2 are expected to be nonpositive. Given are data pairs to be interpolated, (x_0, z_0), (x_1, z_1), and (x_2, z_2), where $z_i > 0$, $i = 0, 1, 2$. Thus, we require $u(x_i) = z_i$.

The function $u(x)$ is not linear in its coefficients, but $v(x) = \ln(u(x))$ is linear in its.

(a) Find a quadratic polynomial $v(x)$ that interpolates appropriately defined three data pairs, and then give a formula for $u(x)$ in terms of the original data.

[This is a pen-and-paper item; the following one should consume much less of your time.]

(b) Write a script to find u for the data $(0, 1)$, $(1, .9)$, $(3, .5)$. Give the coefficients γ_i and plot the resulting interpolant over the interval $[0, 6]$. In what way does the curve behave qualitatively differently from a quadratic?

3. Use the known values of the function $\sin(x)$ at $x = 0, \pi/6, \pi/4, \pi/3$ and $\pi/2$ to derive an interpolating polynomial $p(x)$. What is the degree of your polynomial?

What is the interpolation error magnitude $|p(1.2) - \sin(1.2)|$?

4. Given $n+1$ data pairs $(x_0, y_0), (x_1, y_1), \ldots, (x_n, y_n)$, define for $j = 0, 1, \ldots, n$ the functions $\rho_j = \prod_{i \neq j}(x_j - x_i)$, and let also $\psi(x) = \prod_{i=0}^{n}(x - x_i)$.

(a) Show that

$$\rho_j = \psi'(x_j).$$

(b) Show that the interpolating polynomial of degree at most n is given by

$$p_n(x) = \psi(x) \sum_{j=0}^{n} \frac{y_j}{(x - x_j)\psi'(x_j)}.$$

5. Construct a polynomial $p_3(t)$ of degree at most three *in Lagrange form* that interpolates the data

t	−1.1	1.1	2.2	0.0
y	0.0	6.75	0.0	0.0

6. Given the four data points $(-1, 1), (0, 1), (1, 2), (2, 0)$, determine the interpolating cubic polynomial

- using the monomial basis;
- using the Lagrange basis;
- using the Newton basis.

Show that the three representations give the same polynomial.

7. For the Newton basis, prove that $c_j = f[x_0, x_1, \ldots, x_j]$ satisfies the recursion formula as stated on page 308 in Section 10.4.

[This proof is short but may be challenging.]

8. A secret formula for eternal youth, $f(x)$, was discovered by Dr. Quick, who has been working in our biotech company. However, Dr. Quick has disappeared and is rumored to be negotiating with a rival organization.

   From the notes that Dr. Quick left behind in his hasty exit it is clear that $f(0) = 0$, $f(1) = 2$, and that $f[x_0, x_1, x_2] = 1$ for *any* three points x_0, x_1, x_2. Find $f(x)$.

9. Joe had decided to buy stocks of a particularly promising Internet company. The price per share was \$100, and Joe subsequently recorded the stock price at the end of each week. With the abscissae measured in days, the following data were acquired:
   $(0, 100), (7, 98), (14, 101), (21, 50), (28, 51), (35, 50)$.

   In attempting to analyze what happened, it was desired to approximately evaluate the stock price a few days before the crash.

   (a) Pass a linear interpolant through the points with abscissae 7 and 14. Then add to this data set the value at 0 and (separately) the value at 21 to obtain two quadratic interpolants. Evaluate all three interpolants at $x = 12$. Which one do you think is the most accurate? Explain.

   (b) Plot the two quadratic interpolants above, together with the data (without a broken line passing through the data) over the interval $[0, 21]$. What are your observations?

10. Suppose we are given 137 uniformly spaced data pairs at distinct abscissae: $(x_i, y_i), i = 0, 1, \ldots, 136$. These data are thought to represent a function which is piecewise smooth; that is, the unknown function $f(x)$ which gave rise to these data values has many bounded derivatives everywhere except for a few points where it jumps discontinuously. (Imagine drawing a curve smoothly from left to right, mixed with lifting the pen and moving it vertically a few times.) For each subinterval $[x_{i-1}, x_i]$ we want to pass the hopefully best cubic $p_3(x)$ for an accurate interpolation of $f(x)$ at points $x_{i-1} < x < x_i$. This involves choosing good neighbors to interpolate at.

    Propose an algorithm for this task. Justify.

11. Given a sequence $y_0, y_1, y_2, \ldots$, define the *forward difference operator* Δ by

    $$\Delta y_i = y_{i+1} - y_i.$$

    Powers of Δ are defined recursively by

    $$\Delta^0 y_i = y_i,$$
    $$\Delta^j y_i = \Delta(\Delta^{j-1} y_i), \quad j = 1, 2, \ldots.$$

    Thus, $\Delta^2 y_i = \Delta(y_{i+1} - y_i) = y_{i+2} - 2y_{i+1} + y_i$, etc.

    Consider polynomial interpolation at equispaced points, $x_i = x_0 + ih$, $i = 0, 1, \ldots, n$.

    (a) Show that

    $$f[x_0, x_1, \ldots, x_j] = \frac{1}{j! h^j} \Delta^j f(x_0).$$

    [Hint: Use mathematical induction.]

10.8. Exercises

(b) Show that the interpolating polynomial of degree at most n is given by the *Newton forward difference* formula

$$p_n(x) = \sum_{j=0}^{n} \binom{s}{j} \Delta^j f(x_0),$$

where $s = \frac{x-x_0}{h}$ and $\binom{s}{j} = \frac{s(s-1)\cdots(s-j+1)}{j!}$ (with $\binom{s}{0} = 1$).

12. Given a sequence $y_0, y_1, y_2, \ldots$, define the *backward difference operator* ∇ by

$$\nabla y_i = y_i - y_{i-1}.$$

Powers of ∇ are defined recursively by

$$\nabla^0 y_i = y_i,$$
$$\nabla^j y_i = \nabla(\nabla^{j-1} y_i), \quad j = 1, 2, \ldots.$$

Thus, $\nabla^2 y_i = \nabla(y_i - y_{i-1}) = y_i - 2y_{i-1} + y_{i-2}$, etc.

Consider polynomial interpolation at equispaced points, $x_i = x_0 + ih$, $i = 0, 1, \ldots, n$.

(a) Show that

$$f[x_n, x_{n-1}, \ldots, x_{n-j}] = \frac{1}{j! h^j} \nabla^j f(x_n).$$

[Hint: Use mathematical induction.]

(b) Show that the interpolating polynomial of degree at most n is given by the *Newton backward difference* formula

$$p_n(x) = \sum_{j=0}^{n} (-1)^j \binom{s}{j} \nabla^j f(x_n),$$

where $s = \frac{x_n - x}{h}$ and $\binom{s}{j} = \frac{s(s-1)\cdots(s-j+1)}{j!}$ (with $\binom{s}{0} = 1$).

13. Let $(\hat{x}_0, \hat{x}_1, \ldots, \hat{x}_k)$ be a permutation of the abscissae $(x_0, x_1, \ldots, x_k)$. Show that

$$f[\hat{x}_0, \hat{x}_1, \ldots, \hat{x}_k] = f[x_0, x_1, \ldots, x_k].$$

[Hint: Consider the kth derivative of the unique polynomial of degree k interpolating f at these $k+1$ points, regardless of how they are ordered.]

14. Let the points $x_0, x_1, \ldots, x_n$ be fixed and consider the divided difference $f[x_0, x_1, \ldots, x_n, x]$ as a function of x. (This function appears as part of the expression for the error in polynomial interpolation.)

Suppose next that $f(x)$ is a polynomial of degree m. Show that

- if $m \leq n$, then $f[x_0, x_1, \ldots, x_n, x] \equiv 0$;
- otherwise $f[x_0, x_1, \ldots, x_n, x]$ is a polynomial of degree $m - n - 1$.

[Hint: If $m > n$, show it first for the case $n = 0$. Then proceed by induction, examining the function $g(x) = f[x_1, \ldots, x_n, x]$.]

15. Suppose we want to approximate the function e^x on the interval $[0, 1]$ by using polynomial interpolation with $x_0 = 0, x_1 = 1/2$ and $x_2 = 1$. Let $p_2(x)$ denote the interpolating polynomial.

    (a) Find an upper bound for the error magnitude
    $$\max_{0 \leq x \leq 1} |e^x - p_2(x)|.$$

    (b) Find the interpolating polynomial using your favorite technique.

    (c) Plot the function e^x and the interpolant you found, both on the same figure, using the commands `plot`.

    (d) Plot the error magnitude $|e^x - p_2(x)|$ on the interval using logarithmic scale (the command `semilogy`) and verify by inspection that it is below the bound you found in part (a).

16. For the problem of Exercise 3, find an error bound for the polynomial interpolation on $[0, \pi/2]$. Compare your error bound to the actual interpolation error at $x = 1.2$.

17. Construct two simple examples for any positive integer n, one where interpolation at $n+1$ equidistant points is more accurate than interpolation at $n+1$ Chebyshev points and one where Chebyshev interpolation is more accurate.

    Your examples should be convincing without the aid of any computer implementation.

18. (a) Interpolate the function $f(x) = \sin(x)$ at 5 Chebyshev points over the interval $[0, \pi/2]$. Compare your results to those of Exercises 3 and 16.

    (b) Repeat the interpolation, this time using 5 Chebyshev points over the interval $[0, \pi]$. Plot $f(x)$ as well as the interpolant. What are your conclusions?

19. Interpolate the Runge function of Example 10.6 at Chebyshev points for n from 10 to 170 in increments of 10. Calculate the maximum interpolation error on the uniform evaluation mesh `x = -1:.001:1` and plot the error vs. polynomial degree as in Figure 10.8 using `semilogy`. Observe spectral accuracy.

20. The **Chebyshev extremum points** are close cousins of the Chebyshev points. They are defined on the interval $[-1, 1]$ by
    $$x_i = \xi_i = \cos\left(\frac{i}{n}\pi\right), \quad i = 0, 1, \ldots, n.$$

    See Section 12.4 for more on these points. Like the Chebyshev points, the extremum Chebyshev points tend to concentrate more near the interval's ends.

    Repeat the runs of Exercises 18 and 19, as well as those reported in Example 10.7, interpolating at the $n+1$ Chebyshev extremum points by a polynomial of degree at most n.

    Compare against the corresponding results using the Chebyshev points and show that although the Chebyshev extremum points are slightly behind, it's never by much. In fact, the famed Chebyshev stability and spectral accuracy are also observed here!

21. Interpolate the function $f(x) = \ln(x)$ by passing a cubic through the points $x_i = (0.1, 1, 2, 2.9)$. Evaluate your interpolant at $x = 1.5$ and compare the result against the exact value and against the value of the osculating Hermite cubic through the points $x_i = (1, 1, 2, 2)$, given in Example 10.9.

    Explain your observations by looking at the error terms for both interpolating cubic polynomials.

22. For some function f, you have a table of extended divided differences of the form

i	z_i	$f[\cdot]$	$f[\cdot,\cdot]$	$f[\cdot,\cdot,\cdot]$	$f[\cdot,\cdot,\cdot,\cdot]$
0	5.0	$f[z_0]$			
1	5.0	$f[z_1]$	$f[z_0, z_1]$		
2	6.0	4.0	5.0	-3.0	
3	4.0	2.0	$f[z_2, z_3]$	$f[z_1, z_2, z_3]$	$f[z_0, z_1, z_2, z_3]$

    Fill in the unknown entries in the table.

23. For the data in Exercise 22, what is the osculating polynomial $p_2(x)$ of degree at most 2 that satisfies
    $$p_2(5.0) = f(5.0), \quad p_2'(5.0) = f'(5.0), \quad p_2(6.0) = f(6.0)?$$

24. (a) Write a script that interpolates $f(x) = \cosh(x) = \frac{e^x + e^{-x}}{2}$ with an osculating polynomial that matches both $f(x)$ and $f'(x)$ at abscissae $x_0 = 1$ and $x_1 = 3$. Generate a plot (with logarithmic vertical axis) comparing $f(x)$ and the interpolating polynomial and another plot showing the error in your interpolant over this interval. Use the command semilogy for generating your graphs.

    (b) Modify the code to generate another interpolant that matches $f(x)$ and $f'(x)$ at the abscissae $x_0 = 1$, $x_1 = 2$, and $x_2 = 3$. Generate two more plots: comparison of function to interpolant, and error. Compare the quality of this new polynomial to the quality of the polynomial of part (a).

    (c) Now, modify the code to generate an interpolant that matches $f(x)$, $f'(x)$ and $f''(x)$ at the abscissae $x_0 = 1$ and $x_1 = 3$. Generate two more plots and comment on the quality of this approximation compared to the previous two interpolants.

25. A popular technique arising in methods for minimizing functions in several variables involves a *weak line search*, where an approximate minimum x^* is found for a function in one variable, $f(x)$, for which the values of $f(0)$, $f'(0)$, and $f(1)$ are given. The function $f(x)$ is defined for all nonnegative x, has a continuous second derivative, and satisfies $f(0) < f(1)$ and $f'(0) < 0$. We then interpolate the given values by a quadratic polynomial and set x^* as the minimum of the interpolant.

    (a) Find x^* for the values $f(0) = 1$, $f'(0) = -1$, $f(1) = 2$.

    (b) Show that the quadratic interpolant has a unique minimum satisfying $0 < x^* < 1$. Can you show the same for the function f itself?

26. Suppose we have computed an interpolant for data points $\{(x_i, y_i)\}_{i=0}^n$ using each of the three polynomial bases that we have discussed in this chapter (monomial, Lagrange, and Newton), and we have already constructed any necessary matrices, vectors, interpolation coefficients,

and/or basis functions. Suddenly, we realize that the final data point was wrong. Consider the following situations:

(a) The final data point should have been $(\tilde{x}_n, y_n)$ (that is, y_n is the same as before).

(b) The final data point should have been $(x_n, \tilde{y}_n)$ (x_n is the same as before).

For each of these two scenarios, determine the computational cost of computing the modified interpolants. Give your answers in the $\mathcal{O}$ notation (see page 7).

[A couple of the answers for parts of this question may require some "extra" knowledge of linear algebra.]

10.9 Additional notes

The fundamental role that polynomial interpolation plays in many different aspects of numerical computation is what gave rise to such a long chapter on what is essentially a relatively simple process. Many textbooks on numerical methods and analysis devote considerable space for this topic, and we mention Burden and Faires [11], Cheney and Kincaid [12], and the classical Conte and de Boor [13] for a taste.

Lagrange polynomials are used in the design of methods for numerical integration and solution of differential equations. We will see some of this in Chapters 14, 15, and 16. For much more see, e.g., Davis and Rabinowitz [17], Ascher and Petzold [5], and Ascher [3].

For polynomial interpolation at uniformly spaced (or equispaced) abscissae, where there is a spacing value h such that $x_i - x_{i-1} = h$ for all $i = 1, \ldots, n$, the divided difference table acquires a simpler form. Special names from the past are attached to this: there are the Newton forward and backward difference formulas, as well as Stirling's centered difference formula. See Exercises 11 and 12.

The use of divided differences and the Newton form interpolation in designing methods for complex problems is also prevalent. As elaborated also in some of the exercises, this is natural especially if we add interpolation points one at a time, for instance, because we wish to stay on one side of a discontinuity in the interpolated function (rather than crossing the discontinuity with an infinitely differentiable polynomial). Certain such methods for solving problems with shocks are called essentially nonoscillatory (ENO) schemes; see, e.g., [3].

Chebyshev polynomial interpolation is the most notable exception to the practical rule of keeping the degree n of interpolating polynomials low and using them locally, switching for higher accuracy to the methods of Chapter 11. At least for infinitely smooth functions $f(x)$, high degree Chebyshev polynomial interpolation can yield very high accuracy at a reasonable cost, as Example 10.7 and Exercise 19 indicate. This allows one to relate to smooth functions as objects, internally replacing them by their Chebyshev interpolants in a manner that is invisible to a user. Thus, "symbolic" operations such as differentiation, integration, and even solution of differential equations in one variable, are enabled. For work on this in a MATLAB context by Trefethen and coauthors, see http://www.mathworks.com/matlabcentral/fx_files/23972/10/content/chebfun/guide/html/guide4.html.

Chebyshev polynomials often arise in the numerical analysis of seemingly unrelated methods because of their min-max property. For instance, they feature prominently in the analysis of convergence rates for the famous conjugate gradient method introduced in Section 7.4; see, e.g., Greenbaum [32] or LeVeque [50]. They are also applicable for efficiently solving certain partial differential equations using methods called *spectral collocation*; see, e.g., Trefethen [69].

Chapter 11

Piecewise Polynomial Interpolation

The previous chapter discusses polynomial interpolation, an essential ingredient in the construction of general approximation methods for a variety of problems including function approximation, differentiation, integration, and the solution of differential equations.

However, polynomial interpolation is often not sufficiently flexible to directly yield useful general-purpose procedures; rather, it typically provides an essential building block for other, more general techniques. In the present chapter we develop robust methods for the interpolation of functions which work even if the number of data points is large, or their abscissae locations are not under our control, or the interval over which the function is approximated is long.

> **Note:** In many ways, this chapter is a direct continuation of the previous one. Section 10.1, in particular, is directly relevant here as well.

We start in Section 11.1 by making the case for piecewise polynomial interpolation as the cure for several polynomial interpolation ills. In Section 11.2 we introduce two favorite piecewise polynomial interpolants which are defined entirely locally: each polynomial piece can be constructed independently of and in parallel with the others.

The workhorse of robust interpolation, however, is the *cubic spline* which is derived and demonstrated in Section 11.3. Basis functions for piecewise polynomial interpolation are considered in Section 11.4.

In Section 11.5 we shift focus and introduce parametric curves, fundamental in **computer aided geometric design** (CAGD) and other applications. Despite the different setting, these generally turn out to be a surprisingly simple extension of the techniques of Section 11.2.

Interpolation in more than one independent variable is a much broader, more advanced topic that is briefly reviewed in Section 11.6.

11.1 The case for piecewise polynomial interpolation

We continue to consider interpolation of the $n+1$ data pairs

$$(x_0, y_0), (x_1, y_1), \ldots, (x_n, y_n),$$

looking for a function $v(x)$ that satisfies

$$v(x_i) = y_i, \quad i = 0, 1, \ldots, n.$$

We also continue to assume an underlying function $f(x)$ that is to be approximated on an interval $[a,b]$ containing the abscissae x_i. The function $f(x)$ is unknown in general, except for its values $f(x_i) = y_i$, $i = 0,\ldots,n$. Occasionally we will interpolate function derivative values $f'(x_i)$ as well. Furthermore, although this will not always be obvious in what follows, we consider only interpolants in linear form that can be written as

$$v(x) = \sum_{j=0}^{n} c_j \phi_j(x),$$

where $\phi_j(x)$ are given basis functions and c_j are unknown coefficients to be determined.

Shortcomings of polynomial interpolation

An interpolant of the form discussed in Chapter 10 is not always suitable for the following reasons:

- The error term

$$f(x) - p_n(x) = \frac{f^{(n+1)}(\xi)}{(n+1)!} \prod_{i=0}^{n}(x - x_i)$$

  (see the Interpolation Error Theorem on page 314) may not be small if $\frac{\|f^{(n+1)}\|}{(n+1)!}$ isn't. In Example 10.6 we have already seen that it does not take much for this to happen.

- High order polynomials tend to oscillate "unreasonably."

- Data often are only piecewise smooth, whereas polynomials are infinitely differentiable. The high derivatives $f^{(n+1)}$ may blow up (or be very large) in such a case, which again yields a large error term.

- No locality: changing any one data value may drastically alter the entire interpolant.

Returning to Example 10.6, for that particular function polynomial interpolation at Chebyshev points works well as we have seen. However, polynomial interpolation for various functions at *any* set of fixed points does not always produce satisfactory results, and the quality of the resulting approximation is hard to control. For more examples of "unreasonable" polynomial interpolation see Exercise 15 in this chapter and Exercise 9 in Chapter 10. Such examples are abundant in practice.

Piecewise polynomials

We must find a way to reduce the error term without increasing the degree n. In what follows we do this by decreasing the size of the interval $b - a$. Note that simply rescaling the independent variable x will not help! (Can you see why?) We thus resort to using polynomial pieces only locally. Globally, we use a *piecewise polynomial interpolation*.

Thus, we divide the interval into a number of smaller subintervals (or *elements*) by the partition

$$a = t_0 < t_1 < \cdots < t_r = b$$

and use a (relatively low degree) polynomial interpolation in each of these subintervals $[t_i, t_{i+1}]$, $i = 0,\ldots,r-1$. These polynomial pieces, $s_i(x)$, are then patched together to form a continuous (or C^1, or C^2) global interpolating curve $v(x)$ which satisfies

$$v(x) = s_i(x), \quad t_i \leq x \leq t_{i+1}, i = 0,\ldots,r-1.$$

The points $t_0, t_1, \ldots, t_r$ are called *break points*. See Figure 11.1.

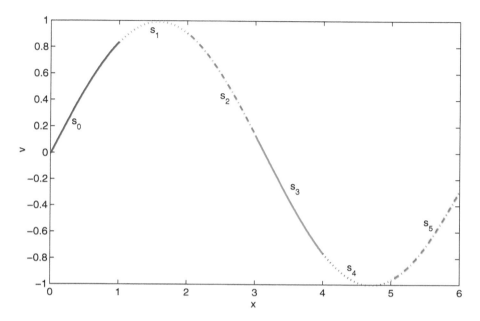

Figure 11.1. *A piecewise polynomial function with break points* $t_i = i$, $i = 0, 1, \ldots, 6$.

11.2 Broken line and piecewise Hermite interpolation

In this section we consider two important general-purpose interpolants that can be constructed entirely locally. That is to say, a polynomial interpolant is constructed on each subinterval $[t_i, t_{i+1}]$, and our interpolant on the entire interval $[a,b]$ is simply the assembly of the local pieces.

Broken line interpolation

The simplest instance of continuous piecewise polynomial interpolation is piecewise linear, or "broken line interpolation." Thus, the polynomial pieces are linear and the piecewise linear interpolant is continuous (but not continuously differentiable) everywhere.

Example 11.1. An instance of broken line interpolation is provided in Figure 11.2. It consists simply of connecting data values by straight lines. By Newton's formula for a linear polynomial interpolant, we can write

$$v(x) = s_i(x) = f(x_i) + f[x_i, x_{i+1}](x - x_i), \quad x_i \leq x \leq x_{i+1}, \ 0 \leq i \leq 4.$$

You can roughly figure out (if you must) the values of $(x_i, f(x_i))$ that gave rise to Figure 11.2. ∎

A great advantage of piecewise linear interpolation, other than its obvious simplicity, is that the maximum and minimum values are at the data points: no new extremum point is "invented" by the interpolant, which is important for a general-purpose interpolation black box routine. MATLAB uses this interpolation as the default option for plotting. (More precisely, a parametric curve based on broken line interpolation is plotted by default—see Section 11.5.)

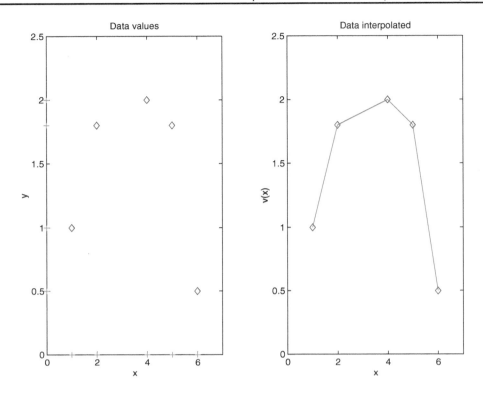

Figure 11.2. *Data and their broken line interpolation.*

Error bound for piecewise linear interpolation

Continuing with the notation of Example 11.1, let $n = r$, $t_i = x_i$, and

$$h = \max_{1 \le i \le r}(t_i - t_{i-1}).$$

It is not difficult to show that the error for piecewise linear interpolation is bounded by

$$|f(x) - v(x)| \le \frac{h^2}{8} \max_{a \le \xi \le b} |f''(\xi)|$$

for any x in the interval $[a,b]$. Indeed, for any $x \in [a,b]$ there is an index i, $1 \le i \le r$, such that $t_{i-1} \le x \le t_i$. The interpolant is linear on this subinterval, so applying the error formula from Section 10.5 for polynomial interpolation to this linear segment, we have

$$f(x) - v(x) = \frac{f''(\xi)}{2!}(x - t_{i-1})(x - t_i).$$

It can be shown (try!) that the maximum value of $|(x - t_{i-1})(x - t_i)|$ is attained at the midpoint $x = \frac{t_{i-1} + t_i}{2}$, and hence

$$|(x - t_{i-1})(x - t_i)| \le \left(\frac{t_i - t_{i-1}}{2}\right)^2.$$

The above error bound is obtained upon noting that $2^2 \cdot 2! = 8$ and applying bounds in an obvious way. ♦

11.2. Broken line and piecewise Hermite interpolation

So, at least for equally spaced data points, as n increases the error decreases, at rate $\mathcal{O}(h^2) = \mathcal{O}(n^{-2})$. Note that n is no longer the degree of the polynomial pieces; the latter is held here at 1.

Piecewise constant interpolation

Often the break points t_i are the given data abscissae x_i, sorted in ascending order, in which case we have $r = n$. Such is the case for the piecewise linear interpolation discussed above. One simple exception is *piecewise constant* interpolation: not knowing anything about the smoothness of f—not even if it is continuous—we may want to construct the simplest approximation, and one such is given by

$$v(x) = s_i(x) = f(x_i), \quad t_i \leq x < t_{i+1},$$

where $t_i = \frac{1}{2}(x_i + x_{i-1})$, $i = 1, 2, \ldots, n$. Set also $t_0 = a \leq x_0$, $t_r = b \geq x_n$, $r = n+1$.

Exercise 1 asks you to find an error bound for such an interpolation scheme. (It clearly should be $\mathcal{O}(h)$ if f has a bounded derivative.)

Piecewise cubic interpolation

But piecewise linear interpolation is often not smooth enough. (Granted, what's "enough" depends on the application.) From Figure 11.2 it is clear that this interpolant has discontinuous first derivatives at data points. To have higher smoothness, say, C^1 or C^2, we must increase the degree of each polynomial piece.

The most popular piecewise polynomial interpolation is with *cubics*. Here we may write

$$v(x) = s_i(x) = a_i + b_i(x - t_i) + c_i(x - t_i)^2 + d_i(x - t_i)^3,$$
$$t_i \leq x \leq t_{i+1}, \ i = 0, \ldots, r-1.$$

Clearly, we have $4r$ unknowns (coefficients). To fix these parameters we need $4r$ algebraic conditions. There are two types of conditions:

- interpolation conditions, and
- continuity conditions.

In the piecewise linear case these conditions read

$$s_i(t_i) = f(t_i), \ s_i(t_{i+1}) = f(t_{i+1}), \quad i = 0, 1, \ldots, r-1,$$

where $t_i = x_i$, $i = 0, 1, \ldots, r$. This gives $2r$ conditions, which exhausts all the freedom we have in the piecewise linear case because we have only $2r$ coefficients to determine. Continuity is implied by $s_i(t_{i+1}) = f(t_{i+1}) = s_{i+1}(t_{i+1})$. But in the piecewise cubic case we can impose $2r$ additional conditions. We now look at a way for doing this, which retains the local construction charm. Another important way for setting those conditions is considered in Section 11.3.

Piecewise cubic Hermite interpolation

The simplest, or cleanest, case is where also values of $f'(t_i)$ are provided. Thus, $n+1 = 2(r+1)$ and the abscissae are

$$(x_0, x_1, x_2, \ldots, x_{n-1}, x_n) = (t_0, t_0, t_1, t_1, \ldots, t_r, t_r).$$

Matching these gives precisely $2r$ more conditions, written as

$$s_i'(t_i) = f'(t_i), \ s_i'(t_{i+1}) = f'(t_{i+1}), \ i = 0, 1, \ldots, r-1.$$

This is the *piecewise cubic Hermite interpolation*. Note that $v(x)$ is clearly in C^1, i.e., both v and v' are continuous everywhere.

This interpolant can be constructed locally, piece by piece, by applying the osculating interpolation algorithm from Section 10.7 for the cubic polynomial in each subinterval separately. Specifically, for any x in the subinterval $[t_i, t_{i+1}]$ the interpolant $v(x)$ coincides with the corresponding Hermite cubic polynomial $s_i(x)$ on this interval, given explicitly by

$$s_i(x) = f_i + (h_i f_i') \tau + \left(3(f_{i+1} - f_i) - h_i(f_{i+1}' + 2f_i')\right)\tau^2$$
$$+ \left(h_i(f_{i+1}' + f_i') - 2(f_{i+1} - f_i)\right)\tau^3,$$

where $h_i = t_{i+1} - t_i$, $f_i = f(t_i)$, $f_i' = f'(t_i)$, and $\tau = \frac{x - t_i}{h_i}$. See Exercise 4.

Note also that changing the data at one data point changes the value of the interpolant only in the two subintervals associated with that point.

Example 11.2. Let us return to Example 10.6 and consider interpolation at 20 equidistant abscissae. Using a polynomial of degree 20 for this purpose led to poor results; recall Figure 10.6 on page 317.

In contrast, interpolating at the same points $x_i = -1 + 2i/19$ using values of $f(x_i)$ and $f'(x_i)$, $i = 0, 1, \ldots, 19$, yields a perfect looking curve with a computed maximum error of .0042. ∎

Error bound for piecewise cubic Hermite interpolation

> **Theorem: Piecewise Polynomial Interpolation Error.**
> Let v interpolate f at the $n+1$ points $x_0 < x_1 < \cdots < x_n$, define $h = \max_{1 \le i \le n} x_i - x_{i-1}$, and assume that f has as many bounded derivatives as appear in the bounds below on an interval $[a, b]$ containing these points.
> Then, using a local constant, linear or Hermite cubic interpolation, for each $x \in [a, b]$ the interpolation error is bounded by
>
> $$|f(x) - v(x)| \le \frac{h}{2} \max_{a \le \xi \le b} |f'(\xi)| \quad \text{piecewise constant,}$$
>
> $$|f(x) - v(x)| \le \frac{h^2}{8} \max_{a \le \xi \le b} |f''(\xi)| \quad \text{piecewise linear,}$$
>
> $$|f(x) - v(x)| \le \frac{h^4}{384} \max_{a \le \xi \le b} |f''''(\xi)| \quad \text{piecewise cubic Hermite.}$$

The error in this interpolant can be bounded directly from the error expression for polynomial interpolation. The resulting expression is

$$|f(x) - v(x)| \le \frac{h^4}{384} \max_{a \le \xi \le b} |f''''(\xi)|.$$

Note that $384 = 2^4 \cdot 4!$, which plays the same role as $8 = 2^2 \cdot 2!$ plays in the error expression for the piecewise linear interpolant.

11.3. Cubic spline interpolation

Example 11.3. Recall Example 10.9. For the function $f(x) = \ln(x)$ on the interval $[1,2]$ we have estimated $|f''''(\xi)| \leq 6$. Therefore, with $h = .25$ the interpolation error upon applying the Hermite piecewise cubic interpolation on four equidistant subintervals will be bounded by

$$|f(x) - v(x)| \leq \frac{6}{384} \times .25^4 \approx 6 \times 10^{-5}.$$

The actual maximum error turns out to be $\approx 3.9 \times 10^{-5}$. Of course, with this error level plotting, the resulting solution yields the same curve as that for $\ln(x)$ in Figure 10.10, as far as the eye can discern. ∎

The error bounds for the various local interpolants that we have seen in this section are gathered in the Piecewise Polynomial Interpolation Error Theorem on the facing page.

Specific exercises for this section: Exercises 1–4.

11.3 Cubic spline interpolation

The main disadvantage of working with Hermite piecewise cubics is that we need values for $f'(t_i)$, which is a lot to ask for, especially when there is no f, just discrete data values! Numerical differentiation comes to mind (see Exercise 9 and Section 14.1), but this does not always make sense and a more direct, robust procedure is desired. Another potential drawback is that there are many applications in which an overall C^1 requirement does not provide sufficient smoothness.

Suppose we are given only data values (x_i, y_i), $i = 0, \ldots, n$, where $x_0 < x_1 < \cdots < x_{n-1} < x_n$ are distinct and $y_i = f(x_i)$ for some function f that may not be explicitly available. Set also $a = x_0$ and $b = x_n$. We identify x_i with the break points t_i, and $n = r$. Having used $2n$ parameters to satisfy the interpolation conditions by a continuous interpolant, we now use the remaining $2n$ parameters to require that $v(x) \in C^2[a,b]$. The result is often referred to as a **cubic spline**.

Thus, the conditions to be satisfied by the cubic spline are

$$s_i(x_i) = f(x_i), \quad i = 0, \ldots, n-1, \tag{11.1a}$$
$$s_i(x_{i+1}) = f(x_{i+1}), \quad i = 0, \ldots, n-1, \tag{11.1b}$$
$$s_i'(x_{i+1}) = s_{i+1}'(x_{i+1}), \quad i = 0, \ldots, n-2, \tag{11.1c}$$
$$s_i''(x_{i+1}) = s_{i+1}''(x_{i+1}), \quad i = 0, \ldots, n-2. \tag{11.1d}$$

See Figure 11.3.[44] Note that the matching conditions are applied only at interior abscissae, not including the first and last points. For this reason there are only $n-1$ such conditions for the first and second derivatives.

Example 11.4. Consider the data

i	0	1	2
x_i	0.0	1.0	2.0
$f(x_i)$	1.1	0.9	2.0

---

[44]It may appear to the naked eye as if the curve in Figure 11.3 is infinitely smooth near $x_i = 0$. But in fact it was constructed by matching at $x = 0$ two cubics that agree in value, first and second derivative, yet differ by ratio 7 : 1 in the values of the third derivative. The apparent smoothness of the combined curve illustrates why the cubic spline is attractive in general.

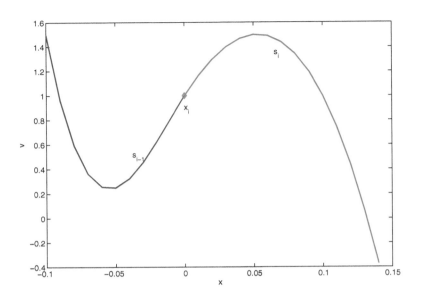

Figure 11.3. *Matching s_i, s_i', and s_i'' at $x = x_i$ with values of s_{i-1} and its derivatives at the same point. In this example, $x_i = 0$ and $y_i = 1$.*

Of course, for these three data points a quadratic polynomial interpolant would do just fine. But our purpose here is to demonstrate the construction of a cubic spline, not to ask if this is a sensible thing to do. For the latter purpose, note that $n = 2$. Thus, we have three data points and just one break point at $x_1 = 1.0$.

The interpolating cubic spline is written as

$$v(x) = \begin{cases} s_0(x) = a_0 + b_0(x-0.0) + c_0(x-0.0)^2 + d_0(x-0.0)^3, & x < 1.0, \\ s_1(x) = a_1 + b_1(x-1.0) + c_1(x-1.0)^2 + d_1(x-1.0)^3, & x \geq 1.0. \end{cases}$$

We now must determine the eight coefficients $a_0, b_0, c_0, d_0, a_1, b_1, c_1$, and d_1.
The interpolation conditions (11.1a) and (11.1b), evaluated at their left ends $i = 0$, give

$$1.1 = v(0.0) = s_0(0.0) = a_0,$$
$$0.9 = v(1.0) = s_1(1.0) = a_1,$$

which determine $a_0 = 1.1$ and $a_1 = 0.9$.
Evaluating these two conditions at $i = 1$, we also have

$$0.9 = v(1.0) = s_0(1.0) = 1.1 + b_0 + c_0 + d_0,$$
$$2.0 = v(2.0) = s_1(2.0) = 0.9 + b_1 + c_1 + d_1.$$

Thus, we have two relations between the remaining six coefficients. Note how equating $s_0(1.0)$ and $s_1(1.0)$ to the same value of $f(1.0)$ implies continuity of the constructed interpolant $v(x)$.

Next, we evaluate (11.1c) and (11.1d) at $i = 0$ (the only value of i for which they are defined here), i.e., we specify that also $s_0'(x_1) = s_1'(x_1)$ and $s_0''(x_1) = s_1''(x_1)$, where $x_1 = 1.0$. This gives

$$b_0 + 2c_0 + 3d_0 = b_1,$$
$$2c_0 + 6d_0 = 2c_1.$$

11.3. Cubic spline interpolation

In total, then, we have four equations for the six coefficients b_0, b_1, c_0, c_1, d_0, and d_1. We require two more conditions to complete the specification of $v(x)$. ∎

Two additional conditions

In total, above there are $4n - 2$ conditions for $4n$ unknowns, so we still have two conditions to specify. We specify one at each of the boundaries x_0 and x_n:

1. One popular choice is that of **free boundary**, giving a **natural spline**:
$$v''(x_0) = v''(x_n) = 0.$$

   This condition is somewhat arbitrary, though, and may cause general deterioration in approximation quality because there is no a priori reason to assume that f'' also vanishes at the endpoints.

2. If f' is available at the interval ends, then it is possible to administer **clamped boundary** conditions, specified by
$$v'(x_0) = f'(x_0), \quad v'(x_n) = f'(x_n).$$

   The resulting interpolant is also known as the **complete spline**.

3. A third alternative is called **not-a-knot**. In the absence of information on the derivatives of f at the endpoints we use the two remaining parameters to ensure third derivative continuity of the spline interpolant at the nearest interior break points, x_1 and x_{n-1}. Unlike the free boundary conditions, this does not require the interpolant to satisfy something that the interpolated function does not satisfy, so error quality does not deteriorate.

Example 11.5. Continuing Example 10.6, discussed on page 317, we use the MATLAB function spline, which implements the not-a-knot variant of cubic spline interpolation, to plot in Figure 11.4 the resulting approximation of $f(x) = 1/(1 + 25x^2)$ at 20 equidistant data points. The interpolant in Figure 11.4 marks a clear improvement over the high degree polynomial interpolant of Figure 10.6.

Computing the maximum absolute difference between this interpolant and $f(x)$ over the mesh -1:.001:1 gives the error .0123, which is only three times larger than that calculated for piecewise Hermite interpolation in Example 11.2, even though twice as much data was used there! ∎

Constructing the cubic spline

The actual construction of the cubic spline is somewhat more technical than anything we have seen so far in this chapter.[45] Furthermore, the cubic spline is not entirely local, unlike the piecewise Hermite cubic. If you feel that you have a good grasp on the principle of the thing from the preceding discussion and really don't need the construction details, then skipping to (11.6) on page 343 is possible.

We begin by completing Example 11.4 and then continue to develop the construction algorithm for the general case.

---

[45] So much so that in the derivation there are a lot of numbered equations, an infrequent event in this book.

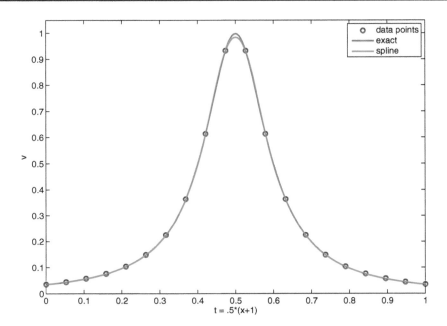

Figure 11.4. *Not-a-knot cubic spline interpolation for the Runge Example 10.6 at 20 equidistant points. The interval has been rescaled to be* $[0, 1]$.

Example 11.6. Let us continue to consider the data of Example 11.4. Four equations have been constructed for the six coefficients b_0, b_1, c_0, c_1, d_0, and d_1. We require two more conditions to complete the specification of $v(x)$.

For instance, suppose we know that $f''(0.0) = f''(2.0) = 0$. Then the "natural" conditions $v''(0.0) = s_0''(0.0) = 0$, $v''(2.0) = s_1''(2.0) = 0$ are justified and yield

$$2c_0 = 0,$$
$$2c_1 + 6d_1 = 0.$$

Solving the resulting six equations yields

$$v(x) = \begin{cases} s_0(x) = 1.1 - 0.525x + 0.325x^3, & x < 1.0, \\ s_1(x) = 0.9 + 0.45(x - 1.0) + 0.975(x - 1.0)^2 - 0.325(x - 1.0)^3, & x \geq 1.0. \end{cases}$$

Figure 11.5 depicts this interpolant. Note again the apparent smoothness at the break point depicted by a red circle, despite the jump in the third derivative. ∎

General construction

We now consider a general $n \geq 2$, so we are patching together n local cubics $s_i(x)$. Having identified the break points with the data points, the spline $v(x)$ satisfies $v(x) = s_i(x)$, $x_i \leq x \leq x_{i+1}$, $i = 0, 1, \ldots, n-1$, where

$$s_i(x) = a_i + b_i(x - x_i) + c_i(x - x_i)^2 + d_i(x - x_i)^3, \tag{11.2a}$$
$$s_i'(x) = b_i + 2c_i(x - x_i) + 3d_i(x - x_i)^2, \tag{11.2b}$$
$$s_i''(x) = 2c_i + 6d_i(x - x_i). \tag{11.2c}$$

Our present task is therefore to determine the $4n$ coefficients a_i, b_i, c_i and d_i, $i = 0, 1, \ldots, n-1$.

11.3. Cubic spline interpolation

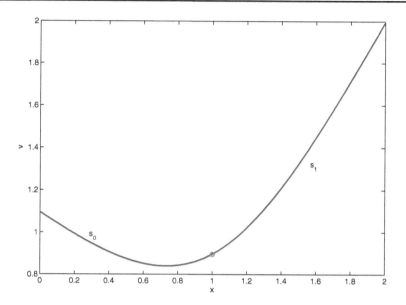

Figure 11.5. *The interpolant of Example* 11.6.

The interpolation conditions (11.1a) at the left end of each subinterval immediately determine

$$a_i = f(x_i), \quad i = 0, \ldots, n-1. \tag{11.3a}$$

Denote

$$h_i = x_{i+1} - x_i, \quad i = 0, 1, \ldots, n-1.$$

Then the interpolation conditions (11.1b) (implying continuity) give

$$a_i + b_i h_i + c_i h_i^2 + d_i h_i^3 = f(x_{i+1}).$$

Plugging the values of a_i, dividing by h_i, and rearranging gives

$$b_i + h_i c_i + h_i^2 d_i = f[x_i, x_{i+1}], \quad i = 0, \ldots, n-1. \tag{11.3b}$$

Now for the smoothness conditions. From (11.1c) and (11.2b) we get for the first derivative continuity the condition

$$b_i + 2h_i c_i + 3h_i^2 d_i = b_{i+1}, \quad i = 0, \ldots, n-2. \tag{11.3c}$$

Likewise, for the second derivative continuity, we get from (11.1d) and (11.2c) the relation

$$c_i + 3h_i d_i = c_{i+1}, \quad i = 0, \ldots, n-2. \tag{11.3d}$$

Note that we may extend (11.3c) and (11.3d) to apply also for $i = n-1$ by defining

$$b_n = v'(x_n), \quad c_n = \frac{1}{2}v''(x_n).$$

Now, we can eliminate the d_i from (11.3d), obtaining

$$d_i = \frac{c_{i+1} - c_i}{3h_i}, \quad i = 0, \ldots, n-1, \tag{11.4a}$$

and then the b_i from (11.3b), which yields

$$b_i = f[x_i, x_{i+1}] - \frac{h_i}{3}(2c_i + c_{i+1}), \quad i = 0, \ldots, n-1. \tag{11.4b}$$

Thus, if we can determine the c_i coefficients, then all the rest are given by (11.4) (and (11.3a)).
Next, let us shift the counter $i \leftarrow i+1$ in (11.3c), so we have

$$b_{i-1} + 2h_{i-1}c_{i-1} + 3h_{i-1}^2 d_{i-1} = b_i, \quad i = 1, \ldots, n-1.$$

Plugging in (11.4b) yields

$$f[x_{i-1}, x_i] - \frac{h_{i-1}}{3}(2c_{i-1} + c_i) + 2h_{i-1}c_{i-1} + h_{i-1}(c_i - c_{i-1}) = f[x_i, x_{i+1}] - \frac{h_i}{3}(2c_i + c_{i+1}),$$

and this, rearranged, finally reads

$$h_{i-1}c_{i-1} + 2(h_{i-1} + h_i)c_i + h_i c_{i+1} = 3(f[x_i, x_{i+1}] - f[x_{i-1}, x_i]),$$
$$i = 1, \ldots, n-1. \tag{11.5}$$

(We could further divide these equations by $h_{i-1} + h_i$, obtaining $3f[x_{i-1}, x_i, x_{i+1}]$ on the right-hand side, but let's not.)

In (11.5) we have a set of $n-1$ linear equations for $n+1$ unknowns. To close this system, we need two more conditions: freedom's just another word for something left to do. This is where the boundary conditions discussed earlier (see page 339) come in. An overview in algorithm form of the cubic spline construction is given on the current page. Next, we concentrate on the important choice of these two extra conditions.

> **Algorithm: Cubic Spline.**
> Given data pairs (x_i, y_i), $i = 0, 1, \ldots, n$:
>
> 1. Identify $f(x_i) \equiv y_i$, $i = 0, 1, \ldots, n$, and set $a_i = y_i$, $i = 0, 1, \ldots, n-1$.
>
> 2. Construct a tridiagonal system of equations for the unknowns $c_0, c_1, \ldots, c_n$ using the $n-1$ equations (11.5) and two more boundary conditions.
>
> 3. Solve the linear system, obtaining the coefficients c_i.
>
> 4. Set the coefficients d_i, $i = 0, 1, \ldots, n-1$, by equations (11.4a); set the coefficients b_i, $i = 0, 1, \ldots, n-1$, by equations (11.4b).
>
> 5. The desired spline $v(x)$ is given by
>
> $$v(x) = s_i(x), \quad x_i \leq x \leq x_{i+1}, \quad i = 0, 1, \ldots, n-1,$$
>
> where s_i are given by equations (11.2a).

Two extra conditions

The free boundary choice, yielding the natural spline, is the easiest; this perhaps accounts for its pedagogical popularity. It translates into simply setting

$$c_0 = c_n = 0.$$

11.3. Cubic spline interpolation

Then we write (11.5) in matrix form, obtaining

$$\begin{pmatrix} 2(h_0+h_1) & h_1 & & & & \\ h_1 & 2(h_1+h_2) & h_2 & & & \\ & \ddots & \ddots & \ddots & & \\ & & h_{n-3} & 2(h_{n-3}+h_{n-2}) & h_{n-2} & \\ & & & h_{n-2} & 2(h_{n-2}+h_{n-1}) \end{pmatrix} \begin{pmatrix} c_1 \\ c_2 \\ \vdots \\ c_{n-2} \\ c_{n-1} \end{pmatrix} = \begin{pmatrix} \psi_1 \\ \psi_2 \\ \vdots \\ \psi_{n-2} \\ \psi_{n-1} \end{pmatrix},$$

where ψ_i is shorthand for

$$\psi_i = 3(f[x_i, x_{i+1}] - f[x_{i-1}, x_i]).$$

The matrix in the above system is nonsingular, because it is strictly diagonally dominant, i.e., in each row, the magnitude of the diagonal element is larger than the sum of magnitudes of all other elements in that row; see Section 5.3. Hence, there is a unique solution for $c_1, \ldots, c_{n-1}$ and thus a unique interpolating natural spline, whose coefficients are found by solving this linear system and then using $c_0 = 0$ and the formulas (11.3a) and (11.4). For those readers who have diligently read Chapter 5, let us add that the matrix is tridiagonal and symmetric positive definite. Hence, solving for the coefficients c_i by Gaussian elimination without pivoting is stable and simple. It requires about $4n$ flops. The entire construction process of the spline therefore costs $\mathcal{O}(n)$ operations.

These matrix equations represent a *global* coupling of the unknowns. Thus, if one data value $f(x_{i_0})$ is changed, all of the coefficients change: a bummer! However, as it turns out, the magnitude of the elements of the inverse of the matrix decay exponentially as we move away from the main diagonal. (This, incidentally, is a consequence of the strict diagonal dominance.) Consequently, c_i depends most strongly upon elements of f at abscissae closest to x_i. So, if $f(x_{i_0})$ is modified, it will cause the largest changes in the c_i which are near $i = i_0$; although all coefficients change, these changes rapidly decay in magnitude as we move away from the spot of modification of f. The nature of the approximation is therefore "almost local."

For clamped boundary conditions we must specify

$$b_0 = f'(x_0), \quad b_n = f'(x_n).$$

This is then plugged into (11.4b) and (11.3c). After simplification we get

$$h_0(2c_0 + c_1) = 3(f[x_0, x_1] - f'(x_0)),$$
$$h_{n-1}(2c_n + c_{n-1}) = 3(f'(x_n) - f[x_{n-1}, x_n]).$$

These relations, together with (11.5), can be written as a tridiagonal system of size $n+1$ for the $n+1$ unknowns $c_0, c_1, \ldots, c_n$.

For the not-a-knot condition we set

$$d_0 = d_1, \quad d_{n-1} = d_{n-2}.$$

This basically says that $s_0(x)$ and $s_1(x)$ together are really one cubic polynomial, and likewise $s_{n-1}(x)$ and $s_{n-2}(x)$ together are just one cubic polynomial. Inserting these conditions into (11.4a) allows elimination of c_0 in terms of c_1 and c_2, and elimination of c_n in terms of c_{n-1} and c_{n-2}. Substituting these expressions in (11.5), the obtained linear system is still tridiagonal and strictly diagonally dominant. The details are left as Exercise 5.

The error in the complete spline approximation (i.e., with clamped ends) satisfies

$$\max_{a \leq x \leq b} |f(x) - v(x)| \leq c \max_{a \leq t \leq b} |f''''(t)| \max_{0 \leq i \leq n-1} h_i^4, \tag{11.6}$$

where $c = \frac{5}{384}$. Equivalently, we use the maximum norm notation (see page 366) to write
$$\|f - v\| \leq c \|f''''\| h^4.$$

This bound is comparable (up to factor 5) with the error *bound* for the Hermite piecewise cubics (see page 336), which is quite impressive given that we are using much less information on f. The error in the not-a-knot approximation is also of similar quality, obeying a similar expression with just a different constant c.

In contrast, the natural spline is generally only second order accurate near the endpoints! Indeed, it can be shown for a general piecewise polynomial approximation that if $\max_{a \leq x \leq b} |f(x) - v(x)| = \mathcal{O}(h^q)$, then the error in the jth derivative satisfies
$$\max_{a \leq x \leq b} |f^{(j)}(x) - v^{(j)}(x)| = \mathcal{O}(h^{q-j}).$$

Thus, if it happens that $f''(x_0) = 1$, say, then the error in the second derivative of the natural spline near x_0 is $\approx 1 = \mathcal{O}(h^0)$, so $q - 2 = 0$ and this translates to an $\mathcal{O}(h^2)$ error in $v(x)$ near x_0. The lesson is that not everything "natural" is good for you.

Having described no less than four favorite piecewise polynomial interpolants, we summarize their respective properties in a table:

Interpolant	Local?	Order	Smooth?	Selling features
Piecewise constant	yes	1	bounded	Accommodates general f
Broken line	yes	2	C^0	Simple, max and min at data values
Piecewise cubic Hermite	yes	4	C^1	Elegant and accurate
Spline (not-a-knot)	not quite	4	C^2	Accurate, smooth, requires only f data

Construction cost

When contemplating taking a relatively large number of subintervals n, it is good to remember that the construction cost of the approximation, i.e., the part of the algorithm that does not depend on a particular evaluation point, is *linear* in n. In fact, the flop count is αn with a small proportionality constant α for all practical piecewise polynomial interpolation methods. This fact is obvious in particular for the methods described in Section 11.2. (Why?)

In contrast, recall that the construction cost for the Chebyshev polynomial interpolation considered in Section 10.6 (the only place in Chapter 10 where we contemplate mildly large polynomial degrees) is proportional to n^2, where now n is the polynomial degree; see Exercise 8.

Specific exercises for this section: Exercises 5–9.

11.4 Hat functions and B-splines

In Section 10.1 we have discussed the representation of an interpolant using basis functions $\phi_j(x)$. The three forms of polynomial interpolation that followed in Sections 10.2–10.4 have all been presented with references to their corresponding basis functions. In the next two chapters basis functions play a central role. But for the piecewise polynomial approximations presented in Sections 11.2

11.4. Hat functions and B-splines

> **Note:** The material in the present section may be studied almost independently from the material in Sections 11.2 and 11.3.

and 11.3 no basis functions have been mentioned thus far. Do such bases exist, and if they do, aren't they relevant?

Let us first hasten to reconfirm that such basis functions do exist. The general linear form for the approximant, written as

$$v(x) = \sum_{j=0}^{n} c_j \phi_j(x) = c_0 \phi_0(x) + \cdots + c_n \phi_n(x),$$

does apply also to piecewise polynomial interpolation. The reason we have not made a fuss about it thus far is that a simpler and more direct determination of such interpolants is often achieved using local representation, building upon polynomial basis functions.

But in other applications, for instance, in the solution of boundary value problems for differential equations, the approximated function is given only indirectly, through the solution of the differential problem, and the need to express the discretization process in terms of basis functions is more prevalent.

In such cases, it is important to choose basis functions that have as narrow a support as possible. The **support** of a function is the subinterval on which (and only on which) it may have nonzero values. If basis functions with narrow support can be found, then this gives rise to local operations in x or, if not entirely local, it typically translates into having a sparse matrix for the linear system of equations that is assembled in the discretization process. This, in turn, gives rise to savings in the computational work required to solve the linear system.

Hat functions

As usual, everything is simple for the case of continuous, piecewise linear approximation. Recall from Section 11.2 that in this case we have

$$(x_0, x_1, x_2, \ldots, x_{n-1}, x_n) = (t_0, t_1, t_2, \ldots, t_{r-1}, t_r),$$

and $r = n$.

For any linear combination of the basis functions ϕ_j to be a continuous, piecewise linear function, the basis functions themselves must be of this type. Insisting on making their support as narrow as possible leads to the famous *hat functions* given by

$$\phi_j(x) = \begin{cases} \frac{x - x_{j-1}}{x_j - x_{j-1}}, & x_{j-1} \leq x < x_j, \\ \frac{x - x_{j+1}}{x_j - x_{j+1}}, & x_j \leq x < x_{j+1}, \\ 0 & \text{otherwise.} \end{cases}$$

The end functions ϕ_0 and ϕ_n are not defined outside the interval $[x_0, x_n]$. See Figure 11.6.

Like the Lagrange polynomials, these basis functions satisfy

$$\phi_j(x_i) = \begin{cases} 0, & i \neq j, \\ 1, & i = j. \end{cases}$$

Therefore, like the Lagrange polynomial representation, the coefficients c_j are none other than the data ordinates themselves, so we write

$$c_j = y_j = f(x_j), \quad j = 0, 1, \ldots, n.$$

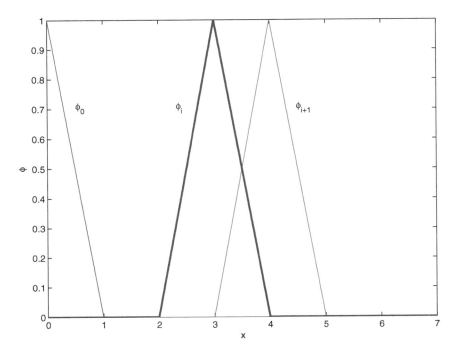

Figure 11.6. *Hat functions: a compact basis for piecewise linear approximation.*

Hat functions appear in almost any text on the **finite element method** (FEM). In that context, representations that satisfy $c_j = f(x_j)$ are called **nodal methods**.

Hat functions form an instance of basis functions with *local support*: the support of each basis function $\phi_j(x)$ spans only a few (two in this instance) mesh subintervals about x_j. Moreover, a change in one data value y_i obviously affects only $c_i \phi_i(x)$ and thus only values of $v(x)$ in the very narrow vicinity of x_i, namely, the subinterval (x_{i-1}, x_{i+1}).

Often in finite element methods and other techniques related to piecewise interpolation, everything is defined and constructed on a particularly simple interval, say, $[-1, 1]$. In the present case the basis functions can in fact be viewed as simple scalings and translations of one *mother hat function*, given by

$$\phi(z) = \begin{cases} 1+z, & -1 \leq z < 0, \\ 1-z, & 0 \leq z < 1, \\ 0 & \text{otherwise.} \end{cases}$$

The two polynomial segments of this mother function, $\psi_1(z) = 1 + z$ and $\psi_2(z) = 1 - z$, are the building blocks for constructing all hat functions.

Hermite cubic basis

Moving on to more challenging piecewise polynomial spaces, let us consider those based on cubic polynomials. The next simplest case is that of Hermite piecewise cubics. Like the piecewise linear approximation, the Hermite interpolation is completely local, as we have seen in Section 11.2. Indeed, the observations of the previous paragraph apply also for the Hermite interpolation, except that

11.4. Hat functions and B-splines

the notation and the notion of "mother function" are more involved. In fact, there are two mothers here!

Let us define four interpolating cubic polynomials (as compared to two in the piecewise linear case) on the interval $[0,1]$ by

$$\psi_1(0) = 1, \quad \psi_1'(0) = 0, \quad \psi_1(1) = 0, \quad \psi_1'(1) = 0,$$
$$\psi_2(0) = 0, \quad \psi_2'(0) = 1, \quad \psi_2(1) = 0, \quad \psi_2'(1) = 0,$$
$$\psi_3(0) = 0, \quad \psi_3'(0) = 0, \quad \psi_3(1) = 1, \quad \psi_3'(1) = 0,$$
$$\psi_4(0) = 0, \quad \psi_4'(0) = 0, \quad \psi_4(1) = 0, \quad \psi_4'(1) = 1.$$

Figuring out the cubic polynomials defined in this way is straightforward based on the Hermite interpolation procedure described in Section 10.7. We obtain $\psi_1(z) = 1 - 3z^2 + 2z^3$, $\psi_2(z) = z - 2z^2 + z^3$, $\psi_3(z) = 3z^2 - 2z^3$, and $\psi_4(z) = -z^2 + z^3$. See Exercise 10.

A global basis function may now be constructed from $\psi_l(z), l = 1, 2, 3, 4$. Recall that for Hermite interpolation we have

$$(x_0, x_1, \ldots, x_n) = (t_0, t_0, \ldots, t_r, t_r),$$

from which it follows that $n = 2r + 1$. We thus need $2r + 2$ basis functions. Define for $0 \le k \le r$ the functions

$$\xi_k(x) = \begin{cases} \psi_1\left(\frac{x - t_k}{t_{k+1} - t_k}\right), & t_k \le x < t_{k+1}, \\ \psi_3\left(\frac{x - t_{k-1}}{t_k - t_{k-1}}\right), & t_{k-1} \le x < t_k, \\ 0 & \text{otherwise}, \end{cases}$$

and

$$\eta_k(x) = \begin{cases} \psi_2\left(\frac{x - t_k}{t_{k+1} - t_k}\right) \cdot (t_{k+1} - t_k), & t_k \le x < t_{k+1}, \\ \psi_4\left(\frac{x - t_{k-1}}{t_k - t_{k-1}}\right) \cdot (t_k - t_{k-1}), & t_{k-1} \le x < t_k, \\ 0 & \text{otherwise}. \end{cases}$$

The functions $\xi_k(x)$ and $\eta_k(x)$ are depicted in Figure 11.7. Clearly, these are piecewise cubic functions in C^1 which have local support. The ξ_k's are all scalings and translations of one mother Hermite cubic, and the η_k's likewise relate to another mother. See Exercise 11.

Our Hermite piecewise interpolant is now expressed as

$$v(x) = \sum_{k=0}^{r} \left(f(t_k) \xi_k(x) + f'(t_k) \eta_k(x) \right).$$

With respect to the standard form $v(x) = \sum_{j=0}^{n} c_j \phi_j(x)$, we have here that $n + 1 = 2(r+1)$, $\phi_{2k}(x) = \xi_k(x)$, and $\phi_{2k+1}(x) = \eta_k(x)$, $k = 0, 1, \ldots, r$. The interpolation properties of the basis functions allow us to identify $c_{2k} = f(t_k)$ and $c_{2k+1} = f'(t_k)$, so in FEM jargon this is a nodal method.

B-splines

While the case for Hermite cubics is almost as easy (in principle!) as for the broken line interpolation, the case for cubic spline interpolation cannot be that simple because the approximation process is no longer entirely local. Rather than looking for a special remedy for cubic splines, we turn instead to a general method for constructing basis functions with local support for piecewise polynomial spaces. The resulting basis functions, called *B-splines*, do not yield a nodal method, but they possess many beautiful properties and have proved useful in CAGD.

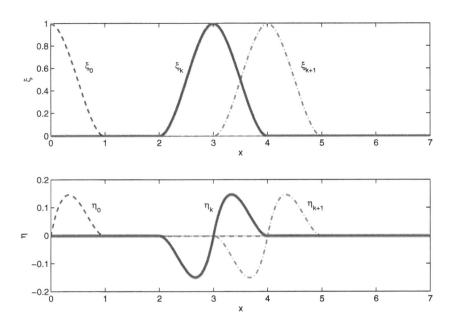

Figure 11.7. *Basis functions for piecewise Hermite polynomials.*

Note: Be warned that the remainder of this section is inherently technical.

Suppose then that we have a sequence of *break points* $t_0, t_1, \ldots, t_r$ and that we wish to form a basis for an approximation space such that in each subinterval $[t_{i-1}, t_i]$ the approximation is a polynomial of degree at most l and such that globally the approximation is in C^m, $m < l$. We form a sequence of *knots* x_i by repeating the values t_j in the sequence, which yields

$$(x_0, x_1, x_2, \ldots, x_n) = \left(\underbrace{t_0, t_0, \ldots, t_0}_{l+1}, \underbrace{t_1, \ldots, t_1}_{l-m}, \ldots, \underbrace{t_{r-1}, \ldots, t_{r-1}}_{l-m}, \underbrace{t_r, \ldots, t_r}_{l+1} \right).$$

The *truncated power function* $w(\tau) = (\tau)_+^k$ is defined as

$$(\tau)_+^k = [\max\{\tau, 0\}]^k.$$

Using this, define

$$M_j(x) = g_x[x_j, \ldots, x_{j+l+1}],$$

where

$$g_x(t) = (t - x)_+^l.$$

(By this notation we mean that each real value x is held fixed while finding the divided difference of g_x in t. See page 308 for the definition of divided differences.) The B-spline of degree l is

$$\phi_j(x) \equiv B_j(x) = (x_{j+l+1} - x_j) M_j(x).$$

These functions form a basis for piecewise polynomials of degree l and global smoothness m. Figure 11.8 depicts the shape of B-spline basis functions for the cubic spline ($m = 2, l = 3$). The basis has a number of attractive features:

11.5. Parametric curves

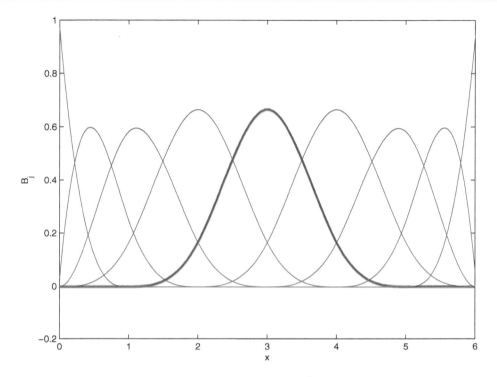

Figure 11.8. *B-spline basis for the C^2 cubic spline.*

- The function B_j is positive on the interval (x_j, x_{j+l+1}) and zero elsewhere, so it has a local support.

- The evaluation of B-splines can be done recursively, a property inherited from the divided difference function used in the definition of M_j above.

- At each point x, $\sum_{j=0}^{n} \phi_j(x) = \sum_j B_j(x) = 1$.

Since the B-spline values are never negative they form a *partition of unity*. Although the value c_j of the obtained interpolant is not equal to the corresponding given data y_j, it is not far from it either. Hence the values of c_j may be used as control points in a CAGD system.

Specific exercises for this section: Exercises 10–12.

11.5 Parametric curves

Suppose we are given a set of data with repeated abscissae and that the data are to be joined in a curve *in that order*—see Figure 11.9. The curve C joining these points cannot be expressed as a function of either coordinate variable. Thus, none of our interpolation techniques thus far can generate C without breaking it up into subproblems, such that on each, y can be expressed as a univalued function of x.

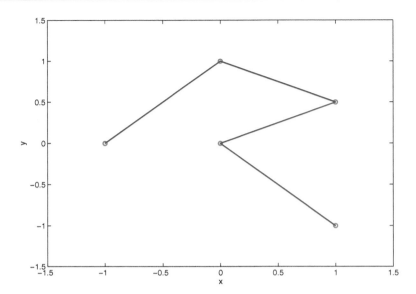

Figure 11.9. *Parametric broken-line interpolation.*

To construct an interpolating approximation in this case, we **parametrize** both abscissa and ordinate. In other words, we find a representation

$$(\tau_0, x_0), (\tau_1, x_1), (\tau_2, x_2), \ldots, (\tau_n, x_n),$$
$$(\tau_0, y_0), (\tau_1, y_1), (\tau_2, y_2), \ldots, (\tau_n, y_n),$$

where $\tau_0 < \tau_1 < \cdots < \tau_n$ is a partition of some interval in the newly introduced common variable τ. For instance, in the absence of a reason not to do so, set

$$\tau_i = i/n, \quad i = 0, 1, \ldots, n.$$

Denote the parametrized curve by

$$C : (X(\tau), Y(\tau)).$$

Then the interpolation conditions are

$$X(\tau_i) = x_i, \qquad Y(\tau_i) = y_i \quad i = 0, 1, \ldots, n,$$

and the curve C is given at any point $\tau \in [\tau_0, \tau_n]$ by $(X(\tau), Y(\tau))$.

For constructing $X(\tau)$ and $Y(\tau)$ we may now use any of the techniques seen so far in this chapter. Different parametric curve fitting techniques differ in the choice of interpolation method used.

Example 11.7. Suppose we are given the data $\{(-1,0), (0,1), (1,0.5), (0,0), (1,-1)\}$. If we issue the MATLAB commands

```
xi = [-1,0,1,0,1];
yi = [0,1,0.5,0,-1];
plot (xi,yi,'go',xi,yi)
axis([-1.5,1.5,-1.5,1.5])
xlabel('x')
ylabel('y')
```

11.5. Parametric curves

then Figure 11.9 is obtained. The part plot(xi,yi,'go') of the plot command merely plots the data points (in green circles), whereas the default plot(xi,yi) does a broken line interpolation (in default blue) of $X(\tau)$ vs. $Y(\tau)$.

Suppose instead that we use polynomial interpolation for each of $X(\tau)$ and $Y(\tau)$. Since we have the two MATLAB functions divdif and evalnewt presented in Section 10.4 already coded, we can obtain Figure 11.10 using the script

```
ti = 0:0.25:1;
coefx = divdif(ti,xi);
coefy = divdif(ti,yi);
tau = 0:.01:1;
xx = evalnewt(tau,ti,coefx);
yy = evalnewt(tau,ti,coefy);
plot (xi,yi,'go',xx,yy)
axis([-1.5,1.5,-1.5,1.5])
xlabel('x')
ylabel('y')
```

Note again that the parametric curves in Figures 11.10 and 11.9 interpolate the same data set! ∎

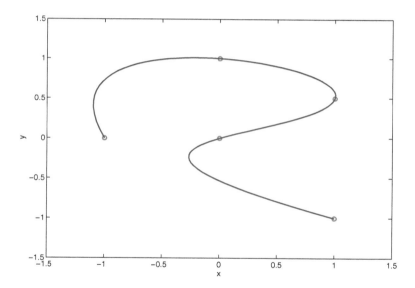

Figure 11.10. *Parametric polynomial interpolation.*

Parametric piecewise cubic hermite polynomials

Computer graphics and CAGD applications make extensive use of parametric curve fitting. These applications require

- rapid generation of curves, and

- easily modified curves. In particular, a modification should be *local*; changing one or two data points should not affect the entire curve.

The second criterion rules out global interpolating polynomials. The natural approach here is to use piecewise cubic Hermite polynomials. In a drawing package, for instance, one may specify, for two consecutive data points (x_0, y_0) and (x_1, y_1), also *guidepoints* $(x_0 + \alpha_0, y_0 + \beta_0)$ and $(x_1 - \alpha_1, y_1 - \beta_1)$. Then the software constructs the two Hermite cubics $X(\tau)$ and $Y(\tau)$ satisfying

$$X(0) = x_0, \ X(1) = x_1, \ X'(0) = \alpha_0, \ X'(1) = \alpha_1,$$
$$Y(0) = y_0, \ Y(1) = y_1, \ Y'(0) = \beta_0, \ Y'(1) = \beta_1.$$

It is easy to come up with an explicit formula for each of $X(\tau)$ and $Y(\tau)$; see Exercise 4.

Common graphics systems use local **Bézier polynomials**,[46] which are cubic Hermite polynomials with a built-in scaling factor of 3 for the derivatives (i.e., $X'(0) = 3\alpha_0$, $Y'(1) = 3\beta_1$). This scaling is transparent to the user and is done to make curves more sensitive to the guidepoint position; also, the mouse does not have to be moved off screen when specifying the guidepoints.[47] Please verify that for $0 \leq \tau \leq 1$ we get the formulas

$$X(\tau) = (2(x_0 - x_1) + 3(\alpha_0 + \alpha_1))\tau^3 + (3(x_1 - x_0) - 3(\alpha_1 + 2\alpha_0))\tau^2 + 3\alpha_0\tau + x_0,$$
$$Y(\tau) = (2(y_0 - y_1) + 3(\beta_0 + \beta_1))\tau^3 + (3(y_1 - y_0) - 3(\beta_1 + 2\beta_0))\tau^2 + 3\beta_0\tau + y_0.$$

Example 11.8. Figure 11.11 shows a somewhat primitive design of a pair of glasses using 11 Bézier polynomials. MATLAB was employed using the function `ginput` to input point coordinates from the figure's screen.

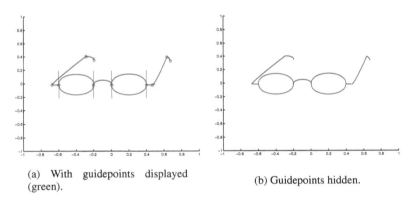

(a) With guidepoints displayed (green).

(b) Guidepoints hidden.

Figure 11.11. *A simple curve design using 11 Bézier polynomials.*

The design was deliberately kept simple so as to make the point that with more dedication you can do much better than this using the Bézier curves. Indeed, the best way to get the hang of this procedure is to try it out interactively, along the lines of Exercise 14. The programming involved is not difficult! ■

Specific exercises for this section: Exercises 13–14.

---

[46] Pierre Bézier came up with these polynomials in the 1970s while designing car shapes for the French auto maker Renault.

[47] Remember that τ is an arbitrarily chosen parametrization. For instance, if we define the parameter $\sigma = c\tau$ for some scaling factor $c \neq 0$, then at any point $\frac{dY/d\tau}{dX/d\tau} = \frac{cdY/d\sigma}{cdX/d\sigma} = \frac{dY/d\sigma}{dX/d\sigma}$.

11.6 *Multidimensional interpolation

All approximating functions that we have seen thus far in Chapter 10 and in the present chapter were essentially functions of one variable: typically, we are looking for

$$v(x) = \sum_{j=0}^{n} c_j \phi_j(x),$$

using some basis functions $\phi_j(x)$. The function $v(x)$ describes a **curve**. In Section 11.5 we have extended the definition of our interpolant $v(x)$, using parametrization to allow the curve to roam more flexibly in the plane (in two dimensions). It is also possible to parametrize a function in one variable so as to describe a curve in a box (in three dimensions).

But there are many applications where the sought interpolating function is a function of more than one independent variable—a **surface**. An example is a digitized black-and-white photograph, which is really a function of two variables describing gray levels at each pixel, with the pixels arranged in an underlying two-dimensional array. A color photo is the same except that there are three such functions—for red, green, and blue. Another example is a body such as a sofa that is made of inhomogeneous material, or a box containing dollar bills buried in the sand. The conductivity or permeability at each point of such an article is a function in three space dimensions. Between these there is the case of the surface of a body in three dimensions, which is a manifold in two dimensions: analogous to the parametrized curve, each point of such a surface is described as a function of three variables, but at least in principle it could be described as a function of two variables using a clever parametrization.

Make no mistake: the passage from one to more independent variables is often far from simple or straightforward. For one thing, while in one dimension the domain of definition (where x roams) is just an interval or, at worst, a union of intervals, in two dimensions the domain need not be a rectangle or even a disk, as a casual glance at the shape of islands and continents on a flattened world map reveals. Moreover, the task of finding appropriate parametrization for surfaces and bodies in more than one variable often becomes prohibitively complex. For instance, people often view the surface of a body (such as part of the human body) as a collection of points in $\mathcal{R}^3$, each having three coordinates, that may or may not be connected by some local neighboring relations to form a *surface mesh*.

There is no one quick approach for all interpolation problems here. The following central questions arise:

- Is the data scattered or located on a grid?

- What is the purpose of the interpolating function (i.e., where is it expected to be evaluated and how often)?

Below we quickly survey but a sample of related techniques and ideas.

Interpolation on a rectangular grid

In general, the interpolation problem in two dimensions, extending Chapter 10 and Sections 11.2–11.4, can be written as seeking a function $v(x, y)$ that satisfies

$$v(x_i, y_i) = f_i, \quad i = 0, 1, \ldots, n.$$

Note that here, unlike before, y_i is the abscissa coordinate in direction y, while the corresponding data value is denoted by f_i. The points (x_i, y_i) for which function (or data ordinate) values are given are contained in some domain Ω that need not be rectangular.

Before getting to this general case, however, let us consider a more special though important one. We assume that there are abscissae in x and y separately, e.g., $x_0 < x_1 < \cdots < x_{N_x}$ and $y_0 < y_1 < \cdots < y_{N_y}$, such that the $n+1 = (N_x+1)(N_y+1)$ data points are given as $(x_i, y_j, f_{i,j})$. So, the interpolation problem is to find $v(x,y)$ satisfying

$$v(x_i, y_j) = f_{i,j}, \quad i = 0, 1, \ldots, N_x, \quad j = 0, 1, \ldots, N_y.$$

In this special case we can think of seeking $v(x,y)$ in the *tensor product* form

$$v(x,y) = \sum_{k=0}^{N_x} \sum_{l=0}^{N_y} c_{k,l} \phi_k(x) \psi_l(y),$$

where ϕ_k and ψ_l are *one-dimensional basis functions*. This is why it is simpler, despite the somewhat cumbersome notation. In fact, extensions of this specific case to more dimensions are in principle straightforward, too. The multidimensional interpolation problem is decoupled into a product of one-dimensional ones.

For a polynomial interpolation we can think, for instance, of a *bilinear* or a *bicubic* polynomial. These are defined on a square or a rectangle, with data typically given at the four corners, and are often considered as a surface *patch*, filling in missing data inside the rectangle. Obviously, without data given inside the rectangle we would need some additional data to determine the bicubic.

Example 11.9 (bilinear interpolation). This example is longish but should be straightforward. Let us consider first a bilinear polynomial interpolation over the unit square depicted on the left of Figure 11.12. The data points are $(0,0, f_{0,0})$, $(1,0, f_{1,0})$, $(0,1, f_{0,1})$, and $(1,1, f_{1,1})$.

We write our interpolant as

$$v(x,y) = c_0 + c_1 x + c_2 y + c_3 xy.$$

Then substituting the data directly gives

$$c_0 = f_{0,0},$$
$$c_0 + c_1 = f_{1,0} \Rightarrow c_1 = f_{1,0} - f_{0,0},$$
$$c_0 + c_2 = f_{0,1} \Rightarrow c_2 = f_{0,1} - f_{0,0},$$
$$c_0 + c_1 + c_2 + c_3 = f_{1,1} \Rightarrow c_3 = f_{1,1} + f_{0,0} - f_{0,1} - f_{1,0}.$$

This completes the construction of the interpolant $v(x,y)$. Next, suppose the task is to interpolate at mid-edges and at the center of the square. We obtain

$$v(.5, 0) = c_0 + .5c_1 = \frac{1}{2}(f_{0,0} + f_{1,0}),$$
$$v(0, .5) = c_0 + .5c_2 = \frac{1}{2}(f_{0,0} + f_{0,1}),$$
$$v(.5, 1) = c_0 + c_2 + .5(c_1 + c_3) = \frac{1}{2}(f_{0,1} + f_{1,1}),$$
$$v(1, .5) = c_0 + c_1 + .5(c_2 + c_3) = \frac{1}{2}(f_{1,0} + f_{1,1}),$$
$$v(.5, .5) = c_0 + .5c_1 + .5c_2 + .25c_3 = \frac{1}{4}(f_{0,0} + f_{1,0} + f_{0,1} + f_{1,1}).$$

(You could have probably guessed these expressions directly. But it's good to know how they are obtained in an orderly manner.)

11.6. *Multidimensional interpolation

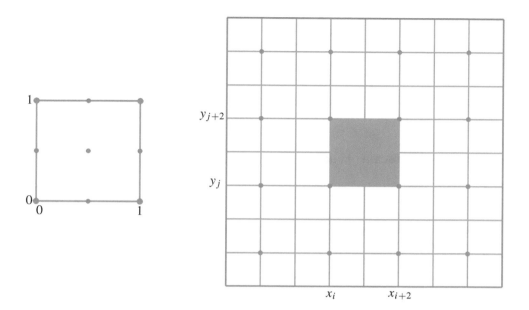

Figure 11.12. *Bilinear interpolation. Left green square: data are given at the unit square's corners (red points), and bilinear polynomial values are desired at mid-edges and at the square's middle (blue points). Right blue grid: data are given at the coarser grid nodes (blue points) and a bilinear interpolation is performed to obtain values at the finer grid nodes, yielding values $f_{i,j}$ for all $i = 0, 1, 2, \ldots, N_x$, $j = 0, 1, 2, \ldots, N_y$.*

Now, suppose we have a rectangular, uniform grid in the (x, y)-plane and data given at every other node of the grid, as depicted in the right part of Figure 11.12. The data value at point (x_i, y_j) is $f_{i,j}$, and we are required to supply values at all points of the finer grid.

Let us pick the particular grid square $[x_i, x_{i+2}] \times [y_j, y_{j+2}]$, whose corner values are given. To this we fit an interpolating bilinear polynomial, which is a scaled version of the one developed above, given by

$$v(x, y) = c_0 + c_1 \frac{(x - x_i)}{(x_{i+2} - x_i)} + c_2 \frac{(y - y_j)}{(y_{j+2} - y_j)} + c_3 \frac{(x - x_i)}{(x_{i+2} - x_i)} \frac{(y - y_j)}{(y_{j+2} - y_j)}.$$

The same formulas as above follow with the obvious notational change. Thus, we evaluate

$$f_{i+1,j} = \frac{1}{2}(f_{i,j} + f_{i+2,j}),$$
$$f_{i,j+1} = \frac{1}{2}(f_{i,j} + f_{i,j+2}),$$
$$f_{i+1,j+2} = \frac{1}{2}(f_{i,j+2} + f_{i+2,j+2}),$$
$$f_{i+2,j+1} = \frac{1}{2}(f_{i+2,j} + f_{i+2,j+2}),$$
$$f_{i+1,j+1} = \frac{1}{4}(f_{i,j} + f_{i+2,j} + f_{i,j+2} + f_{i+2,j+2}).$$

Of course this procedure can (and should) be carried out throughout the grid all at once, using vector operations, rather than one square at a time. The resulting bilinear interpolation procedure from coarse to fine grid has been implemented in our MATLAB multigrid program of Example 7.17 in Section 7.6. ■

More generally, a *bicubic spline* can be constructed from (many) one-dimensional cubic splines such as described in Section 11.3 for a general-purpose smooth interpolant of data on a grid.

Triangular meshes and scattered data

We now consider piecewise polynomial interpolation in the plane and denote $\mathbf{x} = (x, y)^T$. As is the case for one space variable, there are applications where the abscissae of the data are given and others where these locations are ours to choose.

In the latter case the "data," or function values, are often given implicitly, say, as the solution of some partial differential equation on a domain $\Omega \subset \mathcal{R}^2$. The locations of the interpolation points then form part of a discretization scheme for the differential equation, usually a *finite element* or a **finite volume** method, whose complete description is beyond the scope of this text. Suffice it to say here that the domain Ω is divided into nonoverlapping elements (triangles or rectangles), on each of which the sought solution function is a polynomial in two variables. The solution constructed out of these polynomial pieces should maintain an overall degree of smoothness. The union of these elements should form a good approximation of Ω and they each should be small enough in diameter to allow for a high quality approximation overall. Of course, if the domain Ω is square, then we would happily discretize it with a tensor grid as in Examples 7.1 and 11.9. But if the domain has a more complex boundary, then a tensor grid will not do. In more complex geometries, triangles are often better building blocks than rectangles; see Figure 11.13.

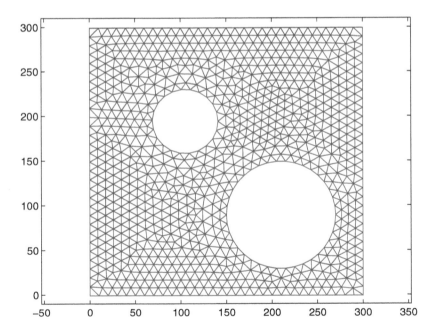

Figure 11.13. *Triangle mesh in the plane. This one is MATLAB's data set* `trimesh2d`.

11.6. *Multidimensional interpolation

In the case where the locations of the data are fixed as part of the given interpolation problem, there are also many situations where we would naturally want to form a piecewise polynomial interpolant. Thus we are led, in either case, to the following basic tasks:

1. Construct a triangulation for a given set of points in the plane.
2. Construct a piecewise polynomial interpolant on the given triangle mesh.

Of the constructed triangulation we want if possible not to have small or large (more than 90°) angles: triangles that are not too far from being equilateral allow for good approximation properties. This leads to *Delaunay triangulation*, a subject from *computational geometry* that again we just mention rather than fully explain here. MATLAB has several commands that deal with point set triangulation in both two and three dimensions: check out `delaunaytri, triplot`.

Now that we have a triangle mesh, we wish to construct an interpolating piecewise polynomial on it. We must ensure the desired number of continuous derivatives, not only across nodes as in one dimension, but also across triangle edges. This can quickly turn hairy.

By far the most popular option is to use linear polynomials, where everything is still relatively simple. On a triangle T_j having vertices $\mathbf{x}_i^j$, $i = 0, 1, 2$, we write

$$v(\mathbf{x}) = v(x, y) = c_0^j + c_1^j x + c_2^j y.$$

If the corresponding data ordinates are f_i^j, then the three interpolation conditions

$$v(\mathbf{x}_i^j) = f_i^j, \quad i = 0, 1, 2,$$

obviously determine the three coefficients c_i^j of the linear polynomial piece. Moreover, continuity across vertices between neighboring triangles is automatically achieved, and so is continuity across edges, because the value of a linear polynomial along an edge is determined by its values at the edge's endpoints (which are the vertices common to the two neighboring triangles). See Figure 11.14. In MATLAB, check out `triscatteredinterp`.

Figure 11.14. *Linear interpolation over a triangle mesh. Satisfying the interpolation conditions at triangle vertices implies continuity across neighboring triangles.*

Smoother, higher order piecewise polynomial interpolants are also possible, and they are discussed in finite element texts. In three dimensions the triangles become tetrahedra while the rectangles become boxes, affording considerably less headache than tetrahedra when tracking them. Thus, engineers often stick with the latter elements after all, at least when the geometry of Ω is not too complicated.

In computer graphics and computer aided geometric design it is popular to construct triangle surface meshes such as in Figure 11.15, both for display and for purposes of further manipulation such as morphing or design.

Radial basis functions for scattered data interpolation

We now leave the land of piecewise polynomials. Given a set of scattered data to interpolate, it may not be worthwhile to triangulate it if the purpose of interpolation is just the display of a pleasingly coherent surface.

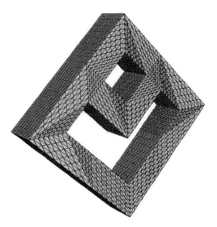

Figure 11.15. *Triangle surface mesh.*

A very popular technique for the latter purpose is radial basis function (RBF) interpolation. Let $|\mathbf{z}|$ denote the ℓ_2-norm of a point $\mathbf{z}$ in two or three dimensions. Given data $(\mathbf{x}_i, f_i)$, $i = 0, 1, \ldots, n$, the values $\phi(|\mathbf{x} - \mathbf{x}_j|)$ of an RBF ϕ measure the distances from a point $\mathbf{x}$ to the data abscissae. Thus, $\phi(\mathbf{z})$ depends only on the radial distance $|\mathbf{z}|$.

In the simplest version of this method we look for an interpolant

$$v(\mathbf{x}) = \sum_{j=0}^{n} w_j \phi(|\mathbf{x} - \mathbf{x}_j|),$$

where the weights w_j are determined by the $n+1$ interpolation conditions

$$f_i = v(\mathbf{x}_i) = \sum_{j=0}^{n} w_j \phi(|\mathbf{x}_i - \mathbf{x}_j|), \quad i = 0, 1, \ldots, n.$$

These are $n+1$ linear equations for $n+1$ unknown weights w_j, so we solve the system once and then construct the interpolant using the formula for $v(\mathbf{x})$.

A good choice for ϕ in two dimensions is the *multiquadric*

$$\phi(r) = \sqrt{r^2 + c^2},$$

where c is a user-determined parameter. In three dimensions the *biharmonic spline*

$$\phi(r) = r$$

plus an extra linear polynomial is recommended. There are other choices in both two and three dimensions.

The algorithm as presented above works reasonably well if n is not too large and the points are reasonably spread out. But for point clouds that contain many points, or very tightly clustered points, modifications must be made. These are well beyond the scope of our quick presentation, so let us just demonstrate the approach by an example.

11.7. Exercises

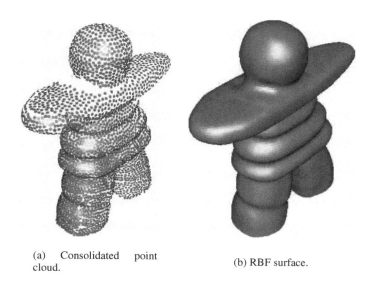

(a) Consolidated point cloud.

(b) RBF surface.

Figure 11.16. *RBF interpolation of an upsampling of a consolidated point cloud.*

Example 11.10. Figure 11.16(a) shows a set of points with directions that was obtained from a raw scanned point cloud of an inukshuk-like three-dimensional model[48] following denoising, removal of outlier points, and association of a normal to each point (thus generating a so-called *surfel*; see Example 4.18). The data was further up-sampled, and the resulting scattered surfel data was interpolated by the FastRBF software of Far Field Technology to form the surface in (b). Note that this is just one still shot of a three-dimensional surface.

We don't expect you to be able to reproduce Figure 11.16 based on the above sketchy description. The purpose, rather, is to show that the simple RBF idea can be made to produce impressive practical results. ∎

11.7 Exercises

0. **Review questions**

    (a) What is a piecewise polynomial?

    (b) State three shortcomings of polynomial interpolation that are improved upon by piecewise polynomial interpolation.

    (c) State at least one advantage and one disadvantage of piecewise constant interpolation.

    (d) State at least two advantages and two disadvantages of broken line interpolation.

    (e) What is a piecewise cubic Hermite?

    (f) In what different ways are the cubic spline and piecewise cubic Hermite useful interpolants?

    (g) Define the different end conditions for cubic spline interpolation, giving rise to the natural, complete, and not-a-knot variants. Which of these is most suitable for general-purpose implementation, and why?

---

[48] The inukshuk is a manmade stone landmark, used by peoples of the Arctic region of North America. It formed the basis of the logo of the 2010 Winter Olympics in Vancouver.

(h) What is a tridiagonal matrix and how does it arise in spline interpolation?

(i) What is the support of a function, and what can be said about the support of basis functions for piecewise interpolation?

(j) How are basis functions for piecewise polynomial interpolation different from basis functions for polynomial interpolation of the sort discussed in Chapter 10?

(k) What is a nodal method?

(l) Define a hat function and explain its use.

(m) In what ways is the B-spline similar to the hat function?

(n) How do parametric curves arise?

(o) Define the Bézier polynomial and explain its use in CAGD.

1. Consider the piecewise constant interpolation of the function

$$f(x) = \sin(x), \quad 0 \leq x \leq 381,$$

at points $x_i = ih$, where $h = 0.1$. Thus, our interpolant satisfies

$$v(x) = \sin(ih), \quad (i - .5)h \leq x < (i + .5)h,$$

for $i = 0, 1, \ldots, 3810$.

(a) Find a bound for the error in this interpolation.

(b) How many leading digits in $v(x)$ are guaranteed to agree with those of $f(x)$, for any $0 \leq x \leq 381$?

2. Suppose we wish to interpolate $n+1$ data points ($n > 2$) with a piecewise *quadratic* polynomial. How many continuous derivatives at most can this interpolant be guaranteed to have without becoming one global polynomial? Explain.

3. Let $f \in C^3[a,b]$ be given at equidistant points $x_i = a + ih$, $i = 0, 1, \ldots, n$, where $nh = b - a$. Assume further that $f'(a)$ is given as well.

(a) Construct an algorithm for C^1 piecewise quadratic interpolation of the given values. Thus, the interpolating function is written as

$$v(x) = s_i(x) = a_i + b_i(x - x_i) + c_i(x - x_i)^2, \quad x_i \leq x \leq x_{i+1},$$

for $i = 0, \ldots, n-1$, and your job is to specify an algorithm for determining the $3n$ coefficients a_i, b_i and c_i.

(b) How accurate do you expect this approximation to be as a function of h? Justify.

4. Verify that the Hermite cubic interpolating $f(x)$ and its derivative at the points t_i and t_{i+1} can be written explicitly as

$$s_i(x) = f_i + \left(h_i f_i'\right)\tau + \left(3(f_{i+1} - f_i) - h_i(f_{i+1}' + 2f_i')\right)\tau^2$$
$$+ \left(h_i(f_{i+1}' + f_i') - 2(f_{i+1} - f_i)\right)\tau^3,$$

where $h_i = t_{i+1} - t_i$, $f_i = f(t_i)$, $f_i' = f'(t_i)$, $f_{i+1} = f(t_{i+1})$, $f_{i+1}' = f'(t_{i+1})$, and $\tau = \frac{x - t_i}{h_i}$.

5. Derive the matrix problem for cubic spline interpolation with the not-a-knot condition. Show that the matrix is tridiagonal and strictly diagonally dominant.

11.7. Exercises

6. The *gamma function* is defined by

$$\Gamma(x) = \int_0^\infty t^{x-1} e^{-t} dt, \quad x > 0.$$

It is known that for integer numbers the function has the value

$$\Gamma(n) = (n-1)! = 1 \cdot 2 \cdot 3 \cdots (n-1).$$

(We define $0! = 1$.) Thus, for example, $(1,1), (2,1), (3,2), (4,6), (5,24)$ can be used as data points for an interpolating polynomial.

   (a) Write a MATLAB script that computes the polynomial interpolant of degree four that passes through the above five data points.

   (b) Write a program that computes a cubic spline to interpolate the same data. (You may use MATLAB's `spline`.)

   (c) Plot the two interpolants you found on the same graph, along with a plot of the gamma function itself, which can be produced using the MATLAB command `gamma`.

   (d) Plot the errors in the two interpolants on the same graph. What are your observations?

7. Suppose you are asked to construct a *clamped* cubic spline interpolating the following set of data:

i	0	1	2	3	4	5	6	7
x	1.2	1.4	1.6	1.67	1.8	2.0	2.1	2.2
f(x)	4.561	5.217	5.634	5.935	6.562	6.242	5.812	5.367

The function underlying these data points is unknown, but clamped cubic splines require interpolation of the first derivative of the underlying function at the end points $x_0 = 1.2$ and $x_7 = 2.2$. Select the formula in the following list that would best assist you to construct the clamped cubic spline interpolating this set of data:

   i. $f'(x_0) = \frac{1}{2h}\left(-f(x_0-h) + f(x_0+h)\right) - \frac{h^2}{6} f^{(3)}(\xi).$

   ii. $f'(x_n) = \frac{1}{2h}\left(f(x_n-2h) - 4f(x_n-h) + 3f(x_n)\right) + \frac{h^2}{3} f^{(3)}(\xi).$

   iii. $f'(x_0) = \frac{1}{12h}\left(f(x_0-2h) - 8f(x_0-h) + 8f(x_0+h) - f(x_0+2h)\right) + \frac{h^4}{30} f^{(5)}(\xi).$

   iv. $f'(x_n) = \frac{1}{12h}\left(3f(x_n-4h) - 16f(x_n-3h) + 36f(x_n-2h) - 48f(x_n-h)\right.$
   $\left. + 25f(x_n)\right) + \frac{h^4}{5} f^{(5)}(\xi).$

   v. $f'(x_0) = \frac{1}{180h}\left(-441 f(x_0) + 1080 f(x_0+h) - 1350 f(x_0+2h)\right.$
   $+ 1200 f(x_0+3h) - 675 f(x_0+4h) + 216 f(x_0+5h)$
   $\left. - 30 f(x_0+6h)\right) + \frac{h^6}{7} f^{(7)}(\xi).$

Provide an explanation supporting your choice of formula.

8. Consider using cubic splines to interpolate the function

$$f(x) = e^{3x} \sin(200x^2)/(1+20x^2), \quad 0 \leq x \leq 1,$$

featured in Figure 10.8.

Write a short MATLAB script using `spline`, interpolating this function at equidistant points $x_i = i/n$, $i = 0, 1, \ldots, n$. Repeat this for $n = 2^j$, $j = 4, 5, \ldots, 14$. For each such calculation record the maximum error at the points `x = 0:.001:1`. Plot these errors against n, using `loglog`.

Make observations in comparison to Figure 10.8.

9. Given function values $f(t_0), f(t_1), \ldots, f(t_r)$, as well as those of $f'(t_0)$ and $f'(t_r)$, for some $r \geq 2$, it is possible to construct the complete interpolating cubic spline.

Suppose that we were instead to approximate $f'(t_i)$ by the divided difference $f[t_{i-1}, t_{i+1}]$, for $i = 1, 2, \ldots, r-1$, and then use these values to construct a Hermite piecewise cubic interpolant.

State one advantage and one disadvantage of this procedure over a complete cubic spline interpolation.

10. Show that the four sets of interpolation conditions given on page 347 define the four cubic polynomials $\psi_1(z), \psi_2(z), \psi_3(z),$ and $\psi_4(z)$ used to construct the Hermite piecewise cubic basis functions.

11. (a) Verify that the functions $\xi_k(x)$ and $\eta_k(x)$ defined in Section 11.4 are C^1 piecewise cubic with local support interval $[t_{k-1}, t_{k+1}]$.

    (b) What are the mother functions $\xi(z)$ and $\eta(z)$ that give rise to the Hermite basis functions by scalings and translations?

12. Derive a B-spline basis representation for piecewise linear interpolation and for piecewise Hermite cubic interpolation.

13. For the script in Example 11.7 implementing parametric polynomial interpolation, plot $X(\tau)$ vs. τ and $Y(\tau)$ vs. τ.

    Is this useful? Discuss.

14. Using parametric cubic Hermite polynomials, draw a crude face in a MATLAB graph. Your picture must include an outline of the head, eyes, ears, a mouth, and a nose. You may use Bézier polynomials to construct the necessary parametric curves. The graph must clearly mark the interpolation and guidepoints used for each curve. Include also a list of the interpolation and guidepoints used for each feature of the face. Finally, include a second plot that shows the curves of your face plot without the guidepoints and interpolation points.

    Hints:

    - You should not have to use too many curves to draw a crude representation of a face. At the same time, you are hereby given artistic license to do better than the bare minimum.
    - Experiment with the parametric curves before you begin to construct your picture.
    - You may want to write two generic MATLAB functions that can be used by `plot` to construct all of the parametric curves.
    - Use the `help` command in MATLAB to check the features of `plot`, `ginput`, and whatever else you find worthwhile.

15. Consider interpolating the data $(x_0, y_0), \ldots, (x_6, y_6)$ given by

x	0.1	0.15	0.2	0.3	0.35	0.5	0.75
y	3.0	2.0	1.2	2.1	2.0	2.5	2.5

Construct the five interpolants specified below (you may use available software for this), evaluate them at the points 0.05 : 0.01 : 0.8, plot, and comment on their respective properties:

(a) a polynomial interpolant;

(b) a cubic spline interpolant;

(c) the interpolant

$$v(x) = \sum_{j=0}^{n} c_j \phi_j(x) = c_0 \phi_0(x) + \cdots + c_n \phi_n(x),$$

where $n = 7$, $\phi_0(x) \equiv 1$, and

$$\phi_j(x) = \sqrt{(x - x_{j-1})^2 + \varepsilon^2} - \varepsilon, \quad j = 1, \ldots, n.$$

In addition to the n interpolation requirements, the condition $c_0 = -\sum_{j=1}^{n} c_j$ is imposed. Construct this interpolant with (i) $\varepsilon = 0.1$, (ii) $\varepsilon = 0.01$, and (iii) $\varepsilon = 0.001$. Make as many observations as you can. What will happen if we let $\varepsilon \to 0$?

11.8 Additional notes

While Chapter 10 lays the foundation for most numerical approximation methods, in the present chapter we reap the rewards: many of the techniques described here see an enormous amount of everyday use by scientists, engineers, and others. Piecewise linear interpolation is the default plotting mode, cubic spline is the default all-purpose data interpolator, and piecewise Hermite cubics are the the default playing units in many CAGD applications.

The complete spline defined in Section 11.3 is a function that approximately minimizes a strain energy functional over all functions (not only piecewise polynomials) which pass through the data. As such, it approximates the trace of a draftsperson's **spline**—a flexible thin beam forced to pass through a set of given pegs. This is the origin of the term "spline" in approximation theory, coined by I. J. Schoenberg in 1946.

B-splines were first shown to be practical by C. de Boor, and we refer to his book [20], where much more theory can be found about polynomial and spline approximation.

Bézier curves and B-splines are used routinely in CAGD and in computer graphics. See, for instance, Chapter 11 in Hill [41]. There is an extension of polynomial bases to rational ones, i.e., bases for *rational interpolation and approximation*. The corresponding extension of B-splines is called NURBs.

In this chapter we are mainly concerned with one-dimensional problems, i.e., curves. But as evident from our brief discussion in Section 11.6 of interpolation in two and three dimensions, the methods and approaches for one dimension may be used as building blocks for the more complex applications. For example, consider methods such as bilinear and bicubic patches, tensor product interpolation, and interpolation on triangular and tetrahedral meshes. This is true for even more dimensions.

The cryptic Example 11.10 is further explained in [45], from which Figure 11.16 has been lifted as well.

Chapter 12
Best Approximation

The quest for *best approximation* is motivated by the same pair of problems as for interpolation in Chapters 10 and 11. These are *data fitting* (discrete approximation problem), where given a set of data points $\{(x_i, y_i)\}_{i=0}^{m}$ we look for a *reasonable* function $v(x)$ that fits the data points, and *approximating known functions*, where a complicated function $f(x)$ (which may be given explicitly or only implicitly) is approximated by a simpler function $v(x)$. See Section 10.1.

The *difference* from interpolation is that we no longer necessarily require the approximation $v(x)$ to *pass through* the data values.

In Chapter 6 we discuss data fitting in the discrete least squares norm. Here we consider the problem of approximating functions, employing the *continuous* least squares norm. We will see that at least the notation becomes simpler and more pleasing in this chapter, not having to deal with discrete data points. In addition, we rely here less heavily on linear algebra, which is why our present discussion is naturally closer to interpolation and numerical integration, while Chapter 6 fits more in the linear algebra sector.

> **Note:** The material introduced in this chapter is classical. Moreover, no efficient computer-oriented algorithm or program is directly introduced. So, the method-hungry crowd may wonder, why not just skip it?!
>
> Well, in a squeeze you could. But then something may be missing in the long run, because here we develop some tools and concepts that, in addition to having their considerable mathematical elegance, are useful for deriving and analyzing practical algorithms in other chapters.

In the previous two chapters we have sought bounds on the interpolation error that would hold at any point in the relevant interval. This necessitates using the *maximum function norm*. Here we employ the *least squares function norm* instead, because there is more structure associated with it, giving rise to both simplicity and beauty. The definition of this norm is given on the following page. In particular, the continuous least squares approximation, discussed in Section 12.1, gives rise to families of *orthogonal* basis functions.

For polynomial approximation we obtain **orthogonal polynomials**. These are developed and discussed in Sections 12.2 and 12.3.

A special class of weighted orthogonal polynomials called *Chebyshev polynomials* is considered in Section 12.4. These polynomial families, rather than being the object of numerical approximation themselves, are useful for developing numerical methods for numerical integration, differential equations, interpolation and linear algebra. We have already seen them in action in Section 10.6.

Other important families of orthogonal basis functions are considered in Chapter 13.

12.1 Continuous least squares approximation

> **Note:** There are many parallels between this section and Section 6.1.

We focus in this section on the problem of fitting known functions in the least squares sense by simpler ones. Given a function $f(x)$ defined on an interval $[a,b]$, we consider an approximation of the same form as before, writing

$$v(x) = c_0 \phi_0(x) + c_1 \phi_1(x) + \cdots + c_n \phi_n(x)$$
$$= \sum_{j=0}^{n} c_j \phi_j(x),$$

where the basis functions $\{\phi_j(x)\}_{j=0}^n$ are linearly independent. The norm of the residual to be minimized is now the L_2 *function norm* defined on the current page. Recall that we have also seen the *maximum* (or *sup*, or L_∞) norm when discussing error bounds in Sections 10.5, 11.2, and 11.3. Here, however, only L_2-norms are considered, so we drop the subscript 2 of the norm notation henceforth.

> **Function Norms and Orthogonality.**
>
> - A **norm** for functions on an interval $[a,b]$, denoted $\|\cdot\|$, is a scalar function satisfying for all appropriately integrable functions $g(x)$ and $f(x)$ on $[a,b]$, (i) $\|g\| \geq 0$; $\|g\| = 0$, if and only if $g(x) \equiv 0$, (ii) $\|\alpha g\| = |\alpha| \|g\|$ for all scalars α, and (iii) $\|f+g\| \leq \|f\| + \|g\|$. The set of all functions whose norm is finite forms a **function space** associated with that particular norm.
>
> - Some popular function norms and corresponding function spaces are
>
> $$L_2: \quad \|g\|_2 = \left(\int_a^b g(x)^2 dx\right)^{1/2} \quad \text{(least squares)},$$
> $$L_1: \quad \|g\|_1 = \int_a^b |g(x)| dx,$$
> $$L_\infty: \quad \|g\|_\infty = \max_{a \leq x \leq b} |g(x)| \quad \text{(maximum)}.$$
>
> - Two square-integrable functions $g \in L_2$ and $f \in L_2$ are **orthogonal** to each other if their inner product vanishes, i.e., they satisfy
>
> $$\int_a^b f(x)g(x)dx = 0.$$

12.1. Continuous least squares approximation

Normal equations for continuous best approximation

We want to find coefficients $c_0,\ldots,c_n$ so as to minimize the norm of the residual $r(x)$. This is written as

$$\min_{\mathbf{c}} \|r\|^2 = \min_{\mathbf{c}} \|f - v\|^2 \equiv \min_{\mathbf{c}} \int_a^b \left[f(x) - \sum_{j=0}^n c_j \phi_j(x) \right]^2 dx.$$

As in the discrete case considered in Section 6.1, denoting $\psi(\mathbf{c}) = \|r\|^2$, the necessary conditions for a minimum are

$$\frac{\partial}{\partial c_k} \psi(\mathbf{c}) = 0, \quad k = 0, 1, \ldots, n.$$

These conditions turn out to be sufficient for a global minimum as well. Here they give

$$-2 \int_a^b \left[f(x) - \sum_{j=0}^n c_j \phi_j(x) \right] \phi_k(x) dx = 0,$$

which is rewritten as

$$\sum_{j=0}^n c_j \int_a^b \phi_j(x)\phi_k(x) dx = \int_a^b f(x)\phi_k(x) dx, \quad k = 0, 1, \ldots, n.$$

The resulting linear conditions are the *normal equations* for the continuous case. They read

$$\tilde{B}\mathbf{c} = \tilde{\mathbf{b}}, \quad \text{where}$$

$$\tilde{B}_{j,k} = \int_a^b \phi_j(x)\phi_k(x) dx, \quad \tilde{b}_j = \int_a^b f(x)\phi_j(x) dx.$$

Notice how the sum in the discrete problem of Section 6.1 has been replaced by an integral in the continuous problem. The corresponding algorithm given on page 370 is to be compared with the discrete least squares algorithm given on page 145.

There are in fact a number of properties of these normal equations that are similar to those associated with the discrete case:

1. The matrix $\tilde{B}$ is symmetric, which is obvious from its definition ($\tilde{B}_{j,k} = \tilde{B}_{k,j}$), and it is positive definite, because for any vector $\mathbf{z} \neq \mathbf{0}$

$$\mathbf{z}^T \tilde{B} \mathbf{z} = \sum_j \sum_k \int z_j z_k \phi_j \phi_k = \int_a^b \left[\sum_{j=0}^n z_j \phi_j(x) \right]^2 dx > 0.$$

   (Recall Section 4.3. The above expression is not 0, and therefore positive, because the basis functions are linearly independent.)

2. Denote the solution of the normal equations by $\mathbf{c}^*$. It is indeed the global minimizer. (This follows, as in the discrete case detailed in Section 6.1, from the fact that $\tilde{B}$ is symmetric positive definite.)

3. The residual (approximation error) $r^*(x) = f(x) - v(x)$ of the best least squares approximation $v(x) = \sum_j c_j^* \phi_j(x)$ is orthogonal to the basis functions, satisfying

$$\int_a^b r^*(x)\phi_k(x)dx = 0, \quad k = 0,\ldots,n.$$

In other words, $v(x)$ is the **orthogonal projection** of $f(x)$ into the space spanned by the basis functions. Recall Figure 6.2 on page 145.

Integration by Parts.
If f and g are both differentiable functions on the interval $[a,b]$, then

$$\int_a^b f'(x)g(x)dx = -\int_a^b f(x)g'(x)dx + \big(f(b)g(b) - f(a)g(a)\big).$$

This formula follows directly from taking integrals of the expression for differentiation given by

$$\big[f(x)g(x)\big]' = f'(x)g(x) + f(x)g'(x).$$

Polynomial best approximation

Of course, as in Chapter 10, Section 6.1, and Example 4.16, polynomials once again come to mind as candidates for best approximation $v(x)$.

With the simplest basis representation on the interval $[0,1]$, namely, the monomials

$$\phi_j(x) = x^j, \quad j = 0, 1, \ldots, n,$$

we obtain

$$\tilde{B}_{j,k} = \int_0^1 x^{j+k}dx = \frac{1}{j+k+1}, \quad 0 \leq j,k \leq n.$$

Thus, $\tilde{B}$ turns out to be the notorious **Hilbert matrix**. For instance, the Hilbert matrix for $n = 4$ reads

$$\tilde{B} = \begin{pmatrix} 1 & \frac{1}{2} & \frac{1}{3} & \frac{1}{4} & \frac{1}{5} \\ \frac{1}{2} & \frac{1}{3} & \frac{1}{4} & \frac{1}{5} & \frac{1}{6} \\ \frac{1}{3} & \frac{1}{4} & \frac{1}{5} & \frac{1}{6} & \frac{1}{7} \\ \frac{1}{4} & \frac{1}{5} & \frac{1}{6} & \frac{1}{7} & \frac{1}{8} \\ \frac{1}{5} & \frac{1}{6} & \frac{1}{7} & \frac{1}{8} & \frac{1}{9} \end{pmatrix}.$$

This matrix gets extremely ill-conditioned as its size n grows; see Exercise 1 and Section 5.8. Recall that the **condition number** is defined by $\kappa(\tilde{B}) = \|\tilde{B}\| \|\tilde{B}^{-1}\|$. If you are not familiar with the material of Chapters 4 and 5, then suffice it to say here that the solution of the normal equations is generally expected to have an error proportional to $\kappa(\tilde{B})n^2\eta$, where η is the rounding unit defined in Section 2.1.

12.1. Continuous least squares approximation

Example 12.1. For $n \leq 4$ the condition number of the Hilbert matrix is not unbearable, so let's see the process at work. In particular, for $n = 4$, $\kappa(\tilde{B}) \approx 5 \times 10^5$. We next compute, for each $n = 0, 1, 2, 3, 4$, the best polynomial approximations of degree $\leq n$ of the function $f(x) = \cos(2\pi x)$.

For this we need the following integrals, which are obtained in sequence using integration by parts (please verify!):

$$\tilde{b}_0 = \int_0^1 \cos(2\pi x)dx = 0, \quad \int_0^1 \sin(2\pi x)dx = 0,$$

$$\tilde{b}_1 = \int_0^1 x\cos(2\pi x)dx = 0, \quad \int_0^1 x\sin(2\pi x)dx = -\frac{1}{2\pi},$$

$$\tilde{b}_2 = \int_0^1 x^2\cos(2\pi x)dx = \frac{1}{2\pi^2}, \quad \int_0^1 x^2\sin(2\pi x)dx = -\frac{1}{2\pi},$$

$$\tilde{b}_3 = \int_0^1 x^3\cos(2\pi x)dx = \frac{3}{4\pi^2}, \quad \int_0^1 x^3\sin(2\pi x)dx = -\frac{1}{2\pi} + \frac{2}{4\pi^3},$$

$$\tilde{b}_4 = \int_0^1 x^4\cos(2\pi x)dx = \frac{1}{\pi^2} - \frac{3}{2\pi^4}.$$

The resulting approximants $p_n(x)$ are plotted in Figure 12.1. Note that here, due to symmetry, $p_1(x) = p_0(x)$ and $p_3(x) = p_2(x)$. The resulting polynomials are very similar to their counterparts obtained by discrete least squares in Example 6.3 given on page 149, because the function $f(x) = \cos(2\pi x)$ is depicted very well there by its values at 21 equidistant points on the interval $[0, 1]$. ∎

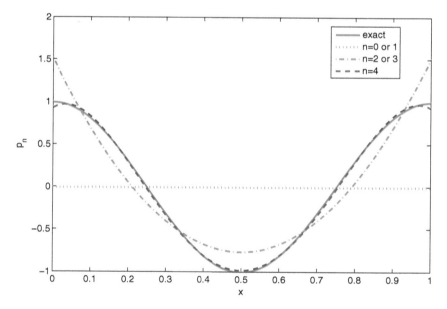

Figure 12.1. *The first five best polynomial approximations to $f(x) = \cos(2\pi x)$. The approximated function is in solid green. Note the similarity to Figure 6.4 of Example 6.3.*

When comparing the computational work involved in continuous vs. discrete least squares, e.g., Example 12.1 vs. Example 6.3, we observe that while in the discrete case of Section 6.1 the bulk of the work is in the construction of the matrix B (think m large there), in the continuous case of the present section it is in the evaluation of the integrals constituting the right-hand-side vector $\tilde{\mathbf{b}}$. The latter is generally done numerically, using numerical integration; see Chapter 15.

The advantage in the continuous case as compared to discrete data fitting (assuming we actually have the choice, i.e., that $f(x)$ with a finite norm is given everywhere on an interval) is that more work can be done in advance and analytically: the Hilbert matrix has been known since the days of, well, D. Hilbert. Likewise, the essential construction work underlying the next two sections is mathematical and can be done with a minimal use of computers: no wonder it was carried out centuries ago.

Specific exercises for this section: Exercises 1–2.

12.2 Orthogonal basis functions

Algorithm: Continuous Least Squares.
Given data function $f(x)$ on an interval $[a,b]$, construct approximation $v(x) = \sum_{j=0}^{n} c_j \phi_j(x)$ by computing the coefficients $\mathbf{c} = (c_0, \ldots, c_n)^T$ as follows:

1. Evaluate $\tilde{B} = (\tilde{B}_{j,k})$, where

$$\tilde{B}_{j,k} = \int_a^b \phi_j(x) \phi_k(x) dx, \quad j,k = 0,1,\ldots,n.$$

   If the basis is orthogonal (we know this in advance), then only the diagonal entries $\tilde{B}_{j,j}$ are evaluated.

2. Evaluate $\tilde{\mathbf{b}} = (\tilde{b}_0, \ldots, \tilde{b}_n)^T$, where

$$\tilde{b}_j = \int_a^b \phi_j(x) f(x) dx.$$

3. Solve the linear system

$$\tilde{B} \mathbf{c} = \tilde{\mathbf{b}}.$$

   In the case of an orthogonal basis, simply set

$$c_j = \tilde{b}_j / \tilde{B}_{j,j}, \quad j = 0, 1, \ldots, n.$$

Because of potential poor conditioning such as that of the Hilbert matrix for higher degree polynomials, it is natural to look for basis functions such that the resulting matrix $\tilde{B}$ is well-conditioned. The best we can do in this sense is to make $\tilde{B}$ the identity matrix, analogous to the Lagrange polynomial basis in Section 10.3, although a diagonal $\tilde{B}$ will do just as well. This not only improves the conditioning, but it also renders the task of solving the linear system of equations $\tilde{B} \mathbf{c} = \tilde{\mathbf{b}}$ trivial.

Thus, we require that the off-diagonal elements of $\tilde{B}$ vanish, which yields the condition

$$\tilde{B}_{j,k} = \int_a^b \phi_j(x) \phi_k(x) dx = 0, \quad j \neq k.$$

12.2. Orthogonal basis functions

If this holds, then we say that the basis functions are **orthogonal**. Alternatively, we say that the basis is orthogonal.

Let us denote
$$d_j = \tilde{B}_{j,j} > 0.$$
Then the solution of the normal equations is simply
$$c_j = \tilde{b}_j/d_j, \quad j = 0, 1, \ldots, n.$$

The entire cost and effort, therefore, is essentially in constructing the integrals of the products of the given function $f(x)$ to be approximated and these orthogonal basis functions.

An additional advantage of orthogonal basis functions is that only $\phi_j(x)$ participates in the determination of the coefficient c_j. Thus, we can pick a positive integer k and determine the best approximation using k basis functions $\phi_0(x), \phi_1(x), \ldots, \phi_{k-1}(x)$; if we then decide to upgrade the approximation by adding also $\phi_k(x)$, then only c_k need be calculated: the previous $c_0, c_1, \ldots, c_{k-1}$ remain the same. For this to work we also need to be able to simply add a new basis function without disturbing the previous ones, a property that is not satisfied by Lagrange polynomials.

Legendre polynomials

An **orthogonal polynomial** basis is provided by the Legendre polynomials. They are defined on the interval $[-1, 1]$ by the three-term recurrence relation

$$\phi_0(x) = 1,$$
$$\phi_1(x) = x,$$
$$\phi_{j+1}(x) = \frac{2j+1}{j+1} x \phi_j(x) - \frac{j}{j+1} \phi_{j-1}(x), \quad j \geq 1.$$

Thus, we compute
$$\phi_2(x) = \frac{1}{2}(3x^2 - 1),$$
$$\phi_3(x) = \frac{5}{6} x (3x^2 - 1) - \frac{2}{3} x = \frac{1}{2}(5x^3 - 3x),$$

and so on.

In the next section we derive the Legendre polynomials and other families of orthogonal polynomials methodically. For now, we ask that you take it on faith that the polynomials defined above indeed form an orthogonal basis. Let us describe a few of their nice properties:

1. *Orthogonality*. Specifically, we have
$$\int_{-1}^{1} \phi_j(x) \phi_k(x) dx = \begin{cases} 0, & j \neq k, \\ \frac{2}{2j+1}, & j = k. \end{cases}$$

So, for a given polynomial least squares approximation problem as before, we can form $\tilde{b}_j$ by integrating $f(x)\phi_j(x)$ and then set $c_j = \frac{2j+1}{2}\tilde{b}_j$.

2. *Calibration*. We have $|\phi_j(x)| \leq 1$, $-1 \leq x \leq 1$, and $\phi_j(1) = 1$.

3. *Oscillation*. The Legendre polynomial $\phi_j(x)$ has degree j (not less). All its j zeros are simple and lie inside the interval $(-1, 1)$. Hence the polynomial oscillates j times in this interval. See Figure 12.2.

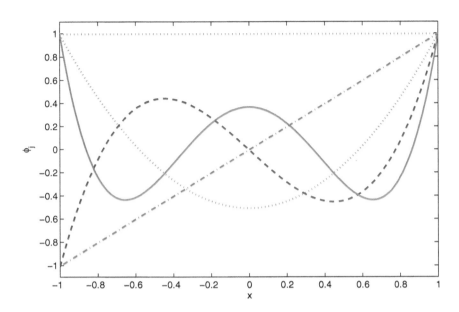

Figure 12.2. *The first five Legendre polynomials. You should be able to figure out which curve corresponds to which polynomial.*

Best approximation on a general interval

Of course, not all approximation problems are given on the interval $[-1, 1]$. What about a general interval $[a, b]$? The answer, as in Section 10.6, is general and simple: we can apply an affine transformation to $x \in [-1, 1]$ which *scales* and *translates* it. That is, we can write a given $t \in [a, b]$ as

$$t = \frac{1}{2}[(b-a)x + (a+b)] \quad \text{or} \quad x = \frac{2t - a - b}{b - a}.$$

Exercise 3 shows for the more general case that

$$d_j = \int_a^b [\phi_j(t)]^2 dt = \frac{b-a}{2j+1}.$$

In Exercise 4 you are asked to reproduce Figure 12.1 using the Legendre polynomials. Here, let us consider another example of this sort.

Example 12.2. Suppose we wish to calculate the quadratic polynomial $v(t)$ that minimizes

$$\int_0^1 \left(e^t - v(t)\right)^2 dt.$$

Here $t = \frac{1}{2}(x+1)$, or $x = 2t - 1$. We get

$$\phi_0(t) = 1,$$
$$\phi_1(t) = 2t - 1,$$
$$\phi_2(t) = \frac{1}{2}(3(2t-1)^2 - 1) = 6t^2 - 6t + 1.$$

12.3. Weighted least squares

This yields

$$\tilde{b}_0 = \int_0^1 e^t dt = e - 1, \quad d_0 = \|\phi_0\|^2 = \int_0^1 1 dt = 1,$$

$$\tilde{b}_1 = \int_0^1 (2t-1)e^t dt = 3 - e, \quad d_1 = \|\phi_1\|^2 = \int_0^1 (2t-1)^2 dt = \frac{1}{3},$$

$$\tilde{b}_2 = \int_0^1 (6t^2 - 6t + 1)e^t dt = 7e - 19, \quad d_2 = \|\phi_2\|^2 = \int_0^1 (6t^2 - 6t + 1)^2 dt = \frac{1}{5}.$$

So, our best approximating quadratic is

$$v(t) = (e-1) + 3(3-e)(2t-1) + 5(7e-19)(6t^2 - 6t + 1).$$

This can obviously be simplified. Plotting the two functions $f(t) = e^t$ and $v(t)$ over $[0,1]$ would reveal a nice agreement between function and approximant. ■

Trigonometric polynomials

As stated earlier, the concept of orthogonality is not restricted to polynomials! An important non-polynomial orthogonal family consists of the basis functions for **trigonometric polynomials**, given on $[-\pi, \pi]$ by

$$\phi_0(x) = \frac{1}{\sqrt{2\pi}},$$

$$\phi_{2l-1}(x) = \frac{1}{\sqrt{\pi}} \sin(lx),$$

$$\phi_{2l}(x) = \frac{1}{\sqrt{\pi}} \cos(lx)$$

for $l = 1, \ldots, \frac{n}{2}$ with n assumed even. Here we have

$$\int_{-\pi}^{\pi} \phi_j(x) \phi_k(x) dx = \delta_{jk} = \begin{cases} 0, & j \neq k, \\ 1, & j = k. \end{cases}$$

Thus, $d_j = 1$, and the basis is called **orthonormal**. We return to trigonometric polynomials in Section 13.1.

Specific exercises for this section: Exercises 3–5.

12.3 Weighted least squares

An important, yet straightforward, generalization of the methods of Sections 12.1 and 12.2 is obtained by introducing a weight function $w(x)$ into the integrals. Often in applications the given data are not equally important everywhere. For example, we may know that the data are more accurate in some regions than in others. The weight function is then used to indicate and take into account where $f(x)$ should be respected more and where it matters less. A weight function on an interval $[a,b]$ must be nonnegative, $w(x) \geq 0$, and it may vanish only at isolated points, i.e., not on a whole subinterval of $[a,b]$. Usually, if $w(x)$ vanishes at all, it would be at the endpoints.

The basic approximation problem is now

$$\min_{\mathbf{c}} \psi(\mathbf{c}) = \int_a^b w(x)(f(x) - v(x))^2 dx,$$

Algorithm: Gram–Schmidt Process.
Given an interval $[a,b]$ and a weight function $w(x)$, set

$$\phi_0(x) = 1,$$
$$\phi_1(x) = x - \beta_1,$$
$$\phi_j(x) = (x - \beta_j)\phi_{j-1}(x) - \gamma_j \phi_{j-2}(x), \quad j \geq 2,$$

where

$$\beta_j = \frac{\int_a^b x w(x)[\phi_{j-1}(x)]^2 dx}{\int_a^b w(x)[\phi_{j-1}(x)]^2 dx}, \quad j \geq 1,$$

$$\gamma_j = \frac{\int_a^b x w(x)\phi_{j-1}(x)\phi_{j-2}(x)dx}{\int_a^b w(x)[\phi_{j-2}(x)]^2 dx}, \quad j \geq 2.$$

where still

$$v(x) = \sum_{j=0}^{n} c_j \phi_j(x).$$

Everything we said before generalizes directly: the *normal equations* read the same, with the obvious modification in inner product, so

$$\tilde{B}\mathbf{c} = \tilde{\mathbf{b}}, \quad \text{where}$$

$$\tilde{B}_{j,k} = \int_a^b w(x)\phi_j(x)\phi_k(x)dx, \quad \tilde{b}_j = \int_a^b w(x)f(x)\phi_j(x)dx.$$

Therefore, to make $\tilde{B}$ diagonal, the orthogonality condition is generalized into

$$\int_a^b w(x)\phi_j(x)\phi_k(x)dx = 0, \quad j \neq k.$$

If orthogonality holds then, as before, $c_j = \tilde{b}_j / d_j$, where now

$$d_j = \int_a^b w(x)\phi_j(x)^2 dx > 0.$$

Gram–Schmidt orthogonalization

In Section 12.2 we have simply written down the Legendre polynomials. But how are they found in the first place?

More generally, we need a way to specify families of orthogonal polynomials for any eligible weight function. Fortunately, there is a simple, general algorithm for this. Given an interval $[a,b]$ and a weight function $w(x)$, we now *construct* the corresponding family of **orthogonal polynomials** by the **Gram–Schmidt process**. The algorithm is given on the current page. Compare it to the (modified) Gram–Schmidt algorithm for the discrete case given on page 159. The proof is by induction.

A useful property of a given family of orthogonal polynomials is that $\phi_n(x)$ *is orthogonal to all polynomials of degree* $< n$. This is because any polynomial $q(x)$ of degree less than n can be

12.3. Weighted least squares

> **Theorem: Continuous Gram–Schmidt Orthogonalization.**
> The Gram–Schmidt orthogonalization algorithm defined on the facing page produces a family $\{\phi_j(x)\}$ of polynomials that are orthogonal to one another with respect to the weight function $w(x)$, satisfying
> $$\int_a^b w(x)\phi_j(x)\phi_k(x)dx = 0, \quad j \neq k.$$

> **Note:** Despite the large number of coefficients with indices, the Gram–Schmidt process is really straightforward to apply: just try it!
> What is far less obvious, at least for us, is to have guessed that such a short algorithm would be at all possible: for *any* weight function on *any* interval an entire class of orthogonal polynomials is constructed using a simple *three-term* recurrence relationship!

written as a linear combination of the previous basis functions, $q(x) = \sum_{j=0}^{n-1} \alpha_j \phi_j(x)$, and hence

$$\int_a^b w(x)q(x)\phi_n(x)dx = \sum_{j=0}^{n-1} \alpha_j \int_a^b w(x)\phi_j(x)\phi_n(x)dx = 0.$$

Here are some important **families of orthogonal polynomials**, given as a sequence of examples.

Example 12.3. For $w(x) \equiv 1$ and $[a,b] = [-1,1]$ we obtain

$\phi_0(x) = 1,$
$$\beta_1 = \int_{-1}^{1} x\, dx \bigg/ \int_{-1}^{1} 1\, dx = 0,$$
$\phi_1(x) = x,$
$$\beta_2 = \int_{-1}^{1} x^3\, dx \bigg/ \int_{-1}^{1} x^2\, dx = 0, \quad \gamma_2 = \int_{-1}^{1} x^2\, dx \bigg/ \int_{-1}^{1} 1\, dx = \frac{1}{3},$$
$\phi_2(x) = x^2 - \frac{1}{3},$

etc.

These are the *Legendre polynomials* introduced in Section 12.2, although they are scaled differently here. Indeed, in an orthogonal basis, unlike an orthonormal basis, the basis functions are determined only up to a constant. Different constants yield different values d_j.

Let us denote the roots of the jth Legendre polynomials by x_i, so

$$-1 < x_1 < \cdots < x_j < 1.$$

Like for all polynomials of degree j we can write $\phi_j(x)$ in terms of its roots, so

$$\phi_j(x) = \beta(x - x_1)(x - x_2) \cdots (x - x_j),$$

where β is a constant. Thus we have for any polynomial $q(x)$ of degree less than j the orthogonality result

$$\int_{-1}^{1} q(x)(x - x_1)(x - x_2) \cdots (x - x_j)dx = 0.$$

This property allows derivation of highly accurate yet basic numerical integration formulas; see Section 15.3. The zeros of the Legendre polynomial are called **Gauss points**. ∎

What if the interval on which we consider our best approximation is the entire real line, or just the nonnegative part of the real line? Even the constant function $f(x) \equiv 1$ obviously does not have a bounded L_2-norm on the interval $[0, \infty)$.

Example 12.4. As for the case of a general finite interval handled in Section 12.2, let us consider a transformation, albeit a more delicate one. Here we get $0 \leq t < 1$ for the variable
$$t = 1 - e^{-x}, \quad 0 \leq x < \infty.$$
Now, $dt = e^{-x} dx$, so for any function $\hat{g}(t)$ that has bounded L_2-norm on $[0,1)$ we have
$$\int_0^\infty e^{-x} g(x)^2 dx = \int_0^1 \hat{g}(t)^2 dt < \infty,$$
where $g(x) = \hat{g}(1 - e^{-x})$.

For $w(x) = e^{-x}$ and $[a, b] \to [0, \infty)$ the **Laguerre polynomials** form the orthogonal family. We obtain
$$\phi_0(x) = 1,$$
$$\phi_1(x) = 1 - x,$$
$$\phi_{j+1}(x) = \frac{2j+1-x}{j+1}\phi_j(x) - \frac{j}{j+1}\phi_{j-1}(x), \quad j = 1, 2, 3, \ldots.$$

Again, multiplying them all by a nonzero constant is possible. ∎

If x roams on the entire real line, not only the nonnegative half, then the natural weighting turns out to be the famous Gaussian function.

Example 12.5. For $w(x) = e^{-x^2}$ and $[a,b] \to (-\infty, \infty)$ we obtain the **Hermite polynomials**
$$\phi_0(x) = 1,$$
$$\phi_1(x) = 2x,$$
$$\phi_{j+1}(x) = 2x\phi_j(x) - 2j\phi_{j-1}(x), \quad j = 1, 2, 3, \ldots.$$

The weighted orthogonality and weighted L_2-boundedness hold even though $\phi_1(x)$, for instance, is unbounded as such.

Note that these polynomials have nothing to do with the Hermite cubics used for interpolation in Sections 10.7, 11.2, and 11.5. ∎

Finally, in this section for $w(x) = \frac{1}{\sqrt{1-x^2}}$ and $[a,b] = [-1,1]$ we obtain the **Chebyshev polynomials**
$$\phi_0(x) = 1,$$
$$\phi_1(x) = x,$$
$$\phi_2(x) = 2x^2 - 1,$$
$$\phi_{j+1}(x) = 2x\phi_j(x) - \phi_{j-1}(x).$$

Note that $w(x) \to \infty$ as $x \to \pm 1$. Still, the process is well-defined.

12.4 Chebyshev polynomials

The Chebyshev polynomials form a particularly interesting family of orthogonal polynomials. They arise as a tool in various instances of numerical analysis. Let us therefore pay particular attention to them next.

Specific exercises for this section: Exercises 6–7.

12.4 Chebyshev polynomials

> **Note:** The polynomials defined in this section will be extensively discussed again in the advanced Section 14.5.

The Chebyshev polynomials[49] just mentioned in the end of the previous section are elegantly defined as

$$\phi_j(x) = T_j(x) = \cos(j \arccos x)$$
$$= \cos(j\theta), \quad \text{where } x = \cos(\theta).$$

In other words, T_j is the expansion of $\cos(j\theta)$ in terms of $\cos(\theta)$.

Recall from Section 12.3 that we have a weight function $w(x) = \frac{1}{\sqrt{1-x^2}}$ associated with these polynomials. To verify orthogonality we substitute the definition, obtaining

$$\int_{-1}^{1} w(x) T_j(x) T_k(x) dx = \int_{-1}^{1} \frac{T_j(x) T_k(x)}{\sqrt{1-x^2}} dx$$
$$= \int_{0}^{\pi} \cos(j\theta) \cos(k\theta) d\theta = \begin{cases} 0, & j \neq k, \\ \frac{\pi}{2}, & j = k > 0. \end{cases}$$

The crucial step above is the transformation from x to θ. Here, $dx = -\sin(\theta)\,d\theta$, and $\sqrt{1-x^2} = \sqrt{1-\cos^2(\theta)} = \sin(\theta)$, so $\sin(\theta)$ cancels in the integrand. Also, $\cos(\pi) = -1$ and $\cos(0) = 1$.

Let us fix our running index as $j = n$. The n zeros (roots) of $T_n(x)$ are called the **Chebyshev points**. From the definition, they are the first n roots of the equation

$$\cos(n\theta) = 0,$$

and hence $\theta_k = \frac{(k-1/2)\pi}{n}$. Consequently, the roots $x_k = \cos(\theta_k)$ are given by

$$x_k = \cos\left(\frac{2k-1}{2n}\pi\right), \qquad k = 1, \ldots, n,$$

and satisfy in particular $1 > x_1 > \cdots > x_n > -1$. Moreover, the $n+1$ *extremum points* of $T_n(x)$ on the interval $[-1, 1]$ (these are the $n-1$ points where the first derivative $T'_n(x)$ vanishes, plus the interval ends) are at

$$\xi_k = \cos\left(\frac{k}{n}\pi\right), \qquad k = 0, 1, \ldots, n,$$

and thus they satisfy

$$T_n(\xi_k) = (-1)^k,$$

see Figure 12.3. These properties are all summarized on the next page.

---

[49] The notation T_j instead of ϕ_j stands for the first letter of Chebyshev's name, when converted (by Westerners) from the Russian as "Tchebycheff." This chapter is indeed loaded with 19th-century history.

Chebyshev Polynomials.
The polynomial of degree n is defined by
$$T_n(x) = \cos(n\theta), \quad \text{where } x = \cos(\theta).$$
Its n zeros are simple, real, lie in the interval $(-1, 1)$, and are given by
$$x_k = \cos\left(\frac{2k-1}{2n}\pi\right), \quad k = 1, \ldots, n.$$
It has $n+1$ extrema on $[-1, 1]$, given by
$$\xi_k = \cos\left(\frac{k}{n}\pi\right), \quad k = 0, 1, \ldots, n.$$
Recurrence relation: $T_0(x) = 1$, $T_1(x) = x$,
$$T_{n+1}(x) = 2xT_n(x) - T_{n-1}(x).$$

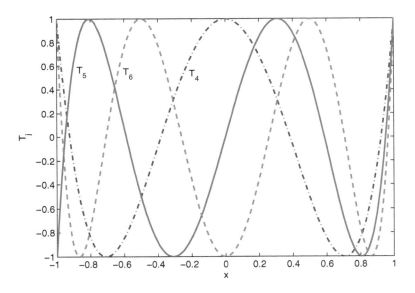

Figure 12.3. *Chebyshev polynomials of degrees* 4, 5, *and* 6.

Min-max property

As in Section 10.6, note that the Chebyshev points tend to concentrate more towards the ends of the interval. Indeed, recall the behavior of the weight function $w(x)$. Also, the maximum norm of T_n satisfies
$$\|T_n\|_\infty = \max_{-1 \leq x \leq 1} |T_n(x)| = 1,$$
and from the recursion formula, the leading coefficient of $T_n(x)$ (what multiplies x^n) equals 2^{n-1}; see Exercise 8. Indeed, for $|x| > 1$ the polynomial $T_n(x)$ grows very rapidly in magnitude.

> **Theorem: Chebyshev Min-Max Property.**
> Let $x_1, \ldots, x_n$ be the Chebyshev points and define $\tilde{T}_n(x) = \prod_{k=1}^{n}(x - x_k)$.
> Then, of all monic polynomials of degree n, $\tilde{T}_n(x)$ uniquely has the smallest maximum magnitude over the interval $[-1, 1]$, and
> $$\max_{-1 \leq x \leq 1} |\tilde{T}_n(x)| = \frac{1}{2^{n-1}}.$$

The **monic** (i.e., polynomial with leading coefficient 1) Chebyshev polynomials are therefore defined as

$$\tilde{T}_0(x) = 1,$$
$$\tilde{T}_n(x) = \frac{T_n(x)}{2^{n-1}} = \prod_{k=1}^{n}(x - x_k), \quad n \geq 1.$$

These polynomials possess the *min-max* property described in the theorem by the same name given on this page.

The min-max property follows from the oscillation of $T_n(x)$ at its extremum points ξ_k, where its first derivative vanishes and

$$\tilde{T}_n(\xi_k) = \frac{1}{2^{n-1}}(-1)^k.$$

For any monic polynomial $p_n(x)$ of degree n, we now show that if $\max_{-1 \leq x \leq 1} |p_n(x)| \leq \frac{1}{2^{n-1}}$, then necessarily $p_n \equiv \tilde{T}_n$.

Thus, let
$$q(x) = \tilde{T}_n(x) - p_n(x).$$

This is a polynomial of degree at most $n - 1$, because both $\tilde{T}_n$ and p_n are monic, hence their subtraction eliminates the nth power of x in $q(x)$. At the points ξ_k we know that

$$q(\xi_k) = \frac{1}{2^{n-1}}(-1)^k - p_n(\xi_k) \begin{cases} \geq 0, & k \text{ even}, \\ \leq 0, & k \text{ odd}, \end{cases}$$

for $k = 0, 1, \ldots, n$. So, $q(x)$ changes sign at least n times on the interval $[-1, 1]$, and therefore it has at least n zeros there. This implies that $q \equiv 0$, and hence $p_n \equiv T_n$. ♦

Specific exercises for this section: Exercises 8–9.

12.5 Exercises

0. **Review questions**

   (a) Distinguish between interpolation and best approximation.

   (b) What is the difference between continuous and discrete least squares approximation?

   (c) Write down the normal equations for least squares approximation.

   (d) Write down the Hilbert matrix of size $n \times n$. When does this matrix arise?

   (e) What are the conditions for a function $v(x)$ to be the orthogonal projection of a function $f(x)$ onto the space spanned by the two functions $\phi_1(x)$ and $\phi_2(x)$?

(f) What are orthogonal polynomials with respect to a weight function $w(x)$? Why are they useful for numerical computations?

(g) What are the Legendre polynomials, and how can they be used to solve the continuous least squares problem?

(h) Describe how to compute a best approximation on a general interval, $[a,b]$, not necessarily equal to the interval $[-1,1]$.

(i) Explain what trigonometric polynomials are. Are they really polynomials? Are they orthogonal?

(j) Describe the Gram–Schmidt process and explain its importance.

(k) What are the Chebyshev polynomials? Why do we hear more about them than about other families of orthogonal polynomials?

(l) What is a min-max property?

(m) The weight function $w(x)$ for Chebyshev polynomials satisfies $w(x) \to \infty$ as $x \to \pm 1$. Is this a problem? Explain.

1. Using the MATLAB instruction `cond`, find the condition numbers of Hilbert matrices for $n = 4, 5, \ldots, 12$. Plot these condition numbers as a function of n using `semilogy`. What are your observations?

2. Construct the second degree polynomial $q_2(t)$ that approximates $g(t) = \sin(\pi t)$ on the interval $[0,1]$ by minimizing
$$\int_0^1 [g(t) - q_2(t)]^2 \, dt.$$

Some useful integrals:

$$\int_0^1 (6t^2 - 6t + 1)^2 \, dt = \frac{1}{5}, \quad \int_0^1 \sin(\pi t) \, dt = \frac{2}{\pi},$$

$$\int_0^1 t \sin(\pi t) \, dt = \frac{1}{\pi}, \quad \int_0^1 t^2 \sin(\pi t) \, dt = \frac{\pi^2 - 4}{\pi^3}.$$

3. The Legendre polynomials satisfy
$$\int_{-1}^1 \phi_j(x) \phi_k(x) \, dx = \begin{cases} 0, & j \neq k, \\ \frac{2}{2j+1}, & j = k. \end{cases}$$

Suppose that the best fit problem is given on the interval $[a,b]$.

Show that with the transformation $t = \frac{1}{2}[(b-a)x + (a+b)]$ and a slight change of notation, we have
$$\int_a^b \phi_j(t) \phi_k(t) \, dt = \begin{cases} 0, & j \neq k, \\ \frac{b-a}{2j+1}, & j = k. \end{cases}$$

4. Redo Example 12.1, reconstructing Figure 12.1, using an orthogonal polynomial basis.

5. (a) Using an orthogonal polynomial basis, find the best least squares polynomial approximations, $q_2(t)$ of degree at most 2 and $q_3(t)$ of degree at most 3, to $f(t) = e^{-3t}$ over the interval $[0,3]$.

12.5. Exercises

[Hint: For a polynomial $p(x)$ of degree n and a scalar $a > 0$ we have $\int e^{-ax} p(x) dx = \frac{-e^{-ax}}{a} (\sum_{j=0}^{n} (\frac{p^{(j)}(x)}{a^j}))$, where $p^{(j)}(x)$ is the jth derivative of $p(x)$. Alternatively, just use numerical quadrature, e.g., the MATLAB function `quad`.]

(b) Plot the error functions $f(t) - q_2(t)$ and $f(t) - q_3(t)$ on the same graph on the interval $[0,3]$. Compare the errors of the two approximating polynomials. In the least squares sense, which polynomial provides the better approximation?

[Hint: In each case you may compute the *norm* of the error, $(\int_a^b (f(t) - q_n(t))^2 dt)^{1/2}$, using the MATLAB function `quad`.]

(c) Without any computation, prove that $q_3(t)$ generally provides a least squares fit, which is never worse than with $q_2(t)$.

6. Let $\phi_0(x), \phi_1(x), \phi_2(x), \ldots$ be a sequence of orthogonal polynomials on an interval $[a,b]$ with respect to a positive weight function $w(x)$. Let $x_1, \ldots, x_n$ be the n zeros of $\phi_n(x)$; it is known that these roots are real and $a < x_1 < \cdots < x_n < b$.

(a) Show that the Lagrange polynomials of degree $n-1$ based on these points are orthogonal to each other, so we can write

$$\int_a^b w(x) L_j(x) L_k(x) dx = 0, \quad j \neq k,$$

where

$$L_j(x) = \prod_{k=1, k \neq j}^{n} \frac{(x - x_k)}{(x_j - x_k)}, \quad 1 \leq j \leq n.$$

[Recall Section 10.3.]

(b) For a given function $f(x)$, let $y_k = f(x_k)$, $k = 1, \ldots, n$. Show that the polynomial $p_{n-1}(x)$ of degree at most $n-1$ that interpolates the function $f(x)$ at the zeros $x_1, \ldots, x_n$ of the orthogonal polynomial $\phi_n(x)$ satisfies

$$\|p_{n-1}\|^2 = \sum_{k=1}^{n} y_k^2 \|L_k\|^2$$

in the weighted least squares norm. This norm is defined by

$$\|g\|^2 = \int_a^b w(x) [g(x)]^2 dx$$

for any suitably integrable function $g(x)$.

7. Prove the Gram–Schmidt Theorem given on page 375.

8. Using the recursion formula for Chebyshev polynomials, show that $T_n(x)$ can be written as

$$T_n(x) = 2^{n-1} (x - x_1)(x - x_2) \cdots (x - x_n),$$

where x_i are the n roots of T_n.

9. Jane works for for a famous bioinformatics company. Last year she was required by Management to approximate an important but complicated formula, $g(x)$, defined on the interval $[-1, 1]$, by a polynomial of degree $n+1$. She did so, and called the result $f(x) = p_{n+1}(x)$.

Last week, Management decided that they really needed a polynomial of degree n, not $n+1$, to represent g. Alas, the original g had been lost by this time and all that was left was $f(x)$. Therefore, Jane is looking for the polynomial of degree n which is closest (in the maximum norm) to f on the interval $[-1, 1]$. Please help her find it.

12.6 Additional notes

This chapter, probably more than any other chapter in our text, connects to many different areas of mathematics and computational applications. The following additional notes will necessarily be brief.

Data fitting in ℓ_2 could just as well belong here or in a basic course on numerical linear algebra. We opted for the latter; see Chapter 6.

The function spaces L_2 and L_1 are special cases of L_p for $1 \leq p < \infty$. Moreover, there is nothing in the definitions given on page 366 that cannot be extended from an interval $[a,b]$ to a set Ω in d dimensions, i.e., $\Omega \subset \mathcal{R}^d$. Thus, we may define

$$\|g\|_{p,\Omega} = \left(\int_\Omega g(\mathbf{x})^p d\mathbf{x} \right)^{1/p}$$

and say that $g \in L_p(\Omega)$ if its norm is finite. Furthermore, the space L_∞ may be defined as consisting of all the functions g on Ω that have a bounded norm

$$\|g\|_{\infty,\Omega} = \lim_{p \to \infty} \|g\|_{p,\Omega}.$$

The *dual space* of L_p is L_q, where $\frac{1}{p} + \frac{1}{q} = 1$. Only L_2 is its own dual space, and this innocent-looking fact has many important consequences, including naturally defined inner product and orthogonality. For much more on function spaces, see, for instance, Rivlin [60] or Rudin [61]. The former, in particular, considers in detail best L_1- and L_∞-approximations.

The zeros of Legendre polynomials, called Gauss points, have found major use in numerical integration (see Section 15.3) and in solving initial and boundary value problems for ordinary differential equations; see, e.g., Ascher and Petzold [5].

Chebyshev polynomials are a particular favorite among analysts because of their min-max properties. Among other things they can also be used to reduce the degree of an approximating polynomial with a minimal loss of accuracy. See, for instance, Exercise 9.

When differential equations are considered, function norms that involve derivatives naturally replace the L_p-norms. This gives rise to **Sobolev spaces**; see, e.g., Adams and Fournier [1]. The theory of **finite element methods** makes extensive use of Sobolev spaces. In particular, the Galerkin approach involves orthogonality requirements with respect to basis functions in a manner that extends the least squares best approximation algorithm given on page 370 in a rather nontrivial fashion; see Section 16.8 for more on this.

Chapter 13

Fourier Transform

The Fourier transform in its many forms is much more than just another way of approximating functions. It decomposes a given integrable function into a sum or an integral of components with varying degrees of oscillations. This representation and its approximations allow for a variety of manipulations, depending on what the input function represents, from image noise and blur removal, through mp3 and jpeg compressions, to solving partial differential equations.

In particular, wave phenomena such as light and sound are naturally described by smooth, periodic functions $f(x)$; see the definition on the current page. A natural approximation basis for such f would then consist of smooth, periodic, wave-like basis functions, and this naturally lends itself to **trigonometric polynomials**. Using these basis functions, briefly introduced already in Section 12.2, leads to the justly famous Fourier transform.

The continuous transform is introduced in Section 13.1, which may be viewed as a continuation of Section 12.1. In practice, though, integrals must be discretized and sums must be finite, and this leads to the **discrete Fourier transform** and the process of *trigonometric polynomial interpolation* discussed in Section 13.2. An efficient way to carry out the discrete transform is discussed in Section 13.3, leading to one of the most celebrated algorithms of all times, the **fast Fourier transform** (FFT). Uses of the FFT for solving differential equations are described in the more advanced Section 14.5.

> **Periodic Function.**
> A function f defined on the real line is *periodic* if there is a positive scalar τ, called the *period*, such that $f(x+\tau) = f(x)$ for all x.

13.1 The Fourier transform

> **Note:** In this chapter we *do not* use polynomials or piecewise polynomials for the approximation process. Rather, we consider trigonometric polynomials: these arise in special but very important practical applications and provide useful function decompositions, not just approximations.

Let us recall the *trigonometric basis functions*. In their simplest real, unnormalized form they can be written as

$$\phi_0(x) = 1, \quad \phi_{2k}(x) = \cos(kx), \quad \phi_{2k-1}(x) = \sin(kx), \quad k = 1, 2, \ldots.$$

These functions are periodic and pairwise orthogonal on the interval $[-\pi,\pi]$, satisfying

$$\int_{-\pi}^{\pi} \phi_i(x)\phi_j(x)dx = 0, \quad i \neq j.$$

Fourier series and transform

Assume that f is square integrable, i.e., that $\int_{-\pi}^{\pi} f^2(x)dx < \infty$. Then it can be shown that f admits the decomposition

$$f(x) = \frac{a_0}{2} + \sum_{k=1}^{\infty} a_k \cos(kx) + b_k \sin(kx),$$

with the coefficients a_k and b_k as defined below in (13.1b) and (13.1c). This is called the **Fourier series**, and the coefficients form the **Fourier transform** of f. The index k is typically referred to as the **frequency**. Thus, the Fourier series provides a neat decomposition of f in terms of frequency-based components.

Best least squares approximation

To compute anything, we must turn the infinite sum in the Fourier series expression into a finite one. For $n = 2l-1$, with l a positive integer (the *cutoff frequency*), we can write the approximation for a given function $f(x)$ on the interval $[-\pi,\pi]$ as

$$v(x) = \sum_{j=0}^{n} c_j \phi_j(x) \equiv \frac{a_0}{2} + a_l \cos(lx) + \sum_{k=1}^{l-1} \Big(a_k \cos(kx) + b_k \sin(kx) \Big). \tag{13.1a}$$

This notation identifies in an obvious way the coefficients c_j with the coefficients a_k and b_k for the cos and sin basis functions. We set by convention $b_l = 0$, because we want n to be odd.

To find these coefficients, let us apply the algorithm for best least squares approximation given on page 370. In Exercise 2 you are asked to verify that, in order for $v(x)$ to be the best approximation for $f(x)$ on $[-\pi,\pi]$ in the least squares sense, we must set

$$a_k = \frac{1}{\pi} \int_{-\pi}^{\pi} f(x)\cos(kx)dx, \quad k = 0,1,\ldots,l, \tag{13.1b}$$

$$b_k = \frac{1}{\pi} \int_{-\pi}^{\pi} f(x)\sin(kx)dx, \quad k = 1,\ldots,l-1. \tag{13.1c}$$

These expressions hold also as we let $l \to \infty$ and thus give the coefficients for the (full) Fourier transform.

Fourier series and trigonometric polynomials are extremely important in **digital signal processing** (DSP) applications and other areas where wave-like phenomena are encountered. The trigonometric basis functions ϕ_j are all periodic as well, with period 2π, so they are naturally suitable for describing a smooth periodic function f, once the variable x is scaled and translated if needed so as to have the period 2π. Thus, we would consider the Fourier series of

$$g(x) = f\left(\frac{\tau}{2\pi}x - \alpha\right)$$

with a suitably chosen translation constant α; see, for instance, Example 13.4.

13.1. The Fourier transform

Example 13.1. As the frequency k increases, $\sin(kx)$ and $\cos(kx)$ oscillate more and more, so if we know that the observed data consist of a true signal plus noise, where the noise oscillates at higher frequencies, i.e., more rapidly,[50] then a best least squares approximation with the first few trigonometric basis functions would naturally **filter out** the noise; see Figure 13.1. This is one form of what is known as a *low pass filter*. Design of data filters is an important part of DSP. ∎

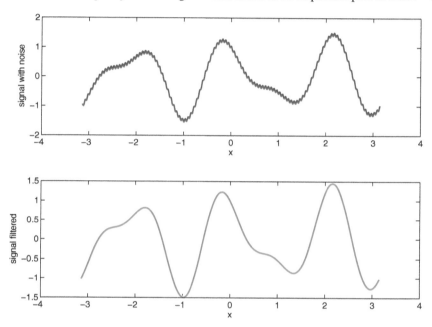

Figure 13.1. *A linear combination of sines and cosines with various k-values up to* 110 *(top panel), filtered by taking its best least squares trigonometric polynomial approximation with* $l = 10$ *(bottom panel). This best approximation simply consists of that part of the given function that involves the first* 20 *basis functions* ϕ_j; *thus the higher frequency contributions disappear.*

Trigonometric Identities.
Here are some useful trigonometric identities for two angles α and β (in radians):

$$\sin(\alpha)\sin(\beta) = \frac{1}{2}[\cos(\alpha - \beta) - \cos(\alpha + \beta)],$$

$$\cos(\alpha)\cos(\beta) = \frac{1}{2}[\cos(\alpha - \beta) + \cos(\alpha + \beta)],$$

$$\sin(\alpha)\cos(\beta) = \frac{1}{2}[\sin(\alpha - \beta) + \sin(\alpha + \beta)].$$

Using complex basis functions

Perhaps contrary to some people's intuition, describing the Fourier transform in terms of complex basis functions is far from being more complex. It is actually simpler or at least neater! We define

$$\phi_k(x) = e^{\iota k x}, \quad k = 0, \pm 1, \pm 2, \ldots,$$

---
[50]Indeed, noise by its nature is typically highly oscillatory.

where $\iota^2 = -1$, and use the Euler identity (4.1) to realize that

$$\phi_k(x) = \cos(kx) + \iota \sin(kx).$$

Thus, for $k = 0$ we have $\phi_0(x) = 1$, and for $k \neq 0$ each $\phi_k(x)$ packs two of the previous real-valued basis functions. See the fact box on this page.

> **Note:** A refresher on complex numbers is provided on page 71. However, if you are allergic to these, then you can safely skip to the beginning of Section 13.2.

Utilizing the frequency domain

Any square integrable function $f(x)$ on $[-\pi, \pi]$ can be written as a *Fourier series*

$$f(x) = \sum_{k=-\infty}^{\infty} c_k \phi_k(x), \tag{13.2a}$$

with the *Fourier transform*

$$c_k = \frac{1}{2\pi} \int_{-\pi}^{\pi} f(x) e^{-\iota kx} dx, \quad k \text{ integer}. \tag{13.2b}$$

See Exercise 4.

> **Complex inner product and orthogonality.**
> Let $v(x)$ be a complex-valued function defined on a real interval $[a,b]$. It is common to denote the function whose values are the conjugates of those of v by $v^*(x)$. Recall that the magnitude of a complex scalar z satisfies $|z|^2 = z\bar{z}$ (and generally $|z|^2 \neq z^2$). Thus, the *inner product* of two complex valued functions u and v is defined by
>
> $$(u,v) = \int_a^b u(x) v^*(x) dx,$$
>
> and the L_2-*norm* of such a function is
>
> $$\|u\|_2 = \sqrt{(u,u)} = \left(\int_a^b |u(x)|^2 dx \right)^{1/2}.$$
>
> For the Fourier basis functions $\phi_k(x) = e^{\iota kx}$, defined for x real and k integer, $\phi_k^*(x) = e^{-\iota kx}$. Hence, on the interval $[a,b] = [-\pi, \pi]$ we have
>
> $$(\phi_j, \phi_k) = \int_{-\pi}^{\pi} e^{\iota kx} e^{-\iota jx} dx = \int_{-\pi}^{\pi} e^{\iota(k-j)x} dx = \begin{cases} 0, & j \neq k, \\ 2\pi, & j = k. \end{cases}$$
>
> In particular, $\|\phi_k\|_2 = \sqrt{2\pi}$ for all integer k. This is an orthogonal basis that can be easily scaled to become orthonormal.

13.1. The Fourier transform

Obtaining $f(x)$ from its Fourier coefficients is called the **inverse Fourier transform**. Moreover, the *Parseval equality* reads

$$\frac{1}{2\pi}\int_{-\pi}^{\pi}|f(x)|^2 dx = \sum_{k=-\infty}^{\infty}|c_k|^2, \tag{13.2c}$$

which is a statement of norm preservation between transform and its inverse transform.

Example 13.2. Here are two important instances where manipulating functions in the frequency domain provides great advantage and simplification. The first instance is that of **convolution**. For two real integrable functions $f(x)$ and $g(x)$ that are periodic on our canonical interval $[-\pi,\pi]$, define their convolution as

$$\psi(x) = \int_{-\pi}^{\pi} g(x-s)f(s)ds.$$

Evaluating such an integral may not seem like a piece of cake. However, writing

$$\psi(x) = \sum_{k=-\infty}^{\infty} c_k \phi_k(x),$$

we have

$$c_k = \frac{1}{2\pi}\int_{-\pi}^{\pi}\psi(x)e^{-\imath kx}dx = \frac{1}{2\pi}\int_{-\pi}^{\pi}\left(\int_{-\pi}^{\pi}g(x-s)f(s)ds\right)e^{-\imath kx}dx$$

$$= \frac{1}{2\pi}\int_{-\pi}^{\pi}\left(\int_{-\pi}^{\pi}g(\xi)e^{-\imath k\xi}d\xi\right)e^{-\imath ks}f(s)ds$$

$$= c_k^g\int_{-\pi}^{\pi}e^{-\imath ks}f(s)ds = 2\pi c_k^g c_k^f,$$

where $\xi = x-s$, the Fourier coefficients c_k^g and c_k^f are those of g and f separately, and we have used the periodicity of g to shift the interval of integration in ξ. Thus, we may apply the Fourier transform to f and g separately and obtain the coefficients for $\psi(x)$ as a simple product in frequency space.

Another very important simplification is obtained for the **derivative** function $g(x) = f'(x)$ of a given differentiable periodic function f. Notice that for such a function we can apply integration by parts (see page 368) to obtain

$$\frac{1}{2\pi}\int_{-\pi}^{\pi}g(x)e^{-\imath kx}dx = (\imath k)\frac{1}{2\pi}\int_{-\pi}^{\pi}f(x)e^{-\imath kx}dx = \imath k c_k. \tag{13.3}$$

Therefore, differentiation in the x-domain translates into a simple multiplication of c_k by $\imath k$ in the frequency domain! ∎

Often in applications, x stands for physical time. People then talk about switching back and forth between the time domain and the frequency domain. However, there are many applications (including the examples in Section 14.5) where x stands for a physical space variable. The frequency is occasionally called a **wave number** then. This does not change the transform or the numerical methods used to approximate it, only the terminology.

Specific exercises for this section: Exercises 1–4.

13.2 Discrete Fourier transform and trigonometric interpolation

Note: In Section 13.2 we stay with real basis functions, not only for the sake of those readers who have an uneasy time imagining (pardon the pun) complex functions but also because we wish to smoothly continue the discussion from previous chapters. The complex basis functions retake center stage in Section 13.3.

For a simple function f, it is possible to calculate the Fourier coefficients a_k and b_k of (13.1) exactly. More generally, however, such integrals must be approximated by discrete sums. This brings us to discrete least squares approximation and interpolation (recall Section 10.1) and to the *discrete Fourier transform*.

It is notationally more convenient here to shift the interval of approximation from $[-\pi, \pi]$ to $[0, 2\pi]$. Thus, consider data $y_i = f(x_i)$ at m *equidistant* abscissae, where m is *even*, $m \geq n+1 = 2l$, so

$$x_i = \frac{2\pi i}{m}, \quad i = 0, 1, \ldots, m-1.$$

Discrete orthogonality and least squares approximation

In general, recalling Section 6.1, we would have to construct and solve the normal equations $B\mathbf{c} = \boldsymbol{\beta}$, where $B_{jk} = \sum_{i=0}^{m-1} \phi_j(x_i)\phi_k(x_i)$ and $\beta_k = \sum_{i=0}^{m-1} y_i \phi_k(x_i)$ for $j, k = 0, 1, \ldots, n$. But here it is possible to show that *discrete orthogonality* holds: we have

$$\frac{2}{m} \sum_{i=0}^{m-1} \cos(kx_i)\cos(jx_i) = \begin{cases} 0, & k \neq j, \\ 1, & 0 < k = j < m/2, \\ 2, & k = j = 0, \text{ or } k = j = m/2, \end{cases}$$

$$\frac{2}{m} \sum_{i=0}^{m-1} \sin(kx_i)\sin(jx_i) = \begin{cases} 0, & k \neq j, \\ 1, & 0 < k = j < m/2, \end{cases}$$

and likewise, $\frac{2}{m} \sum_{i=0}^{m-1} \sin(kx_i)\cos(jx_i) = 0$ for any relevant integers j and k; see Exercise 5.

So, just as with orthogonal polynomials in the continuous case of Section 12.2, the matrix B is diagonal here, and the solution process simplifies considerably, yielding $c_j = \beta_j / B_{jj}$, $j = 0, 1, \ldots, n$. For $2l < m$ we can write the best approximating trigonometric polynomial as

$$p_n(x) = \frac{a_0}{2} + a_l \cos(lx) + \sum_{k=1}^{l-1} a_k \cos(kx) + b_k \sin(kx),$$

$$a_k = \frac{2}{m} \sum_{i=0}^{m-1} y_i \cos(kx_i), \quad k = 0, 1, \ldots, l,$$

$$b_k = \frac{2}{m} \sum_{i=0}^{m-1} y_i \sin(kx_i), \quad k = 1, \ldots, l-1.$$

The discrete Fourier transform

In general, the number of data points must be large enough compared to the number of basis functions so that $n+1 = 2l \leq m$. The case $m = n+1$ yields **trigonometric polynomial interpolation**. In this case, with the same expressions for the coefficients a_k and b_k, we must divide a_l by 2 (see Exercise 6) obtaining the algorithm given on the current page.

> **Note:** For the rest of the present chapter we assume that there is a positive integer l such that
> $$m = n+1 = 2l.$$
> In particular, the underlying approximation process is trigonometric polynomial interpolation.

Obtaining the $2l$ trigonometric polynomial coefficients $a_0, \ldots, a_l, b_1, \ldots, b_{l-1}$ from the $2l$ data values **y** is called the *discrete Fourier transform* (DFT).

> **Algorithm: Discrete Fourier Transform (DFT).**
> Given equidistant data (x_i, y_i), $i = 0, 1, \ldots, n$, with $n = 2l - 1$ and $x_i = \frac{\pi i}{l}$, i.e., $2l$ input values y_i, the DFT defines the $2l$ coefficients
> $$a_k = \frac{1}{l} \sum_{i=0}^{n} y_i \cos(k x_i), \quad k = 0, 1, \ldots, l,$$
> $$b_k = \frac{1}{l} \sum_{i=0}^{n} y_i \sin(k x_i), \quad k = 1, \ldots, l-1.$$
> The interpolating trigonometric polynomial is then given by
> $$p_n(x) = \frac{1}{2}(a_0 + a_l \cos(lx)) + \sum_{k=1}^{l-1} [a_k \cos(kx) + b_k \sin(kx)].$$
> Furthermore, the *inverse DFT* gives back the data values in terms of the Fourier coefficients as
> $$y_i = p_n(x_i) = \frac{1}{2}(a_0 + a_l \cos(l x_i)) + \sum_{k=1}^{l-1} [a_k \cos(k x_i) + b_k \sin(k x_i)]$$
> for $i = 0, 1, \ldots, n$.

Periodic grid

Let us emphasize that the DFT setting assumes periodicity. In fact, it is useful to imagine the points $\{x_i\}_{i=0}^n$ as sitting on a circle, rather than on a straight line, having x_{n+1} identified with x_0. Thus, x_{n+2} is identified with x_1, x_{-1} is identified with x_n, and so on. See Figure 13.2. In MATLAB type `help circshift` for a potentially useful command.

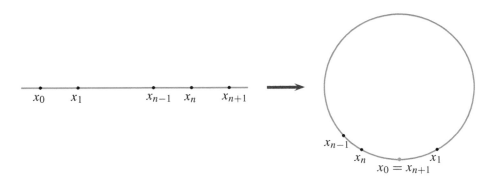

Figure 13.2. *For the discrete Fourier transform, imagine the red mesh points located on a blue circle with $x_0 = x_{n+1}$ closing the ring. Thus, x_n is the left (clockwise) neighbor of x_0, x_1 is the right (counterclockwise) neighbor of x_{n+1}, and so on.*

Note: As mentioned in Section 13.1, the Fourier transform provides much more than yet another interpolation method. By the same token, for various purposes and uses it is also important to assess its approximation qualities in a variety of situations. This latter aspect is what we concentrate on for the rest of the present section, using a set of examples.

Example 13.3. For $l = 2$ and the interval $[0, 2\pi]$, the interpolation abscissae are $x_i = (0, \frac{\pi}{2}, \pi, \frac{3\pi}{2})$. Consider the approximation for the "hat function"

$$f(x) = \begin{cases} x, & 0 \leq x \leq \pi, \\ 2\pi - x, & \pi < x \leq 2\pi. \end{cases}$$

Since this function is symmetric about $x = \pi$, which is the middle of the interval, we expect $b_k = 0$ for all k. Indeed, upon carrying out the calculations we obtain

$$a_0 = \frac{2}{4}\left(\frac{\pi}{2} + \pi + \frac{\pi}{2}\right) = \pi,$$

$$a_1 = \frac{2}{4}(0 - \pi + 0) = -\frac{\pi}{2},$$

$$b_1 = \frac{2}{4}\left(\frac{\pi}{2} + 0 - \frac{\pi}{2}\right) = 0,$$

$$a_2 = \frac{2}{4}\left(-\frac{\pi}{2} + \pi - \frac{\pi}{2}\right) = 0.$$

The function f and its approximation are plotted in Figure 13.3.

Note that the interpolation at the extra point $x = 2\pi$ is due to periodicity of $p_n(x)$ coupled with the fact that $f(2\pi) = f(0)$. On the flip side, we would have had $p_n(2\pi) = p_n(0)$ even if $f(2\pi) \neq f(0)$, because $x_5 = 2\pi$ is identified with $x_0 = 0$ by the algorithm, so beware of using such interpolation absentmindedly! We return to this point following Example 13.7.

13.2. Discrete Fourier transform and trigonometric interpolation

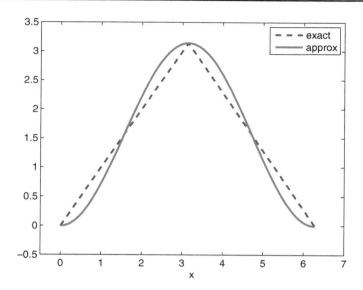

Figure 13.3. *Trigonometric polynomial interpolation for the hat function with $p_3(x)$.*

The quality of this approximation is not high, particularly because of the discontinuity in f' at $x = \pi$, a property that stands in stark contrast to the infinite smoothness of the approximant $p_n(x)$, written as $p_n \in C^\infty[0, 2\pi]$. If we increase n, then the maximum trigonometric polynomial interpolation error is expected to decrease for a while only like $1/n$.

The approximation accuracy can be much more impressive if f is smooth and (almost) periodic, as the next two examples demonstrate. ∎

Simple DFT code

The following two short MATLAB functions implement the algorithm for DFT given on page 389 in a straightforward way:

```
function [a0,a,b] = dft1e(y)
%
%  function [a0,a,b] = dft1e(y)
%
%  Given m values of y, m even, these values are interpreted as
%  function values at m equidistant points on [0, 2*pi).
%  Construct the DFT

y = y(:); m = length(y); l = m / 2;
pi2 = 2*pi; pi2m = pi2/m;
x = 0: pi2m: (m-1)*pi2m;

a0 = sum(y)/l;
for k=1:l
  co = cos(k*x); si = sin(k*x);
  a(k) = (co*y)/l;
  if k < l
    b(k) = (si*y)/l;
  end
```

```
end

function yy = dft2e(xx,a0,a,b)
%
%    function yy = dft2e(xx,a0,a,b)
%
%    Given Fourier coefficients, evaluate trigonometric
%    polynomial at xx.
%
l = size(a,2);
yy = a0/2 * ones(size(xx));
for k=1:l-1
   yy = yy + a(k)*cos(k*xx) + b(k)*sin(k*xx);
end
yy = yy + a(l)/2 * cos(l*xx);
```

Although the efficiency of this DFT code is not optimal, it has the great advantage of simplicity and transparency.

Example 13.4. Consider the function

$$g(t) = t^2(t+1)^2(t-2)^2 - e^{-t^2}\sin^2(t+1)\sin^2(t-2)$$

on the interval $[-1,2]$. To apply the discrete Fourier transform we must first rescale and translate the independent variable to $x = \frac{2\pi}{3}(t+1)$ so that $x \in [0, 2\pi]$. Then in the above notation, $f(x) = g(t)$, where $t = \frac{3}{2\pi}x - 1$. So, for a given $n = 2l - 1$ we have the abscissae $x_i = \frac{\pi i}{l}$ (hence $t_i = \frac{3}{2\pi}x_i - 1 = \frac{3i}{2l} - 1$) and data $y_i = g(\frac{3i}{2l} - 1)$. Here is a MATLAB program that does the job. The printing and plotting instructions are sufficiently primitive so that you can safely concentrate on other aspects.

```
% prepare data y
l = 16; % input parameter
n = 2*l-1; pi2 = 2*pi;
x = 0 : pi/l : n*pi/l;    % abscissae on [0,pi2]
t = 3/pi2*x - 1;          % abscissae on [-1,2]
y = t.^2 .* (t+1).^2 .* (t-2).^2 - ...
    exp(-t.^2) .* sin(t+1).^2 .* sin(t-2).^2;

% find dft real coefficients
[a0,a,b] = dft1e(y);

% interpolate on fine mesh, plot and find max interpolation error
xx = 0:.01*pi:pi2; tt = 3/pi2*xx - 1;
yexact = tt.^2 .* (tt+1).^2 .* (tt-2).^2 - ...
         exp(-tt.^2) .* sin(tt+1).^2 .* sin(tt-2).^2;
yapprox = dft2e(xx,a0,a,b);

% display results
plot(tt,yexact,'b--',tt,yapprox)
axis([-1 2 -2 6])
legend('exact','approx')
xlabel('t')
err_max = max(abs(yexact-yapprox)./(abs(yexact)+1))
```

13.2. Discrete Fourier transform and trigonometric interpolation

The results using $l = 2, 4, 8$, and 16 are plotted in Figure 13.4. The improvement as the number of basis functions and data increases is evident. The error in maximum norm decreases like n^{-2} over this range, and faster if we continue to increase n. ∎

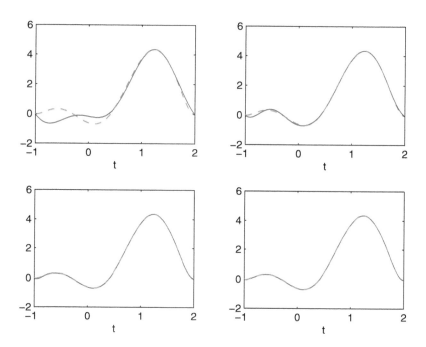

Figure 13.4. *Trigonometric polynomial interpolation for a smooth function with $p_3(t)$ (top left), $p_7(t)$ (top right), $p_{15}(t)$ (bottom left), and $p_{31}(t)$ (bottom right). The approximated function is plotted in dashed green.*

Spectral accuracy

If the approximated function $f(x)$, or $g(t)$, is infinitely smooth and periodic, then it can be shown that the convergence of trigonometric polynomial interpolation is very fast: as n increases, the error decreases faster than n^{-q} for *any integer q*. This is called **spectral accuracy**.

Example 13.5. Continuing Example 13.4, with the prospect of spectral accuracy it is natural to ask why, then, does the error in Figure 13.4 behave like that of a second order method?!

To understand this we have to ask first if $g(t)$ is really a periodic function. In fact, so far it has been defined (and used) only on the interval $[-1, 2]$, where it satisfies $g(-1) = g(2)$, $g'(-1) = g'(2)$. If g is to be periodic, then we must extend it to the entire real line by replicating it. Thus, we define $g(t) = g(t - 3)$ for $t \in [2, 5]$; $g(t) = g(t + 3)$ for $t \in [-4, -1]$; and so on. The extended g is indeed a periodic function with period $\tau = 3$. But now, consider the smoothness of the extended g at the point $t = 2$. Approaching from the right, $t \downarrow 2$, we have by construction $g^+(2) = g^+(-1)$, and the same holds for any derivative $\frac{d^p g}{dt^p}$. So, the number of continuous derivatives that g has equals the largest $q - 1$ such that $\frac{d^p g}{dt^p}(2) = \frac{d^p g}{dt^p}(-1)$, $p = 0, 1, \ldots, q - 1$. In Example 13.4 we have $q = 2$, and this yields the observed second order convergence for small n. If q is larger, then the relative approximation error for the same n will be smaller.

For instance, replacing g of Example 13.4 by

$$g(t) = t^2(t+1)^r(t-2)^r - e^{-t^2}\sin^r(t+1)\sin^r(t-2), \quad -1 \leq t \leq 2,$$

and repeating the same program, we obtain the errors recorded in the following table:

n	Error ($r=2$)	Error ($r=10$)
31	9.8e-3	2.0e-5
63	2.4e-3	2.2e-8
127	3.9e-4	2.5e-11

Note from the script in Example 13.4 that these values are mixed relative-absolute maximum errors and that in this example, $q = r$, meaning g and its first $r-1$ derivatives match at the endpoints. Thus, replacing $r = 2$ by $r = 10$ yields a much faster convergence, where the error decreases like n^{-10} before roundoff error starts to dominate. For an infinitely smooth periodic function, q is infinitely large (meaning, we reach roundoff error kingdom very rapidly). ∎

Fourier and Chebyshev interpolation schemes

From the families of interpolants considered in Chapters 10 and 11, the closest to the trigonometric polynomial interpolation considered here is *polynomial interpolation at Chebyshev points*. Like here, this interpolant is global and infinitely smooth, and it also exhibits spectral accuracy under appropriate conditions. How does it fare on the examples we have seen thus far in this chapter?

Example 13.6. We have applied our programs from Section 10.6 to obtain Chebyshev interpolants for the functions f or g of Examples 13.1, 13.3, 13.4, and 13.5. Here is what transpires.

Going in reverse order of the examples, the corresponding results to those in the table of Example 13.5 are as follows:

n	Error ($r=2$)	Error ($r=10$)
31	1.3e-13	6.5e-6
63	2.1e-15	2.5e-14
127	3.7e-15	6.4e-14

These results are much better than those obtained with the DFT, and the spectacular spectral accuracy of Chebyshev interpolation shows through. Mild roundoff error already dominates the values for $n = 127$. Indeed, the main selling point of Chebyshev interpolation vs. Fourier is that the former does not require the approximated function to be periodic! Thus, the results for larger r are not better here than those for the smaller r; in fact, they are slightly worse because the derivatives of the interpolated function are larger.

For the hat function of Example 13.3, the results using Chebyshev polynomial interpolation are comparable to and even slightly worse than those of the Fourier interpolant. The error decreases like $1/n$ here, too. This is to be expected because we are still trying to approximate a function with a kink (a discontinuous first derivative) by an infinitely smooth interpolant.

Finally, applying Chebyshev interpolation at $n+1 = 10, 20, \ldots, 350$ points to the function

$$f(x) = \cos(3x) - .5\sin(5x) + .05\cos(104x)$$

of Example 13.1 produces two observations. The first is that for intermediate values such as 50, 100, or 200 points, the maximum error remains almost constant at about .09, suddenly descending quickly into rounding unit (machine precision) level around $n = 350$. This sudden improvement occurs when the high frequencies of $f(x)$ finally get resolved, just like in Example 10.7.

The other observation is possibly of even more interest: at $n + 1 = 10$ we observe "noise filtering," where the obtained polynomial approximates $\hat{f}(x) = \cos(3x) - .5\sin(5x)$ much more closely than it does $f(x)$. Chebyshev interpolation is in this sense close to the optimal Fourier interpolation! ∎

> **Note:** Let us summarize the comparison of Fourier and Chebyshev interpolation schemes.
> If the function $f(x)$ to be interpolated is smooth and periodic, and certainly if it is only given as sampled on a *uniform mesh*, then a trigonometric polynomial interpolation (Fourier) is preferable.
> On the other hand, if $f(x)$ is smooth and given anywhere on the interval of interest, but it is not periodic, then Chebyshev interpolation wins.

Discontinuities and nonperiodic functions

We have seen in Example 13.3 that a discontinuity in the derivative of the approximated function implies that Fourier interpolation is not very accurate. Not surprisingly, this gets worse when the interpolated function itself is discontinuous, as the following example demonstrates.

Example 13.7. Let $f(x)$ be the square wave function given by

$$f(x) = \begin{cases} 1, & \pi - 1 \leq x \leq \pi + 1, \\ 0, & \text{otherwise.} \end{cases}$$

This function is periodic but not smooth: it is discontinuous at $x = \pi \pm 1$.

The Fourier interpolant for $l = 64$ is depicted in Figure 13.5. There are spurious oscillations throughout most of the interval, and their amplitude is particularly large near the jump discontinuities. These overshoots are often referred to as the **Gibbs phenomenon**. Of course, approximating a discontinuity with infinitely differentiable basis functions is bound to yield some trouble, which here unfortunately looks like spurious waves rather than random error.

As n is increased further the approximation error does decrease, but slowly. The overshoots are still visible to the naked eye at $n = 10,001$, where the ℓ_2-error divided by $\sqrt{n}$ is significantly smaller than the maximum error. ∎

Suppose next that we are given a set of values $\{(x_i, y_i)\}_{i=0}^{n}$ with equidistant abscissae to interpolate. Denote $x_i = a + ih$, where $h = (b-a)/(n+1)$ for some interval $[a,b]$. Suppose further that $y_i = f(x_i)$, $i = 0, 1, \ldots, n$, for some smooth function $f(x)$. Now, if we are unlucky and $f(b) \neq f(a)$ (think, for example, of $f(b) = f(a) + 1$), then a Fourier interpolant $p_n(x)$ of the given data would try to follow the smooth $f(x)$ until just before the right end, but then it would also need to satisfy $p_n(b) = p_n(a) = f(a) \neq f(b)$. The resulting jump would generate spurious waves in $p_n(x)$, as in Example 13.7, polluting the approximation also in the interpolation range $[a, b-h]$.

There is a simple fix to this problem: an even extension. See Exercise 9. That would get rid of the bulk of the spurious oscillations, but the quality of the resulting approximation is still not very high.

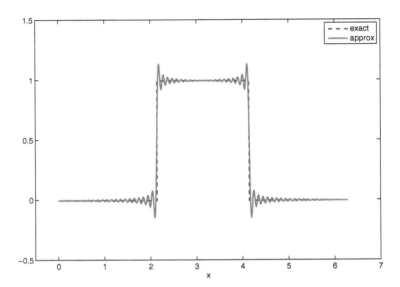

Figure 13.5. *Trigonometric polynomial interpolation for the square wave function on* $[0, 2\pi]$ *with* $n = 127$.

Inverse transform

Retrieving the $2l$ values y_i back from the Fourier coefficients as specified in the algorithm given on page 389 is called the **inverse DFT**. Using our function dft2e to obtain the inverse DFT is straightforward: simply identify xx there with x defined in our DFT function dft1e.

Note that the same elements $\cos(kx_i)$ and $\sin(kx_i)$ appear in both the transform and its inverse. One can take the transform of a given function or data in the space of x, manipulate the coefficients meaningfully in frequency space, and then transform back to the original space. Filtering a noisy signal as in Figure 13.1 is just one example of this powerful idea. Convolution and function differentiation are two others. This will become more transparent in Sections 13.3 and 14.5.

Specific exercises for this section: Exercises 5–9.

13.3 Fast Fourier transform

As noted before, the discrete Fourier transform is very popular in practical applications. Often, it must be carried out for large values of n, say, $n = 1023$, in order to capture rapidly varying phenomena.[51] However, carrying out the discrete Fourier transform, as we have done above, clearly costs $\mathcal{O}(n^2)$ operations.

Thus, even though our function dft1e which implements the DFT algorithm given on page 389 is short and simple to understand, it is too inefficient in many situations. A revolution in the field of signal processing resulted when Cooley and Tukey developed in 1965 an algorithm which costs only $\mathcal{O}(n \log_2 n)$ operations. For $n = 1023$, for example, this roughly cuts 1 million operations to about 3000. Here we must sound the usual warning about how misleading operation counts can generally be for a true assessment of efficiency in a modern computing environment, and yet, like for the basic direct linear algebra algorithms, this simplified complexity assessment again picks the

---

[51] As you will soon see, it is particularly convenient in this section to set $n = 2^k - 1$ with k integer.

13.3. Fast Fourier transform

winner. The resulting fast Fourier transform (FFT) is without a doubt one of the most important algorithms in scientific computing.

> **Note:** Be warned that the derivation below is somewhat technical, not to say occasionally mysterious. However, the resulting algorithm is short and simple.

DFT in complex basis functions

We continue to restrict n to be odd and write $m = n+1 = 2l$. In fact, it is convenient to assume here that m is a power of 2, i.e., that there is a positive integer p such that $m = 2^p$. For instance, think of $m = 8$. Then $p = 3$, $n = 7$, and $l = 4$. Furthermore, we continue to use the notation

$$x_i = \frac{2\pi i}{m}, \quad i = 0, 1, \ldots, n.$$

Explaining the FFT algorithm is much simpler and more intuitive if we switch back to the complex basis functions $\phi_k(x) = e^{\iota k x}$ defined in Section 13.1. In fact, along the lines of (13.2a)–(13.2b) it is easily seen that the DFT can be neatly expressed as

$$c_j = \frac{1}{m} \sum_{i=0}^{m-1} y_i e^{-\iota j x_i}, \quad -l \leq j \leq l-1.$$

Using these coefficients the interpolant can be written as

$$p_n(x) = \sum_{j=-l}^{l-1} c_j e^{\iota j x}.$$

In particular, the discrete inverse transform is given by

$$y_i = p_n(x_i) = \sum_{j=-l}^{l-1} c_j e^{\iota j x_i}, \quad i = 0, 1, \ldots, m-1.$$

For the purpose of deriving the FFT it is customary and more convenient to write the transform as

$$\hat{y}_k = \sum_{i=0}^{m-1} y_i e^{-\iota k x_i}, \quad 0 \leq k \leq m-1.$$

The shift in the index relates the coefficients c_j to the computed values $\hat{y}_k$ by

$$(c_{-l}, c_{-l+1}, \ldots, c_{-1}, c_0, c_1, \ldots, c_{l-1}) = \frac{1}{m}(\hat{y}_l, \hat{y}_{l+1}, \ldots, \hat{y}_{m-1}, \hat{y}_0, \hat{y}_1, \ldots, \hat{y}_{l-1}).$$

The inverse transform is given by

$$y_i = \frac{1}{m} \sum_{k=0}^{m-1} \hat{y}_k e^{\iota k x_i}, \quad i = 0, 1, \ldots, m-1.$$

Roots of unity

Let us define the *m*th *root of unity* by

$$\omega = \omega_m = e^{-\imath 2\pi/m}.$$

There are m different complex numbers ω^j, $j = 0, 1, \ldots, m-1$, all satisfying $(\omega^j)^m = 1$. Note also that

$$(\omega_m)^2 = e^{-\imath 2\pi/(m/2)} = \omega_l, \quad 2l = m.$$

Example 13.8. Figure 13.6 displays the eight roots of unity for the case $m = 8$.

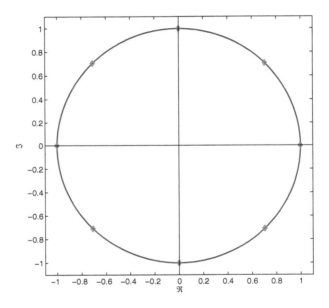

Figure 13.6. *The unit circle in the complex plane is where the values of $e^{\imath\theta}$ reside. The $m = 8$ roots of the polynomial equation $\theta^8 = 1$, given by $e^{-\imath 2j\pi/8}$, $j = 0, 1, \ldots, 7$, are displayed as red diamonds.*

Note that we could have defined the roots as $e^{\imath 2j\pi/8}$ instead of $e^{-\imath 2j\pi/8}$, obtaining the same eight values in a different order. ∎

Observe that $\omega^{qm+r} = \omega^r$ for any integer q and $0 \leq r \leq m-1$. Thus, any number of the form $e^{\pm \imath k x_i}$, where x_i is given as above and k is an integer, must equal one of these m roots. We have

$$e^{-\imath k x_i} = e^{-\imath 2\pi k i/m} = \omega_m^{ki}.$$

So, what multiplies y_i in the transform sum for $\hat{y}_k$ is equal to ω_m raised to the power ki.

13.3. Fast Fourier transform

We can therefore write the DFT as

$$\begin{pmatrix} \hat{y}_0 \\ \hat{y}_1 \\ \vdots \\ \hat{y}_{m-2} \\ \hat{y}_{m-1} \end{pmatrix} = W \begin{pmatrix} y_0 \\ y_1 \\ \vdots \\ y_{m-2} \\ y_{m-1} \end{pmatrix},$$

where the elements of the $m \times m$ matrix W are all integer powers, from among $0, 1, \ldots, m-1$, of ω. There are only m different such values in total.

Moreover, from the form of the inverse transform, i.e., the formula for y_i above, we see that W^{-1} can be explicitly written (thus the techniques of Chapters 5 and 7 are irrelevant here), and its elements are also all from among the m integer powers of ω. Indeed, a simple variation of whatever algorithm that is used for the DFT can be applied for its inverse transform, too.

A divide-and-conquer approach

The FFT algorithm is based on a *divide-and-conquer* approach. From the definition of ω and the DFT expression, we can write

$$\hat{y}_k = \sum_{i=0}^{m-1} y_i \omega_m^{ik} = \sum_{j=0}^{l-1} \left(y_{2j} \omega_m^{(2j)k} + y_{2j+1} \omega_m^{(2j+1)k} \right)$$

$$= \sum_{j=0}^{l-1} y_{2j} \omega_l^{jk} + \omega_m^k \sum_{j=0}^{l-1} y_{2j+1} \omega_l^{jk}.$$

We have therefore decomposed the sum into two smaller, similar sums for the even and the odd data components, separately. Let us write this as

$$\hat{y}_k = \tilde{y}_k^{\text{even}} + \omega_m^k \tilde{y}_k^{\text{odd}}.$$

The problems for the odd and even components are only of size $l = m/2$. Note that $l = 2^{p-1}$.

But the good news does not end here. Since $\omega_l^{jl} = 1$ with any integer j, we have for each $0 \leq \tilde{k} < l$ that

$$\tilde{y}_{l+\tilde{k}}^{\text{even}} = \sum_{j=0}^{l-1} y_{2j} (\omega_l^{jl}) \omega_l^{j\tilde{k}} = \tilde{y}_{\tilde{k}}^{\text{even}},$$

and similarly for $\tilde{y}_{l+\tilde{k}}^{\text{odd}}$. Finally, observe that $\omega_m^{l+\tilde{k}} = -\omega_m^{\tilde{k}}$, because $\omega_m^l = e^{-\iota \pi} = -1$. Therefore, we can write

$$\hat{y}_{\tilde{k}} = \tilde{y}_{\tilde{k}}^{\text{even}} + \omega_m^{\tilde{k}} \tilde{y}_{\tilde{k}}^{\text{odd}},$$
$$\hat{y}_{\tilde{k}+l} = \tilde{y}_{\tilde{k}}^{\text{even}} - \omega_m^{\tilde{k}} \tilde{y}_{\tilde{k}}^{\text{odd}},$$

for $\tilde{k} = 0, 1, \ldots, l-1$.

This reasoning is repeated recursively, until at the bottom level the obtained subproblems have a sufficiently small size to be dealt with directly using a (small and) constant number of flops. There are $\mathcal{O}(m)$ bottom level subproblems. Such a recursion must be applied essentially p times, i.e., there are $\mathcal{O}(\log_2 m)$ recursion levels. The resulting operation count for the complete calculation of the Fourier coefficients is $\mathcal{O}(mp) = \mathcal{O}(m \log_2 m) = \mathcal{O}(n \log_2 n)$.

Here is an implementation of the above derivation:

```
function yh = myfft(y)
%
%   function yh = myfft(y)
%
%   Apply FFT, assuming the length m of y is a power of 2 and even.
m = length(y);
if m == 2
  % directly define coefficients
  yh(1) = y(1)+y(2); yh(2) = y(1)-y(2);
else
  % recursion step
  l = m/2;
  yeven = y(1:2:m-1); yodd = y(2:2:m);
  ceven = myfft(yeven); codd = myfft(yodd);
  omegam = exp(-i*2*pi/m); omegap = omegam.^(0:1:l-1);
  yh(1:l) = ceven + omegap .* codd;
  yh(l+1:m) = ceven - omegap .* codd;
end
```

Interpolating with the FFT

MATLAB of course has a highly optimized FFT function, called `fft`. Its input and output are the same as our `myfft`, but the number of data values m may be any positive integer and the canned routine can be much more efficient than ours.

Suppose that a call to `fft` or `myfft` results in a coefficient array `yh`. To retrieve the real coefficients necessary for the representation of Section 13.2, use the instructions

```
a0 = yh(1)/l;
a  = (yh(2:l)+yh(m:-1:l+2))/m;
b  = i*(yh(2:l)-yh(m:-1:l+2))/m;
a(l) = yh(l+1)/l;
```

The calculated coefficients can then be utilized as input for our routine `dft2e` presented in Example 13.4 to evaluate the interpolating trigonometric polynomial; see Exercises 10 and 11. This latter evaluation is not highly efficient anymore. (Why?)

The IFFT

Let us write the m data values as a vector $\mathbf{y}$ and the corresponding m Fourier coefficients as a vector $\hat{\mathbf{y}}$. Recall that the operations described earlier for evaluating the Fourier coefficients $\hat{y}_k$ can be compactly written as

$$\hat{\mathbf{y}} = W \mathbf{y}$$

for a corresponding $m \times m$ matrix W. This matrix has a special form, but it is *not* sparse, unlike matrices discussed in Sections 5.6 and 5.7 and throughout Chapter 7. The FFT can then be regarded as a **fast matrix-vector product** technique, where the matrix-vector multiplication is achieved in much fewer than the usual m^2 operations.

13.3. Fast Fourier transform

The inverse operation, written as
$$\mathbf{y} = W^{-1}\hat{\mathbf{y}},$$
corresponds to the *inverse Fourier transform*. As we have seen it can be realized using similar techniques that do not involve matrix inversion; see Exercise 12. This is useful in cases where the coefficients c_k are first manipulated in frequency space and then the corresponding solution values y_i in state space are recovered. An instance arises when differentiating a function. The latter topic is further discussed in Section 14.5. Needless to say, there is a fast matrix-vector multiplication algorithm for W^{-1} as well, called IFFT. In MATLAB, use `ifft`.

More than one independent variable

The discussion in this chapter so far, as in preceding chapters, is associated with approximating a function of one variable, $f(x)$. But in many applications the data relates to values of a function in more than one variable, say, $f(x, y)$.

If the given data consists of values in a two-dimensional array that correspond to a function defined on a rectangular domain, then extending the Fourier transform is conceptually simple: we apply the transform in one direction, say, for each row of the data separately, and then similarly in the other direction. In MATLAB the command `fft2` achieves this.

However, if the underlying domain has a significantly more complicated geometry (see Section 11.6), then Fourier methods lose much of their attraction.

Example 13.9 (image deblurring). This example describes a case study involving several weighty issues, and full details of the following description are well beyond the scope of our book. We concentrate to the extent possible on what is relevant in the present context, namely, the evaluation of a convolution in two space dimensions using FFT.

Recall from Example 13.2 that one operation which greatly simplifies in the Fourier frequency domain is **convolution**. In blurred images there is typically a blurring kernel $K(x, y)$ (say, because the camera is in motion) convolving the clear image $u(x, y)$. The observed image can be written as

$$K(x, y) * u(x, y) = \int_0^1 \int_0^1 K(x - x', y - y') u(x', y') dx' dy'.$$

So, an observed blurred image $b(x, y)$ is related to the ideal sharp image $u(x, y)$ by $b = K * u + \epsilon$, where $\epsilon = \epsilon(x, y)$ is an ever-present noise. The task is then to "invert" the blurring kernel and at the same time remove the noise while still retaining, indeed potentially enhancing, image edges. See Figure 13.7.

The image is given on a square or rectangular grid of pixels, and we discretize the convolution integral on this grid: the details are unimportant in the present context, but let us say in passing that they relate to techniques considered in Section 15.2. The result of discretizing $K(x, y) * u(x, y)$ is written as $J\mathbf{v}$, where $\mathbf{v}$ is an array containing our approximations for grayscale levels $u_{i,j}$ at the grid pixels, reshaped as a vector of unknowns, and J is a matrix corresponding to the discretization of the kernel K. The images in Figure 13.7 are 270×270; hence J is a dense (*not* sparse, i.e., unlike in Section 5.6 and Example 7.1) $m \times m$ matrix with $m = 270^2$. Therefore, a matrix-vector product involving J could be costly if not performed efficiently. Moreover, for an image taken by a common SLR digital camera nowadays, m is so large that we cannot even store J in fast memory.

The actual algorithm used to deblur the image in Figure 13.7 solves the optimization problem

$$\min_{\mathbf{v}} \frac{1}{2} \|J\mathbf{v} - \mathbf{b}\|_2^2 + R(\mathbf{v}).$$

Here $\mathbf{b}$ is the observed data, reshaped as a vector like $\mathbf{v}$. The first term is thus a measure of data fitting in the ℓ_2-norm, squared for convenience. In the second term, R is a **regularization** operator.

(a) Blurred, noisy image.   (b) Deblurred image.

Figure 13.7. *Example* 13.9*: an observed blurred image b, and the result of a deblurring algorithm applied to this data. (Images courtesy of H. Huang [43].)*

It is there for several reasons: (i) solving $J\mathbf{v} = \mathbf{b}$ is an ill-conditioned problem (recall Sections 1.3 and 5.8); (ii) there is noise to handle by ensuring that the recovered image does not look rough everywhere; and (iii) smoothing across edges should be avoided. The full development of this regularization term again falls outside the scope of our presentation. Suffice it to say that R is not quite quadratic in $\mathbf{v}$, and it functions to smooth the image anisotropically, or selectively.

The solution process of this minimization problem (see Sections 9.2 and 7.4) involves at each iteration several matrix-vector multiplications of the form $J\mathbf{w}$ or $J^T\mathbf{w}$ for various encountered vectors $\mathbf{w}$. We thus concentrate below on achieving the effect of multiplying a vector by J, both rapidly and without ever forming J.

Fortunately, J arises as a discretization of a convolution process. Reverting to the pixel grid rather than a vector notation, we can write

$$(K * u)_{i,j} = \sum_{k=0}^{\nu-1}\sum_{l=0}^{\nu-1} K_{i-k,j-l} u_{k,l}, \quad 0 \leq i, j < \nu,$$

where $\nu = \sqrt{m}$. So in Figure 13.7, $\nu = 270$. In general, the discrete $\{K_{i,j}\}_{i,j=0}^{\nu-1}$ is two-dimensional periodic, defined as

$$K_{i,j} = K_{i',j} \quad \text{whenever} \quad i = i' \bmod \nu,$$
$$K_{i,j} = K_{i,j'} \quad \text{whenever} \quad j = j' \bmod \nu.$$

This allows us to consider $K \in \mathcal{R}^{\nu \times \nu}$, because by periodic extension we can easily get (ν, ν)-periodic arrays K^{ext} for which

$$K_{i,j}^{ext} = K_{i,j} \quad \text{whenever} \quad 0 \leq i, j < \nu.$$

Thus, the double FFT algorithm implemented in `fft2` may be applied. Denoting its operation by $\mathcal{F}$ and the operation of the inverse transform implemented in `ifft2` by $\mathcal{F}^{-1}$, we obtain the effect of the product $J\mathbf{v}$ by calculating

$$K^{ext} * u = \nu \mathcal{F}^{-1}\{\mathcal{F}(K) .* \mathcal{F}(u)\},$$

13.3. Fast Fourier transform

where .∗ denotes componentwise matrix-matrix multiplication. This multiplication corresponds to convolution in the frequency domain; recall Example 13.2.

The procedure just outlined in effect circumvents the use of the large matrix J, and it reduces the cost of forming $J\mathbf{v}$, which is $\mathcal{O}(m^2)$, by employing an algorithm that requires only $\mathcal{O}(m \log_2 m)$ operations. A fast, storage-efficient overall method results! ∎

The discrete cosine transform

If the function to be approximated or manipulated is real and there is no operation in the frequency domain that involves complex variables, then it makes sense to attempt to construct the Fourier transform based on cosine functions alone: the imaginary part of $e^{\imath kx}$ is discarded. This leads to the discrete cosine transform (DCT),[52] a highly popular variant of DFT. In MATLAB, check out `dct`.

> **Algorithm: Discrete Cosine Transform (DCT).**
> Given m data values y_i, $i = 0, 1, \ldots, n$, set $m = n+1$ and define
>
> $$x_i = \frac{\pi(i+1/2)}{m}, \quad i = 0, 1, \ldots, n.$$
>
> The DCT defines the m coefficients
>
> $$a_k = \frac{2}{m} \sum_{i=0}^{n} y_i \cos(k x_i), \quad k = 0, 1, \ldots, n.$$
>
> The *inverse DCT* gives back the data values in terms of the transform coefficients as
>
> $$y_i = \frac{1}{2}a_0 + \sum_{k=1}^{n} a_k \cos(k x_i), \quad i = 0, 1, \ldots, n.$$
>
> The interpolating trigonometric polynomial is given by
>
> $$p_n(x) = \frac{1}{2}a_0 + \sum_{k=1}^{n} a_k \cos(k x).$$

We have seen in Example 13.3 that sine basis functions are dropped automatically when the approximated function is *even* (see Figure 13.3). Indeed, an important difference between DFT and DCT is that the former assumes a periodic boundary condition, so a given data sequence is automatically extended periodically, whereas the latter assumes an **even boundary condition**, meaning the given data series is automatically extended to describe an even function.

In the most popular version of this algorithm, on which we concentrate, the given data values $\{y_i\}_{i=0}^{n}$ are associated with abscissae

$$x_i = \frac{\pi(i+1/2)}{m}, \quad i = 0, 1, \ldots, n, \text{ where } m = n+1.$$

The extension to an even function is about $x_{-1/2}$ and $x_{n+1/2}$, corresponding to $x = 0$ and $x = \pi$, respectively. The algorithm is given on this page.

---
[52]This chapter is particularly rich in acronyms, which unfortunately are all frequently used in the literature.

Example 13.10. Let the data consist of the values $y_i = \ln(1 + 2\pi(i + 1/2)/32)$, $i = 0, 1, \ldots, 31$. The results may then be compared against those of Exercise 9. Applying the DCT algorithm leads to Figure 13.8.

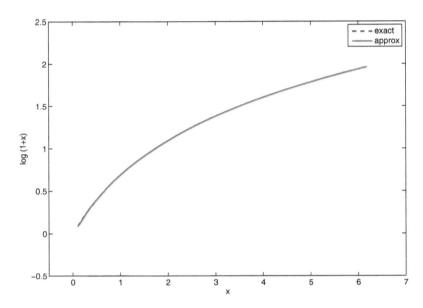

Figure 13.8. *Cosine basis interpolation for the function* $\ln(x + 1)$ *on* $[0, 2\pi]$ *with* $n = 31$.

The maximum absolute interpolation error is a decent 9.61e-3, even though the two end values of the data do not match. ∎

It is important to understand that the DCT effect can be achieved using DFT with twice as large data and transform arrays. There is a **fast cosine transform** version, too. In general, in appropriate circumstances there are no disadvantages to using DCT over DFT, and this makes it a method worthy of our additional discussion. Moreover, the DCT is closely related to the Chebyshev polynomials, fondly remembered from Sections 12.4 and 10.6, a connection that we will not explore further here.

Example 13.11. Continuing with Example 13.10, assume now that the same function is sampled at 1001 uniformly spaced points $x_i = ih$, $i = 0, \ldots, 1000$, with $h = .001 \times 2\pi$. Now, the cosine basis functions corresponding to high frequencies do not have a significant effect in the decomposition of such a smooth $f(x) = \ln(1 + x)$, i.e., the magnitudes of these high-frequency components will be small.

We next apply the DCT and then threshold in frequency space, setting the coefficients whose magnitude is less than 1% of the total to zero. This leaves only 45 nonzero components out of the original 1001, yielding a **sparse representation**. Transforming the modified component vector back using the inverse DCT results in a maximum difference of only 1.2e-3 between the obtained values and the original $\ln(1 + x_i), i = 0, 1, \ldots, 1001$. ∎

When digitizing images or music, an important question is that of **data compression**. For instance, a good SLR camera may produce a picture that consumes 10^7 pixels, and we want to compress it into an image that requires only 10^6 or even 10^5 pixels (it's only the neighbor's dog,

13.4 Exercises

after all) while still looking very close to the original. A straightforward SVD compression (see Figures 8.5 and 8.6) does not cut it because the nature of the compression exercise is not taken into account there. Transform techniques of the type considered in the present chapter do much better, for the reasons described and demonstrated in Example 13.11 as well as in the introduction and in Section 13.1. In particular, variants of the DCT are used in popular compression schemes such as mp3 and jpeg.

Specific exercises for this section: Exercises 10–13.

13.4 Exercises

0. Review questions

    (a) Define a periodic function. If f is periodic over an interval $[a,b]$, how can we scale and translate it to be periodic over $[-\pi,\pi]$?

    (b) What are trigonometric polynomials? What is their period?

    (c) Define the continuous Fourier series and Fourier transform and explain their importance in terms of frequencies.

    (d) Explain why a trigonometric basis approximation may not be a good idea if the approximated function $f(x)$ is not periodic.

    (e) Why should the interpolation data abscissae x_i in Section 13.2 *not* include both $i = 0$ and $i = 2l$?

    (f) What is the computational cost (in $\mathcal{O}$ terms) of computing the DFT?

    (g) What is the importance of the fast Fourier transform?

    (h) What are the roots of unity for $m = 16$?

    (i) Describe a divide and conquer approach.

    (j) When is the DCT preferable over DFT?

1. Verify that the trigonometric polynomial functions (13.1a) form an orthogonal basis, and find their normalizing constants, i.e., find $\int_{-\pi}^{\pi} \cos^2(kx)\,dx$ and $\int_{-\pi}^{\pi} \sin^2(kx)\,dx$.

2. Verify the expressions (13.1b)–(13.1c) for Fourier coefficients a_k and b_k given on page 384 for the case $n = 2l - 1$.

3. Find the continuous Fourier transform for $l = 50$, i.e., the coefficients of (13.1b)–(13.1c), for the function $f(x) = \cos(3x) - .5\sin(5x) + .05\cos(54x)$ on the interval $[-\pi,\pi]$. Let $v(x)$ denote the corresponding approximation.

    Note that most of the Fourier coefficients vanish for this example. Plot $f(x)$ and $v(x)$ and make observations.

    [Hint: Either you find this exercise easy or you are on the wrong track.]

4. Show that the formula (13.2a) holds upon defining the Fourier coefficients by (13.2b).

5. Verify the discrete orthogonality identities given on page 388 at the beginning of Section 13.2 by writing a short program that evaluates these expressions for even m and running it for $m = 4, 6,$ and 8.

6. Explain why we must replace a_l by $a_l/2$ when moving from approximation with $m > 2l$ to interpolation, where $m = 2l$ on page 389.

7. Write a short program to interpolate the hat function of Example 13.3 by trigonometric polynomials for $l = 2, 4, 8, 16,$ and 32. Measure maximum absolute error on the mesh `0:.01*pi:2*pi` and observe improvement that is only linear in l (or $n = 2l - 1$).

8. Following the procedure outlined in Example 13.4, construct and plot interpolating trigonometric polynomials for the Chebyshev polynomials $T_5(x)$ and $T_6(x)$ defined in Section 10.6. (Thus, the Chebyshev polynomials are not the interpolants; rather, they are being interpolated here, for a change.) Try using $l = 8$ and $l = 16$. What are your observations? Why is the error for the more oscillatory T_6 smaller than the error for T_5?

9. Consider using a DFT to interpolate the function $f(x) = \log(x+1)$ on the interval $[0, 2\pi]$ as in the examples of Section 13.2.

    (a) Construct and plot the interpolant on $[0, 2\pi]$ for $l = 16$ and $l = 32$.

    Explain why the results look unsatisfactory.

    (b) Consider an even extension of $f(x)$, defining

    $$g(t) = \begin{cases} f(t), & 0 \le t < 2\pi, \\ f(4\pi - t), & 2\pi \le t < 4\pi. \end{cases}$$

    Apply DFT interpolation to $g(t)$ and plot the results on $[0, 2\pi]$. Find maximum errors for $l = 16$ and $l = 32$. Are they better than before? Why?

10. Write down the connection between the coefficients c_k in Section 13.3 and the coefficients for the real representation a_k and b_k of Section 13.2.

11. Repeat the calculations of Examples 13.4 and 13.5 using the MATLAB function `fft`.

12. Write a short MATLAB function called `myifft` that has the same input (in simplest form) and output as the built-in `ifft`, restricting the number of data m to be a power of 2. Thus, for any given data array $\mathbf{y}$ of such a size fed as input to `myfft`, which yields $\hat{\mathbf{y}}$, your function should take as input $\hat{\mathbf{y}}$ and deliver the original $\mathbf{y}$ as output.

    Test your program on the function $g(t) = t^2(t+1)^2(t-2)^2 - e^{-t^2}\sin^2(t+1)\sin^2(t-2)$ of Example 13.4 sampled uniformly at 32 points on the interval $[-1, 2)$.

13. Compare the results reported in Example 13.10 and in particular Figure 13.8 to those obtained in Exercise 9. Explain similarities and differences.

13.5 Additional notes

Trigonometric interpolation and the Fourier transform appear in many different branches of science and engineering. For use in computer vision, see Horn [42]. For time series data analysis, see Brillinger [10]. For computerized tomography, see Natterer [56]. A theoretical computer science view of the FFT appears in Cormen, Leiserson, and Rivest [14]. None of these references lends itself to an easy, leisurely reading, though. The treatment in Vogel [73] of FFT use in convolution and image deblurring is more accessible.

Indeed, there are some pedagogical problems in exposing the material in this chapter. The concepts and programs presented in Sections 13.1 and especially 13.2 are straightforward (we think!), but the fast algorithm that everyone uses in practice is more opaque. This perhaps is not unlike the general algorithms for eigenvalues and SVD in Section 8.3; however, even presenting the principle of FFT in Section 13.3 requires an increased level of attention to indices. And then, making this into

13.5. Additional notes

a working program for any number of data points (not just $m = 2^p$), or extending to non-equally spaced data, or "windowing" to localize the effect of the transform, are undertakings leading to many technicalities. For details, see, for instance, Press et al. [59]. Most users treat the MATLAB function fft as a *black box* in a more fundamental sense than, say, the functions svd or spline, occasionally with detrimental consequences.

The discrete Fourier transform is intimately connected with the notion of *circulant matrices* and can indeed be represented and treated in traditional linear algebra ways, involving spectral decompositions and so on. See Davis [18] for a thorough description of circulant matrices. Example 13.9 puts such matrices into action; its details are described in [44].

Fourier transform techniques are also used in the numerical solution of partial differential equations, both for analysis (see, e.g., Ascher [3]) and in order to form highly accurate numerical methods called *spectral methods*. We will say more in Section 14.5; for now let us merely refer to Fornberg [26] and Trefethen [68].

An essential property of trigonometric polynomials is that their wavy, or oscillatory, behavior is uniform throughout the real line. This is not always what we want. The desire to have a better, or at least a more natural, localization of rapidly varying phenomena in the function to be approximated has led to another highly popular approximation basis, *wavelets*, which give rise to the **wavelet transform**. These basis functions are naturally used for **multiresolution** techniques in image processing and have successfully competed with Fourier transform techniques in that arena. For sources on wavelets see, e.g., Mallat [52] or Daubechies [16]. Be warned that these books are heavy.

Chapter 14
Numerical Differentiation

The need to differentiate a function $f(x)$ arises often in mathematical modeling. This chapter is devoted to techniques for approximating these operations numerically.

Although we all remember from calculus how to analytically evaluate derivatives of a function $f(x)$, there are reasons to do this numerically. One is that derivatives of very complicated functions may be required, and the process of obtaining them by hand may be prone to errors. Another reason is that the function $f(x)$ may not be known explicitly. For instance, formulas from numerical differentiation are frequently used in the numerical solution of differential equations.

The techniques considered in this chapter are at the heart of approximation methods for solving differential and integral equations, which are important and active areas of research.

To derive formulas for numerical (or approximate) differentiation, the basic tool used is *polynomial approximation*. If a polynomial $p_n(x)$ approximates a given function $f(x)$, then $p'_n(x)$ approximates $f'(x)$. In Section 14.1 we use Taylor series expansions at nearby points to derive basic numerical differentiation formulas. Some of these will probably be familiar to you from an earlier course.

Section 14.2 describes a general *extrapolation* technique that builds high order (more accurate) derivative approximations from low order ones in a simple, disciplined way.

The main drawback of the Taylor series approach is that it is ad hoc, in a sense that we promise to explain further on. In Section 14.3 we therefore consider an alternative derivation approach based on Lagrange polynomial interpolation.

There is a fundamental difficulty with numerical differentiation, in that roundoff error and other high-frequency noise are actually *enlarged* by the process, occasionally significantly so, regardless of which method from the first three sections is used. This is analyzed in Section 14.4.

The process of differentiation at a particular value of x is local. But if we do this for points on an entire mesh, then a global process of approximating the function f' on an interval results. This gives rise to *differentiation matrices* and global approximations, considered in the more advanced Section 14.5.

14.1 Deriving formulas using Taylor series

From calculus, the derivative of a function f at x_0 is defined by

$$f'(x_0) = \lim_{h \to 0} \frac{f(x_0+h) - f(x_0)}{h}.$$

This uses values of f near the point of differentiation, x_0. To obtain a formula approximating a first derivative, we therefore typically choose equally spaced points nearby (e.g., $x_0 - h, x_0, x_0 + h, \ldots$, for a small, positive h) and construct an approximation from values of the function $f(x)$ at these points.

The resulting discretization error, also called **truncation error**, shrinks as h shrinks, provided that the function $f(x)$ is *sufficiently smooth*. Throughout this section, indeed throughout this chapter and others, we obtain estimates for truncation errors which depend on some higher derivatives of $f(x)$. We will assume, unless specifically noted otherwise, that such high derivatives of f exist and are bounded: this is what we mean by "sufficiently smooth."

What determines an actual value for h depends on the specific application. If we really aim only at obtaining a good approximation at a particular point and f is available everywhere, then only roundoff error considerations would stop us from taking exceedingly small values for h (recall Example 1.3). But in the context of deriving formulas for use in the numerical solution of differential equations, an interval $[a,b]$ of a fixed size has to be traversed. This requires $\frac{b-a}{h}$ such h-steps, so the smaller the step size h the more computational effort is necessary. Thus, *efficiency* considerations would limit us from taking $h > 0$ to be incredibly small.

> **Note:** Throughout this chapter we make extensive use of the order notation defined on page 7.

Formula derivation

The local nature of the differentiation operator, in the sense that for a particular argument the value of the derivative depends on the value of the function in the vicinity of that argument, gives rise to the prominence of Taylor expansions (see page 5) as a viable approach for approximate differentiation. Note that, as in Section 10.7, we can think of the Taylor expansion merely as an osculating polynomial interpolant whose interpolation data are all given at one particular point.

The virtue of the Taylor series expansion approach is simplicity and ease of use. Its detractor is the lack of orderliness or generality: it is an ad hoc approach. Therefore we introduce it by examples, or instances. We start off with three instances of an approximation of the first derivative of a function. Then a fourth instance is provided for the approximation of a second derivative.

1. **Two-point formulas**

    Expanding $f(x)$ in a Taylor series about x_0 yields the **forward** and **backward** difference formulas, which both fall in the category of *one-sided* formulas.

    For instance, letting $x = x_0 - h$ we have

    $$f(x_0 - h) = f(x_0) - hf'(x_0) + \frac{h^2}{2}f''(\xi), \quad x_0 - h \leq \xi \leq x_0$$
    $$\Rightarrow f'(x_0) = \frac{f(x_0) - f(x_0 - h)}{h} + \frac{h}{2}f''(\xi).$$

    The formula $\frac{f(x_0) - f(x_0 - h)}{h}$ is first order accurate; i.e., it provides an approximation to $f'(x_0)$ such that the associated truncation error is $\mathcal{O}(h)$. This is the *backward difference formula*, since we use $x_0 - h$ as an argument for the formula, which can be thought of as proceeding backwards from the point we evaluate the derivative at, namely, x_0. The *forward* formula is the one derived in Example 1.2, and it reads $\frac{f(x_0 + h) - f(x_0)}{h}$.

14.1. Deriving formulas using Taylor series

2. **Three-point formulas**

    (a) A **centered** formula for $f'(x_0)$ is obtained by expanding about $x = x_0$ at both $x = x_0 + h$ and $x = x_0 - h$, obtaining

    $$f(x_0 + h) = f(x_0) + hf'(x_0) + \frac{h^2}{2}f''(x_0) + \frac{h^3}{6}f'''(\xi_1),$$

    $$f(x_0 - h) = f(x_0) - hf'(x_0) + \frac{h^2}{2}f''(x_0) - \frac{h^3}{6}f'''(\xi_2).$$

    Then subtracting the second from the first of these two expressions and solving for $f'(x_0)$ yields

    $$f'(x_0) = \frac{f(x_0 + h) - f(x_0 - h)}{2h} - \frac{h^2}{6}f'''(\xi), \quad x_0 - h \leq \xi \leq x_0 + h.$$

    The truncation error term is obtained from $\frac{h^2}{12}(f'''(\xi_1) + f'''(\xi_2))$ by using the Intermediate Value Theorem (see page 10) to find a common ξ between ξ_1 and ξ_2. This three-point formula (namely, the first term on the right-hand side) is second order accurate, so its accuracy order is higher than the order of the previous ones, although it also uses only two evaluations of f. However, the points are more spread apart than for the forward and backward formulas.

    (b) A higher order one-sided formula can be obtained using similar techniques and can be written as

    $$f'(x_0) = \frac{1}{2h}(-3f(x_0) + 4f(x_0 + h) - f(x_0 + 2h)) + \frac{h^2}{3}f'''(\xi),$$

    where $\xi \in [x_0, x_0 + 2h]$. A similar formula using $-h$ instead of h can readily be obtained as well.

    Once again, this formula is second order. Note that it is expected to be less accurate than its centered comrade, in that the constant of the $\mathcal{O}(h^2)$ error term is larger, even though they are both second order.

3. **Five-point formula**

    Higher order approximations of the first derivative can be obtained by using more neighboring points. For instance, we derive a centered fourth order formula by writing

    $$f(x_0 \pm h) = f(x_0) \pm hf'(x_0) + \frac{h^2}{2}f''(x_0) \pm \frac{h^3}{6}f'''(x_0)$$
    $$+ \frac{h^4}{24}f^{(iv)}(x_0) \pm \frac{h^5}{120}f^{(v)}(x_0) + \frac{h^6}{720}f^{(vi)}(x_0) + \mathcal{O}(h^7),$$

    $$f(x_0 \pm 2h) = f(x_0) \pm 2hf'(x_0) + 2h^2 f''(x_0) \pm \frac{8h^3}{6}f'''(x_0)$$
    $$+ \frac{16h^4}{24}f^{(iv)}(x_0) \pm \frac{32h^5}{120}f^{(v)}(x_0) + \frac{64h^6}{720}f^{(vi)}(x_0) + \mathcal{O}(h^7).$$

    Subtracting the pair $f(x_0 \pm h)$ from each other and likewise for the pair $f(x_0 \pm 2h)$ leads to two centered second order approximations to $f'(x_0)$, with the truncation error for the first being four times smaller than for the second. (Can you see why without any further expansion?) Some straightforward algebra subsequently verifies that the formula

$$f'(x_0) \approx \frac{1}{12h}(f(x_0-2h) - 8f(x_0-h) + 8f(x_0+h) - f(x_0+2h))$$

has the truncation error

$$e(h) = \frac{h^4}{30} f^{(v)}(\xi).$$

4. **Three-point formula for the second derivative**

The same approach can be used to generate approximations to higher order derivatives of f. For example, if we add together the two expansions

$$f(x_0+h) = f(x_0) + hf'(x_0) + \frac{h^2}{2}f''(x_0) + \frac{h^3}{6}f'''(x_0) + \frac{h^4}{24}f^{(iv)}(\xi_1),$$

$$f(x_0-h) = f(x_0) - hf'(x_0) + \frac{h^2}{2}f''(x_0) - \frac{h^3}{6}f'''(x_0) + \frac{h^4}{24}f^{(iv)}(\xi_2),$$

then all the odd powers of h cancel out and we have

$$f(x_0+h) + f(x_0-h) = 2f(x_0) + h^2 f''(x_0) + \frac{h^4}{24}\left(f^{(iv)}(\xi_1) + f^{(iv)}(\xi_2)\right).$$

Solving for $f''(x_0)$ gives a famous **centered formula for the second derivative**, written as

$$f''(x_0) = \frac{1}{h^2}(f(x_0-h) - 2f(x_0) + f(x_0+h)) - \frac{h^2}{12} f^{(iv)}(\xi).$$

Once again, $x_0 - h \leq \xi \leq x_0 + h$ and the formula consisting of the first term on the right-hand side is second order accurate.

Unlike for the first derivative, where we have seen several equally important basic discretizations, there is really only one useful formula of order ≤ 2 for the second derivative. A five-point centered formula of order 4 will be derived in Section 14.2.

Example 14.1. Table 14.1 lists errors obtained for the step sizes $h = 10^{-k}$, $k = 1, \ldots, 5$, using three methods of orders 1, 2, and 4 for approximating $f'(x_0)$, applied to $f(x) = e^x$ at $x = 0$. Note how the truncation error for a method of order q decreases by a factor of roughly 10^{-q} when h is refined by a factor of 10.

Figure 14.1 displays the errors for values $h = 10^{-k}$, $k = 1, \ldots, 8$. On the log-log scale of the figure, different orders of accuracy appear as different slopes of straight lines. Where the straight lines terminate (going from right to left in h) is where roundoff error takes over. We address this further in Section 14.4. Note that before roundoff error dominates, the higher order methods perform very well. ∎

Suppose now that we are given the values of $f(x)$ at the points of a uniform *mesh*, or *grid*, $x_0 \pm jh$, $j = 0, 1, 2, \ldots$, and that these values are to be used to generate an approximation to the kth derivative of f at x_0. Since the lowest polynomial degree to have a nontrivial kth derivative is k, we must use at least $k+1$ mesh points in the interpolation approach in order to generate such an approximation. The forward and backward differences for f' and the centered difference for f'' use a minimal number of neighboring mesh points. The second order schemes for f' do not; the centered scheme for f', in particular, is sometimes referred to as a *long difference*.

Specific exercises for this section: Exercises 1–3.

14.2. Richardson extrapolation

Table 14.1. *Errors in the numerical differentiation of $f(x) = e^x$ at $x = 0$ using methods of order 1, 2, and 4.*

h	$\frac{e^h-1}{h} - 1$	$\frac{e^h-e^{-h}}{2h} - 1$	$\frac{-e^{2h}+8e^h-8e^{-h}+e^{-2h}}{12h} - 1$
0.1	5.17e-2	1.67e-3	3.33e-6
0.01	5.02e-3	1.67e-5	3.33e-10
0.001	5.0e-4	1.67e-7	−4.54e-14
0.0001	5.0e-5	1.67e-9	−2.60e-13
0.00001	5.0e-6	1.21e-11	−3.63e-12

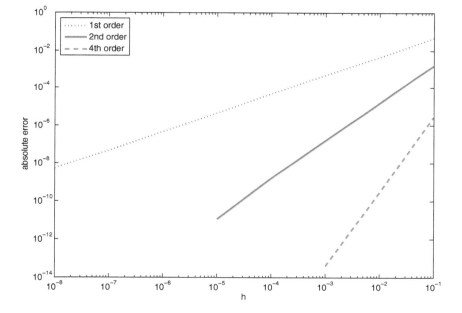

Figure 14.1. *Actual error using the three methods of Example 14.1. Note the log-log scale of the plot. The order of the methods is therefore indicated by the slope of the straight line (note that h is decreased from right to left).*

14.2 Richardson extrapolation

Richardson extrapolation is a simple, effective mechanism for generating higher order numerical methods from lower order ones. Given two methods of the same order, we can exploit the relationship between their leading error terms in order to eliminate such a term. Let us demonstrate this using a simple example.

Example 14.2. In Section 14.1 we have used the Taylor expansions for $f(x_0 \pm h)$ and for $f(x_0 \pm 2h)$ to obtain a fourth order formula for the first derivative. Here we use the same expansions in a different way to obtain a fourth order formula for the second derivative of $f(x)$.

At first, let us keep more terms of the expansions for $f(x_0 \pm h)$ to write the three-point formula for $f''(x_0)$ derived on page 412 as

$$f''(x_0) = \frac{1}{h^2}(f(x_0-h) - 2f(x_0) + f(x_0+h))$$
$$- \frac{h^2}{12} f^{(iv)}(x_0) - \frac{h^4}{360} f^{(vi)}(x_0) + \mathcal{O}(h^5).$$

Next, we do exactly the same using $f(x_0 \pm 2h)$. There is no reason to repeat the entire derivation process: we merely apply the same formula with the spacing $2h$ replacing h. This gives

$$f''(x_0) = \frac{1}{4h^2}(f(x_0-2h) - 2f(x_0) + f(x_0+2h))$$
$$- \frac{4h^2}{12} f^{(iv)}(x_0) - \frac{16h^4}{360} f^{(vi)}(x_0) + \mathcal{O}(h^5).$$

The formula based on the spacing $2h$ has the leading error term $\frac{4h^2}{12} f^{(iv)}(x_0)$, which is precisely four times larger than that in the preceding formula.

This naturally suggests the extrapolation idea of combining the two formulas in a way so as to eliminate the leading error term, thus obtaining a higher order approximation! Multiplying the formula based on the closer neighbors by 4, subtracting the other one, and dividing by 3, we obtain the expression

$$f''(x_0) = \frac{1}{12h^2}(-f(x_0-2h) + 16f(x_0-h) - 30f(x_0) + 16f(x_0+h) - f(x_0+2h))$$
$$+ \frac{h^4}{90} f^{(vi)}(x_0) + \mathcal{O}(h^5).$$

Thus, the five-point difference formula consisting of the first term on the right-hand side is fourth order accurate and has the truncation error

$$e(h) = \frac{h^4}{90} f^{(vi)}(\xi)$$

for some ξ, $x_0 - 2h \leq \xi \leq x_0 + 2h$. ∎

The extrapolation principle demonstrated in Example 14.2 is general. All we need is function smoothness and an error expansion, and the latter is usually provided by the Taylor series. Similar processes can be launched for other derivative approximations in an obvious way. The idea can be applied more than once to obtain even higher order methods (Exercise 5), and it may be used also to accelerate convergence of methods for numerical integration (see Section 15.5) and for solving differential equations.

Error estimation

It is often desirable to obtain a practical estimate for the error committed in a numerical differentiation formula without resorting to precise knowledge of the higher derivatives of $f(x)$. Such an error estimate can be obtained in a manner akin to the extrapolation philosophy. Let us again demonstrate by example.

14.3. Deriving formulas using Lagrange polynomial interpolation

Example 14.3. Instead of combining the two second order expressions in Example 14.2 to obtain a higher order formula, we could use them to obtain an *error estimate*. Thus, subtracting one from the other and ignoring $\mathcal{O}(h^4)$ terms, we clearly obtain an expression for $-\frac{3h^2}{12} f^{(iv)}(x_0)$ in terms of computable values of f at $x_0, x_0 \pm h$, and $x_0 \pm 2h$. Dividing by 3, this yields the computable error estimate

$$e(h) \approx \frac{1}{3} \left[\frac{1}{4h^2} (f(x_0 - 2h) - 2f(x_0) + f(x_0 + 2h)) \right.$$
$$\left. - \frac{1}{h^2} (f(x_0 - h) - 2f(x_0) + f(x_0 + h)) \right],$$

applicable for the classical three-point formula.

Notice, however, that there is a choice to be made here: the computable error estimate applies to the second order formula, not the fourth order one, so using the two second order expressions we can obtain either a fourth order formula *or* a second order one with an error estimate. Obtaining a fourth order method *and* an error estimate requires more; see Exercise 6. ∎

Advantages and disadvantages of extrapolation

The simplicity and generality of the Richardson extrapolation method have made it a favorable approach in practice. However, this approach is not without faults. One disadvantage is that the resulting formulas are typically noncompact: they use more neighboring points involving a wider spacing than perhaps necessary. In principle, a more compact higher order formula could be obtained using another approach, although in actual instances this is not always the case.

Another potential disadvantage is that the explicit reliance on the existence of higher and higher *nicely bounded* derivatives of f, so that neglecting higher order error terms is enabled, exposes vulnerability when this assumption is violated. Of course, any formula approximating $f''(x_0)$, say, will fall apart if low derivatives of f are discontinuous around x_0, but some formulas collapse more gracefully in such a case than those based on extrapolation.

For the numerical solution of ODEs, for instance, methods based on extrapolation are no longer in vogue. They really don't lag far behind the best ones, but they are never better than today's favorites either.

Perhaps the best thing about Richardson's extrapolation is that it enables the conclusion that the task of producing derivative approximations of higher and higher order for sufficiently smooth functions is under control in a simple, even if not always optimal, manner.

Specific exercises for this section: Exercises 4–6.

14.3 Deriving formulas using Lagrange polynomial interpolation

> **Note:** The approach considered in Section 14.3 has several merits, but it is also heavier, so if you must for some reason miss one of these sections, make it this one.

As mentioned before, the Taylor series approach of Section 14.1 can become difficult to apply when the going gets tough. One instance of such complication is when the desired order is high (do, e.g., Exercise 3). Another is when the points used are not equally spaced. The complication of using Taylor expansions quickly increases, especially if a high order method is desired in such a case.

A *general approach* for deriving approximate differentiation formulas at some point x_0 is to choose a few nearby points, interpolate by a polynomial, and then differentiate the interpolant.

Example 14.4. Let us rederive the formula of Example 1.2 using polynomial interpolation. We construct the linear interpolant $p_1(x)$ to $f(x)$ using abscissae x_0 and $x_1 = x_0 + h$ (where $x_0, x_1 \in [a,b]$). This reads

$$p_1(x) = f(x_0) + \frac{f(x_1) - f(x_0)}{h}(x - x_0).$$

By the polynomial interpolation error formula of Section 10.5 we have

$$f(x) = p_1(x) + (x - x_0)(x - x_1) f[x_0, x_1, x],$$

so

$$f'(x) = \frac{f(x_1) - f(x_0)}{h} + ((x - x_0) + (x - x_1)) f[x_0, x_1, x]$$
$$+ (x - x_0)(x - x_1) \frac{d}{dx} f[x_0, x_1, x].$$

To make the dreadful rightmost term disappear we need to evaluate f at one of the data points x_i. Setting $x = x_0$, we obtain

$$f'(x_0) = \frac{f(x_0 + h) - f(x_0)}{h} - \frac{h}{2} f''(\xi), \qquad x_0 \leq \xi \leq x_0 + h.$$

The approximation method is $\frac{f(x_0+h) - f(x_0)}{h}$, and the term $-h/2 f''(\xi)$ is the truncation error.[53] The method is first order accurate.

The approximation formula we have just obtained is the good old **forward difference** formula. In a similar way, it is easy to obtain the **backward difference** formula

$$f'(x_0) = \underbrace{\frac{f(x_0) - f(x_0 - h)}{h}}_{\text{approximation}} + \underbrace{\frac{h}{2} f''(\xi)}_{\text{truncation error}}, \qquad x_0 - h \leq \xi \leq x_0,$$

which is also first order accurate, as mentioned in Section 14.1. ∎

Differentiation using Lagrange interpolation

We can easily generalize and mechanize the interpolation and differentiation methodology of Example 14.4. This assumes that you have read and understood Section 10.3. Since we expect to obtain formulas in terms of values of $f(x_i)$, the Lagrange interpolation form is particularly convenient here.

Consider the derivation of a formula involving the points $x_0, x_1, \ldots, x_n$, not ordered in any particular fashion. The interpolating polynomial of degree at most n is

$$p(x) = \sum_{j=0}^{n} f(x_j) L_j(x),$$

---

[53] Of course, we do not know what exactly ξ is, only in which small interval it resides.

14.3. Deriving formulas using Lagrange polynomial interpolation

where the Lagrange polynomials are

$$L_j(x) = \frac{(x-x_0)\cdots(x-x_{j-1})(x-x_{j+1})\cdots(x-x_n)}{(x_j-x_0)\cdots(x_j-x_{j-1})(x_j-x_{j+1})\cdots(x_j-x_n)}.$$

Taking the derivative of $p(x)$ and substituting $x = x_0$ yields the *general formula for numerical differentiation*

$$p'(x_0) = \sum_{j=0}^{n} f(x_j) L'_j(x_0).$$

This formula does not require the points $x_0, x_1, \ldots, x_n$ to be equidistant!

Differentiation using equidistant points

To obtain cleaner expressions, let us next assume equidistant spacing. We will get to the more general case later. Thus, assume that the points x_i are distributed around x_0 and given as $x_0 - lh, x_0 - (l-1)h, \ldots, x_0 - h, x_0, x_0 + h, \ldots, x_0 + uh$, where l and u are nonnegative integers and $n = l + u$. There is an obvious shift in index involved, from 0 to $-l$, because we want to emphasize that the points $x_i = x_0 + ih$ are generally on both sides of x_0, where we seek to approximate $f'(x_0)$.

Evaluating the weights of the formula, we get

$$L'_0(x_0) = \sum_{\substack{k=-l \\ k \neq 0}}^{u} \frac{1}{x_0 - x_k} = \frac{1}{h}\sum_{\substack{k=-l \\ k\neq 0}}^{u}\left(\frac{1}{-k}\right),$$

$$L'_j(x_0) = \frac{1}{x_j - x_0} \prod_{\substack{k=-l \\ k\neq 0 \\ k\neq j}}^{u} \frac{x_0 - x_k}{x_j - x_k} = \frac{1}{jh}\prod_{\substack{k=-l \\ k\neq 0 \\ k\neq j}}^{u}\left(\frac{-k}{j-k}\right) \quad \text{for } j \neq 0.$$

Note that

$$a_j = h L'_j(x_0), \quad j = -l, \ldots, u,$$

are independent of both h and f. We calculate these weights once and for all and obtain the formula

$$p'(x_0) = h^{-1} \sum_{j=-l}^{u} a_j f(x_j).$$

How accurate is this formula? Recall that the interpolation error is given by

$$f(x) - p_n(x) = f[x_{-l}, x_{-l+1}, \ldots, x_u, x] \prod_{k=-l}^{u}(x - x_k).$$

Therefore the error in the numerical differentiation formula at $x = x_0$ is

$$f'(x_0) - p'_n(x_0) = \frac{d}{dx}\left\{ f[x_{-l}, x_{-l+1}, \ldots, x_u, x] \prod_{k=-l}^{u}(x - x_k) \right\}_{x=x_0}$$

$$= f[x_{-l}, x_{-l+1}, \ldots, x_u, x_0] \prod_{\substack{k=-l \\ k\neq 0}}^{u}(x_0 - x_k)$$

$$= \left[\frac{f^{(n+1)}(\xi)}{(n+1)!} l! u! \right] h^n$$

for some ξ, $x_{-l} \leq \xi \leq x_u$.

In particular, *using $n+1$ points yields a method of accuracy order n.*

Numerical Differentiation.
Based on the points $x_i = x_0 + ih$, $i = -l, \ldots, u$, where $l + u = n$, an nth order formula approximating $f'(x_0)$ is given by

$$f'(x_0) \approx \frac{1}{h} \sum_{j=-l}^{u} a_j f(x_j),$$

where

$$a_j = \begin{cases} -\sum_{\substack{k=-l \\ k \neq 0}}^{u} \left(\frac{1}{k}\right), & j = 0, \\ \frac{1}{j} \prod_{\substack{k=-l \\ k \neq 0 \\ k \neq j}}^{u} \left(\frac{k}{k-j}\right), & j \neq 0. \end{cases}$$

The truncation error is estimated by

$$\left| f'(x_0) - \frac{1}{h} \sum_{j=-l}^{u} a_j f(x_j) \right| \leq \frac{l! u!}{(n+1)!} \|f^{(n+1)}\|_\infty h^n.$$

Example 14.5. Take $l = 1$, $u = 1$. Then $n = 2$, and

$$L'_0(x_0) = \frac{1}{h}(1-1) = 0,$$
$$L'_{-1}(x_0) = \frac{-1}{h} \cdot \frac{-1}{-2} = -\frac{1}{2h},$$
$$L'_1(x_0) = \frac{1}{h} \cdot \frac{1}{2} = \frac{1}{2h}.$$

The symmetric, three-point formula is therefore

$$f'(x_0) \approx \frac{f(x_0 + h) - f(x_0 - h)}{2h}.$$

The error is estimated by

$$f'(x_0) - p'(x_0) = -\frac{f'''(\xi)}{6} h^2.$$

This is a second order formula, even though f is evaluated at only two points, because $L'_0(x_0) = 0$.

Likewise, taking $l = 2$, $u = 2$, yields the formula

$$f'(x_0) \approx \frac{1}{12h} (f(x_0 - 2h) - 8f(x_0 - h) + 8f(x_0 + h) - f(x_0 + 2h)),$$

with truncation error

$$e(h) = \frac{h^4}{30} f^{(v)}(\xi).$$

It is easy to write a short MATLAB function to find these coefficients. ∎

14.3. Deriving formulas using Lagrange polynomial interpolation

Nonuniformly spaced points

Recall a major advantage of the current more general but complicated approach, in that the points need not be equally spaced. We now give an example for this more general case.

Example 14.6. Suppose that the points x_{-1}, $x_0 = x_{-1} + h_0$, and $x_1 = x_0 + h_1$ are to be used to derive a second order formula for $f'(x_0)$ which holds even when $h_0 \neq h_1$. Such a setup arises often in practice.

A rushed approach, which has actually been used in the research literature, would suggest the formula
$$f'(x_0) \approx \frac{f(x_1) - f(x_{-1})}{h_0 + h_1},$$
which quickly generalizes the centered three-point formula for the case $h_0 = h_1$.

Instead, consider $p_2'(x_0) = \sum_{j=-1}^{1} f(x_j) L_j'(x_0)$, where p_2 is the interpolating polynomial in Lagrange form. We have
$$L_0'(x_0) = \frac{1}{h_0} - \frac{1}{h_1} = \frac{h_1 - h_0}{h_0 h_1},$$
$$L_{-1}'(x_0) = -\frac{h_1}{h_0(h_0 + h_1)},$$
$$L_1'(x_0) = \frac{h_0}{h_1(h_0 + h_1)},$$
hence
$$f'(x_0) \approx \frac{h_1 - h_0}{h_0 h_1} f(x_0) + \frac{1}{h_0 + h_1} \left(\frac{h_0}{h_1} f(x_1) - \frac{h_1}{h_0} f(x_{-1}) \right).$$

For $h_0 = h_1 = h$ this, too, becomes the familiar centered three-point formula. The added advantage here is that we have operated in a non ad hoc manner, within the solid framework of polynomial interpolation, and thus we have a good understanding of how to obtain the formula and what the expected error is.

We now repeat the calculations of Example 14.1 with $h_1 = h$ and $h_0 = 0.5h$. We use e_g to denote the error in the more elaborate formula obtained from the polynomial interpolation and e_s to denote the error in the simpler formula preceding it. The results are gathered in Table 14.2. Note the second order accuracy of e_g. The simple-minded method is necessarily first order only (see Exercise 7), and it is significantly inferior to the more carefully derived method. ■

Table 14.2. *Errors in the numerical differentiation of $f(x) = e^x$ at $x = 0$ using three-point methods on a nonuniform mesh, Example 14.6. The error e_g in the more elaborate method is second order, whereas the simpler method yields first order accuracy e_s.*

h	e_g	e_s
0.1	8.44e-4	2.63e-2
0.01	8.34e-6	2.51e-3
0.001	8.33e-8	2.50e-4
0.0001	8.35e-10	2.50e-5
0.00001	1.46e-11	2.50e-6

Formulas for Higher Derivatives

Polynomial interpolation in Lagrange form may be used also to obtain formulas for approximating higher derivatives of f. Indeed, for the kth derivative of $f(x)$ at $x = x_0$ we interpolate at $x_{-l},\ldots,x_u$ as before, obtaining $p(x)$, and then take k derivatives to obtain the approximation formula

$$f^{(k)}(x_0) \approx p^{(k)}(x_0) = \sum_{j=-l}^{u} f(x_j) L_j^{(k)}(x_0).$$

Choosing the (distinct) points x_j to be at most at an $\mathcal{O}(h)$ distance away from x_0 produces an accuracy order of *at least* $n+1-k$, i.e., the error obeys

$$|f^{(k)}(x_0) - p^{(k)}(x_0)| \leq C h^{n+1-k}$$

for some constant C. However, in some cases the order of accuracy may be higher than $n+1-k$, so this is only a lower bound.

With this approach it also becomes clear that we should never take $n < k$, because the kth derivative of the polynomial vanishes then, and no meaningful approximation results.

The case $n = k$ is of particular interest, and corresponding differentiation formulas are sometimes referred to as being **compact**. The formula is particularly easy to derive because

$$L_j^{(n)} = \frac{n!}{\prod_{\substack{i=-l \\ i \neq j}}^{u}(x_j - x_i)}.$$

For instance, setting $n = k = 2$, $x_{\pm 1} = x_0 \pm h$, immediately produces the centered three-point formula for the second derivative (see Exercise 9). First order accuracy is guaranteed in general, even if the interpolation points are not equally spaced, but in the case of the centered second derivative formula the order is actually 2, thus providing an example where the general guaranteed order of accuracy is indeed only a lower bound for the actual order of accuracy.

Specific exercises for this section: Exercises 7–9.

14.4 Roundoff and data errors in numerical differentiation

We have seen in Examples 1.3 and 14.1 that decent approximations may be obtained using numerical differentiation of smooth functions, and yet for very small values of the discretization parameter h, roundoff error increases. Let us now examine the situation more carefully.

> **Note:** Section 14.4 provides one of those rare occasions where roundoff error is allowed to take center stage.

We will see below how the choice of step size h affects the relative magnitudes of truncation and roundoff errors. The special thing about roundoff error is actually that it is not special at all (rather, truncation error is). A similar phenomenon of error magnification by numerical differentiation occurs also for noise in given data, and we consider instances of this as well. Our discussion commences with a simple example.

Example 14.7. Let us concentrate on the three-point centered difference approximation of Example 14.1, which reads

$$f'(x_0) \simeq \frac{f(x_0+h) - f(x_0-h)}{2h} = \frac{e^h - e^{-h}}{2h}.$$

14.4. Roundoff and data errors in numerical differentiation

We write $D_h = (e^h - e^{-h})/(2h)$ and denote by $\bar{D}_h$ the calculated value for D_h. Using the default arithmetic in MATLAB leads to the corresponding error curve in Figure 14.1.

Specifically, at first the accuracy improves by about two decimal digits each time we decrease h by a factor of 10, as befits an $\mathcal{O}(h^2)$ error. However, at about $h = 10^{-5}$ the error stops decreasing, and for smaller h it even starts increasing like h^{-1}.

The default arithmetic in MATLAB is double precision. Using single precision instead (see Section 2.4), we get the following values:

h	0.1	0.01	0.001	0.0001	0.00001
$\bar{D}_h$	1.00167	1.00002	1.00002	1.00017	1.00136

So, the same phenomenon occurs, but the error bottoms out much sooner, at about $h = 10^{-3}$. ∎

How small can the total error be?

The phenomenon demonstrated in Examples 14.1 and 14.7 is not specific to those instances. We must remember that in addition to the truncation error associated with this approximation, there is roundoff error, too. For h small, the expression approximating $f'(x_0)$ exhibits cancellation error, magnified by the division by h. Let $\bar{f}(x) \equiv \text{fl}(f(x))$ be the floating point approximation for $f(x)$. So, at each value of the argument x we have

$$\bar{f}(x) = f(x) + e_r(x),$$

where $e_r(x)$ is the roundoff error term, assumed bounded by ϵ, $|e_r(x)| \leq \epsilon$. The value of ϵ directly depends on the rounding unit[54] η.

For transparency of exposition, we stick with the centered difference formula of Example 14.7. Then we actually have $\bar{D}_h$ given by

$$\bar{D}_h = \frac{\bar{f}(x_0+h) - \bar{f}(x_0-h)}{2h},$$

assuming for simplicity that the floating point subtraction and division operations are exact in this formula. So, a bound on the roundoff error associated with our floating point approximation is derived as

$$|\bar{D}_h - D_h| = \left| \frac{\bar{f}(x_0+h) - \bar{f}(x_0-h)}{2h} - \frac{f(x_0+h) - f(x_0-h)}{2h} \right|$$

$$= \left| \frac{e_r(x_0+h) - e_r(x_0-h)}{2h} \right|$$

$$\leq \left| \frac{e_r(x_0+h)}{2h} \right| + \left| \frac{e_r(x_0-h)}{2h} \right| \leq \frac{\epsilon}{h}.$$

The key point is that the roundoff error function $e_r(x)$ is not smooth. So, although $e_r(x_0+h)$ and $e_r(x_0-h)$ are both bounded by a small number, they are not close to each other in the relative sense! Indeed, their signs may differ.

---

[54] Recall that the rounding unit, or machine precision, for the IEEE standard (double precision) word is

$$\eta = 2^{-53} \approx 1.1 \times 10^{-16}.$$

Note further that $\epsilon \sim \eta |f(x)|$.

Assuming $|f'''(\xi)| \leq M$ in $[x_0 - h, x_0 + h]$, the actual error we have in our approximation can be bounded by

$$|f'(x_0) - \overline{D}_h| = |(f'(x_0) - D_h) + (D_h - \overline{D}_h)|$$
$$\leq |f'(x_0) - D_h| + |D_h - \overline{D}_h| \leq \frac{h^2 M}{6} + \frac{\epsilon}{h}.$$

Looking at this total error bound, we see that as h is reduced, the truncation component of the error bound, which is the first term in the last expression, *shrinks* in magnitude like h^2, while the roundoff error component *grows* in magnitude like h^{-1}. If h is very small, the roundoff error term dominates the calculations. See Figure 14.2.

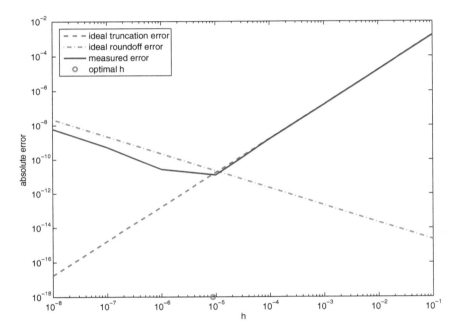

Figure 14.2. *The measured error roughly equals truncation error plus roundoff error. The former decreases but the latter grows as h decreases. The "ideal roundoff error" is proportional to η/h. Note the log-log scale of the plot. A red circle marks the "optimal h" value for Example 14.1.*

There is an optimal value of sorts for h; from calculus, if we define $E(h) = h^2 M/6 + \epsilon/h$, then

$$\frac{dE}{dh} = \frac{hM}{3} - \frac{\epsilon}{h^2} = 0 \Rightarrow h = h_* = \left(\frac{3\epsilon}{M}\right)^{\frac{1}{3}}.$$

(To verify that this is a minimizer, check that $\frac{d^2 E}{dh^2}(h_*) > 0$.) At $h = h_*$ the bound on the total error is

$$E(h_*) = h_*^2 M/6 + \epsilon/h_* = \frac{h_*}{2}\frac{\epsilon}{h_*^2} + \frac{\epsilon}{h_*} = \frac{3\epsilon}{2h_*}.$$

(If all this looks suspiciously familiar to you, it may be because you have diligently solved Exercise 2.14 on page 34.)

14.4. Roundoff and data errors in numerical differentiation

The actual choice of h therefore also depends on the rounding unit, as well as on the order of the truncation error and the bound on the appropriate derivative of f. The latter as well as ϵ are rarely known in practice. Moreover, there are additional reasons to want the truncation error to *dominate* rather than roughly equal the roundoff error! For this reason we refrain from postulating a theorem here.

The main lesson from this analysis is not to take h too close to the rounding unit η. A rough rule of thumb is to keep h well above $\eta^{1/(q+1)}$ for a method of accuracy order q.

A general intuitive explanation

It is impossible to avoid severe cancellation errors with a very small h in the case of numerical differentiation of the sort we have seen thus far in this chapter, although in a restricted setting there is the method considered in Exercises 10–11. Perhaps the following explanation will be useful for intuition. (Simply skip the next paragraph if it is not working for you!)

Consider the error in a polynomial approximation $p(x)$ to $f(x)$, say, $e(x) = f(x) - p(x)$. If we use $p'(x)$ to approximate $f'(x)$, then the error is $e'(x)$. Now, imagine expanding $e(x)$ in a Fourier series, recalling Sections 13.1 and 13.3. We are looking at

$$e(x) = \sum_{j=-\infty}^{\infty} c_j e^{\iota j x}.$$

The error for the derivative approximation is then

$$e'(x) = \iota \sum_{j=-\infty}^{\infty} j c_j e^{\iota j x}.$$

So, the amplitude $|c_j|$ for the the jth frequency is replaced by $|j c_j|$ in the derivative expansion. The higher the frequency, the larger the amplification $|j|$. Now, truncation error is smooth: the high-frequency components in the Fourier expansion have vanishing amplitudes. But roundoff error is not smooth (recall Figure 2.2), and it is rich in high-frequency components. As $j \to \pm\infty$, corresponding to $h \to 0$, the magnification of these error components is unbounded.

Accurate numerical differentiation

Let us distinguish between the two uses of numerical differentiation techniques mentioned at the beginning of this chapter:

- Formulas from numerical differentiation are frequently used in the numerical solution of differential equations.

  In this case h and η must be such that truncation error dominates roundoff error (i.e., it is considerably larger in magnitude). This is why double precision (long word in the IEEE standard) is the default option in scientific computing. Also, h is usually not extremely small for reasons of efficiency and the degree of accuracy (or, tolerance of error) required in practice.

- If an approximate value for a function derivative is desired, then h is typically much smaller than what is employed in the numerical solution of differential equations. Here the roundoff problem is more immediately prominent.

Concentrating on the second item above, a practical cure is to use a sufficiently small rounding unit η. But although there are software packages for extended floating point precision, this cure may have a high computational price tag attached to it.

Alternative techniques to numerical differentiation called **automatic differentiation** have been devised as well, but these are beyond the scope of our text.

Numerical differentiation of noisy data

The "cure" of decreasing the unit roundoff is simply not available when the function $f(x)$ to be differentiated is acquired by noisy measurements. If the given $f(x)$ contains a component of high-frequency noise, then this noise will essentially be magnified by a factor of $1/h$ upon applying numerical differentiation. If numerical differentiation is applied to approximately determine the kth derivative, then the factor of noise amplification will be $1/h^k$, so the higher the derivative, the worse our situation is. The following simple example illustrates the point.

Example 14.8. We simulate noisy data by sampling the function $\sin(x)$ on the interval $[0, 2\pi]$ using a step size $h = 0.01$, large enough so as to ignore roundoff error effects, then adding 1% of Gaussian noise. This is achieved by the MATLAB commands

```
x = 0:.01:2*pi;
l = length(x);
sinx = sin(x);
sinp = (1+.01*randn(1,1)).*sinx;
```

Next, we apply the same centered second order formula as in Examples 14.1 and 14.7 to the clean data in the array `sinx` and to the noisy (or perturbed) data in the array `sinp`:

```
cosx = (sinx(3:l)-sinx(1:l-2))/.02;
cosp = (sinp(3:l)-sinp(1:l-2))/.02;
```

To see how far `cosp` is from `cosx` relative to where they came from, we calculate the maximum differences and also plot their values:

```
err_f = max(abs(sinx-sinp))
err_fp = max(abs(cosx-cosp))
subplot(1,2,1)
plot(x,sinp,x,sinx,'r')
subplot(1,2,2)
plot(x(2:l-1),cosp,x(2:l-1),cosx,'r')
```

The results are displayed in Figure 14.3; the error values are $\text{err}_f = 0.0252$, $\text{err}_{fp} = 1.8129$. The error is significantly magnified, which demonstrates that the practical significance of the results of such an algorithm is severely limited. ∎

The lesson is clear: numerical differentiation should not be directly applied to data with non-trivial noise, in general. As we have seen in Example 14.8, it does not take much to obtain meaningless results this way. Methods exist which filter out the high frequency noise in various ways,

14.4. Roundoff and data errors in numerical differentiation

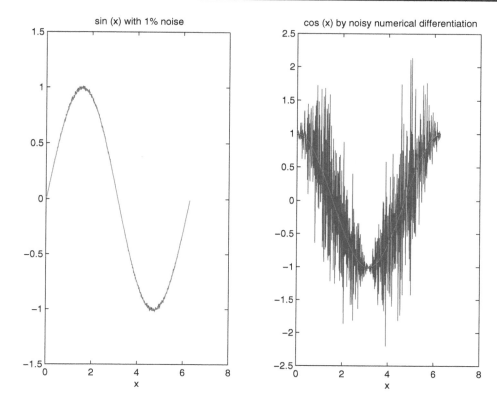

Figure 14.3. *Numerical differentiation of noisy data. On the left panel, $\sin(x)$ is perturbed by 1% noise. On the right panel, the resulting numerical differentiation is a disaster.*

thus obtaining more meaningful results. This **regularizes** the problem by appropriately altering it slightly.

Example 14.9. For a more practical example of the same sort consider Figure 14.4. The unsmoothed image on the left is constructed from satellite-measured depth data of a mountainous terrain near St. Mary Lake, British Columbia, Canada.

If we were to estimate terrain slopes in some direction, say, along one of the horizontal axes, by applying numerical differentiation to these depth (or rather, height) data, then numerical error as in Example 14.8 would frustrate our effort. It is better to smooth the image first, resulting in the image on the right.[55] Numerical differentiation may then cautiously be applied to the right image. ∎

Specific exercises for this section: Exercises 10–13.

---

[55] The smoothing was done using a *thin plate spline* (never mind what precisely that is) by David Moulton and Ray Spiteri as a term project in a graduate course.

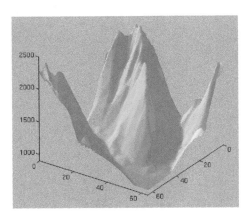 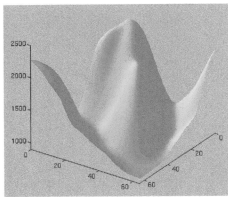

(a) Unsmoothed image.   (b) Smoothed image.

Figure 14.4. *An image with noise and its smoothed version. Numerical differentiation must not be applied to the original image.*

14.5 *Differentiation matrices and global derivative approximation

Throughout most of the present chapter thus far we have been concerned with approximating a function derivative at one point x_0. An exception is Example 14.8, where an approximation of the derivative function $f'(x)$ over the interval $[0, 2\pi]$ is constructed for $f(x) = \sin(x)$ as well as for a noisy version of this $f(x)$. In this section we continue the pursuit of approximating an entire *derivative function* on an interval $[a, b]$.

In a discrete version of this quest, we are given function values $f(x_0), f(x_1), \ldots, f(x_n)$ as input and expect approximations to $f'(t_0), f'(t_1), \ldots, f'(t_m)$ as output. Here, the input abscissae $\{x_i\}_{i=0}^{n}$ and the output abscissae $\{t_i\}_{i=0}^{m}$ are related, but they do not have to coincide. Their selection must allow at least for a faithful reconstruction on $[a, b]$ of f and f', respectively.

Example 14.10. In Example 14.8 we have for the input points $n \approx 200\pi$, $h = .01$, and $x_i = ih$, $i = 0, \ldots, n$. Then for the output points we have the input points except for the first and the last ones. This gives $m = n - 2$ and $t_i = x_{i+1}$, $i = 0, \ldots, m$.

Alternatively, we can have $m = n - 1$ using a short rather than a long difference formula for the first derivative. The script

```
h = .01;
xx = 0:h:2*pi;
np1 = length(xx);
n = np1 - 1;
sinx = sin(xx);
cosx = (sin(2:np1)-sin(1:n))./h;
```

yields first order approximations to $\cos(x)$ at $x_1, \ldots, x_n$. Thus, $t_i = x_{i+1}$.

Occasionally more interesting is to note that the same values obtained by the above script are second order approximations to $\cos(x)$ at the *midpoints* of the input mesh, i.e., at $t_i = x_i + h/2$, $i = 0, 1, \ldots, m$. This is so because these t_i are the points about which the short difference approximations are centered. It is an example of a **staggered mesh**. ∎

14.5. *Differentiation matrices and global derivative approximation

Finite difference differentiation matrices

Writing the above in the form of a matrix-vector product, we have for Example 14.8 the system
$$\mathbf{f}' = D\mathbf{f}, \quad \text{where}$$

$$\mathbf{f}' \approx \begin{pmatrix} f'(x_1) \\ f'(x_2) \\ \vdots \\ f'(x_{n-1}) \end{pmatrix}, \quad D = \frac{1}{2h} \begin{pmatrix} -1 & 0 & 1 & & & \\ & -1 & 0 & 1 & & \\ & & \ddots & \ddots & \ddots & \\ & & & -1 & 0 & 1 \end{pmatrix}, \quad \mathbf{f} = \begin{pmatrix} f(x_0) \\ f(x_1) \\ f(x_2) \\ \vdots \\ f(x_{n-1}) \\ f(x_n) \end{pmatrix}.$$

The matrix D is a simple instance of a *differentiation matrix*. Note, moreover, that because D is very sparse the matrix-vector product yielding $\mathbf{f}'$ can be carried out efficiently. Indeed, in Example 14.8 we did not bother to construct D as such and wrote an instruction resulting in this matrix-vector product directly.

However, this brings up a concern, because the matrix D is not square. Shouldn't it be square, both for symmetry's sake between function and derivative and because we may want to invert it?

Of course, f and f' do not play symmetric roles! Indeed, f' is uniquely obtained from f but not vice versa. To obtain f from f', one additional side condition must be specified, as this is an albeit trivial ordinary differential equation. (Indeed, recall that integrating a function can only be done up to an additive constant, whereas differentiating a function does not entail any such difficulty.) Thus, in this sense, ideally we should have $m = n - 1$. For the compact discretization in Example 14.10, if we know an initial value $f(x_0) = 0$, say, then we can approximately obtain $f(x_1), \ldots, f(x_n)$ from the values $f'(x_1), \ldots, f'(x_n)$ by solving a linear system of equations

$$\begin{pmatrix} f'(x_1) \\ f'(x_2) \\ \vdots \\ f'(x_n) \end{pmatrix} = \frac{1}{h} \begin{pmatrix} 1 & & & \\ -1 & 1 & & \\ & \ddots & \ddots & \\ & & -1 & 1 \end{pmatrix} \begin{pmatrix} f_1 \\ f_2 \\ \vdots \\ f_n \end{pmatrix},$$

where the first row in the matrix corresponds to the initial condition $f_0 = f(x_0) = 0$. The obtained matrix now is lower triangular, nonsingular, and sparse.

For the noncompact long difference of Example 14.8 we would still have one extra f value, but there are simple methods to fix this. Moreover, generally if we insist, as we will below, on obtaining all of $f'(x_0), \ldots, f'(x_n)$, i.e., that $t_i = x_i$, $i = 0, 1, \ldots, n$, then *one-sided* second order formulas defined on page 411 may be used in an obvious way for approximating the end values $f'(x_0)$ and $f'(x_n)$.

Chebyshev differentiation matrix

> **Note:** Before reading this part, you may benefit from a quick review of Section 12.4.

Suppose we are given $n+1$ values $\{f(x_i)\}_{i=0}^n$ and we want to approximate the derivative values $\{f'(x_i)\}_{i=0}^n$ *at the same points*. Thus, the output locations $\{t_i\}$ coincide with the input locations $\{x_i\}$.

We can think of interpolating the f-values by a polynomial of degree at most n, differentiating the obtained polynomial and evaluating the result at the points x_i. In fact, this slightly generalizes the procedure already developed in Section 14.3 using Lagrange polynomials; see in particular page 417. Here we do not assume the data locations x_i to be equispaced, and we evaluate the derivative at all (distinct) $n+1$ points, not just one. The resulting formula and differentiation matrix D are given on the current page.

Polynomial Differentiation Matrix.
Given data $(x_0, f(x_0)), \ldots, (x_n, f(x_n))$, an approximation to $f'(x_i)$ is given by

$$p'_n(x_i) = \sum_{j=0}^{n} f(x_j) L'_j(x_i), \quad 0 \le i \le n,$$

where

$$L'_j(x_i) = \begin{cases} \displaystyle\sum_{\substack{k=0 \\ k \ne j}}^{n} \left(\frac{1}{x_j - x_k}\right), & i = j, \\ \displaystyle\frac{1}{x_j - x_i} \prod_{\substack{k=0 \\ k \ne i \\ k \ne j}}^{n} \left(\frac{x_i - x_k}{x_j - x_k}\right), & i \ne j. \end{cases}$$

In matrix form, $\mathbf{f'} = D\mathbf{f}$, where D is an $(n+1) \times (n+1)$ matrix given by

$$d_{i,j} = L'_j(x_i).$$

Regardless of which points are chosen, the resulting matrix D is almost singular, because the derivative of the interpolation polynomial, $p'_n(x)$, is a polynomial of degree $< n$ that gets evaluated at $n+1$ points: an overdetermination. It is safer to use D than D^{-1}, then.

Of course, if n is large, then we would not get far with equidistant data. However, if we are free to sample f at any location in the interval $[a,b]$, then we choose an appropriately scaled translation of the *Chebyshev extremum points*

$$x_i = \cos\left(\frac{i}{n}\pi\right), \quad i = 0, \ldots, n.$$

These points are introduced in Exercise 20 of Chapter 10 and discussed in the context of Chebyshev polynomials in Section 12.4; see pages 328 and 377. Applying the differentiation procedure described above for the Chebyshev extremum points yields the *Chebyshev differentiation matrix*, denoted D_C.

Note that the Chebyshev extremum points are cosines of the same points (only differently scaled) used for the DFT in Sections 13.2 and 13.3, plus the last one. The latter is not needed in the DFT case due to the periodicity assumed there, while here we are working essentially on an extension to the more general, nonperiodic case. Moreover, it turns out to be possible (indeed, the resulting program is pleasantly short) to use FFT for a *fast evaluation of matrix-vector products* involving D_C. This is an important advantage.

14.5. *Differentiation matrices and global derivative approximation

Example 14.11. For the function $f(x) = e^x \sin(10x)$ on $[0,\pi]$ we check the error in $D_C \mathbf{f}$ and $D_C^2 \mathbf{f}$ as approximations for the first and second derivatives at the Chebyshev extremum points. See Figure 14.5.

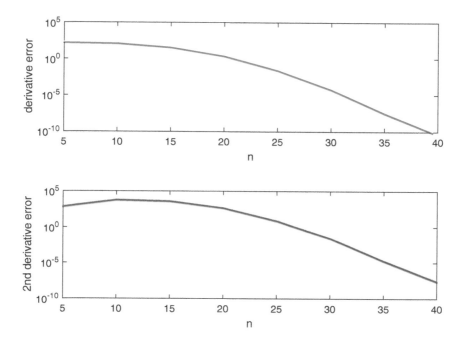

Figure 14.5. *Maximum absolute errors for the first and second derivatives of $f(x) = e^x \sin(10x)$ on the interval $[0,\pi]$ at the Chebyshev extremum points, i.e., using the Chebyshev differentiation matrix.*

Spectral accuracy is again demonstrated. At $n = 25$ we are already at the roundoff error level which, especially for the second derivative, is considerably higher than the rounding unit (machine precision).

To be duly impressed by Figure 14.5, check the accuracy of the centered three-point finite difference method using 25 (or 50, or 100, or 200) equidistant points for this function! ∎

Thus, it is not surprising to learn that the Chebyshev differentiation matrix is popular, especially in smooth, classroom-type environments where clean model problems are solved to incredibly high accuracy. When partial differential equations are solved the resulting method is called **pseudospectral**, and also **Chebyshev collocation**.

Fourier differentiation matrix

If the function $f(x)$ to be differentiated is periodic, then Fourier transform techniques suggest themselves. The mesh $\{x_i\}_{i=0}^n$ here is equispaced and is best considered *periodic*: on the interval $[0, 2\pi]$, $x_i = \frac{2\pi i}{n+1}$, and x_{n+1} coincides with x_0. Recall Figure 13.2 on page 390.

Indeed, we have seen in (13.3) (page 387) that differentiation in the x-space translates into multiplication of each coefficient c_k by $\imath k$. For the discrete Fourier transform, with $2l = m = n + 1$, we similarly have (see Section 13.3) that

$$c_k = \frac{1}{m}\sum_{i=0}^{m-1} y_i e^{-\imath k x_i}, \quad -l \leq k \leq l-1,$$

$$p_n(x) = \sum_{k=-l}^{l-1} c_k e^{\imath k x}, \quad \text{hence}$$

$$y_i' = p_n'(x_i) = \sum_{k=-l}^{l-1} \imath k c_k e^{\imath k x_i}, \quad i = 0, 1, \ldots, m-1.$$

This means that in frequency space the differentiation matrix is *diagonalized*! Indeed, writing this process in matrix notation as before we have $\mathbf{f}' = D_F \mathbf{f}$, where the $m \times m$ differentiation matrix D_F is given by

$$D_F = \mathcal{F}^{-1}\left[\imath \, \text{diag}\{-l, -l+1, \ldots, l-1\}\right]\mathcal{F}.$$

Here $\mathcal{F}$ stands for the matrix realizing the discrete Fourier transform.

Great ease in practice may be realized in this way. Note, in particular, that inverting the differentiation matrix is as straightforward as generating it, utilizing the inverse Fourier transform.

Of course, rather than generating the differentiation matrix we use FFT as in Section 13.3 to rapidly obtain its action. In MATLAB, assuming that f is a column array containing the values $f(x_i)$ on the interval $[a, b]$ and that a corresponding column of rth derivatives, call it g, is the desired outcome, let $m2 = 0$ if r is odd and $m2 = l$ otherwise. Then the script

```
f_hat = fft(f);
g_hat = (i)^r * ([0:l-1 m2 -l+1:-1].^r)' .* f_hat;
g_hat = (2*pi/(b-a))^r * g_hat;
g = real(ifft(g_hat));
```

does the job. The deal with m2 has to do with the convention MATLAB uses to order these coefficients.

Example 14.12. The function $f(x) = e^{-5x^2}$ is "very close" to being periodic on the interval $[-2\pi, 2\pi]$, being almost flat at the interval ends. For this function Fourier differentiation is significantly more accurate than Chebyshev; see Figure 14.6. ∎

Spectral methods

Wave phenomena often satisfy a system of partial differential equations in time and space, subject to periodic boundary conditions in space and initial conditions in time. For linear problems of this sort that have smooth coefficients, the use of FFT differentiation results in a *spectral method*, having spectral accuracy if all goes well.

> **Note:** Here we turn the heat up by a notch or two and consider, until the end of this section, solving special partial differential equations by spectral methods.

Example 14.13. Where spectral methods are truly superior is when considering a linear partial differential equation (PDE) having *constant coefficients* and periodic boundary conditions.

14.5. *Differentiation matrices and global derivative approximation

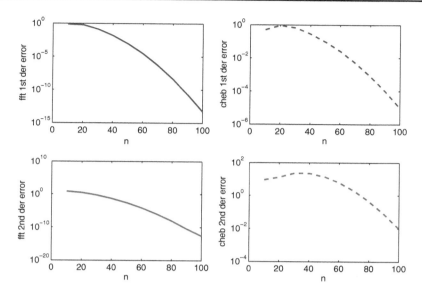

Figure 14.6. *Maximum absolute errors for the first and second derivatives of $f(x) = e^{-5x^2}$ on the interval $[-2\pi, 2\pi]$. The two left subplots use FFT, and the right subplots use the Chebyshev differentiation matrix. The errors are evaluated at the corresponding data locations (also called collocation points).*

Consider, for instance, the coupled system

$$\frac{\partial v}{\partial t} = -\frac{\partial^2 w}{\partial x^2},$$

$$\frac{\partial w}{\partial t} = \frac{\partial^2 v}{\partial x^2},$$

defined for all $a \leq x \leq b$ and time $t \geq 0$. In addition to the assumed periodicity in the space variable x there are initial conditions, so $v(t=0,x)$ and $w(t=0,x)$ are given.

A neat representation for this PDE is obtained using complex arithmetic, defining

$$u(t,x) = v(t,x) + \imath w(t,x),$$

with $\imath = \sqrt{-1}$. This yields the PDE

$$\frac{\partial u}{\partial t} = \imath \frac{\partial^2 u}{\partial x^2}.$$

The complex-valued function u (note that t and x stay real!) satisfies the same periodic boundary conditions, as well as the initial condition

$$u(t=0,x) = v(t=0,x) + \imath w(t=0,x).$$

To solve the PDE for $u(t,x)$, consider applying the continuous Fourier transform in x (not in t: there is no periodicity in time here). This is done as in Section 13.1; see in particular the equations (13.2). Assuming for a moment that $[a,b] = [-\pi, \pi]$, we obtain for each Fourier coefficient c_k

at any fixed t, in a manner similar to (13.3), the two equations

$$\frac{1}{2\pi}\int_{-\pi}^{\pi}\imath\frac{\partial^2 u}{\partial x^2}(t,x)e^{-\imath kx}dx = (-\imath k^2)\frac{1}{2\pi}\int_{-\pi}^{\pi}u(t,x)e^{-\imath kx}dx = -\imath k^2 c_k$$

and

$$\frac{1}{2\pi}\int_{-\pi}^{\pi}\frac{\partial u}{\partial t}(t,x)e^{-\imath kx}dx = \frac{d}{dt}c_k.$$

The PDE then decouples in Fourier space into the set of trivial ODEs

$$\frac{d}{dt}c_k = -\imath k^2 c_k, \quad -\infty < k < \infty,$$

with appropriate initial conditions obtained from the corresponding transform of the initial value function $u(t=0,x)$. The solution is

$$c_k(t) = c_k(0)e^{-\imath k^2 t},$$

and the inverse Fourier transform may then be applied at any desired time level t to obtain the *exact solution* $u(t,x)$.

Of course, infinitely many coefficients c_k are a lot to ask for from a mere computer. However, moving from Fourier to discrete Fourier as in Sections 13.2 and 13.3, we cheaply obtain, *at any time t*, an approximate solution with spectral accuracy. With `l = m/2`, $h = (b-a)/m$ and `u0` $= (u(0,a),u(0,a+h),\ldots,u(0,b-h))^T$ already in place, the corresponding MATLAB script reads

```
scale = 2*pi/(b-a);
expfac = exp(-i*t* scale^2* [0:l -l+1:-1]'.^2);
u0_hat = fft(u0);
ut_hat = expfac .* u0_hat;
u = ifft(ut_hat);
```

Startlingly short, this yields roundoff-level accuracy in $\mathcal{O}(m\log m)$ flops with a usually moderate m. Note that t need not be small. ∎

Essentially, then, initial value PDEs with constant coefficients and periodic boundary conditions may be assumed to have known solutions, obtained at will using the method described in Example 14.13. Of course (or perhaps *therefore*), in practical computations such simple problems rarely arise. Still, Example 14.13 emphasizes the availability of an important tool. Here is another example, demonstrating use of this tool for a more interesting case study.

Example 14.14 (nonlinear Schrödinger). The famous cubic nonlinear Schrödinger equation[56] in one space dimension is given by

$$\frac{\partial \psi}{\partial t} = \imath \left(\frac{\partial^2 \psi}{\partial x^2} + |\psi|^2 \psi\right)$$

for a complex-valued wave function $\psi(t,x)$. This PDE typically comes equipped with periodic boundary conditions as well as initial conditions.

Nonlinearity usually gives spectral methods the creeps: a full explanation and discussion of remedies belongs to another, more specialized and advanced text. Here we consider a simple technique called **operator splitting**.

---

[56] So famous is this equation that it is routinely referred to by the acronym NLS by researchers who do not have the habit of dealing with nonlinear least squares.

14.5. *Differentiation matrices and global derivative approximation

Observe that the right-hand side of our equation consists of two terms of a rather different nature. The first gives the PDE

$$\frac{\partial u}{\partial t} = \iota \frac{\partial^2 u}{\partial x^2},$$

which is precisely what we have dealt with so successfully in Example 14.13. The second part reads

$$\frac{\partial w}{\partial t} = \iota |w|^2 w,$$

which for each fixed x is just an ODE. In fact, we even know the exact solution for this one, because it so happens that $|w|$ is independent of t, leading to

$$w(t_1) = w(t_0)\, e^{\iota(t_1-t_0)|w|^2}$$

for any $0 \le t_0 \le t_1$.

This suggests a method whereby we alternately solve the problems for u and for w, using the solution from the previous stage as the initial value function for the next. Of course we cannot expect the result to be precisely the same as when solving the complete problem simultaneously. But over a step Δt in time the error is $\mathcal{O}(\Delta t^2)$, and it can be made $\mathcal{O}(\Delta t^3)$ using a simple staggering trick, reminiscent of what we have observed about finite difference differentiation (see, e.g., Exercise 15). Again we skip the details of this, except to say that these errors do accumulate. To reach a fixed time t_f from the initial level $t = 0$ requires $t_f/\Delta t$ steps, so the global error with the staggering trick is $\mathcal{O}(\Delta t^2)$, while without it the error is only $\mathcal{O}(\Delta t)$.

As in Example 14.13, the MATLAB code is amazingly short (although not *that* short). With notation as in Example 14.13 and $\texttt{dt} = \Delta t$, let us define

```
expfac = exp(-i*dt* scale^2* [0:l -l+1:-1]'.^2);
```

once. Looping in time, assume that at the beginning of the current time step from t to $t + \Delta t$, w approximates $\psi(t,x)$ at the spatial mesh points $x_0,\ldots,x_{m-1}$. The code for carrying out one such time step then reads

```
% advance dt for iu_xx
v_hat = fft(w);
u_hat = expfac .* v_hat;
u = ifft(u_hat);
% advance nonlinear term exactly
w = u.* exp(i*dt* abs(u).^2);
```

For a concrete example, we set $a = -20$, $b = 80$, and use the initial value function

$$\psi(0,x) = e^{\iota x/2}\mathrm{sech}(x/\sqrt{2}) + e^{\iota(x-25)/20}\mathrm{sech}((x-25)/\sqrt{2}).$$

(Don't worry about the hyperbolic secant function sech: there is a built-in MATLAB function for it.) As it happens this unintuitive initial value profile results in two pulses, called **solitons**, that propagate to the right at different speeds as time advances. They occasionally briefly merge and then split again with their original shapes intact. When a pulse reaches the right end $x = 80$ from the left it dutifully reappears at $x = -20$ as befits a periodic, or circular, grid. This is depicted by plotting the magnitude $|\psi(t,x)|$ in Figure 14.7. For this purpose we have used $m = 10{,}000$ and $\Delta t = .01$, and in order not to exceed fast memory capacity we have plotted solution curves every 100th spatial point at every 200th time level.

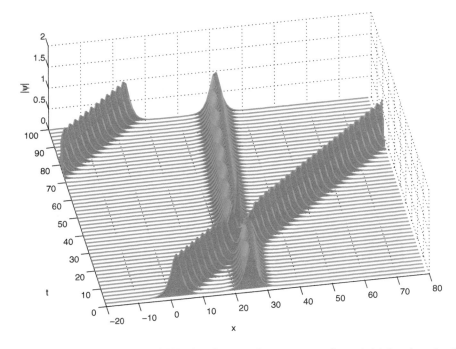

Figure 14.7. *This "waterfall" plot depicts the progress of two initial pulses in time; see Example 14.14. The two solitons merge and then split with their form intact. Rest assured that they will meet (and split) again many times in the future.*

The splitting method performs very well, both in terms of efficiency and in terms of the quality of the obtained solution. Note the clean shapes of the solitons in Figure 14.7, even after 10,000 time steps.

However, let us also caution that if we were to continue this integration for much longer, an instability would eventually show up! Details are again omitted here: suffice it to say that to ensure stability for long runs with this value of m it turns out that we must take $\Delta t < .0001/\pi$. Such a prospect is no longer as appetizing. ∎

Note: In numerical computing, just like in artificial intelligence and computer graphics, if something looks too good to be true, then chances are that it is.

Specific exercises for this section: Exercises 14–17.

14.6 Exercises

0. **Review questions**

   (a) How does numerical differentiation differ from manual or symbolic differentiation?

   (b) Define order of accuracy.

   (c) Explain briefly the Richardson extrapolation process and give two potential advantages and two potential disadvantages.

14.6. Exercises

(d) What possible advantage does the formula derivation using Lagrange polynomial interpolation have over using Taylor expansions?

(e) What sort of error frustrates attempts to numerically differentiate a given smooth function to arbitrary accuracy? What is the step size h with smallest overall error?

(f) Why is the numerical differentiation of noisy data a risky affair?

(g) What is a differentiation matrix?

1. Derive a difference formula for the fourth derivative of f at x_0 using Taylor's expansions at $x_0 \pm h$ and $x_0 \pm 2h$. How many points will be used in total and what is the expected order of the resulting formula?

2. Let $f(x)$ be a given function that can be evaluated at points $x_0 \pm jh$, $j = 0, 1, 2, \ldots$, for any fixed value of h, $0 < h \ll 1$.

    (a) Find a second order formula (i.e., truncation error $\mathcal{O}(h^2)$) approximating the third derivative $f'''(x_0)$. Give the formula, as well as an expression for the truncation error, i.e., not just its order.

    (b) Use your formula to find approximations to $f'''(0)$ for the function $f(x) = e^x$ employing values $h = 10^{-1}, 10^{-2}, \ldots, 10^{-9}$, with the default MATLAB arithmetic. Verify that for the larger values of h your formula is indeed second order accurate. Which value of h gives the closest approximation to $e^0 = 1$?

    (c) For the formula that you derived in (a), how does the roundoff error behave as a function of h, as $h \to 0$?

    (d) How would you go about obtaining a fourth order formula for $f'''(x_0)$ in general? (You don't have to actually derive it: just describe in one or two sentences.) How many points would this formula require?

3. Using Taylor's expansions derive a sixth order method for approximating the second derivative of a given sufficiently smooth function $f(x)$.

    Use your method to approximate the second derivative of $f(x) = e^x \sin(3x)$ at $x_0 = .4$, verifying its order by running your code for $h = .1$, $.05$, and $.025$.

4. Let

$$g = \frac{1}{h^2}(f(x_0 - h) - 2f(x_0) + f(x_0 + h)),$$

$$\hat{g} = \frac{1}{4h^2}(f(x_0 - 2h) - 2f(x_0) + f(x_0 + 2h)).$$

Show that if f has three bounded derivatives in a neighborhood of x_0 that includes $[x_0 - 2h, x_0 + 2h]$, then the computable expression

$$\hat{e} = \frac{g - \hat{g}}{3}$$

provides an estimation for the error in $f''(x_0) - g$ accurate up to $\mathcal{O}(h)$.

Note that here g is the actual approximation used, and $\hat{g}$ is an auxiliary expression employed in computing the error estimation for g.

5. Let
$$g = \frac{1}{h^2}(f(x_0-h)-2f(x_0)+f(x_0+h)),$$
$$\hat{g} = \frac{1}{4h^2}(f(x_0-2h)-2f(x_0)+f(x_0+2h)),$$
$$\tilde{g} = \frac{1}{9h^2}(f(x_0-3h)-2f(x_0)+f(x_0+3h)).$$

Assuming that f is smooth, use these three approximations for $f''(x_0)$ to obtain a sixth order formula.

Compare with Exercise 3.

6. Use the same setting as in Exercise 5 to obtain an error estimate for a fourth order formula.

7. It is apparent that the error e_s in Table 14.2 is only first order. But why is this *necessarily* so? More generally, let $f(x)$ be smooth with $f''(x_0) \neq 0$. Show that the truncation error in the formula
$$f'(x_0) \approx \frac{f(x_1)-f(x_{-1})}{h_0+h_1}$$
with $h_1 = h$ and $h_0 = h/2$ must decrease linearly, and not faster, as $h \to 0$.

8. Consider the derivation of an approximate formula for the second derivative $f''(x_0)$ of a smooth function $f(x)$ using three points x_{-1}, $x_0 = x_{-1}+h_0$, and $x_1 = x_0+h_1$, where $h_0 \neq h_1$.

   Consider the following two methods:

   (i) Define $g(x) = f'(x)$ and seek a *staggered mesh*, centered approximation as follows:
   $$g_{1/2} = (f(x_1)-f(x_0))/h_1; \quad g_{-1/2} = (f(x_0)-f(x_{-1}))/h_0;$$
   $$f''(x_0) \approx \frac{g_{1/2}-g_{-1/2}}{(h_0+h_1)/2}.$$

   The idea is that all of the differences are short (i.e., not long differences) and centered.

   (ii) Using the second degree interpolating polynomial in Newton form, differentiated twice, define
   $$f''(x_0) \approx 2f[x_{-1},x_0,x_1].$$

   Here is where you come in:

   (a) Show that the above two methods are one and the same.

   (b) Show that this method is only first order accurate in general.

   (c) Run the two methods for the example depicted in Table 14.2 (but for the second derivative of $f(x) = e^x$). Report your findings.

9. Show that setting $n = k = 2$, $x_{\pm 1} = x_0 \pm h$, in the expression for $L_j^{(n)}$ defined towards the end of Section 14.2, produces the centered three-point formula for the second derivative of a given function f at x_0.

10. If you are a fan of complex arithmetic, then you will like this exercise.

    Suppose that $f(z)$ is infinitely smooth on the complex plane and $f(z)$ is real when z is real. We wish to approximate $f'(x_0)$ for a given real argument x_0 as usual.

(a) Let $h > 0$, assumed small. Show by a Taylor expansion of $f(x_0 + \imath h)$ about x_0 that

$$f'(x_0) = \Im[f(x_0 + \imath h)]/h + \mathcal{O}(h^2).$$

Thus, a second order difference formula is obtained that does not suffer the cancellation error that plagues all the methods in Sections 14.1–14.3.

(b) Show that, furthermore, the leading term of the truncation error is the same as that of the centered formula that stars in Examples 14.1 and 14.7.

11. (a) Continuing Exercise 10, implement the complex variable formula for the problem of Examples 1.2–1.3. Implement also the real centered difference formula and generate error curves as in Figure 1.3 for both methods on the same graph. What are your conclusions? [The programming here requires but a small variation of the code given in Example 1.3. Note that $\Im$ in MATLAB is imag.]

(b) Contemplate how the technique of Exercises 10 and the current one may be applied or extended to the problem depicted in Figure 14.3 or in more than one variable. That should sober us up.

12. Let us denote $x_{\pm 1} = x_0 \pm h$ and $f(x_i) = f_i$. It is known that the difference formula

$$fpp_0 = (f_1 - 2f_0 + f_{-1})/h^2$$

provides a second order method for approximating the second derivative of f at x_0, and also that roundoff error increases like h^{-2}.

Write a MATLAB script using default floating point arithmetic to calculate and plot the actual total error for approximating $f''(1.2)$, with $f(x) = \sin(x)$. Plot the error on a log-log scale for $h = 10^{-k}$, k = 0:.5:8. Observe the roughly V shape of the plot and explain it. What is (approximately) the observed optimal h?

13. Continuing with the notation of Exercise 12, one could define

$$g_{1/2} = (f_1 - f_0)/h \quad \text{and} \quad g_{-1/2} = (f_0 - f_{-1})/h.$$

These approximate to second order the first derivative values $f'(x_0 + h/2)$ and $f'(x_0 - h/2)$, respectively. Then define

$$fpp_0 = (g_{1/2} - g_{-1/2})/h.$$

All three derivative approximations here are centered (hence second order), and they are applied to first derivatives and hence have roundoff error increasing proportionally to h^{-1}, not h^{-2}. Can we manage to (partially) cheat the hangman in this way?!

(a) Show that in exact arithmetic fpp_0 defined above and in Exercise 12 are one and the same.

(b) Implement this method and compare to the results of Exercise 12. Explain your observations.

14. Using the centered difference $h^{-2}(f(x_{i+1}) - 2f(x_i) + f(x_{i-1}))$, construct an $(n-2) \times n$ differentiation matrix, D^2, for the second derivative of $f(x) = e^x \sin(10x)$ at the points $x_i = ih, i = 1, 2, \ldots, n-1$, with $h = \pi/n$. Record the maximum absolute error in $D^2\mathbf{f}$ for $n = 25, 50, 100$, and 200. You should observe $\mathcal{O}(n^{-2})$ improvement.

Compare these results against those obtained using the Chebyshev differentiation matrix, as recorded in Figure 14.5.

15. Consider the numerical differentiation of the function $f(x) = c(x)e^{x/\pi}$ defined on $[0,\pi]$, where
$$c(x) = j, \quad .25(j-1)\pi \le x < .25j\pi,$$
for $j = 1,2,3,4$.

   (a) Contemplating a difference approximation with step size $h = \pi/n$, explain why it is a very good idea to ensure that n is an integer multiple of 4, $n = 4l$.

   (b) With $n = 4l$, show that the expression $h^{-1}c(t_i)(e^{x_{i+1}/\pi} - e^{x_i/\pi})$ provides a second order approximation (i.e., $\mathcal{O}(h^2)$ error) of $f'(t_i)$, where $t_i = x_i + h/2 = (i+1/2)h$, $i = 0,1,\ldots,n-1$.

16. Let us continue with the problem and notation of Exercise 15.

   (a) Explain why both Chebyshev and Fourier differentiation matrices constructed over the interval $[0,\pi]$ are bound to yield poor accuracy here.

   (b) How would you construct a Chebyshev-type approximation with spectral accuracy for this problem?

17. Continuing with the notation of Exercise 15, consider approximating $q(x_i)$ in terms of values of $c(x)$ and $g(x)$, where
$$q(x) = -[c(x)g'(x)]'$$
is known to be square integrable (but not necessarily differentiable) on $[0,\pi]$. The function g is assumed given, and it has some jumps of its own to offset those of $c(x)$ so as to create a smoother function $\phi(x) = c(x)g'(x)$. The latter is often termed the *flux function* in applications.

   (a) Convince yourself yet again that Chebyshev and Fourier differentiations on the entire interval $[0,\pi]$ are not the way to go.

   (b) Evaluate the merits (or lack thereof) of the difference approximation
   $$h^{-1}\left[\frac{c_{i+1/2}(g_{i+1}-g_i)}{h} - \frac{c_{i-1/2}(g_i - g_{i-1})}{h}\right],$$
   with g_i, $g_{i\pm 1}$ and $c_{i\pm 1/2}$ appropriately defined for $i = 1,\ldots,n-1$.

   (c) We could instead write $q(x) = -c(x)g''(x) - c'(x)g'(x)$ and discretize the latter expression. Is this a good idea? Explain.

14.7 Additional notes

This chapter, more than its immediate predecessors, may be considered as "layered" in its development, addressing potentially different audiences along the way. The material in Sections 14.1 and 14.2 is often considered a precursor to finite difference or finite volume methods for solving partial differential equations; see, e.g., LeVeque [50] or Ascher [3]. Many practitioners know of the material covered in this chapter nothing more than these first two sections, and do not experience void as a result.

But numerical differentiation has more uses. Another relevant topic, spanning an entire area of research, deals with *symbolic computing*, where exact, symbolic formulas for differentiation are obtained. Symbolic computing (as implemented, for instance, in Maple) yields more information than a numerical value when applicable and is often very useful for mathematical analysis. But for practically differentiating very complicated functions, as well as appearing in formulas for simulating differential equations, the numerical option is the only practical one, and this is what we pursue here.

14.7. Additional notes

In the first three sections of the present chapter, good and useful formulas are devised and all seems well. But when the desire for high accuracy increases, things begin to fall apart, as described in Section 14.4. Finally, we end up with a disaster in the case of differentiating noisy data. The essential reason for this is that in the limit numerical differentiation is an *ill-posed* problem. See, e.g., Vogel [73] for much more than our brief, intuitive explanation.

A way around the hazards of numerical differentiation is *automatic differentiation*; see the relevant chapter in Nocedal and Wright [57] or the dedicated Griewank [33]. These methods are much faster than differentiation using symbolic computation, and they do not amplify noise. Their usage is limited, though, and should not be seen as potentially replacing all the basic methods introduced in this chapter.

Differentiation matrices are developed in Trefethen [68] in the context of *spectral methods*. For the latter, see also Fornberg [26].

Chapter 15

Numerical Integration

The need to integrate a function $f(x)$ arises often in mathematical modeling. This chapter is devoted to techniques for approximating this operation numerically. Such techniques are at the heart of approximation methods for solving differential and integral equations, which are important and active areas of research.

> **Note:** The material covered here is classical, but we do not settle for just presenting a sequence of integration rules. Rather, we take the opportunity to introduce several important concepts and approaches that extend well beyond numerical integration without having to go into complicating or confusing detail. This is the main reason for the present chapter's considerable length.

We concentrate on definite integrals. Thus, unless otherwise noted we seek to approximate

$$I_f = \int_a^b f(x)dx \approx \sum_{j=0}^n a_j f(x_j) \tag{15.1}$$

for a given finite interval $[a,b]$ and an integrable function f. The numerical integration formula, often referred to as a **quadrature rule**, has *abscissae* x_j and *weights* a_j.

In Section 15.1 we start off by deriving basic rules for numerical integration based on, guess what, polynomial interpolation. Then the arguments that led in previous chapters from polynomial interpolation to piecewise polynomial interpolation apply again: these basic rules are good locally but often not globally when the integration interval $[a,b]$ is not necessarily short. The composite formulas developed in Section 15.2 are similar in spirit to those of Chapter 11, although they are simpler here.

Sections 15.3–15.5 introduce more advanced concepts that arise, in a far more complex form, in numerical methods for differential equations. When designing basic rules for numerical integration, it is generally possible to achieve a higher accuracy order than is possible for numerical differentiation based on the same points. This is exploited in Section 15.3, where clever choices of abscissae are considered, resulting in **Gaussian quadrature**. Another approach called **Romberg integration** for obtaining high order formulas, extending the extrapolation methods of Section 14.2, is developed in Section 15.5.

In Section 15.4 an essential premise changes. We no longer seek just formulas that behave well, in principle, when some parameter h is "small enough." Rather, a general-purpose program is designed that delivers approximations for definite integrals that are within a user-specified tolerance of the exact value.

Finally, numerical integration in more than one dimension is briefly considered in the advanced Section 15.6. When there are many dimensions, the corresponding numerical methods change radically.

15.1 Basic quadrature algorithms

Basic quadrature rules are based on low degree polynomial interpolation. Given a function $f(x)$ on a short interval $[a,b]$, we can choose a set of nodes $x_0, x_1, \ldots, x_n \in [a,b]$ and construct a polynomial interpolant $p_n(x)$. From this it follows that

$$\int_a^b p_n(x)dx \quad \text{approximates} \quad \int_a^b f(x)dx.$$

> **Note:** Sections 15.1 and 15.2 cover the essentials for numerical integration. Following that, Sections 15.3–15.5 are less crucial but more advanced and possibly more fun.

Deriving basic rules

As in Section 14.3 on numerical differentiation, the Lagrange interpolation form suggests itself, here even more strongly. Suppose that abscissae $x_0, x_1, \ldots, x_n$ have been specified somehow. Then the interpolating polynomial in Lagrange form is

$$p_n(x) = \sum_{j=0}^n f(x_j) L_j(x)$$

(recall Section 10.3), where

$$L_j(x) = \frac{(x-x_0)\cdots(x-x_{j-1})(x-x_{j+1})\cdots(x-x_n)}{(x_j-x_0)\cdots(x_j-x_{j-1})(x_j-x_{j+1})\cdots(x_j-x_n)} = \prod_{\substack{k=0 \\ k \neq j}}^n \frac{(x-x_k)}{(x_j-x_k)}.$$

This leads to

$$\int_a^b f(x)dx \approx \int_a^b p_n(x)dx = \int_a^b \sum_{j=0}^n f(x_j) L_j(x) dx$$

$$= \sum_{j=0}^n f(x_j) \int_a^b L_j(x) dx.$$

We therefore set

$$a_j = \int_a^b L_j(x) dx.$$

These *quadrature weights* are precomputed *once and for all* as part of constructing a quadrature rule (and not each time we wish to evaluate a particular integral!).

15.1. Basic quadrature algorithms

Example 15.1. Set $n = 1$ and interpolate at the ends $x_0 = a$ and $x_1 = b$. Then

$$L_0(x) = \frac{x-b}{a-b}, \quad L_1(x) = \frac{x-a}{b-a}.$$

Thus, we have the weights

$$a_0 = \int_a^b \frac{x-b}{a-b} dx = \frac{b-a}{2},$$
$$a_1 = \int_a^b \frac{x-a}{b-a} dx = \frac{b-a}{2}.$$

The resulting **trapezoidal rule** (recall the notation in (15.1)) is

$$I_f \approx I_{trap} = \frac{b-a}{2}\Big[f(a) + f(b)\Big].$$

Now, instead of a linear interpolant at the interval ends, let us consider interpolating also at the middle using a quadratic. Thus, $n = 2$, $x_0 = a$, $x_1 = \frac{b+a}{2}$, $x_2 = b$. Forming the three Lagrange quadratic polynomials and integrating them yields the **Simpson rule**, given by

$$I_f \approx I_{Simp} = \frac{b-a}{6}\left[f(a) + 4f\left(\frac{b+a}{2}\right) + f(b)\right].$$

Figure 15.1 depicts the approximations to the area under the curve of $f(x)$ that these quadrature rules provide. ∎

The trapezoidal and the Simpson rules are instances of **Newton–Cotes** formulas. These are formulas based on polynomial interpolation at equidistant abscissae. If the endpoints a and b are included in the abscissae $x_0, \ldots, x_n$, then the formula is *closed*. Otherwise it is an *open* Newton–Cotes formula. The simplest example of an open formula of this sort is the **midpoint rule**, given by

$$I_f \approx I_{mid} = (b-a) f\left(\frac{a+b}{2}\right),$$

which uses a constant interpolant at the middle of the interval of integration. See Figure 15.1 and recall again the notation in (15.1).

Basic quadrature error

What is the error in these basic quadrature rules? Is it reasonably small? Is it controllable?
Recall from Section 10.5 that the error in a polynomial interpolation at $n + 1$ points is given by

$$f(x) - p_n(x) = f[x_0, x_1, \ldots, x_n, x] \prod_{i=0}^{n}(x - x_i).$$

The divided difference function $f[\cdot, \ldots, \cdot]$ at the $n+2$ points $x_0, x_1, \ldots, x_n$ and x is really just a function of the variable x, with all the other arguments x_i held fixed. The quadrature error is the

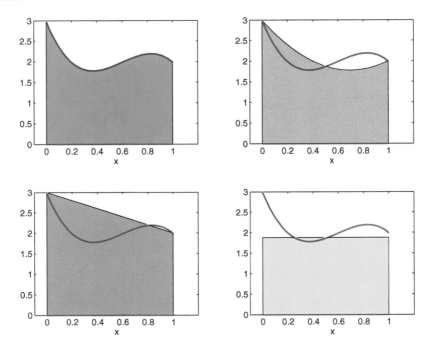

Figure 15.1. *Area under the curve. Top left (cyan): for $f(x)$ that stays nonnegative, I_f equals the area under the function's curve. Bottom left (green): approximation by the trapezoidal rule. Top right (pink): approximation by the Simpson rule. Bottom right (yellow): approximation by the midpoint rule.*

integral of this interpolation error, so

$$E(f) = \int_a^b f(x)dx - \sum_{j=0}^n a_j f(x_j)$$

$$= \int_a^b f[x_0, x_1, \ldots, x_n, x](x-x_0)(x-x_1)\cdots(x-x_n)dx.$$

From here on the going gets a tad tougher than when deriving truncation error expressions for numerical differentiation as in Section 14.1, and we proceed to examine our three basic rules one by one. If you are not interested in the details, then at least read the results, summarized in a table on the facing page.

Error in the trapezoidal rule

For the trapezoidal rule we get

$$E(f) = \int_a^b f[a,b,x](x-a)(x-b)dx.$$

Note that the function $\psi_1(x) = (x-a)(x-b)$ is nonpositive (since $x - a \geq 0$ and $x - b \leq 0$) for all $a \leq x \leq b$. Because $\psi_1(x)$ does not change its sign, there is an intermediate value ξ such that

$$E(f) = f[a,b,\xi] \int_a^b \psi_1(x)dx.$$

15.1. Basic quadrature algorithms

Below we summarize the three basic quadrature rules developed in Section 15.1 along with their error estimates.

Quadrature	Rule	Error
Midpoint	$(b-a)f(\frac{a+b}{2})$	$\frac{f''(\xi_1)}{24}(b-a)^3$
Trapezoidal	$\frac{b-a}{2}[f(a)+f(b)]$	$-\frac{f''(\xi_2)}{12}(b-a)^3$
Simpson	$\frac{b-a}{6}[f(a)+4f(\frac{b+a}{2})+f(b)]$	$-\frac{f''''(\xi_3)}{90}(\frac{b-a}{2})^5$

See Exercise 2. Further, there is an intermediate value η such that $f[a,b,\xi] = \frac{f''(\eta)}{2}$, and the integral of ψ_1 evaluates to $-\frac{(b-a)^3}{6}$. Thus, the basic trapezoidal rule error is estimated by

$$E(f) = -\frac{f''(\eta)}{12}(b-a)^3.$$

Error in the Simpson rule

For the Simpson rule we get

$$E(f) = \int_a^b f\left[a, \frac{a+b}{2}, b, x\right](x-a)\left(x - \frac{a+b}{2}\right)(x-b)dx.$$

Here, the function $\psi_2(x) = (x-a)(x - \frac{a+b}{2})(x-b)$ does not retain one sign throughout the interval $[a,b]$. On the other hand, ψ_2 is symmetric about the middle point $\frac{a+b}{2}$. Thus, its integral evaluates to 0. Expanding $f[a, \frac{a+b}{2}, b, x]$ in Taylor series up to the linear term, written as

$$f\left[a, \frac{a+b}{2}, b, x\right] = f\left[a, \frac{a+b}{2}, b, \frac{a+b}{2}\right] + \left(x - \frac{a+b}{2}\right)\frac{d}{dx}f\left[a, \frac{a+b}{2}, b, x\right]\bigg|_{x=\xi},$$

we obtain that the first, constant term $f[a, \frac{a+b}{2}, b, \frac{a+b}{2}]$ integrates against $\psi_2(x)$ to 0. The second term times $\psi_2(x)$ evaluates (believe us!) to the basic Simpson rule error satisfying

$$E(f) = -\frac{f''''(\zeta)}{90}\left(\frac{b-a}{2}\right)^5.$$

Error in the midpoint rule

The error estimate in the midpoint rule is similar in form to that of the trapezoidal rule and is given by

$$E(f) = \frac{f''(\xi)}{24}(b-a)^3.$$

See Exercise 3. This is so despite the apparently poorer approximation in Figure 15.1 as compared to that of the trapezoidal rule in the same figure.

Example 15.2. Let $f(x) = e^x$ and consider two integrals.

1. Set $a = 0$, $b = 1$.

   Then
   $$I_f = \int_0^1 e^x dx = e - 1 = 1.7183\ldots,$$
   $$I_{trap} = \frac{1}{2}(1 + e) = 1.8591\ldots,$$
   $$I_{Simp} = \frac{1}{6}(1 + 4e^{1/2} + e) = 1.7189\ldots,$$
   $$I_{mid} = e^{1/2} = 1.6487\ldots.$$

   Clearly the Simpson rule I_{Simp} gives the best approximation, at the highest cost of evaluation, to the exact value I_f. At the same time, none of these approximations is very accurate.

2. Set $a = 0.9$, $b = 1$.

   Then
   $$I_f = \int_{0.9}^1 e^x dx = e - e^{0.9} = .2586787173\ldots,$$
   $$I_f - I_{trap} = I_f - \frac{0.1}{2}(e^{0.9} + e) = -2.2 \times 10^{-4},$$
   $$I_f - I_{Simp} = I_f - \frac{0.1}{6}(e^{0.9} + 4e^{0.95} + e) = -9.0 \times 10^{-9},$$
   $$I_f - I_{mid} = I_f - 0.1 \times e^{0.95} = 1.1 \times 10^{-4}.$$

   All three rules perform much more accurately on the shorter interval, and the Simpson rule is particularly accurate. This is no surprise: we have $b - a = 0.1$ and this value is raised to higher powers in the error terms for these quadrature rules.

This simple example indicates that basic quadrature rules can be adequate on sufficiently short intervals of integration. ∎

The **precision** (also called *degree of accuracy*) of a quadrature formula is the largest integer ρ such that $E(p_n) = 0$ for all polynomials $p_n(x)$ of degree $n \leq \rho$. Thus, the trapezoidal and midpoint rules have precision 1, and the Simpson rule has precision 3. In Section 15.3 we further examine the precision of basic quadrature rules and develop methods of higher precision.

Specific exercises for this section: Exercises 1–4.

15.2 Composite numerical integration

Example 15.3, as well as Figure 15.1, clearly demonstrates that even for a very smooth integrand the basic quadrature rules may be ineffective when the integration is performed over a long interval. More sampling of $f(x)$ is intuitively required in such circumstances.

Increasing the order of the Newton–Cotes formulas is one way to go about this, but high-precision formulas of this sort suffer the same problems that high degree polynomial interpolation experiences over long intervals; recall Section 11.2.

15.2. Composite numerical integration

Instead, we proceed with the approach that is equivalent to approximating $f(x)$ with piecewise polynomials. Unlike splines and Hermite piecewise cubics, though, there is no challenge here to patch together the polynomial pieces in order to achieve some overall degree of smoothness. The resulting quadrature formulas, called *composite rules*, or **composite quadrature methods**, are the techniques most often used in practice.

Thus, in its simplest form we divide the interval $[a,b]$ into r equal subintervals of length $h = \frac{b-a}{r}$ each. Then

$$\int_a^b f(x)dx = \sum_{i=1}^r \int_{a+(i-1)h}^{a+ih} f(x)dx = \sum_{i=1}^r \int_{t_{i-1}}^{t_i} f(x)dx,$$

where $t_i = a + ih$. Now, apply one of the basic rules we saw earlier in Section 15.1 to each of the integrals on the right-hand side.

The associated error is obviously the sum of errors committed on each of the subintervals. If the error in the basic rule is written as $\tilde{K}(b-a)^{q+1}$ for some positive integer q and constant $\tilde{K}$, then we can write the error on the ith subinterval of length h as $K_i h^{q+1}$. Summing up and noting that $\sum_{i=1}^r h = \sum_{i=1}^r t_i - t_{i-1} = b - a$, we obtain

$$E(f) = \sum_{i=1}^r K_i h^{q+1} = K(b-a)h^q$$

for an appropriate constant K.

Let us next see what specific composite methods look like.

The composite trapezoidal method

The trapezoidal rule applied to the subinterval $[t_{i-1}, t_i]$ yields

$$\int_{t_{i-1}}^{t_i} f(x)dx \approx \frac{h}{2}[f(t_{i-1}) + f(t_i)].$$

So, the *composite trapezoidal method* is

$$\int_a^b f(x)dx \approx \frac{h}{2} \sum_{i=1}^r \left[f(t_{i-1}) + f(t_i) \right]$$

$$= \frac{h}{2}[f(t_0) + 2f(t_1) + 2f(t_2) + \cdots + 2f(t_{r-1}) + f(t_r)]$$

$$= \frac{h}{2}[f(a) + 2f(t_1) + 2f(t_2) + \cdots + 2f(t_{r-1}) + f(b)].$$

See Figure 15.2, where $r = 5$ and $t_i = \frac{i}{5}$, $i = 0, 1, \ldots, 5$.

The associated error is given by

$$E(f) = \sum_{i=1}^r \left(-\frac{f''(\eta_i)}{12} h^3 \right) = -\frac{f''(\eta)}{12}(b-a)h^2$$

for some $t_{i-1} \leq \eta_i \leq t_i$ and $a \leq \eta \leq b$. The last equality comes about by applying the Intermediate Value Theorem given on page 10 to the values $f''(\eta_i)$. The composite trapezoidal method is therefore a second order accurate method, meaning that its truncation error decreases like $\mathcal{O}(h^2)$.

Let us delay presenting a numerical example until after introducing the composite Simpson method (but you can have a peek at Figure 15.3 now if you must).

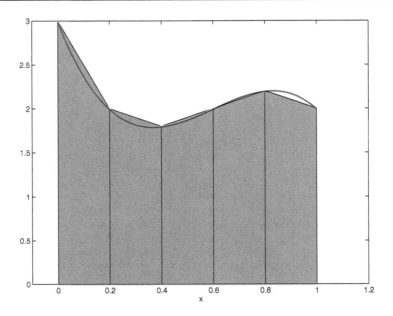

Figure 15.2. *Composite trapezoidal quadrature with $h = 0.2$ for the integral of Figure* 15.1.

The quadrature abscissae

Let us make a general distinction between the points $x_0, x_1, \ldots, x_n$ used by the chosen basic quadrature rule over a generic interval, say, $[-1, 1]$, and the points $t_0, t_1, \ldots, t_r$ which are used in the composite method to subdivide the interval $[a, b]$ of the given integration problem. The points (argument values) where the integrand $f(x)$ gets evaluated are the basic rule points $x_0, x_1, \ldots, x_n$ mapped into each subinterval, or *panel*, $[t_{i-1}, t_i]$. These points are

$$t_{i,k} = \frac{t_{i-1} + t_i}{2} + \frac{t_i - t_{i-1}}{2} x_k, \quad k = 0, 1, \ldots, n, \quad i = 1, \ldots, r.$$

There appear to be $r(n + 1)$ such arguments. For the composite trapezoidal method, where $n = 1$, this would imply $2r$ arguments, hence $2r$ function evaluations, whereas in the formula we saw there were only $r + 1$ function evaluations. The reason is that many of these arguments coincide: for the basic trapezoidal rule over the generic interval, $x_0 = -1$ and $x_1 = 1$. Thus, the rightmost point of the ith panel, $t_{i,1} = \frac{t_{i-1}+t_i}{2} + \frac{t_i - t_{i-1}}{2} x_1 = t_i$, coincides with the leftmost point of the next, $(i+1)$st panel, $t_{i+1,0} = \frac{t_i + t_{i+1}}{2} + \frac{t_{i+1} - t_i}{2} x_0 = t_i$, for $i = 1, 2, \ldots, r - 1$.

On the other hand, for the **composite midpoint method** the points $t_{i,k} = (t_i + t_{i-1})/2$ where the integrand f gets evaluated do not coincide and are all distinct. We return to this method later on, towards the end of this section.

The composite Simpson method

The composite Simpson method is one of the most commonly used general-purpose quadrature methods. We let r be *even* and consider the subintervals in *pairs*, thus forming double subintervals of length $2h$ each, $[t_{2k-2}, t_{2k}]$, $k = 1, 2, \ldots, r/2$. On each subinterval $[t_{2k-2}, t_{2k}]$ we interpolate the integrand $f(x)$ by a quadratic, then integrate these quadratics and sum up the results.

15.2. Composite numerical integration

So, on each of these double subintervals we have the basic rule

$$\int_{t_{2k-2}}^{t_{2k}} f(x)dx \approx \frac{2h}{6}[f(t_{2k-2})+4f(t_{2k-1})+f(t_{2k})].$$

Summing these up then yields the celebrated formula

$$\int_a^b f(x)dx \approx \frac{h}{3}\left[f(a)+2\sum_{k=1}^{r/2-1} f(t_{2k})+4\sum_{k=1}^{r/2} f(t_{2k-1})+f(b)\right].$$

Example 15.3. To get the hang of these composite methods, consider the approximation of

$$I = \int_0^1 e^{-x^2} dx \approx 0.746824133\ldots$$

by subdividing the interval $[0,1]$ into four equal panels. Thus, $t_i = ih$, $i = 0, 1, 2, 3, 4$, and $h = 0.25$. The composite trapezoidal and Simpson formulas yield

$$I_{trap} = .125[e^0 + 2e^{-h^2} + 2e^{-(2h)^2} + 2e^{-(3h)^2} + e^{-(4h)^2}] = 0.\underline{742}984\ldots,$$

$$I_{Simp} = \frac{.25}{3}[e^0 + 4e^{-h^2} + 2e^{-(2h)^2} + 4e^{-(3h)^2} + e^{-(4h)^2}] = 0.\underline{746855}\ldots.$$

The correct digits are underlined. Evidently, the composite Simpson method is more accurate here.

Figure 15.3 records maximum errors for various values of $r = 1/h$. The error in the Simpson method behaves like $\mathcal{O}(h^4)$ and is significantly better than the $\mathcal{O}(h^2)$ error of the trapezoidal method. ■

Composite Quadrature Methods.
With $rh = b - a$, where r is a positive integer (must be even in the Simpson case), we have the formulas

$$\int_a^b f(x)dx \approx \frac{h}{2}\left[f(a)+2\sum_{i=1}^{r-1} f(a+ih)+f(b)\right], \quad \text{trapezoidal}$$

$$\approx \frac{h}{3}\left[f(a)+2\sum_{k=1}^{r/2-1} f(t_{2k})+4\sum_{k=1}^{r/2} f(t_{2k-1})+f(b)\right], \quad \text{Simpson}$$

$$\approx h\sum_{i=1}^{r} f(a+(i-1/2)h), \quad \text{midpoint.}$$

The associated error for the composite Simpson formula is the sum of errors committed on each of the subintervals, and hence it is given by

$$E(f) = -\frac{f''''(\zeta)}{180}(b-a)h^4$$

for some $a \leq \zeta \leq b$. Thus, the error in the Simpson method is $\mathcal{O}(h^4)$, i.e., the order of accuracy is 4.

Observe that both for the trapezoidal rule and for the Simpson rule the order of accuracy of the composite method is one higher than the precision. Please convince yourself that this is no coincidence!

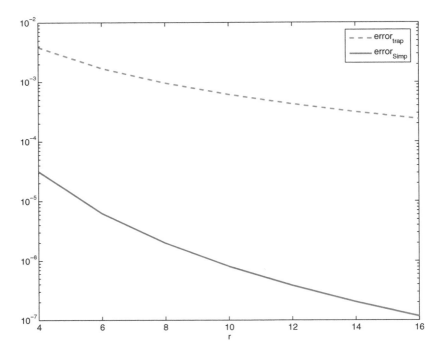

Figure 15.3. *Numerical integration errors for the composite trapezoidal and Simpson methods; see Example* 15.3.

Example 15.4. If the integrand $f(x)$ happens to be periodic on the interval $[a,b]$, implying in particular that $f(b) = f(a)$, then the trapezoidal method reads

$$\int_a^b f(x)dx \approx h \sum_{i=0}^{r-1} f(t_i).$$

As it turns out, the composite trapezoidal method is particularly accurate for periodic functions. The reason for this is beyond the scope of this chapter; it has to do with best ℓ_2-approximation and trigonometric polynomials. There are intriguing connections between this very simple composite quadrature method and the discrete Fourier transform (DFT) described in Sections 13.2 and 13.3. Here, let us just give a numerical example.

Thus, consider the same integrand as in Example 15.3, but on the interval $[-10, 10]$. At both interval ends this function and many of its derivatives are close to 0, so it is "almost periodic." Figure 15.4 depicts the performance of the composite trapezoidal and Simpson methods. The trapezoidal method yields spectacular results here.

We hasten to caution, though, that the typical picture would be more like Figure 15.3 than Figure 15.4. ∎

The computational cost of a quadrature method

Using composite quadrature methods we can drive the discretization error to be arbitrarily small by taking h smaller and smaller.[57] The price, however, is an increased computational cost. How

---

[57] We still assume that the discretization error dominates roundoff error. So, although the total error does not really tend to 0 in the pure sense as $h \to 0$, we can still concentrate on a practical range of values of h where the error in a method of order q decreases proportionally to h^q.

15.2. Composite numerical integration

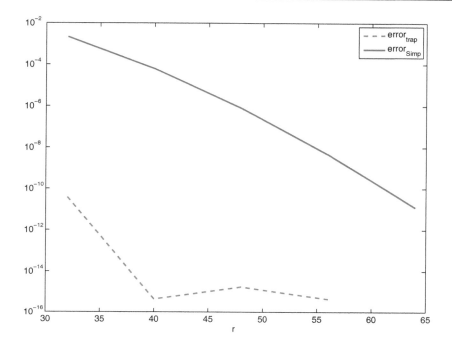

Figure 15.4. *Numerical integration errors for the composite trapezoidal and Simpson methods; see Example* 15.4.

can the algorithm's cost be quickly estimated? Typically, we imagine integrating a function $f(x)$ which is relatively expensive to evaluate and measure the approximate cost of a particular quadrature method by the number of evaluations of the function f that are required. In other words, we assume that function evaluations are significantly more costly than summation and the other arithmetic operations involved in quadrature. (You may recall that we use a similar approach in assessing computational cost of other algorithms; see, for example, Chapter 3.) We see that for the same value of h the costs of the composite trapezoidal method and the composite Simpson method are comparable.

Example 15.5. Approximate $\int_0^1 e^{-x^2} dx$ with an absolute error less than 10^{-5}.

Using the composite Simpson method we have

$$\int_0^1 e^{-x^2} dx = \frac{h}{3}\left[1 + 2\sum_{k=1}^{r/2-1} e^{-t_{2k}^2} + 4\sum_{k=1}^{r/2} e^{-t_{2k-1}^2} + e^{-1}\right] - \frac{f''''(\zeta)}{180}(b-a)h^4,$$

where $t_k = kh$. Here, since $f(x) = e^{-x^2}$ is particularly simple, we can calculate and bound the fourth derivative, obtaining

$$f''(x) = (-2 + 4x^2)e^{-x^2},$$
$$f''''(x) = (12 - 48x^2 + 16x^4)e^{-x^2},$$
$$\|f''''\|_\infty = \max_{0 \le x \le 1} |f''''(x)| = 12.$$

The last equality is valid because the maximum value of $f''''(x)$ on the interval $[0,1]$ is attained at $x = 0$ and evaluates to 12. So, we have a solid bound on the error, given by

$$|E(f)| \leq \frac{12h^4}{180} = \frac{1}{15r^4}.$$

The requirement on the error is therefore satisfied if we choose n such that

$$\frac{1}{15r^4} < 10^{-5}.$$

The first even r that satisfies this inequality is $r = 10$, i.e., $h = 0.1$. Thus, we evaluate

$$\int_0^1 e^{-x^2} dx \approx \frac{1}{30}\left[1 + 2\sum_{k=1}^{4} e^{-(\frac{2k}{10})^2} + 4\sum_{k=1}^{5} e^{-(\frac{2k-1}{10})^2} + e^{-1}\right] = 0.7468249\ldots.$$

Recall from Example 15.3 that the exact value is $\int_0^1 e^{-x^2} dx = 0.746824133\ldots$, so the absolute error is indeed less than its bound.

If we want to use the composite trapezoidal method instead of Simpson, then we must choose r so that the bound is satisfied for its error. This involves bounding the second derivative of f. Here, it is clear that $\max_{0 \leq x \leq 1} |f''(x)| = 2$. Thus, the error obeys

$$|E(f)| \leq \frac{2h^2}{12} = \frac{1}{6r^2}.$$

To bound this by 10^{-5} we must choose $r \geq 129$.

Clearly, then, for this problem the composite trapezoidal method would require significantly more computational effort than the composite Simpson method to achieve a comparable accuracy. ∎

Generally speaking, if the integrand is smooth and the required accuracy is high (error tolerance very small), then higher order methods are more efficient. For rough integrands the picture changes, though, and lower order methods may perform better. Note that the typical error bound for a composite method of order q reads

$$|E(f)| \leq c\|f^{(q)}\|h^q$$

for some moderate constant c. So, the higher the method's order, the higher the derivative of the integrand which must be bounded by a constant of moderate size for the method to be effective.

The composite midpoint method

A specialty of the closed Newton–Cotes rules, as compared to the open ones, is that in the composite formulas the last abscissa from one subinterval is the first abscissa of the next. We have seen this demonstrated with both the trapezoidal rule and the Simpson rule. This cuts down the number of function evaluations, which is one per mesh point for the composite trapezoidal method. For the midpoint rule, which is an open formula, this does not happen. The points where the integrand gets evaluated are

$$\frac{t_{i-1} + t_i}{2}, \quad i = 1, 2, \ldots, r,$$

so there are r of them, as compared to $r + 1$ for the composite trapezoidal method. (See Exercise 5.) Thus, although in Example 15.1 the midpoint rule achieves in one function evaluation essentially

15.2. Composite numerical integration

Theorem: Quadrature Errors.
Let f be sufficiently smooth on $[a,b]$, and consider a composite method using a mesh $a = t_0 < t_1 < \cdots < t_r = b$ with $h_i = t_i - t_{i-1}$. Denote $h = \max_{1 \leq i \leq r} h_i$. In the case of the Simpson method assume that $h_{i+1} = h_i$ for all i odd.

Then the error in the composite trapezoidal method satisfies

$$|E(f)| \leq \frac{\|f''\|_\infty}{12}(b-a)h^2,$$

the error in the composite midpoint method satisfies

$$|E(f)| \leq \frac{\|f''\|_\infty}{24}(b-a)h^2,$$

and the error in the composite Simpson method satisfies

$$|E(f)| \leq \frac{\|f''''\|_\infty}{180}(b-a)h^4.$$

what the trapezoidal rule achieves in two, in the composite formulas when $r \gg 1$ the performance of the midpoint and the trapezoidal methods is comparable, both in terms of error size and in terms of evaluation cost.

Stability of numerical integration

Generally speaking, for humans (at least those humans who have taken a calculus course) it is easier to differentiate a given function than to integrate it: the recipes for differentiation are more automatic and require less ingenuity.

For computers employing floating point arithmetic, however, there are certain aspects of integration which make it in some sense *easier* to deal with than differentiation. Indeed, differentiation may be considered as a *roughing* operation whereas integration (finding the primitive function) is a *smoothing* operation.

Example 15.6. Let
$$e(x) = \epsilon \cos(\omega x), \quad x \geq 0,$$
where $0 < \epsilon \ll 1$ and $1 \ll \omega$ are two parameters. Thus, the magnitude of e, $\|e\|_\infty = \epsilon$, is small, but $e(x)$ oscillates rapidly, just like a component of a roundoff error in the floating point representation of a smooth function (recall Figure 2.2).

Now, differentiating and integrating we have
$$e'(x) = -\epsilon \omega \sin(\omega x),$$
$$E(x) = \int^x e(t)dt = \frac{\epsilon}{\omega}\sin(\omega x).$$

So, the maximum magnitude of $e'(x)$ has grown significantly, whereas the magnitude of the indefinite integral, $E(x)$, has decreased significantly, as compared to that of $e(x)$. ■

The innocent observation about the smoothing property of integration is one worth remembering if you find yourself deriving numerical methods for differential equations.

Specific exercises for this section: Exercises 5–7.

15.3 Gaussian quadrature

In Sections 15.1 and 15.2 we have seen that the error in quadrature rules has a particular dependence on higher derivatives of the integrand $f(x)$. Specifically, a composite rule of order q has an error bound of the form
$$|E(f)| \leq C h^q \|f^{(q)}\|,$$
where C is some constant. Now, if $f(x)$ happens to be any polynomial of degree less than q, then $f^{(q)} \equiv 0$, and hence the quadrature error vanishes. Therefore, *a basic rule which has* precision $p = q - 1$ *leads to a composite method of accuracy order* q.

Recall that basic rules are based on polynomial interpolation and have the general form
$$\int_a^b f(x)dx \approx \sum_{j=0}^n a_j f(x_j).$$

A particular rule is determined upon specifying the abscissae $x_0, x_1, \ldots, x_n$. Can we choose these points wisely?

In the Newton–Cotes formulas, including all methods we have seen thus far, the abscissae were chosen equidistant, e.g., $x_0 = a$, $x_1 = \frac{a+b}{2}$, and $x_2 = b$ for the Simpson rule. This is often a convenient choice in practice. But now let us widen the search and consider *any* choice of the abscissae $\{x_i\}_{i=0}^n$. Our goal is to get as high a precision as possible; i.e., we want the rule to be exact on as large a class of polynomials as possible by selecting these abscissae judiciously. This then would lead to as high an order as possible for the corresponding composite method.

> **Note:** Sections 15.3–15.5 may be treated as stand-alone sections. They are essentially independent of one another, and it is possible to skip any of them without an immediate loss of continuity. If you have the time, however, we do not recommend skipping any of them because they bring to the fore several interesting and more advanced concepts.

Zeros of the Legendre polynomial

Let the polynomial $p_n(x)$ interpolate $f(x)$ at $n+1$ points satisfying
$$a \leq x_0 < x_1 < \cdots < x_n \leq b.$$

As before, the basic rule is
$$I_f = \int_a^b f(x)dx \approx \int_a^b p_n(x)dx,$$

with the corresponding error
$$E(f) = \int_a^b (f(x) - p_n(x))dx = \int_a^b f[x_0, x_1, \ldots, x_n, x] \prod_{i=0}^n (x - x_i)dx.$$

With the $n+1$ points $x_0, \ldots, x_n$ fixed, the divided difference expression $f[x_0, x_1, \ldots, x_n, x]$ is just a function of the independent variable x. Suppose that $f(x)$ is a polynomial of degree m. If $m \leq n$, then
$$f[x_0, x_1, \ldots, x_n, x] = \frac{f^{(n+1)}(\xi)}{(n+1)!} = 0,$$

15.3. Gaussian quadrature

no matter what the points $\{x_i\}$ *(or* ξ*) are*. Thus, any basic rule using $n+1$ points has precision of at least n. For instance, the trapezoidal rule has the minimal precision n with $n=1$.

If we are now allowed to choose these $n+1$ abscissae, then we have $n+1$ degrees of freedom to play with. Intuition suggests that we may be able, if lucky, to increase the precision by $n+1$ from the basic n to $2n+1$. We have already seen higher precision than n for both the midpoint and the Simpson rules. How can we get the precision as high as possible for a given n, in general?

Zeros of Legendre Polynomials.
To fully understand the derivation of Gauss points you would want to read about Legendre polynomials in Section 12.2. But merely for the mechanical aspect of obtaining their values it suffices to note that the Gauss points are the roots of the Legendre polynomials, and that the latter are given on the interval $[-1,1]$ by

$$\phi_0(x) = 1, \quad \phi_1(x) = x,$$
$$\phi_{j+1}(x) = \frac{2j+1}{j+1} x \phi_j(x) - \frac{j}{j+1} \phi_{j-1}(x), \quad j \geq 1.$$

If $f(x)$ is a polynomial of degree m, $m > n$, then $f[x_0, x_1, \ldots, x_n, x]$ is a polynomial of degree $m-n-1$; see Exercise 10.14 on page 327. But now recall from Section 12.2 that for a class of orthogonal polynomials $\phi_0(x), \phi_1(x), \ldots, \phi_{n+1}(x)$ that satisfy

$$\int_a^b \phi_i(x) \phi_j(x) dx = 0, \quad i \neq j,$$

we have

$$\int_a^b g(x) \phi_{n+1}(x) dx = 0$$

for any polynomial $g(x)$ of degree $\leq n$. So, we can choose the points $x_0, x_1, \ldots, x_n$ as the zeros (roots) of the *Legendre polynomial* $\phi_{n+1}(x)$, appropriately scaled and shifted from the canonical interval $[-1,1]$ to our present $[a,b]$. See Figure 12.2 on page 372. Then, for any polynomial $f(x)$ of degree $2n+1$ or less, the divided difference $f[x_0, x_1, \ldots, x_n, x]$ is a polynomial of degree n or less. Since we can write $\phi_{n+1}(x) = c_{n+1} \prod_{i=0}^{n}(x-x_i)$, we conclude from orthogonality that the quadrature error $E(f)$ vanishes for any such polynomial f. So, the precision is $2n+1$.

The resulting family of highest precision methods is called *Gaussian quadrature*, and the roots of the Legendre polynomial are called **Gauss points**. Once we are set on the Gauss points of some degree, the corresponding quadrature weights $\{a_j\}$ are obtained by integrating the Lagrange polynomials, as we did for the other basic rules considered in Section 15.1.

Example 15.7. Let $a = -1$ and $b = 1$. Then, according to the recursion formula given on the current page, the Legendre polynomials are

$$\phi_0(x) = 1, \ \phi_1(x) = x, \ \phi_2(x) = \frac{1}{2}(3x^2 - 1), \ \phi_3(x) = \frac{1}{2}(5x^3 - 3x), \ldots.$$

1. For $n = 0$, we obtain the *midpoint rule* with $x_0 = 0$ and $a_0 = 2$, yielding the formula

$$\int_{-1}^{1} f(x) dx \approx 2 f(0).$$

The precision is 1, the same as for the trapezoidal rule.

2. For $n = 1$, we get the roots $\pm\sqrt{1/3}$, yielding $x_0 = -\sqrt{1/3}$, $x_1 = \sqrt{1/3}$, and $a_0 = a_1 = 1$. The resulting rule is
$$\int_{-1}^{1} f(x)dx \approx f\left(-\sqrt{1/3}\right) + f\left(\sqrt{1/3}\right).$$
The precision is 3, the same as for the Simpson rule.

3. For $n = 2$, we find the roots $x_0 = -\sqrt{3/5}$, $x_1 = 0$, $x_2 = \sqrt{3/5}$. The weights are $a_0 = a_2 = \frac{5}{9}$, $a_1 = 2 - 2a_0 = \frac{8}{9}$. The formula reads
$$\int_{-1}^{1} f(x)dx \approx \frac{1}{9}\left(5f\left(-\sqrt{3/5}\right) + 8f(0) + 5f\left(\sqrt{3/5}\right)\right).$$
The precision is 5, which suggests that a composite form of this rule will have order of accuracy 6. ∎

Gaussian Quadrature.
On the canonical interval $[-1, 1]$ for a given nonnegative integer n:

- The *Gauss points* are the zeros of the Legendre polynomial of degree $n+1$, $\phi_{n+1}(x)$.

- The corresponding quadrature weights are
$$a_j = \frac{2(1-x_j^2)}{[(n+1)\phi_n(x_j)]^2}, \quad j = 0, 1, \ldots, n.$$

- The corresponding error is
$$\int_{-1}^{1} f(x)dx - \sum_{j=0}^{n} a_j f(x_j) = \frac{2^{2n+3}((n+1)!)^4}{(2n+3)!((2n+2)!)^2} f^{(2n+2)}(\xi).$$

It can be shown that the weights a_j are given, for the interval $[-1, 1]$, by
$$a_j = \frac{2(1-x_j^2)}{[(n+1)\phi_n(x_j)]^2}, \quad j = 0, 1, \ldots, n.$$
It is important to understand that these quadrature weights are independent of the integrand and are calculated once and for all, not each time we encounter a new integration problem.

Gaussian quadrature rules for a general interval

For a general interval $[a, b]$, we apply the affine transformation
$$t = \frac{b-a}{2}x + \frac{b+a}{2}, \quad -1 \leq x \leq 1.$$
Then $dt = \frac{b-a}{2}dx$, so
$$\int_{a}^{b} f(t)dt = \int_{-1}^{1} f\left(\frac{b-a}{2}x + \frac{b+a}{2}\right) \frac{b-a}{2} dx.$$

15.3. Gaussian quadrature

Thus, we get the formula

$$\int_a^b f(t)dt \approx \sum_{j=0}^n b_j f(t_j),$$

where

$$t_j = \frac{b-a}{2}x_j + \frac{b+a}{2},$$
$$b_j = \frac{b-a}{2}a_j.$$

These formulas for $t_0, \ldots, t_n$ and $b_0, \ldots, b_n$ depend only on the interval of integration, not the integrand, and are calculated once for a given quadrature.

The *error* in basic Gaussian rules can be shown to be

$$E_n(f) = \frac{(b-a)^{2n+3}((n+1)!)^4}{(2n+3)!((2n+2)!)^2} f^{(2n+2)}(\xi).$$

Example 15.8. Approximate the integral

$$I_f = \int_0^1 e^{-t^2} dt \approx 0.74682413281243$$

(see Examples 15.5, 15.13, and 15.17), this time using basic Gaussian quadrature with $n = 1$ and $n = 2$. Here we have the transformation

$$t_j = \frac{1}{2}(x_j + 1), \quad b_j = \frac{1}{2}a_j$$

for $j = 0, 1, \ldots, n$.

1. For $n = 1$ we have $t_0 = \frac{\sqrt{3}-1}{2\sqrt{3}}$, $t_1 = \frac{\sqrt{3}+1}{2\sqrt{3}}$, $b_0 = b_1 = \frac{1}{2}$. We calculate

$$I_{Gauss2} = 0.\underline{746}594689.$$

2. For $n = 2$ we have $t_0 = \frac{\sqrt{5}-\sqrt{3}}{2\sqrt{5}}$, $t_1 = \frac{1}{2}$, $t_2 = \frac{\sqrt{5}+\sqrt{3}}{2\sqrt{5}}$, $b_0 = b_2 = \frac{5}{18}$, $b_1 = \frac{4}{9}$. We calculate

$$I_{Gauss3} = 0.\underline{7468}14584.$$

The correct digits in these approximations are underlined.

These results, using but two and three function evaluations, respectively, are definitely impressive. With the basic Simpson rule we get the value $0.74718\ldots$, which is somewhat less accurate than I_{Gauss2}. As it turns out, obtaining an error comparable to I_{Gauss3} using the Simpson rule requires five function evaluations; see Example 15.13. ∎

Weighted Gaussian quadrature rules

The Legendre polynomials are not the only family of polynomials that are used for the choice of nodes for Gaussian-type quadrature. Variants exist that depend not only on the interval, but also on the choice of *weight function*, $w(x)$, in a way that is closely related to the weighted least squares problem considered in Section 12.3.

The general form of such a quadrature rule is

$$\int_a^b f(x)w(x)dx = \sum_{j=0}^n a_j f(x_j).$$

The parameters x_j and a_j are determined, as usual, by maximizing the precision of the integral.

It should come as no surprise (assuming you have read Section 12.3) that with the weight function $w(x) = \frac{1}{\sqrt{1-x^2}}$ on the interval $(-1,1)$, quadrature at Chebyshev points is obtained. Note that this weight function is unbounded when approaching the interval's endpoints from within.

More generally, an elegant way to find the nodes is by using the Gram–Schmidt algorithm described on page 374 in Section 12.3 to construct the required orthogonal polynomials with respect to the weight function, and then find their roots.

Example 15.9. Consider approximating the integral

$$I_f = \int_0^1 \frac{e^x}{\sqrt{x}} dx.$$

The integral is well-defined, even though the integrand is undefined for $x = 0$ and $\frac{e^x}{\sqrt{x}} \to \infty$ as $x \to 0$. The fact that it is well-defined can be observed upon using integration by parts, which gives $I_f = 2(e - \int_0^1 \sqrt{x} e^x dx)$. But let's approximate the given form of I_f directly.

Using the weighted Gaussian quadrature rule with $w(x) = \frac{1}{\sqrt{x}}$, the function $f(x) = e^x$ is well approximated by a polynomial. Thus, a weighted Gaussian quadrature suggests itself. For $n = 1$ we get in this way the value

$$I_f \approx a_0 e^{t_0} + a_1 e^{t_1} = 2.9245\ldots.$$

See Exercise 9. The exact result is $2.9253\ldots$. So, we obtain three correct decimal digits: not bad for two function evaluations and a singular integrand! This is an example where weighted Gaussian quadrature fares better than composite closed Newton–Cotes formulas, due to the singularity at the origin and the large values of the integrand in the immediate neighborhood of the origin. ∎

Composite Gaussian formulas

As with the other basic rules we saw, the way to use Gaussian quadrature in practice is often as a composite method, as discussed in Section 15.2. However, here there is no apparent trick for saving on function evaluations: the given interval is divided into r subintervals, $a = t_0 < t_1 < \cdots < t_r = b$, and the $n+1$ Gauss points are scaled and translated into each subinterval to form a set of $(n+1)r$ distinct points where the integrand is evaluated. Thus, the distinct points where f gets evaluated are

$$t_{i,k} = t_{i-1} + \frac{t_i - t_{i-1}}{2}(x_k + 1).$$

We have already mentioned in Section 15.1 how the composite midpoint method looks like. For the next ($n = 1$) member of the Gaussian quadrature family of rules we have to evaluate the integrand at the points

$$\frac{t_{i-1} + t_i}{2} \pm \frac{t_i - t_{i-1}}{2\sqrt{3}}, \quad i = 1, \ldots, r.$$

(Recall Example 15.7 and verify that you know what the composite rule for $n = 2$ looks like.)

15.3. Gaussian quadrature

Let the mesh be uniform, $t_i = ih$, with $h = \frac{b-a}{r}$. The error in the obtained **composite Gaussian rule** is estimated by

$$E_{n,h}(f) = \frac{(b-a)((n+1)!)^4}{(2n+3)!((2n+2)!)^2} f^{(2n+2)}(\xi) h^{2n+2}.$$

Counting function evaluations, we see that the two-point composite Gaussian method is comparable to the composite Simpson method, both in terms of order of accuracy and in terms of number of function evaluations. The Simpson method has the important added simplicity of equal subintervals. The Gauss method has a somewhat better error constant, requires one less function evaluation, and, most importantly, shows more flexibility with discontinuities in the integrand. The following, somewhat long, example demonstrates the latter point.

Example 15.10. Consider the integral

$$I_f = \int_0^{2\pi} f(x) dx,$$

where

$$f(x) = \begin{cases} \sin(x), & 0 \leq x \leq \pi/2, \\ \cos(x), & \pi/2 < x \leq 2\pi. \end{cases}$$

As evident from Figure 15.5 the integrand has a discontinuity at $x = \pi/2$. Nonetheless, there is no difficulty in carrying out the integration exactly, and we obtain

$$I = \int_0^{\pi/2} \sin(x) \, dx + \int_{\pi/2}^{2\pi} \cos(x) \, dx = 1 - 1 = 0.$$

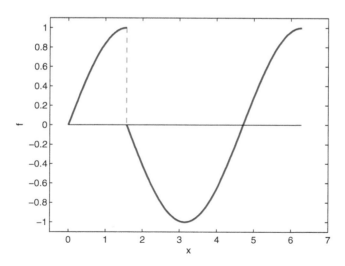

Figure 15.5. *The discontinuous integrand of Example* 15.10.

Let us see how our methods fare for this integral.

At first, note that any basic rule would yield an $\mathcal{O}(1)$ error, because such rules are all based on approximating the discontinuous function $f(x)$ by an infinitely smooth polynomial. For instance, the basic midpoint rule gives $I_{mid} = -2\pi$, which bears little resemblance to the exact value 0.

Likewise, composite methods will not work well if we partition the interval of integration $[0, 2\pi]$ in such a way that no partition point t_i coincides with the point of discontinuity $\pi/2$. For instance, applying the Simpson method to each of the subintervals $[0, \pi]$ and $[\pi, 2\pi]$ and adding the results gives the value $I_{Simpc} = 1.57$, which is again unacceptable.

So, we partition the interval of integration to 4 and to 8 uniform subintervals and apply our composite integration methods. The underlying approximation is now a piecewise polynomial with a breakpoint at the discontinuity location $x = \pi/2$, so we should be better off.

Using the composite methods we list some numerical results in the following table:

Method	# subintervals	Fevals	Error
Midpoint	4	4	5.23×10^{-16}
Trapezoidal	4	5	7.85×10^{-1}
Trapezoidal	8	9	3.93×10^{-1}
Gauss, $n = 1$	4	8	4.56×10^{-16}
Simpson	4	9	2.62×10^{-1}
Simpson	8	17	1.31×10^{-1}

The errors using Gaussian quadrature are but a few multiples of the rounding unit. The discretization error is, in fact, precisely 0. This is not general and is due to the symmetry in the present integrand that is matched by symmetry in the Gauss points. For instance, the basic abscissae $x_0 = -0.4$, $x_1 = 0.4$, which do not stand out in any obvious way, also make the discretization error vanish under the same circumstances here. These results are too good to be generally representative.

But the poor performance of the composite trapezoidal and Simpson methods is, unfortunately, typical and representative. These methods, upon taking for the rightmost basic point in one panel the same value as for the leftmost basic point in the next panel, tacitly assume that the integrand is continuous. An error of magnitude 1 is therefore generated at the point of discontinuity, and this error gets multiplied by h. Thus, the discretization error for both the trapezoidal method and the Simpson method is only first order accurate in h, as is evident in the error values listed in the above table.

Of course, if we know where the discontinuity in $f(x)$ is and assign this location as a breakpoint t_i, then we could apply two function evaluations there, one from the left and one from the right, and thus resurrect the high order of the trapezoidal and Simpson rules, but that leads to an overhead which is elegantly avoided by the Gaussian quadrature schemes. ■

Despite Example 15.10, the composite Simpson is the more usual choice for quadrature, because of the equal spacing of its evaluation points. This yields simplicity, particularly in computing error estimates (see Section 15.4). Moreover, the Simpson derivation avoids the extra notation and does not require the additional understanding that Gaussian quadrature requires. But when discontinuities appear or when solving differential equations, the methods introduced in this section are more popular.

Gauss–Radau quadrature

In some important situations the choice of the abscissae $x_0, x_1, \ldots, x_n$ in a given integration interval $[a, b]$ is not completely free. Basic quadrature rules are then determined where those abscissae that must be fixed are kept at their prescribed values, while the rest are chosen, as in Gaussian quadrature, to maximize precision and hence the order of accuracy of the corresponding composite rule.

15.3. Gaussian quadrature

One such family of quadrature rules is based on the **Radau points**. Here, $x_n = b$ is held fixed. The obtained precision with $n+1$ points is $2n$, so the order of the composite method is $2n+1$.

Example 15.11. Let $a = -1$ and $b = 1$, and apply the Gauss–Radau rule for a general function $f(x)$:

1. For $n = 0$ we set $x_0 = 1$ and get the rule

$$\int_{-1}^{1} f(x)dx \approx 2f(1).$$

   The precision is 0 and the corresponding composite method is first order accurate.

2. For $n = 1$ we set $x_1 = 1$, then calculate x_0 from

$$\int_{-1}^{1} (x - x_0)(x - x_1)dx = 0.$$

   The result is $x_0 = -1/3$.

   The quadrature weights subsequently evaluate to $a_0 = 3/2$, $a_1 = 1/2$. So, the formula is

$$\int_{-1}^{1} f(x)dx \approx \frac{1}{2}\left[f(1) + 3f(-1/3)\right].$$

   The composite form of this rule is third order accurate.

In quadrature there is little use for these methods. But in the solution of *stiff differential equations* (Section 16.5) they see major use, so let's not bury them just yet. ∎

Gauss–Lobatto quadrature

Another family of quadrature rules is based on the **Lobatto points**. Here, both $x_0 = a$ and $x_n = b$ are held fixed. The obtained precision with $n+1$ points is $2n-1$, so the order of the composite method is $2n$. The MATLAB function `quadl` is an adaptive quadrature routine based on Lobatto points.

Example 15.12. Let $a = -1$ and $b = 1$, and apply the Gauss–Lobatto rule for a general function $f(x)$:

1. We must have $n \geq 1$. For $n = 1$ we set $x_0 = -1$ and $x_1 = 1$. This yields the trapezoidal rule. Indeed, $2n = 2$ and the composite trapezoidal method is second order accurate, as we have seen.

2. For $n = 2$ we have $x_0 = -1$ and $x_2 = 1$. The choice of x_1 which maximizes orthogonality is $x_1 = 0$, and this yields the Simpson rule. Indeed, this rule has precision $2n - 1 = 3$ and the accuracy order of the composite method is 4.

3. Values of n that are higher than 2 do yield other, higher order Lobatto rules which we have not seen before in this chapter. The corresponding abscissae are no longer uniformly distributed.

   Like the Gauss points, the Lobatto points of any order are placed symmetrically about 0. ∎

In general, the Lobatto rule based on $n+1$ points is comparable in both precision and cost to the Gauss rule based on n points, so long as the integrand is continuous. For discontinuous integrands one is better off with Gaussian quadrature, as described in Example 15.10. Unlike the Gauss points and the Lobatto points, the Radau points are not placed symmetrically in the interval of integration. They provide for one-sided formulas.

Specific exercises for this section: Exercises 8–10.

15.4 Adaptive quadrature

Suppose we are to write a general-purpose program that evaluates definite integrals. The user of such a program would not be interested in questions such as which method we use, or what value of h is chosen. Rather, the user wants to input the interval ends a and b and the function $f(x)$ to be integrated. Our quadrature routine should subsequently produce a number Q_f such that

$$|Q_f - I_f| < \text{tol},$$

where $I_f = \int_a^b f(x)dx$, and `tol` is an error tolerance which is either specified (possibly reluctantly) by the user or set to a default value by our program.[58] How can such a quadrature routine be written? Let us address this question now.

Computing an error estimate

The first tool we must have is an error estimate. Such an estimate can be obtained, for instance, from two evaluations of the same integral using different values of h, as in Section 14.2. It can be shown for all the quadrature methods we have seen that the error in the composite rule can be written as

$$E(f) = E(f;h) = Kh^q + \mathcal{O}(h^{q+1}),$$

where q is the order of the method and K is a constant independent of h. This K can be estimated by two approximations, and thus the *principal error term*, Kh^q, can be estimated as well.

For instance, let us apply the composite trapezoidal rule twice, for h and for $h/2$, giving

$$R_1 = \frac{h}{2}[f(a) + 2f(a+h) + \cdots + 2f(b-h) + f(b)],$$
$$R_2 = \frac{h}{4}[f(a) + 2f(a+h/2) + 2f(a+h) + \cdots + 2f(b-h) + 2f(b-h/2) + f(b)].$$

Since the error in the first approximation is $I_f - R_1 \approx Kh^2$ and the error in the second approximation is $I_f - R_2 \approx K(h/2)^2$, we have

$$I_f - R_2 \approx \frac{1}{4}(I_f - R_1).$$

Thus, also

$$I_f - R_1 = (I_f - R_2) + (R_2 - R_1)$$
$$\approx \frac{1}{4}(I_f - R_1) + (R_2 - R_1),$$

---

[58] In MATLAB a call to such a quadrature function has the form `Q = quad('f',a,b,tol)`, where `tol` can be a combination of absolute and relative error tolerances. Let us not worry about this further complication in the tolerance interpretation: it is straightforward but rather tedious to write down.

15.4. Adaptive quadrature

so

$$I_f - R_1 \approx \frac{4}{3}(R_2 - R_1),$$
$$I_f - R_2 \approx \frac{1}{3}(R_2 - R_1).$$

Since R_1 and R_2 have been computed, this gives us a computable *error estimate*. Such an error estimate is sometimes called *a posteriori error estimate*, in contrast to error expressions such as $E(f) = -\frac{f''(\eta)}{12}(b-a)h^2$, which is an *a priori error estimate*. In the a priori expression nothing is based on the computed approximation, but there is no realistic value for the actual error either.

Likewise, if we use the Simpson rule (which is fourth order accurate; remember, $2^4 = 16$) and generate approximations S_1 by the composite rule with step h and S_2 by the same rule with step $h/2$, then

$$I_f - S_1 \approx \frac{16}{15}(S_2 - S_1),$$
$$I_f - S_2 \approx \frac{1}{15}(S_2 - S_1).$$

With such an error estimate at hand we can check whether our computed quadrature value is approximately within a specified tolerance to the exact integral value.

The advantage in both trapezoidal and Simpson rules, as compared to applying Gaussian quadrature rules in the same way, is that the function evaluations required for R_1 or S_1 are all used also for R_2 or S_2, respectively. Thus, if our goal were to evaluate R_2 or S_2, then the corresponding coarser approximations, and thus the error estimates, come essentially for free!

Adaptive subdivision

Let us proceed with the composite Simpson rule.

We construct the approximation Q_f as a sum of contributions of approximations Q_i over subintervals of $[a,b]$. The great beauty and simplicity of quadrature is that these local contributions do not mix: they simply add up, as opposed to the situation with piecewise interpolation or solving differential equations. Thus, if we have an arbitrary *mesh* (or *grid*) of points

$$a = t_0 < t_1 < \cdots < t_r = b,$$

where $h_i = t_i - t_{i-1} > 0$, and if we have quadrature values $Q_i \approx I_i = \int_{t_{i-1}}^{t_i} f(x)dx$, such that

$$|I_i - Q_i| < \frac{h_i}{b-a}\texttt{tol}, \quad i = 1, 2, \ldots, r,$$

then

$$Q_f = \sum_{i=1}^{r} Q_i$$

is within the prescribed error tolerance and as such is an acceptable return value to the user of our general-purpose program.

Next, observe that a general integrand $f(x)$ may vary more in some parts of the interval than in others. Thus, we construct an **adaptive** process where we *locally* refine the grid on which the composite Simpson rule is evaluated. The resulting divide-and-conquer procedure is simple and recursive. In pseudocode it reads

```
function Q = quadsimp (a, b, tol)

% evaluate the basic Simpson rule S_1(a,b)
% and the one based on one subdivision S_2(a,b);
% estimate the error E_2 = 1 / 15 | S_2(a,b) - S_1(a,b) |;

if E_2 < tol
   Q = S_2(a,b);
else
   Q1 = quadsimp (a, (a+b)/2, tol/2);
   Q2 = quadsimp ((a+b)/2, b, tol/2);
   Q = Q1 + Q2;
end
```

The reason this works even though the tolerances $\frac{h_i}{b-a}$ tol shrink upon subdivision is that the error in the Simpson rule decreases much faster than linearly in h.

Below we give a MATLAB program to carry out adaptive quadrature as outlined above. Before that, some implementation notes:

1. The technique is based on the assumption that the error estimate is valid, which in turn holds approximately only if f'''' does not vary violently. To compensate for overoptimism, let us insert a safety factor by defining

$$E_2 = \frac{1}{10}|S_2(a,b) - S_1(a,b)|.$$

2. The above recursion can be translated into an iterative algorithm at the price of loss of some conciseness and clarity. Importantly, the recursion is also wasteful in function evaluations. For instance, $f(a)$ is repeatedly evaluated. It is better to pass $f(a), f(b)$, and $f(\frac{a+b}{2})$ as input arguments to quade below. Then only the new $f(\frac{3a+b}{4})$ and $f(\frac{a+3b}{4})$ are evaluated in addition.

3. Problems can be constructed for which the tolerance is never met. See Example 15.16 below. In this unusual eventuality the recursion will never stop. Appropriate bells and whistles must be added here, too. Specifically, we limit the recursion depth level to at most 10.

Here is our set of three functions for adaptive quadrature:

```
function [Q,mesh,fevals] = quads(a, b, tol)
%
% function [Q,mesh,fevals] = quads(a, b, tol)
%
% adaptively evaluate Q - an approximation to the
% integral from a to b of func(x), to within tolerance tol.
%
% mesh is the resulting mesh where f was evaluated
% fevals is the number of function evaluations required.

% initialize

maxlevel = 10;
fa = func(a); fb = func(b); fab2 = func((a+b)/2);
sab = srule (a,b,fa,fab2, fb);
```

15.4. Adaptive quadrature

```
  fevals = 3;
  mesh(1) = a; mesh(2) = b; mesh(3) = (a+b)/2;

  % Evaluate the integral

  [Q,mesh,fevals] = quade (a, b, tol, fa, fab2, ...
fb, sab, maxlevel, fevals, mesh);

  % sort mesh in ascending order

  mesh = sort(mesh);


  function [Q,mesh,fevals] = quade(a, b, tol, fa, ...
fab2, fb, sab, level, fevals, mesh)
  %
  % adaptively evaluate Q - one recursive step.
  % called by quads
  %
  % mesh is the resulting mesh where f was evaluated
  % fevals is the number of function evaluations required.

  fa3b1 = func((3*a+b)/4);      fa1b3 = func((a+3*b)/4);
  mesh(fevals+1) = (3*a+b)/4;  mesh(fevals+2) = (a+3*b)/4;
  fevals = fevals + 2;
  ab2 = (a+b)/2;
  saab = srule (a,ab2,fa,fa3b1,fab2);
  sabb = srule (ab2,b,fab2,fa1b3,fb);

  % compare error estimate to tolerance

  if abs ( sab - (saab+sabb)) < 10*tol

     Q = saab+sabb;

  else

  %    must subdivide further
  %    first check if max recursion levels reached

     if level == 0
       fprintf('max recursion levels reached. ')
       fprintf('Current level = %d\n',level)
       Q = saab+sabb;

     else

        [Q1,mesh,fevals] = quade(a, ab2, tol/2, fa, fa3b1, ...
fab2, saab, level-1, fevals, mesh);

        [Q2,mesh,fevals] = quade(ab2, b, tol/2, fab2, fa1b3, ...
fb, sabb, level-1, fevals, mesh);
```

```
        Q = Q1 + Q2;
    end
end


function val = srule (a,b,fa,fab2, fb)
%
% function val = srule (a,b,fa,fab2, fb)
%
% evaluate basic Simpson rule
%
val = (b-a)/6 * (fa + 4*fab2 + fb);
```

Example 15.13. Let us test our implementation by computing, yet again, approximations to the integral

$$I_f = \int_0^1 e^{-x^2} dx.$$

We experiment with `tol` $= 10^{-3}, 10^{-5}, 10^{-7}, 10^{-9}$, and 10^{-11}. Using calls such as

`[Q,mesh,fevals] = quads(0,1,1.e-11);`

we obtain a very close agreement ($< 10^{-12}$) between the results with `tol` $= 10^{-11}$ and `tol` $= 10^{-13}$. Taking the value with `tol` $= 10^{-13}$ as exact, $I_f = 0.74682413281243$, we measure "exact errors" and record them in the following table:

tol	Error	Fevals
10^{-3}	3.1×10^{-5}	5
10^{-5}	1.2×10^{-7}	17
10^{-7}	7.8×10^{-9}	33
10^{-9}	6.1×10^{-11}	121
10^{-11}	2.8×10^{-13}	417

Our program appears somewhat conservative for this example, but that's fine: this integrand is particularly well behaved.

The meshes obtained by the adaptive codes for this problem instance are uniform, so the results are directly comparable to those of Example 15.5, which also uses the composite Simpson method on a uniform mesh. ∎

Example 15.14. On to a more challenging example, we use our adaptive routine to approximate the integral

$$\int_{1.5}^4 f(x)dx = \int_{1.5}^4 \left(\frac{200}{2x^3 - x^2}\right)(5\sin(20/x))^2 dx$$

to within `tol` $= 5 \times 10^{-3}$. Using 113 function evaluations, we find

$$I_f = \int_{1.5}^4 f(x)dx \approx 281.08 = Q_f.$$

15.4. Adaptive quadrature

A similar call with `tol` $= 10^{-5}$ returns, after 537 function evaluations, a value for the approximate integral which suggests that all the above displayed digits for Q_f are correct.

Figure 15.6 displays the integrand $f(x)$ and marks for each node t_i at which the adaptive routine evaluates $f(x)$. In MATLAB, the command

```
plot(mesh,zeros(size(mesh)),'+')
```

places blue "+" marks where mesh values are on the x-axis.

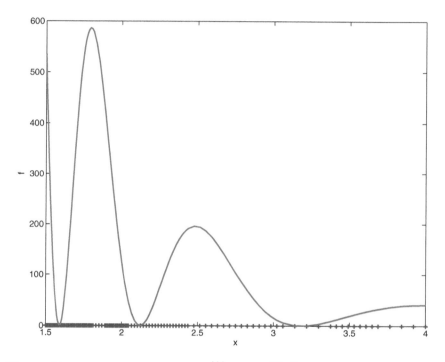

Figure 15.6. *The integrand* $f(x) = \frac{200}{2x^3-x^2}(5\sin(\frac{20}{x}))^2$ *with quadrature mesh points along the x-axis for* `tol` $= 5 \times 10^{-3}$.

Looking at the plot of the function and the clustering of the integration nodes, it is clear that the subintervals (panels) are smallest where the function $f(x)$ is changing most rapidly. The subintervals are widest where $f(x)$ is changing most slowly. ∎

Example 15.15. Next, we use our routine to approximate the integral

$$\int_0^2 e^{2x} \sin(6x)dx$$

to within 10^{-3}. We recommend that you write a 3-line script to picture this integrand. The program uses 41 function evaluations and finds the approximate value

$$I_f \approx -8.2239.$$

Let us compare this performance of the adaptive routine to the performance of the nonadaptive methods considered in Section 15.2. Using the composite trapezoidal method with equal size panels (or a uniform mesh), the error is

$$E(f) = -\frac{(b-a)}{12}h^2 f''(\eta), \quad \eta \in [a,b], \quad h = \frac{b-a}{r}.$$

In this case, $a = 0$ and $b = 2$, so $h = 2/r$, and

$$f'(x) = e^{2x}(2\sin(6x) + 6\cos(6x)),$$
$$f''(x) = e^{2x}(-32\sin(6x) + 24\cos(6x)).$$

This yields

$$\max_{\eta \in [0,2]} |f''(\eta)| \simeq 2.0432 \times 10^3 \text{ at } \eta = 2,$$

and hence

$$|E(f)| \leq \frac{2}{12}\left(\frac{2}{r}\right)^2 (2.0432 \times 10^3).$$

If $|E(f)| < 10^{-3}$ is desired, then we must choose r so that

$$\frac{2}{3r^2}(2.0432 \times 10^3) < 10^{-3}.$$

This implies that

$$r \geq 1168,$$

so the composite trapezoidal method requires at least $(1168 + 1)$ function evaluations to achieve this kind of accuracy, which is much more than the adaptive routine needs.

Using the composite Simpson method on a uniform mesh, the error is

$$E(f) = -\frac{(b-a)}{180}h^4 f^{(4)}(\eta) \quad \left(\eta \in [a,b], h = \frac{b-a}{r}\right).$$

In this case, $a = 0$ and $b = 2$, so $h = 2/r$. The third and fourth order derivatives of f are

$$f'''(x) = e^{2x}(-208\sin(6x) - 144\cos(6x)),$$
$$f^{(4)}(x) = e^{2x}(448\sin(6x) - 1536\cos(6x)).$$

This yields

$$\max_{\eta \in [0,2]} |f^{(4)}(\eta)| \simeq 8.39 \times 10^4 \text{ at } \eta = 2,$$

and hence

$$|E(f)| \leq \frac{1}{90}\left(\frac{2}{r}\right)^4 (8.39 \times 10^4).$$

The requirement $|E(f)| < 10^{-3}$ leads to choosing r so that

$$\frac{16}{90r^4}(8.39 \times 10^4) < 10^{-3},$$

which implies

$$r \geq 63.$$

So, the composite Simpson method using a uniform step size requires at least $(64 + 1)$ (r must be even) function evaluations to achieve this kind of accuracy. This is again more than the adaptive routine requires, although not by much.

15.5. Romberg integration

In summary, to approximate $\int_0^2 e^{2x}\sin(6x)dx$ to within an error of 10^{-3}, the three methods considered require the following number of function evaluations:

Method	Fevals
Adaptive `quads`	41
Composite Simpson	65
Composite Trapezoidal	1169

The composite Simpson approach, adaptive or not, handles this problem well. ∎

It is also possible to merge adaptive quadrature with Romberg integration, described next in Section 15.5, where not only the step size h but also the order of the method used (related to the number of extrapolations) is adaptive and may be varied locally, achieving a higher order method only where the integrand appears to vary smoothly and slowly. There are some successful implementations of this idea.

Example 15.16. It is not difficult to construct an example for which the a posteriori error estimate fails. For instance, the function $\sin(4\pi x)$ vanishes at all points $0, 1, 1/2, 1/4$, and $3/4$. Therefore, so does the function
$$f(x) = e^{-x}\sin(4\pi x).$$
For this integrand, we have the definite integral value
$$I_f = \int_0^1 e^{-x}\sin(4\pi x) \neq 0,$$
and yet both $S_1(0,1) = 0$ and $S_2(0,1) = 0$.

Two wrongs do not make a right in this case, and the error estimate $\frac{1}{15}(0-0) = 0$ is wrong as well. Our adaptive algorithm will stop with the false value $Q_f = 0$.

On the other hand, using the same adaptive algorithm to calculate approximations up to $\text{tol} = 10^{-7}$ for $\int_0^{0.7} e^{-x}\sin(4\pi x)dx$ and for $\int_{0.7}^1 e^{-x}\sin(4\pi x)dx$ (call them Q_1 and Q_2, respectively), we get
$$Q_f = Q_1 + Q_2 = 0.049986,$$
which as an approximation to I_f is accurate to the number of digits shown. No "bad luck" this time. Here is an example where an arbitrary modification, discreetly implemented in a general black box code, can make the chance of catching the software off-guard more remote. ∎

Adaptive quadrature is not always the automatically winning approach. In some situations where error control is less important and effective utilization of parallelism is, a uniform predetermined mesh on which to evaluate the integrand can have strong merits.

Specific exercises for this section: Exercises 11–12.

15.5 Romberg integration

The composite trapezoidal method has an attractive simplicity. Moreover, it can be shown (Exercise 15) that its error term, $E(f) = -\frac{f''(\eta)}{12}(b-a)h^2$, can be further expanded in *even powers of h*, written as

$$E(f;h) = E(f) = K_1 h^2 + K_2 h^4 + \cdots + K_s h^{2s} + \mathcal{O}(h^{2s+1}),$$

where K_i are constants depending on higher and higher derivatives of f, but not on h.[59]

The main drawback of the trapezoidal method as such is its low order accuracy (except for functions $f(x)$ that are periodic on the interval of integration $[a,b]$). This, and the above form of the error, invite extending the **extrapolation** process introduced for numerical differentiation in Section 14.2 to numerical integration. In the present context of quadrature the resulting method is called *Romberg integration*.

One extrapolation step

Let us demonstrate the first step first. For an integral value $I_f = \int_a^b f(x)dx$, we set

$$h_1 = h = b - a, \quad h_2 = \frac{h}{2}.$$

Then we calculate the two composite trapezoidal approximations

$$R_{1,1} = \frac{h_1}{2}(f(a) + f(b)),$$

$$R_{2,1} = \frac{h_2}{2}(f(a) + 2f(a+h_2) + f(b)).$$

The corresponding errors are

$$E_{1,1} = I_f - R_{1,1} = K_1 h^2 + K_2 h^4 + K_3 h^6 + \cdots,$$

$$E_{2,1} = I_f - R_{2,1} = K_1 (h/2)^2 + K_2 (h/2)^4 + K_3 (h/2)^6 + \cdots.$$

Now, note that $E_{1,1} = 4 E_{2,1} + \mathcal{O}(h^4)$. Specifically, we have

$$E_{1,1} - 4 E_{2,1} = \frac{3}{4} K_2 h^4 + \frac{15}{16} K_3 h^6 + \cdots.$$

But also $E_{1,1} - 4 E_{2,1} = (4 R_{2,1} - R_{1,1}) - 3 I_f$. Hence

$$I_f = \frac{4 R_{2,1} - R_{1,1}}{3} - \frac{1}{4} K_2 h^4 - \frac{5}{16} K_3 h^6 + \cdots,$$

so

$$R_{2,2} = \frac{4 R_{2,1} - R_{1,1}}{3} = R_{2,1} + \frac{R_{2,1} - R_{1,1}}{3}$$

is an $\mathcal{O}(h^4)$ accurate approximation to I_f, because the second order term has been eliminated.

The general algorithm

The above extrapolation step is generalized in an obvious fashion. To obtain a formula accurate to $\mathcal{O}(h^{2s})$, we simply generate composite trapezoidal approximations with $h, h/2, \ldots, h/2^{s-1}$ and then apply the Richardson extrapolation procedure just described $s - 1$ times. This is captured in an algorithm given on the facing page.

---

[59] Typically, K_i depends on $\|f^{(2i)}\|$. Deriving such an error expansion in general is not simple, but it can nevertheless be done.

15.5. Romberg integration

Algorithm: Romberg Integration.
Given a smooth integrand $f(x)$ on an interval $[a,b]$, and utilizing a starting step $h = (b-a)/r$ and a positive integer s, construct an $\mathcal{O}(h^{2s})$ approximation to $I_f = \int_a^b f(x)dx$ as follows:

1. Evaluate the starting trapezoidal formula

$$R_{1,1} = \frac{h}{2}\left[f(a)+f(b)+2\sum_{k=1}^{r-1} f(a+kh)\right].$$

2. For $j = 1,2,\ldots,s-1$

   (a) set $h = h/2$, and calculate the trapezoidal approximation on the finer mesh by

   $$R_{j+1,1} = \frac{1}{2}R_{j,1} + h \sum_{k=1}^{r*2^{j-1}} f(a+(2k-1)h).$$

   (b) calculate

   $$R_{j+1,k} = R_{j+1,k-1} + \frac{R_{j+1,k-1}-R_{j,k-1}}{4^{k-1}-1}, \quad k = 2,3,\ldots,j+1.$$

Then $R_{j,j}$ provides an $\mathcal{O}(h^{2j})$-accurate approximation for I_f, $1 \leq j \leq s$.

In this algorithm, generating all the values $R_{j,1}$ for $j = 1,\ldots,s$ requires in total only $r*2^{s-1} + 1$ evaluations of the integrand f. The integrand is evaluated at the same points as for the composite trapezoidal rule with $r*2^{s-1}$ panels, $R_{s,1}$.

The computations can be ordered in a triangular table:

$\mathcal{O}(h^2)$	$\mathcal{O}(h^4)$	$\mathcal{O}(h^6)$	$\ldots$	$\mathcal{O}(h^{2s})$
$R_{1,1}$				
$R_{2,1}$	$R_{2,2}$			
$R_{3,1}$	$R_{3,2}$	$R_{3,3}$		
$\vdots$	$\vdots$	$\vdots$	$\ddots$	
$R_{s,1}$	$R_{s,2}$	$R_{s,3}$	$\ldots$	$R_{s,s}$

Note that this table can be generated one row at a time. Thus, we do not have to determine in advance what order of method we want. If the function $f(x)$ is really expensive to evaluate, then most of the cost of generating the high order approximation $R_{s,s}$ goes into generating the second order approximation $R_{s,1}$.

Example 15.17. Let us generate the Romberg table for the (by now famous) integral

$$I_f = \int_0^1 e^{-x^2} dx.$$

For $r = 1$ and $s = 4$ in the Romberg integration algorithm, the obtained values (to 6 digits) are

$$
\begin{array}{llll}
.683940 & & & \\
.731370 & .747180 & & \\
.742984 & .746855 & .746834 & \\
.745866 & .746826 & .746824 & .746824
\end{array}
$$

The corresponding errors $I_f - R_{j,k}$ are

$$
\begin{array}{llll}
6.29\text{e-}2 & & & \\
1.55\text{e-}2 & -3.56\text{e-}4 & & \\
3.84\text{e-}3 & -3.12\text{e-}5 & -9.58\text{e-}6 & \\
9.59\text{e-}4 & -1.99\text{e-}6 & -3.71\text{e-}8 & 1.14\text{e-}7
\end{array}
$$

The errors down the first column decrease by a factor of roughly 4 each time h is cut in half, as they should. Down the second column the decrease is by a factor of a bit less than the expected $2^4 = 16$. In the third column we see an improvement factor of 25 instead of $2^6 = 64$. The value of $R_{4,4}$ is actually slightly worse than $R_{4,3}$, although both are pretty good for 9 function evaluations: compare to the composite Simpson method in Example 15.5. ■

Assessment of the Romberg method

As Example 15.17 shows, the Romberg integration can be quite effective. Moreover, an *error estimate* can be obtained for each approximation in the $(s-1) \times (s-1)$ Romberg table in a manner similar to that described in Sections 15.4 and 14.2 (see Exercise 16).

In practice, however, other effects such as larger high derivatives of f and roundoff errors may intervene if the starting step size h is too coarse or the extrapolation is carried out too many times.

In general, Romberg integration would be particularly effective if carried out for very smooth integrands and if the accuracy requirements are high. But it is not always the most robust method, especially not for integrands with discontinuities or other rough behavior. The subtle point here is that the starting h must already be fine enough so that the error in the trapezoidal method exhibit the expansion on which the entire procedure is built. For instance, $K_1 h^2$ should be significantly larger than $K_2 h^4$ for the actual, fixed values of h, K_1, and K_2, without relying on a limit estimate as $h \to 0$; and so on.

Specific exercises for this section: Exercises 13–16.

15.6 *Multidimensional integration

Most of the present chapter is devoted to one-dimensional integration. The concepts and methods introduced provide a precursor for similar but more challenging tasks in the more involved context of solving differential equations.

More complex problems also arise when considering integration in several independent variables. These often arise in mathematical and statistical modeling. A natural question then is if and how the methods we have seen thus far in the present chapter extend to d-dimensional problems.

As it turns out, from the numerical point of view it is natural to split multidimensional integration problems into two classes: those with only a few (say, $d = 2$ or 3) independent variables and

15.6. *Multidimensional integration

those with many. For the first class the variables typically correspond to physical space variables (as in the few PDEs that occasionally arise in this text), and we attempt to extend the methods we have studied thus far.

Problems of the second class typically arise where the arguments of the integrand are independent, varying parameters, of which there can be hundreds and more. Such integrations require a completely different treatment, as briefly described in what follows.

Integration with a few arguments

There are some parallels between the development in this chapter and that of Chapters 10 and 11: the basic quadrature rules correspond to polynomial interpolation and composite rules correspond to piecewise polynomial interpolation. Extending this line, we now consider integration over a domain $\Omega \subset \mathcal{R}^d$ that, as in Section 11.6, may not be simply a square or a cube. Fortunately, however, the situation is simpler here than in Section 11.6 in the same way that the methods of Section 15.2 are simpler than those of Sections 11.2–11.4. Here, unlike what is depicted in Figure 11.13, we need not worry about global continuity across Ω of the approximate integrand. In fact, if $C_1, C_2, \ldots, C_N$ are mutually exclusive subdomains in $\mathcal{R}^d$ such that $\Omega = \cup_{i=1}^{N} C_i$, then

$$I_f = \int_\Omega f(\mathbf{x})d\mathbf{x} = \sum_{i=1}^{N} \int_{C_i} f(\mathbf{x})d\mathbf{x},$$

where the integrand f is a function of $\mathbf{x} = (x_1, x_2, \ldots, x_d)$.

Often the subdomain Ω is simple enough that the multiple integral I_f can be written as an **iterated integral** of the form

$$I_f = \int_{l_0}^{u_0} dx_1 \int_{l_1(x_1)}^{u_1(x_1)} dx_2 \cdots \int_{l_{d-1}(x_1,\ldots,x_{d-1})}^{u_{d-1}(x_1,\ldots,x_{d-1})} f(x_1,\ldots,x_d)\, dx_d.$$

Despite the uninviting notation this is great news: it means that we can perform the integration recursively, one dimension at a time. Thus, we have in the case $d = 2$, with $x = x_1$ and $y = x_2$, the integral

$$I_f = \int_{l_0}^{u_0} \left[\int_{l_1(x)}^{u_1(x)} f(x,y) dy \right] dx.$$

The approximation of $\int_{l_0}^{u_0} g(x)dx$, where $g(x) = \int_{l_1(x)}^{u_1(x)} f(x,y)dy$, can be carried out, for instance, using composite Simpson or basic Gaussian quadrature. Then, for each point x where $g(x)$ is required for the quadrature in x, we evaluate g by a similar one-dimensional quadrature in y. See Figure 15.7.

Example 15.18. Let us calculate the area of the unit circle defined by $x^2 + y^2 \leq 1$. The exact solution is $I_f = \pi$.

An integral of $f(x, y)$ over the unit circle can be written as

$$I_f = \int_{-1}^{1} \left[\int_{-\sqrt{1-x^2}}^{\sqrt{1-x^2}} f(x,y) dy \right] dx.$$

Also, in this simple example, $f \equiv 1$.

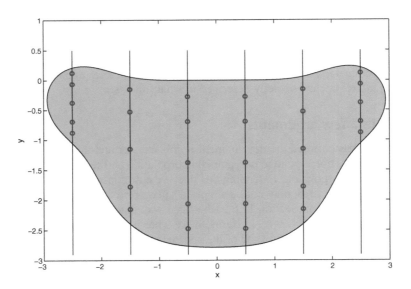

Figure 15.7. *Approximating an iterated integral in two dimensions. For each of the quadrature points in x we integrate in y to approximate g(x).*

The integral in the square brackets is thus given by $g(x) = 2\sqrt{1-x^2}$. This value for each x is obtained exactly by the simplest basic quadrature rules in this chapter, because these methods all reproduce integrals of constants. For the integration of $g(x)$ we now use our routine quads given in Section 15.4 with tolerance 1.e-6. This yields a value with error 2.11e-7, consuming 345 function evaluations. ∎

Integration in many dimensions

The methods that are briefly mentioned here involve random sampling of the integrand $f(\mathbf{x}) = f(x_1, \ldots, x_d)$. In its simplest form, the **Monte Carlo** method evaluates the integrand at a random sample of points from the domain Ω and averages the results.

The implementation of this method is therefore extremely simple: we draw d numbers for $(x_1, \ldots, x_d)$ from an appropriately scaled uniform distribution (i.e., from a bounding box of Ω) and require only a mechanism to tell whether the point so obtained is inside Ω or not.

Example 15.19. Returning to Example 15.18, we can define the integrand as having the value 1 inside the unit circle and 0 outside. So, we draw N samples (x, y) from a uniform distribution on $(-1, 1) \times (-1, 1)$ (using the MATLAB function rand), checking at each sample point whether $x^2 + y^2 < 1$. Multiplying the fraction of positive responses by 4 (the area of the unit square), we obtain the following absolute errors in approximating $I_f = \pi$: for $N = 100$, error = .2984; for $N = 10,000$, error = 1.56e-2; and for $N = 1,000,000$, error = 6.43e-4. The error is seen to be going down slowly, roughly like $\sqrt{N^{-1}}$. ∎

Compared to the results of Example 15.18, the results of Example 15.19 look laughable. Even if we needed 300^2, not just 300, function evaluations to achieve 5 significant digits with the former approach, this is still much more accurate than what the Monte Carlo method yields in many more function evaluations. However, in more dimensions, i.e., larger d, the relative comparison changes and we laugh no more. While deterministic methods that extend one-dimensional algorithms through iterated integrals, or by any other means, essentially require $\mathcal{O}(J^d)$ function evaluations if J evaluations are utilized per dimension, the Monte Carlo method, according to the statistical law of large numbers, requires $\mathcal{O}(\sqrt{N})$ evaluations for a constant error reduction regardless of dimensionality d. Thus, if we modestly set $J = 10$, then the previous approaches are all limited in practice at best to $d \leq 10$, say. When rough integration in hundreds of dimensions is desired, Monte Carlo–type methods, and other special methods, are the only methods of choice.

The level of complexity of methods that are based on random sampling can increase significantly if the function to be integrated has regions with sharp peaks and in other situations. We close the discussion at this somewhat arbitrary juncture by noting that it is not difficult to generate examples in many variables such that no practical integration method is available.

Specific exercises for this section: Exercises 17–18.

15.7 Exercises

0. **Review questions**

    (a) Define quadrature rule and precision.

    (b) In what basic way is numerical integration easier than numerical differentiation?

    (c) Why is polynomial interpolation a convenient tool for deriving quadrature rules, and why in Lagrange form and not Newton form?

    (d) Compare the properties of the basic trapezoidal and midpoint rules.

    (e) Define a composite quadrature method.

    (f) How come the composite trapezoidal and midpoint rules cost about the same?

    (g) What are the accuracy orders of the composite Simpson, midpoint, and trapezoidal methods?

    (h) What is the accuracy order of the composite Gaussian quadrature method using k points in each mesh subinterval (so, $k = 1$ for midpoint)?

    (i) Define Gauss, Radau, and Lobatto points and explain their importance in quadrature.

    (j) Give two advantages that adaptive quadrature has over using a composite method on a fixed number of uniform panels. Are there potential disadvantages?

    (k) What is the Romberg integration and how does it relate to the Richardson extrapolation?

1. Show in detail how to obtain the Simpson rule
$$\int_a^b f(x)dx \approx \frac{b-a}{6}\left[f(a) + 4f\left(\frac{a+b}{2}\right) + f(b)\right].$$

2. Prove the **mean value theorem for integrals**, stated below.

    Assume that $g \in C[a,b]$ and that ψ is an integrable function that is either nonnegative or nonpositive throughout the interval $[a,b]$. Then there is a point $\xi \in [a,b]$ such that
$$\int_a^b g(x)\psi(x)dx = g(\xi)\int_a^b \psi(x)dx.$$

    [Hint: Bound g below and above by its minimum and maximum values on the interval, respectively, then bound the desired integral value likewise and use the Intermediate Value Theorem given on page 10.]

3. The basic midpoint rule and its associated error are described in Section 15.1.

   Derive the formula for the error in the basic midpoint rule.

   [Hint: Although the formula looks like half the error of the trapezoidal rule, its derivation is more like a simpler version of Simpson's rule.]

4. The basic trapezoidal rule for approximating $I_f = \int_a^b f(x)dx$ is based on linear interpolation of f at $x_0 = a$ and $x_1 = b$. The Simpson rule is likewise based on quadratic polynomial interpolation. Consider now a cubic Hermite polynomial, interpolating both f and its derivative f' at a and b. The osculating interpolation formula gives

   $$p_3(x) = f(a) + f'(a)(x-a) + f[a,a,b](x-a)^2 + f[a,a,b,b](x-a)^2(x-b),$$

   and integrating this yields (after some algebra)

   $$I_f \approx \int_a^b p_3(x)dx = \frac{b-a}{2}[f(a) + f(b)] + \frac{(b-a)^2}{12}[f'(a) - f'(b)].$$

   This formula is called the **corrected trapezoidal rule**.

   (a) Show that the error in the basic corrected trapezoidal rule can be estimated by

   $$E(f) = \frac{f''''(\eta)}{720}(b-a)^5.$$

   (b) Use the basic corrected trapezoidal rule to evaluate approximations for $\int_0^1 e^x dx$ and $\int_{0.9}^1 e^x dx$. Compare errors to those of Example 15.2. What are your observations?

5. (a) Derive a formula for the *composite midpoint rule*. How many function evaluations are required?

   (b) Obtain an expression for the error in the composite midpoint rule. Conclude that this method is second order accurate.

6. Let us continue Exercise 4.

   (a) Derive the **composite corrected trapezoidal method**. How does it relate to the composite trapezoidal method? Use both composite trapezoidal and corrected composite trapezoidal to evaluate approximations for $I_f = \int_0^1 e^{-x^2} dx$ with $r = 10$ subintervals. What are your observations? [The exact value is $I_f = 0.746824133...$.]

   (b) Show that the error in the *uncorrected* composite trapezoidal method can be written as

   $$E(f) = I_f - I_{tr} = K_1 h^2 + \mathcal{O}(h^4),$$

   where K_1 is independent of h.

7. Write a program that computes an integral numerically, using the composite trapezoidal, midpoint, and Simpson methods. Input for the program consists of the integrand, the ends of the integration interval, and the number of (uniform) subintervals. Apply your program to the two following integrals:

   (a) $\int_0^1 \frac{4}{1+x^2} dx$,

   (b) $\int_0^1 \sqrt{x}\, dx$.

15.7. Exercises

Use $r = 2, 4, 8, 16, 32$ subintervals. For each of the integrals, explain the behavior of the error by observing how it is reduced each time the number of subintervals is doubled, and how large it is, compared to analytical bounds on the error. As much as you can, compare the performance of the methods to each other, taking into consideration both the error and the number of function evaluations. The exact values of the above integrals are π and $2/3$, respectively.

8. (a) Using Gaussian quadrature with $n = 2$ (i.e., three function evaluations in the basic rule), approximate π employing the integral identity

$$\pi = \int_0^1 \left(\frac{4}{1+x^2}\right) dx.$$

(b) Divide the interval $[0, 1]$ into two equal subintervals and approximate π by applying the same Gaussian rule to each subinterval separately. Repeat with three equal subintervals. (These are examples of composite Gaussian quadrature.) Compare the accuracy of the three Gaussian quadrature prescriptions.

9. Find the values of t_0, t_1, a_0, and a_1 in the weighted Gaussian formula used in Example 15.9.

10. Write a short MATLAB program that will find the $n + 1$ Gauss points on the interval $[-1, 1]$ for each n, $n = 0, 1, 2, \ldots, 9$. (You may find the MATLAB function `roots` useful.) Describe a composite quadrature method of order 20.

    Display these points in one plot. It should look essentially like Figure 15.8.

11. Invent an example (i.e., an integrand $f(x)$, an interval $[a, b]$, and a tolerance `tol`) for which the composite Simpson rule on a uniform mesh requires over 100 times more function evaluations than our adaptive program `quads` to compute an approximation for the integral within the tolerance.

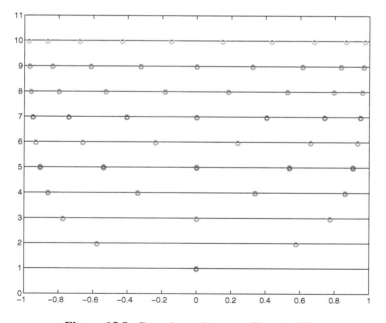

Figure 15.8. *Gaussian points; see Exercise 10.*

To figure out how many evaluations of the integrand function are required by the nonadaptive method, you can use either the technique of Example 15.15 or trial and error.

12. How would you go about writing an adaptive quadrature program based on the trapezoidal rule? Describe the details.

13. Suppose that the interval of integration $[a,b]$ is divided into equal subintervals of length h each such that $r = (b-a)/h$ is even. Denote by R_1 the result of applying the composite trapezoidal method with step size $2h$ and by R_2 the result of applying the same method with step size h. Show that one application of Richardson extrapolation, reading

$$S = \frac{4R_2 - R_1}{3},$$

yields the composite Simpson method.

14. Using Romberg integration, compute π to 8 digits (i.e., 3.xxxxxxx) by obtaining approximations to the integral

$$\pi = \int_0^1 \left(\frac{4}{1+x^2}\right) dx.$$

Describe your solution approach and provide the appropriate Romberg table.

Compare the computational effort (function evaluations) of Romberg integration to that using the adaptive routine developed in Section 15.4 with `tol` $= 10^{-7}$.

You may find for some rows of your Romberg table that only the first step of extrapolation improves the approximation. Explain this phenomenon.

[Hint: Reconsider the assumed form of the composite trapezoidal method's truncation error and the effects of extrapolation for this particular integration.]

15. Show that for a sufficiently smooth integrand $f(x)$ the error in the trapezoidal method can be expanded as

$$E(f;h) = E(f) = K_1 h^2 + K_2 h^4 + \cdots + K_s h^{2s} + \mathcal{O}(h^{2s+1}),$$

where K_i are constants depending on higher and higher derivatives of f, but not on h.

16. Use the values of $R_{j,k}$, $k = 1, \ldots, j$, $j = 1, \ldots, 4$, given in Example 15.17 to estimate the errors in $R_{j,k}$, $k = 1, \ldots, j$, $j = 1, \ldots, 3$, by exploiting for each pair (j,k) the relationship between $R_{j,k}$ and $R_{j+1,k}$.

Compare to the actual errors listed in Example 15.17.

17. Write a short program to approximately integrate

$$I_f = \int_0^1 \int_0^1 e^{x-y} \sin(x+y) dx dy$$

using one, four, and nine points consisting respectively of one, two, and three Gauss points placed symmetrically in both x and y. What are the corresponding quadrature weights?

Use the MATLAB function `quad2d` with strict tolerances to find a tight approximation for I_f and use this value to find the errors in the basic Gaussian rules. What are your observations?

18. Write a short program to approximately integrate

$$I_f = \int_0^1 \int_0^1 \int_0^1 \int_0^1 \left[k\cos(w) - 7kw\sin(w) - 6kw^2\cos(w) + kw^3\sin(w)\right] dx_1 dx_2 dx_3 dx_4,$$

where $w = kx_1x_2x_3x_4$ and k is a parameter. Use a method of your choice.

For the value $k = \pi/2$, the exact solution is $I_f = 1$. Your task is to find an approximation to this value to within 10^{-5} using fewer than 800 function evaluations.

15.8 Additional notes

Numerical integration is an old subject, as evidenced by various words such as "precision" and "rule" which are not used in other similar contexts (a bit like the rich vocabulary that the English language offers in the context of sailing and boats). The vast majority of this chapter has been devoted to approximating definite integrals on a finite interval, using this relatively simple class of problems to develop and demonstrate methods and concepts that are applicable to far wider and more complex problem areas.

An excellent reference for many aspects of numerical integration is the classic by Davis and Rabinowitz [17]. Some of our exercises are solved there, too. A relatively recent paper that discusses delicate issues of error tolerance and more is Gander and Gautschi [27].

An extension to the class of problems considered here is when one or both of a and b is infinite. Such is the case for the intervals on which the Laguerre and Hermite families of orthogonal polynomials are defined; see Section 12.3. Another extension is for cases where the function f is not smooth, or it has singularities. This has been considered briefly in Section 15.3, for cases where the singularity can simply be described by a weight function and integrated away. There is much more on these issues in [17].

Indefinite integrals are often obtained using different techniques of *symbolic computing*, where exact, symbolic formulas for the primitive function are obtained. Symbolic computing (as implemented, for instance, in Maple) yields more information than just a numerical value when applicable. But for very complicated functions, as well as for deriving formulas for simulating differential and integral equations, numerical integration is the only practical option, and this is what we pursue here.

While the subject of numerical integration in one dimension appears to be well under control, the question of how to integrate efficiently multidimensional integrals, especially in many dimensions, is not satisfactorily settled. Such a problem is at the heart of the difficulties in applying statistical Bayesian methods in inverse problems, computer vision, and robotics, to name but a few areas of application. For a description of Monte Carlo methods see, for instance, Hastie, Tibshirani, and Friedman [38] as well as [17].

Chapter 16

Differential Equations

A vast number of mathematical models in various areas of science and engineering involve differential equations. This chapter provides a starting point for a journey into the branch of scientific computing that is concerned with the simulation of differential problems.

We shall concentrate mostly on developing methods and concepts for solving initial value problems for ordinary differential equations: this is the simplest class (although, as you will see, it can be far from being simple), yet a very important one. It is the only class of differential problems for which, in our opinion, up-to-date numerical methods can be learned in an orderly and reasonably complete fashion within a first course text. Section 16.1 prepares the setting for the numerical treatment that follows in Sections 16.2–16.6 and beyond. It also contains a synopsis of what follows in this chapter.

Numerical methods for solving boundary value problems for ordinary differential equations receive a quick review in Section 16.7. More fundamental difficulties arise here, so this section is marked as advanced.

Most mathematical models that give rise to differential equations in practice involve partial differential equations, where there is more than one independent variable. The numerical treatment of partial differential equations is a vast and complex subject that relies directly on many of the methods introduced in various parts of this text. An orderly development belongs in a more advanced text, though, and our own description in Section 16.8 is downright anecdotal, relying in part on examples introduced earlier.

16.1 Initial value ordinary differential equations

Consider the problem of finding a function $y(t)$ that satisfies the *ordinary differential equation* (ODE)

$$\frac{dy}{dt} = f(t, y), \quad a \leq t \leq b.$$

The function $f(t, y)$ is given, and we denote the derivative of the sought solution by $y' = \frac{dy}{dt}$ and refer to t as the *independent variable*.

Previous chapters have dealt with the question of how to numerically approximate, differentiate, or integrate an explicitly known function. Here, similarly, f is given and the sought result is different from f but relates to it. The main difference though is that f depends on the unknown y to be recovered. We would like to be able to compute the function $y(t)$, possibly for all t in the interval $[a, b]$, given the ODE which characterizes the relationship between y and some of its derivatives.

Example 16.1. The function $f(t,y) = -y+t$ defined for $t \geq 0$ and any real y gives the ODE

$$y' = -y+t, \quad t \geq 0.$$

You can verify directly that for any scalar α the function

$$y(t) = t - 1 + \alpha e^{-t}$$

satisfies this ODE. So, the solution is not unique without further conditions.

Suppose next that a value c is given in addition such that at the left end of the interval $y(0) = c$. Then $c = 0 - 1 + \alpha e^0$, and hence $\alpha = c + 1$, and the unique solution is

$$y(t) = t - 1 + (c+1)e^{-t}.$$

Such a solution function, seen as if spawned by the initial value, is sometimes referred to as a **trajectory**. ∎

ODE systems

It will be convenient for presentation purposes in what follows to concentrate on a scalar ODE, because this makes the notation easier when introducing numerical methods. We should note though that scalar ODEs rarely appear in practice. ODEs almost always arise as systems, and these systems may sometimes be large. In the case of systems of ODEs we use vector notation and write our prototype ODE system as

$$\mathbf{y}' \equiv \frac{d\mathbf{y}}{dt} = \mathbf{f}(t,\mathbf{y}), \quad a \leq t \leq b. \tag{16.1a}$$

We shall assume that $\mathbf{y}$ has m components, like $\mathbf{f}$.

In general, as in Example 16.1, there is a family of solutions depending on m parameters for this ODE system. The solution becomes unique, under some mild assumptions, if m *initial values* $\mathbf{c}$ are specified, so that

$$\mathbf{y}(a) = \mathbf{c}. \tag{16.1b}$$

This is then an **initial value problem**.

Example 16.2. Consider a tiny ball of mass 1 attached to the end of a rigid, massless rod of length $r = 1$. At its other end the rod's position is fixed at the origin of a planar coordinate system; see Figure 16.1.

Denoting by θ the angle between the pendulum and the negative vertical axis, the friction-free motion is governed by the ODE

$$\frac{d^2\theta}{dt^2} \equiv \theta'' = -g\sin(\theta),$$

where g is the scaled constant of gravity, e.g., $g = 9.81$, and t is time. This is a simple, nonlinear ODE for θ. The initial position and velocity configuration translates into values for $\theta(0)$ and $\theta'(0)$.

We can write this ODE as a first order system: let

$$y_1(t) = \theta(t), \quad y_2(t) = \theta'(t).$$

16.1. Initial value ordinary differential equations

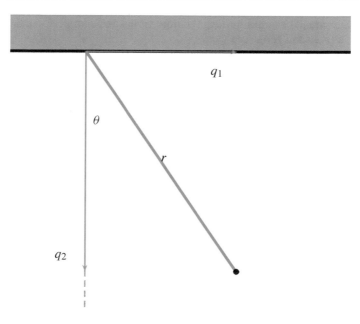

Figure 16.1. *A simple pendulum.*

Then $y_1' = y_2$ and $y_2' = -g\sin(y_1)$. The problem is then written in the form (16.1) with $a = 0$, where

$$\mathbf{y} = \begin{pmatrix} y_1 \\ y_2 \end{pmatrix}, \quad \mathbf{f}(t,\mathbf{y}) = \begin{pmatrix} y_2 \\ -g\sin(y_1) \end{pmatrix}, \quad \mathbf{c} = \begin{pmatrix} \theta(0) \\ \theta'(0) \end{pmatrix}.$$

Note that even in this simple example, which seems to appear in just about any introductory text on ODEs, the dependence of $\mathbf{f}$ on $\mathbf{y}$ is *nonlinear*.

Moreover, the dependence of $\mathbf{f}$ on t is only through $\mathbf{y}(t)$, and not explicitly or directly, so in this example $\mathbf{f} = \mathbf{f}(\mathbf{y})$. The ODE system is called **autonomous** in such a case. ∎

In the previous few chapters we have denoted the independent variable by x, but here we denote it by t. This is because it is convenient to think of initial value problems as depending on **time**, as in Example 16.2, thus also justifying the word "initial" in their name. However, t does not have to correspond to physical time in a particular application.

Going back to scalar ODE notation, note that in principle we can write the initial value ODE in integral form: integrating both sides of the ODE from a to t we have

$$y(t) = c + \int_a^t f(s, y(s))ds, \quad a \le t \le b.$$

This highlights both similarities to and differences from the problem of integration considered in Sections 15.1–15.5. Specifically, on the one hand both prototype problems involve integration; indeed, choosing f not to depend on y and setting $t = b$ and $c = 0$ reveals integration as a special case of the problem considered here. But on the other hand, it is a very special case, because here we are recovering a function of t rather than a value as in definite integration, and even more importantly when deriving numerical methods, the integrand f depends on the unknown function. Numerical methods for ODEs are in general significantly more varied and complex than those considered before for quadrature, as we shall soon see.

Synopsis: Numerical methods described in this chapter

The simplest method for numerically integrating an initial value ODE is *forward Euler*. In fact, when teaching it we often have the feeling that everyone has seen it before. We nonetheless devote Section 16.2 to it, because it is a great vehicle for introducing a variety of important concepts that are relevant for other, more sophisticated methods as well.

One deficiency of the forward Euler method is its low accuracy order. Both Sections 16.3 and 16.4 are devoted to higher order families of methods. Particular members of these families are the methods of choice in almost any general-purpose ODE code today.

Issues of stability of the numerical method are considered in Section 16.2 and, more generally, in Section 16.5. For many problems, just about any reasonable method (including of course all those considered in this chapter) is basically stable. But for an important class of *stiff* problems there is an issue of *absolute stability* restriction which makes some methods preferable in a significant sense over others. Let us say no more at this early stage.

Of course, as in Section 15.4, writing a general-purpose code involves also estimating and controlling the error, and Section 16.6 provides a brief introduction to this topic. Things quickly get much messier here than for the case of numerical integration considered in Section 15.4.

Boundary value ODEs

Sections 16.2–16.6 are all concerned with methods for initial value ODE systems of the form (16.1), where $\mathbf{y}(a)$ is given. Other problems where all of the solution components $\mathbf{y}$ are given at one point can basically be converted to an initial value problem. For example, the *terminal value problem* where $\mathbf{y}(b)$ is given is converted to an initial value problem by a change of variable $\tau = b - t$, which transforms the interval of definition from $[a, b]$ to $[0, b-a]$ with $\mathbf{y}(0)$ now given. But there are other, *boundary value problems* for the same prototype ODE system (16.1a), where some component of $\mathbf{y}$ is given at a different value of t than another. For instance, in Example 16.2 we may be given two position values for $\theta(0)$ and $\theta(1)$ and none on the velocity θ'. See also Examples 4.17 and 9.3 for instances of boundary value ODEs. Methods for this more difficult class of ODE problems that are more general than in Example 9.3 are briefly considered in Section 16.7.

Partial differential equations

Most differential problems that arise in practice depend on more than one independent variable, giving rise to a *partial differential equation* (PDE). For instance, Example 7.1 considers the Poisson equation

$$-\left(\frac{\partial^2 u}{\partial x^2} + \frac{\partial^2 u}{\partial y^2}\right) = g(x, y), \quad 0 < x, y < 1,$$

with the solution values $u(x, y)$ given on the boundary of the unit square. The independent variables here are x and y.

Reminiscent of the situation with interpolation and integration briefly described in Sections 11.6 and 15.6, respectively, everything becomes more complicated in a hurry when moving from ODEs to PDEs. Indeed, one reason for concentrating in this chapter on the easiest case of initial value ODEs is that it gives significant taste and provides important expertise buildup necessary for PDE computations. In Section 16.8 we briefly review PDEs and some corresponding numerical methods.

16.2 Euler's method

Euler's method, which is also known as the **forward Euler** method (to distinguish it from its *backward Euler* counterpart, to be discussed later), is the simplest numerical method for approximately solving initial value ODEs. Here it is used as a vehicle for studying several important, basic notions.

In this section we introduce in the context of the forward Euler method the following **general concepts** for numerical ODE methods:

- method derivation,
- explicit vs. implicit methods,
- local truncation error and global error,
- order of accuracy,
- convergence, and
- absolute stability and stiffness.

Method derivation

We first consider finding an approximate solution for a scalar initial value ODE at equidistant abscissae. Thus, define the points

$$t_0 = a, \quad t_i = a + ih, \quad i = 0, 1, 2, \ldots, N,$$

where $h = \frac{b-a}{N}$ is the step size. Denote the approximate solution for $y(t_i)$ by y_i. The step size h may vary in principle, $h = h_i$, but unless otherwise noted we keep it uniform (constant) in this section and consider varying it wisely in Section 16.6.

Recall from Example 1.2 and Section 14.1 the forward difference formula

$$y'(t_i) = \frac{y(t_{i+1}) - y(t_i)}{h} - \frac{h}{2} y''(\xi_i).$$

By the ODE, $y'(t_i) = f(t_i, y(t_i))$, so

$$y(t_{i+1}) = y(t_i) + h f(t_i, y(t_i)) + \frac{h^2}{2} y''(\xi_i).$$

This is satisfied by the exact solution $y(t)$ of the ODE. Dropping the truncation term, we obtain the *forward Euler method*, which defines the approximate solution $\{y_i\}_{i=0}^{N}$ by

$$y_0 = c,$$
$$y_{i+1} = y_i + h f(t_i, y_i), \quad i = 0, 1, \ldots, N-1.$$

See Figure 16.2. We omit the details of the problem for which the graph in this figure was produced, since the demonstrated issues are not specific to one particular ODE problem.

This simple formula allows us to march forward in t. Assuming that the various parameters and the function f are specified, the following MATLAB script does the job:

```
t = [a:h:b];
y(1) = c;
for i=1:N
   y(i+1) = y(i) + h * f(t(i),y(i));
end
```

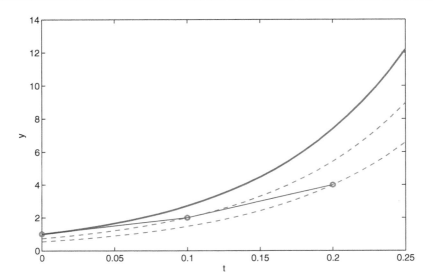

Figure 16.2. *Two steps of the forward Euler method. The exact solution is the curved solid line. The numerical values obtained by the Euler method are circled and lie at the nodes of a broken line that interpolates them. The broken line is tangential at the beginning of each step to the ODE trajectory passing through the corresponding node (dashed lines).*

This produces an array of abscissae and ordinates, (t_i, y_i), which is good for plotting. In other applications, fewer output points may be required. For instance, if only the approximate value of $y(b)$ is desired, then we can save storage by the script

```
t = a; y = c;
for i=1:N
  y = y + h * f(t,y);
  t = t + h;
end
```

This script works also if c, y, and f are arrays of the same size, corresponding to an ODE system.

Example 16.3. Consider the simple, linear initial value ODE problem given by

$$y' = y, \quad y(0) = 1.$$

The exact solution is $y(t) = e^t$.

This ODE is autonomous, $f(t, y) = y$. Euler's method reads

$$y_0 = 1,$$
$$y_{i+1} = y_i + hy_i = (1+h)y_i, \quad i = 0, 1, 2, \ldots.$$

We obtain the results listed in Table 16.1. The errors clearly reduce by a factor of roughly $1/2$ when h is cut by the same factor from 0.2 to 0.1.

Note also that the absolute error increases as t increases in this particular example. Even the relative error increases here with t, although not as fast. ∎

Table 16.1. *Absolute errors using the forward Euler method for the ODE $y' = y$. The values $e_i = y(t_i) - y_i$ are listed under Error.*

		$h = 0.2$		$h = 0.1$	
t_i	$y(t_i)$	y_i	Error	y_i	Error
0	1.000	1.000	0.0	1.000	0.0
0.1	1.105			1.100	0.005
0.2	1.221	1.200	0.021	1.210	0.011
0.3	1.350			1.331	0.019
0.4	1.492	1.440	0.052	1.464	0.028
0.5	1.694			1.611	0.038
0.6	1.822	1.728	0.094	1.772	0.051

Explicit vs. implicit methods

What happens if we replace the forward difference formula that leads to the forward Euler method by the *backward* formula

$$y'(t_{i+1}) \approx \frac{y(t_{i+1}) - y(t_i)}{h},$$

which in the context of Section 14.1 is the most innocent modification imaginable? Here this leads to the **backward Euler** method, given by

$$y_0 = c,$$
$$y_{i+1} = y_i + hf(t_{i+1}, y_{i+1}), \quad i = 0, 1, \ldots, N-1.$$

It might be tempting to think of the backward Euler method as a minor variation, not much different from its forward counterpart. But there is a rather substantial difference here: in the backward Euler method, the computation of y_{i+1} depends *implicitly* on y_{i+1} itself! This leads to the important notion of *explicit* and *implicit* methods. If the evaluation of y_{i+1} involves the evaluation of f at the unknown value y_{i+1} itself, then the method is implicit. Carrying out an integration step requires solving a usually nonlinear equation for y_{i+1}. If on the other hand the evaluation of y_{i+1} involves the evaluation of f only at known points, i.e., values such as y_i obtained in previous steps or stages, then the method is explicit. Hence, the forward Euler method is explicit, whereas the backward Euler method is implicit.

Explicit methods are significantly easier to implement, and carrying out an integration step is typically much faster. This, and its general simplicity, is what has given the forward Euler method its great popularity in numerous areas of application. The backward Euler method, in contrast, while easy to understand conceptually, requires a deeper understanding of numerical issues and cannot be programmed as easily and seamlessly as the forward Euler method. Nonetheless, as we shall see later in this section and in Section 16.5, backward Euler and other implicit methods have numerical properties that in certain cases, and for certain applications, make them superior to explicit methods.

Local truncation error, order, and global error

The *local truncation error*, d_i, is the amount by which the exact solution fails to satisfy the difference equation, written in divided difference form, at integration step i. The concept is general and applies not just for Euler's method.

The method has *order of accuracy q* if q is the lowest positive integer such that for any sufficiently smooth exact solution $y(t)$ we have

$$\max_i |d_i| = \mathcal{O}(h^q).$$

The *global error* of the method is defined by

$$e_i = y(t_i) - y_i, \quad i = 0, 1, \ldots, N.$$

Let us demonstrate these concepts on the forward Euler method. Here we have

$$d_i = \frac{y(t_{i+1}) - y(t_i)}{h} - f(t_i, y(t_i)).$$

From the method's derivation, $d_i = \frac{h}{2} y''(\xi_i)$. This expression is linear in h, i.e., $d_i = \mathcal{O}(h)$. Hence the method is first order accurate, meaning $q = 1$.

The same holds for the backward Euler method, where $d_i = -\frac{h}{2} y''(\xi_i)$. This method is also first order accurate.

More generally, designing a numerical discretization of order q would involve manipulating Taylor series expansions (see page 5) to achieve an appropriately small local truncation error. For the methods of Section 16.3 this can be a hair-raising job, more so than anything encountered in Chapter 14. But the principle, and for the Euler methods also the execution, are simple.

Convergence

The method is said to *converge* if the maximum global error tends to 0 as h tends to 0, provided the exact solution exists and is reasonably smooth.

In general, if nothing goes wrong we expect the global error $e_i = y(t_i) - y_i$ to be of the same order as the local truncation error. So, for the forward Euler method we expect

$$e_i = \mathcal{O}(h) \quad \forall \, i.$$

We have seen such behavior in Example 16.3. Let us show that this is true in general under very mild conditions on $f(t, y)$. Specifically, we assume the conditions specified in the Forward Euler Convergence Theorem given on the next page.

Since the local truncation error satisfies

$$d_i = \frac{y(t_{i+1}) - y(t_i)}{h} - f(t_i, y(t_i)), \quad \text{and also}$$

$$0 = \frac{y_{i+1} - y_i}{h} - f(t_i, y_i),$$

we can subtract the two expressions and obtain for the error the difference formula

$$d_i = \frac{e_{i+1} - e_i}{h} - [f(t_i, y(t_i)) - f(t_i, y_i)].$$

The assumption that f satisfies a Lipschitz condition with a constant L implies that we can write the error difference equation as

$$|e_{i+1}| = |e_i + h[f(t_i, y(t_i)) - f(t_i, y_i)] + h d_i|$$
$$\leq |e_i| + h L |e_i| + h d,$$

16.2. Euler's method

Theorem: Forward Euler Convergence.
Let $f(t, y)$ have bounded partial derivatives in a region $\mathcal{D} = \{a \leq t \leq b, |y| < \infty\}$.
Note that this implies Lipschitz continuity in y: there exists a constant L such that for all (t, y) and $(t, \hat{y})$ in $\mathcal{D}$ we have

$$|f(t, y) - f(t, \hat{y})| \leq L|y - \hat{y}|.$$

Then Euler's method converges and its global error decreases linearly in h.
Moreover, assuming further that

$$|y''(t)| \leq M, \quad a \leq t \leq b,$$

the global error satisfies

$$|e_i| \leq \frac{Mh}{2L}[e^{L(t_i - a)} - 1], \quad i = 0, 1, \ldots, N.$$

where d is a bound on the local truncation errors, $d \geq \max_{0 \leq i \leq N-1} |d_i|$. Thus, if we know

$$M = \max_{a \leq t \leq b} |y''(t)|,$$

then set

$$d = \frac{M}{2}h.$$

It follows that

$$|e_{i+1}| \leq (1 + hL)|e_i| + hd$$
$$\leq (1 + hL)[(1 + hL)|e_{i-1}| + hd] + hd = (1 + hL)^2|e_{i-1}| + (1 + hL)hd + hd$$
$$\leq \cdots \leq (1 + hL)^{i+1}|e_0| + hd \sum_{j=0}^{i}(1 + hL)^j$$
$$\leq d\left[e^{L(t_{i+1} - a)} - 1\right]/L$$
$$\leq \frac{Mh}{2L}\left[e^{L(t_{i+1} - a)} - 1\right].$$

Above, to arrive at the one inequality before last we have used $e_0 = 0$ and a bound for the geometric sum of powers of $1 + hL$. ◆

This provides a proof of convergence for the forward Euler method as stated in the theorem on this page. Note that the global error in Euler's method is indeed first order in h, as observed in Example 16.3.

Note also that the error bound may grow with t. This is realistic to expect when the exact solution grows; the relative error is more meaningful then. But when the exact solution decays, then the above bound on the absolute error becomes too pessimistic, in that the actual global error will be much smaller than its bound.

Example 16.4. A more challenging problem than that in Example 16.3 originates in plant physiology and is defined by the following MATLAB script:

```
function f = hires(t,y)
%
% f = hires(t,y)
%
% High irradiance response function arising in plant physiology
f = y;
f(1) = -1.71*y(1) + .43*y(2) + 8.32*y(3) + .0007;
f(2) = 1.71*y(1) - 8.75*y(2);
f(3) = -10.03*y(3) + .43*y(4) + .035*y(5);
f(4) = 8.32*y(2) + 1.71*y(3) - 1.12*y(4);
f(5) = -1.745*y(5) + .43*y(6) + .43*y(7);
f(6) = -280*y(6)*y(8) + .69*y(4) + 1.71*y(5) - .43*y(6) + .69*y(7);
f(7) = 280*y(6)*y(8) - 1.81*y(7);
f(8) = -280*y(6)*y(8) + 1.81*y(7);
```

This ODE system (which has $m = 8$ components) is to be integrated from $a = 0$ to $b = 322$ starting from $\mathbf{y}(0) = \mathbf{y}_0 = (1,0,0,0,0,0,0,.0057)^T$. The script

```
h = .005; t = 0:h:322;
y = y0 * ones(1,length(t));
for i = 1:length(t)-1
  y(:,i+1) = y(:,i) + h*hires(t(i),y(:,i));
end
plot(t,y(6,:))
```

(plus labeling) produces Figure 16.3. To gauge accuracy of the Euler method we repeat the solution process with a more accurate method from Section 16.3 below and regard the difference between these approximate solutions as the error for the less accurate forward Euler method. Based on this the maximum absolute error occurs after 447 steps at $t_* = 2.235$, where $y_{447}^{(6)} \approx .4483$ and $|y^{(6)}(t_*) - y_{447}^{(6)}| \approx 4.67 \times 10^{-4}$. ∎

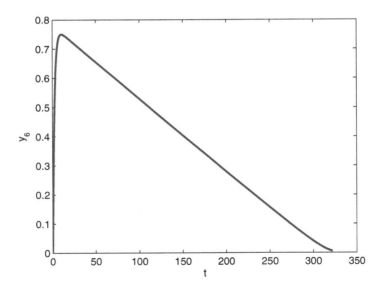

Figure 16.3. *The sixth solution component of the HIRES model.*

16.2. Euler's method

> **Note:** Since deriving discretization methods for differential equations involves numerical differentiation, there is an unavoidable roundoff error that behaves like $\mathcal{O}(h^{-1})$; recall the discussion in Section 14.4.
>
> However, roundoff error is typically a secondary concern here, so long as the method is absolutely stable as explained next. This is so both because h is usually not incredibly small (for reasons of efficiency and modeling limitations) and because y, and not y', is the function for which the approximations are sought.

Absolute stability and stiffness

Let us consider next a simple scalar ODE known as the **test equation** and given by

$$y' = \lambda y,$$

where λ is a real constant. The exact solution is $y(t) = e^{\lambda t} y(0)$. So, the exact solution increases if $\lambda > 0$ and decreases otherwise.

Euler's method gives

$$y_{i+1} = y_i + h\lambda y_i = (1 + h\lambda) y_i = \cdots = (1 + h\lambda)^{i+1} y(0).$$

If $\lambda > 0$, then the approximate solution grows, and so does the absolute error, yet the relative error remains reasonably small. But if $\lambda < 0$, then the exact solution decays, so we must require at the very least that the approximate solution not grow. We then demand

$$|y_{i+1}| \leq |y_i|.$$

For the forward Euler method this corresponds to insisting that

$$|1 + h\lambda| \leq 1 \Rightarrow h \leq \frac{2}{|\lambda|}.$$

It is a requirement of *absolute stability*: regardless of accuracy considerations, the constant step size h must not exceed a certain bound that depends on the problem which is being approximately solved.

Example 16.5. For the initial value ODE problem

$$y' = -1000(y - \cos(t)) - \sin(t), \quad y(0) = 1,$$

the solution is $y(t) = \cos(t)$. A broken line interpolation of $y(t)$, which is what MATLAB uses by default for plotting, looks good already for $h = 0.1$. But stability decrees that we must look at the test equation $y' = -1000y$. Thus, for Euler's method we need

$$h \leq \frac{1}{500}$$

for *any* reasonable accuracy after the first few steps!

For instance, using $h = .0005\pi$ and evaluating the numerical solution at $t = \pi/2$ (where the exact solution vanishes), we get the error $y_{1000} = 3.7\text{e-}10$. But using $h = .001\pi$ instead, which is only double the step size but no longer satisfies the absolute stability bound, we get $y_{500} = -3.7\text{e+}159$, meaning that essentially a blowup has occurred. This can be annoying. ■

An ODE such as that of Example 16.5, where stability considerations make us take a much smaller step size for the forward Euler method than accuracy considerations would otherwise dictate, is called **stiff**.

Let us make two essential comments:[60]

- You may wonder how it is possible that the local truncation error would change so much upon making such a small change in h in Example 16.5. Well, it doesn't. The error that accumulates wildly upon violation of the absolute stability condition is the unruly *roundoff error*!

- There is a fundamental difference between the error concepts introduced earlier and the absolute stability requirement: the former all ask what happens when $h \to 0$ for a fixed ODE problem, whereas the latter is concerned with the situation for a fixed and not necessarily small step size or, more precisely, with a bound involving $z = h\lambda$.

When deriving the forward Euler method, we have also introduced the implicit, *backward Euler* method. At the time it did not seem that anything good could come out of this method, being so much harder to use than its equally accurate forward Euler counterpart. But now it comes to life: applied to the test equation we get $y_{i+1} = y_i + h\lambda y_{i+1}$, and hence

$$y_{i+1} = \frac{1}{1-h\lambda} y_i.$$

Therefore, $|y_{i+1}| \leq |y_i|$ for any $h > 0$ and $\lambda < 0$. There is no annoying absolute stability restriction here!

Example 16.6. For the problem of Example 16.5, applying the backward Euler method yields the following results:

1. Using $h = .0005\pi$ and evaluating the numerical solution at $t = \pi/2$, we get the error $y_{1000} = -1.2\text{e-}9$.

2. Using $h = .001\pi$, we get the error $y_{500} = -3.2\text{e-}9$.

The global error behaves as expected from local truncation error considerations alone. ∎

Let us return to forward Euler and ask what happens for a more general ODE system (16.1a). For the simplest case of interest of such a system we have $\mathbf{f}(\mathbf{y}) = A\mathbf{y}$, where A is a constant $m \times m$ matrix, further assumed to have eigenvalues $\lambda_1, \lambda_2, \ldots, \lambda_m$ and be diagonalizable (see Section 4.1). Thus, there is a transformation matrix T such that $T^{-1}AT$ is diagonal with the eigenvalues on its main diagonal. Then it is not difficult to see that for the transformed unknowns $\mathbf{x} = T^{-1}\mathbf{y}$ the system decouples to give the scalar ODEs

$$x'_j = \lambda_j x_j, \quad j = 1, 2, \ldots, m.$$

The absolute stability requirement for the forward Euler method thence translates into the requirement that

$$|1 + h\lambda_j| \leq 1, \quad j = 1, 2, \ldots, m.$$

The complication is, however, in that eigenvalues need not be real! In general we must therefore consider the test equation with *complex* λ, and we do so in Section 16.5.

Specific exercises for this section: Exercises 1–5.

---

[60]Do not be misled by the seeming simplicity of these comments: they involve complex, nontrivial issues.

16.3 Runge–Kutta methods

Euler's method is only first order accurate. This implies inefficiency for many applications, as the step size must be taken rather small, and thus N becomes rather large, to achieve satisfactory accuracy.

To describe higher order methods, consider the integration of our prototype scalar ODE $y' = f(t, y)$ over one step, from t_i to t_{i+1}. Thus, we assume that an approximation y_i to $y(t_i)$ is known at the point t_i and develop ways to obtain an approximation y_{i+1} to $y(t_{i+1})$ at the next point t_{i+1}.

To obtain higher order methods there are extensions of the forward Euler method in several directions. A very common approach described in this section is to use only information in the current step $[t_i, t_{i+1}]$. Thus, a *Runge–Kutta* (RK) method is a **one-step method** in which repeated function evaluations are used to achieve a higher order; see Figure 16.4.

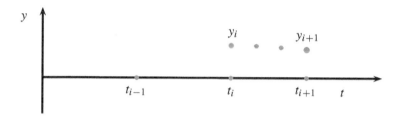

Figure 16.4. *RK method: repeated evaluations of the function f in the current mesh subinterval $[t_i, t_{i+1}]$ are combined to yield a higher order approximation y_{i+1} at t_{i+1}. (Reprinted from Ascher [3].)*

Below we proceed to derive explicit and implicit RK methods, starting from simple and intuitive ones and gradually becoming more general.

Two simple RK methods

Our purpose next is to derive a particular, second order accurate, explicit RK method as an example, to get the hang of it.

Integrating from t_i to t_{i+1} we can write the ODE as

$$y(t_{i+1}) = y(t_i) + \int_{t_i}^{t_{i+1}} f(t, y(t)) dt.$$

Now, consider applying the *trapezoidal* quadrature rule as in Sections 15.1 and 15.2. This gives

$$y_{i+1} = y_i + \frac{h}{2}(f(t_i, y_i) + f(t_{i+1}, y_{i+1})),$$

which is referred to as the **implicit trapezoidal** method. By considering the local truncation error

$$d_i = \frac{y(t_{i+1}) - y(t_i)}{h} - \frac{1}{2}[f(t_i, y(t_i)) + f(t_{i+1}, y(t_{i+1}))],$$

you should be able to quickly convince yourself that $d_i = \mathcal{O}(h^2)$, i.e., this is a second order method.

However, a serious setback is that this method is implicit. Thus, the evaluation of y_{i+1} requires solving a generally nonlinear equation. As discussed in Section 16.2, this is a major difference between quadrature and numerical integration of differential equations. For an ODE system of size m we get a possibly nonlinear system of m algebraic equations to solve for a vector $\mathbf{y}_{i+1}$ at each step *i*. One would like, if possible, to avoid this expense and complication (although not at any cost, as becomes clear in Section 16.5).

Using the only explicit method we know thus far (yes, that would be forward Euler), we can approximate y_{i+1} at first by

$$Y = y_i + hf(t_i, y_i).$$

Then plug this into the implicit trapezoidal formula, obtaining the **explicit trapezoidal** method

$$y_{i+1} = y_i + \frac{h}{2}(f(t_i, y_i) + f(t_{i+1}, Y)).$$

Note that there are two function evaluations involved here at the *i*th step, namely, of $f(t_i, y_i)$ and $f(t_{i+1}, Y)$: this is a two-stage, explicit RK method.

To ascertain that this method is second order accurate we would have to plug in the exact solution into the difference equation, obtaining

$$d_i = \frac{y(t_{i+1}) - y(t_i)}{h} - \frac{1}{2}[f(t_i, y(t_i)) + f(t_{i+1}, \hat{Y})],$$
$$\hat{Y} = y(t_i) + hy'(t_i).$$

Taylor expansions follow, and Exercise 6 completes the job. This gets a bit technical, but the result is intuitively clear, because the first order accurate intermediate step Y gets multiplied by h in the formula for y_{i+1}.

Two more simple RK methods

Applying the trapezoidal rule in the way demonstrated above is perhaps the simplest way to obtain an explicit RK method of order 2. But using the midpoint rule instead is not much more complex.

A direct application of the midpoint quadrature rule gives the **implicit midpoint** method

$$y_{i+1} = y_i + hf(t_{i+1/2}, y_{i+1/2}),$$

where

$$t_{i+1/2} = \frac{t_i + t_{i+1}}{2} = t_i + h/2, \quad y_{i+1/2} = \frac{y_i + y_{i+1}}{2}.$$

Then using forward Euler to approximate $y_{i+1/2}$ yields the two-stage, **explicit midpoint** method, which is an RK method of order 2 given by

$$y_{i+1} = y_i + hf(t_{i+1/2}, Y), \quad \text{where}$$
$$Y = y_i + \frac{h}{2} f(t_i, y_i).$$

At each step i we evaluate Y and then y_{i+1}. This requires two function evaluations per step, so the method is roughly twice as expensive as the forward Euler method per step. The explicit midpoint method is comparable to the explicit trapezoidal method, both in order and in expense.

16.3. Runge–Kutta methods

An explicit RK method of order 4

The **classical RK method** is based on the Simpson quadrature rule and uses four explicit stages, hence four function evaluations per step, to achieve $\mathcal{O}(h^4)$ accuracy. It is given by

$$Y_1 = y_i,$$
$$Y_2 = y_i + \frac{h}{2} f(t_i, Y_1),$$
$$Y_3 = y_i + \frac{h}{2} f(t_{i+1/2}, Y_2),$$
$$Y_4 = y_i + h f(t_{i+1/2}, Y_3),$$
$$y_{i+1} = y_i + \frac{h}{6} \left(f(t_i, Y_1) + 2 f(t_{i+1/2}, Y_2) + 2 f(t_{i+1/2}, Y_3) + f(t_{i+1}, Y_4) \right).$$

Showing that this formula is actually fourth order accurate is not a simple matter, unlike for the composite Simpson quadrature, and will not be discussed further in this text.

Here is a simple MATLAB function that implements the classical RK method using a fixed step size. It is written for an ODE system, with the extension from the scalar ODE method requiring almost no effort. Note that instead of storing the Y_j's we evaluate and store $K_j = f(t_j, Y_j)$.

```
function [t,y] = rk4(f,tspan,y0,h)
%
% function [t,y] = rk4(f,tspan,y0,h)
%
% A simple integration routine to solve the
% initial value ODE   y' = f(t,y),  y(a) = y0,
% using the classical 4-stage Runge-Kutta method
% with a fixed step size h.
% tspan = [a b] is the integration interval.
% Note that y and f can be vector functions

y0 = y0(:);            % make sure y0 is a column vector
m = length(y0);        % problem size
t = tspan(1):h:tspan(2);  % output abscissae
N = length(t)-1;       % number of steps
y = zeros(m,N+1);
y(:,1) = y0;    % initialize

% Integrate
for i=1:N
  % Calculate the four stages
  K1 = feval(f, t(i),y(:,i)     );
  K2 = feval(f, t(i)+.5*h, y(:,i)+.5*h*K1);
  K3 = feval(f, t(i)+.5*h, y(:,i)+.5*h*K2);
  K4 = feval(f, t(i)+h,    y(:,i)+h*K3    );

  % Evaluate approximate solution at next step
  y(:,i+1) = y(:,i) + h/6 *(K1+2*K2+2*K3+K4);
end
```

A script that employs our function `rk4` is given in Example 16.8.

Note: The appearance of methods of different orders in this section motivates exploring the question of testing whether a particular method at least comes close to reflecting its order in a given application. A way to go about this is the following. If the error at fixed t is $e(h) \approx \gamma h^q$, with γ depending on t but not on h, then with step size $2h$, $e(2h) \approx \gamma (2h)^q \approx 2^q e(h)$. Thus, calculate

$$\text{rate}(h) = \log_2 \left(\frac{e(2h)}{e(h)} \right).$$

This **observed order**, or **rate**, is compared to the predicted order q of the given method. See also Exercise 9 for a generalization.

Example 16.7. Consider the scalar problem

$$y' = -y^2, \qquad y(1) = 1.$$

The exact solution is $y(t) = \frac{1}{t}$. We compute and list absolute errors at $t = 10$ in Table 16.2.

Table 16.2. *Errors and calculated observed orders (rates) for the forward Euler, the explicit midpoint (RK2), and the classical Runge–Kutta (RK4) methods.*

h	Euler	Rate	RK2	Rate	RK4	Rate
0.2	4.7e-3		3.3e-4		2.0e-7	
0.1	2.3e-3	1.01	7.4e-5	2.15	1.4e-8	3.90
0.05	1.2e-3	1.01	1.8e-5	2.07	8.6e-10	3.98
0.02	4.6e-4	1.00	2.8e-6	2.03	2.2-11	4.00
0.01	2.3e-4	1.00	6.8e-7	2.01	1.4e-12	4.00
0.005	1.2e-4	1.00	1.7e-7	2.01	8.7e-14	4.00
0.002	4.6e-5	1.00	2.7e-8	2.00	1.9e-15	4.19

By using the approach described above for computing observed orders, we see that indeed the three methods introduced demonstrate orders 1, 2, and 4, respectively. Note that for h very small, roundoff error effects show up in the error with the more accurate formula RK4. Given the cost per step, a fair comparison with roughly equal computational effort would be of Euler with $h = .005$, RK2 with $h = .01$, and RK4 with $h = .02$. Clearly the higher order methods are better if an accurate approximation (say, error below 10^{-7}) is sought.

This example notwithstanding, let us bear in mind that for lower accuracy requirements, or for rougher ODEs, lower order methods may become more competitive. ∎

Example 16.8. Next, we unleash our function `rk4` on the ODE problem given by

$$y_1' = .25 y_1 - .01 y_1 y_2, \quad y_1(0) = 80,$$
$$y_2' = -y_2 + .01 y_1 y_2, \quad y_2(0) = 30.$$

Integrating from $a = 0$ to $b = 100$ with step size $h = 0.01$, the resulting solution is displayed in Figure 16.5.

16.3. Runge–Kutta methods

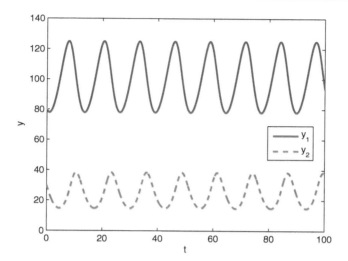

Figure 16.5. *Predator-prey model:* $y_1(t)$ *is number of prey,* $y_2(t)$ *is number of predator.*

Here is the MATLAB script used for this example, yielding both Figure 16.5 and Figure 16.6.

```
y0 = [80,30]';     % initial data
tspan = [0,100];   % integration interval
h = .01;           % constant step size

[tout,yout] = rk4(@func,tspan,y0,h);

figure(1)
plot(tout,yout)
xlabel('t')
ylabel('y')
legend('y_1','y_2')

figure(2)
plot(yout(1,:),yout(2,:))
xlabel('y_1')
ylabel('y_2')

function f = func(t,y)

a = .25; b = -.01; c = -1; d = .01;
f(1) = a*y(1) + b*y(1)*y(2);
f(2) = c*y(2) + d*y(1)*y(2);
```

This is a simple *predator-prey* model, originally considered independently by A. Lotka and V. Volterra. There is one prey species whose number at any given time is $y_1(t)$. The number of prey grows unboundedly in time if unhunted by the predator. There is only one predator species whose number at any given time is $y_2(t)$. The number of predators would shrink to extinction if they do not encounter prey. But they do, and thus a life cycle forms. Note the way the peaks and lows in $y_1(t)$ are related to those of $y_2(t)$.

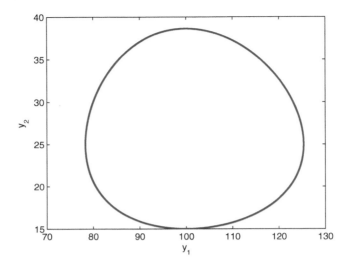

Figure 16.6. *Predator-prey solution in phase plane: the curve of $y_1(t)$ vs. $y_2(t)$ yields a limit cycle.*

In Figure 16.6 we plot y_1 vs. y_2. A *limit cycle* is obtained, suggesting that at least according to this rather simple model neither species will become extinct or grow unboundedly at any future time. ∎

> **Note:** It is often convenient to suppress the explicit dependence of $\mathbf{f}(t,\mathbf{y})$ on t and to write the ODE (16.1a) as
> $$\mathbf{y}' = \mathbf{f}(\mathbf{y}).$$
> Indeed, there are many problem instances, such as those in Examples 16.2–16.8, where there is no explicit dependence on t in $\mathbf{f}$.
> Even if there is such dependence, it is possible to imagine t as yet another dependent variable, i.e., define a new independent variable $x = t$, and let $\tilde{\mathbf{y}}^T = (\mathbf{y}^T, t)$ and $\tilde{\mathbf{f}}^T = (\mathbf{f}^T, 1)$. Then $\mathbf{y}' = \mathbf{f}(t,\mathbf{y})$ becomes
> $$\frac{d\tilde{\mathbf{y}}}{dx} = \tilde{\mathbf{f}}(\tilde{\mathbf{y}}).$$
> We are not proposing to actually carry out such a transformation, only to imagine it in case you don't know how to handle t when implementing a particular discretization formula.

General *s*-stage RK methods

In the explicit RK methods we have seen so far, each internal stage Y_j depends on the previous Y_{j-1}. More generally, each internal stage Y_j depends on all previously computed stages Y_k, so an **explicit, *s*-stage RK** method looks like

$$y_{i+1} = y_i + h \sum_{j=1}^{s} b_j f(Y_j), \quad \text{where}$$

$$Y_j = y_i + h \sum_{k=1}^{j-1} a_{jk} f(Y_k).$$

16.3. Runge–Kutta methods

Here we assume no explicit dependence of f on the independent variable t, to save on notation. For instance, the explicit midpoint method is written in this form with $s = 2$, $a_{11} = a_{12} = a_{22} = 0$, $a_{21} = 1/2$, $b_1 = 0$, and $b_2 = 1$.

The even more general **implicit, s-stage RK** method is written as

$$y_{i+1} = y_i + h \sum_{j=1}^{s} b_j f(Y_j), \quad \text{where}$$

$$Y_j = y_i + h \sum_{k=1}^{s} a_{jk} f(Y_k).$$

The implicit midpoint method is an instance of this, with $s = 1$, $a_{11} = 1/2$, and $b_1 = 1$. Implicit RK methods become interesting when considering stiff problems; see Section 16.5.

It is common to gather the coefficients of an RK method into a *tableau*

$$\begin{array}{c|cccc} c_1 & a_{11} & a_{12} & \cdots & a_{1s} \\ c_2 & a_{21} & a_{22} & \cdots & a_{2s} \\ \vdots & \vdots & \vdots & \ddots & \vdots \\ c_s & a_{s1} & a_{s2} & \cdots & a_{ss} \\ \hline & b_1 & b_2 & \cdots & b_s \end{array}$$

where $c_j = \sum_{k=1}^{s} a_{jk}$ for $j = 1, 2, \ldots, s$. Thus, the explicit midpoint method is written as

$$\begin{array}{c|cc} 0 & 0 & 0 \\ \frac{1}{2} & \frac{1}{2} & 0 \\ \hline & 0 & 1 \end{array}$$

and the implicit midpoint method is

$$\begin{array}{c|c} \frac{1}{2} & \frac{1}{2} \\ \hline & 1 \end{array}$$

In Exercise 10 you are asked to construct corresponding tableaus for the explicit and the implicit trapezoidal methods, both of which satisfy $s = 2$. In general the method is explicit if and only if $a_{j,k} = 0$ for all $j \leq k$. (Why?)

The definitions of local truncation error and order, as well as the proof of convergence and global error bound given in Section 16.2, readily extend for any RK method.

Some of the advantages and disadvantages of RK methods are evident by now. Advantages include:

- Simplicity in concept and in starting the integration process.
- Flexibility in varying the step size.
- Flexibility in handling discontinuities in $f(t, y)$ and other events (e.g., collision of bodies whose motion is being simulated).

Disadvantages of RK methods include:

- The number of function evaluations required for higher order RK methods—our rough measure of work expense—is relatively high, compared to multistep methods.
- Deriving higher order RK methods and proving their order of accuracy can be challenging.
- There are difficulties with adaptive order variation. So, in practice people settle on one method of a certain order and vary only the step size to achieve error control, as described in Section 16.6.
- More involved and possibly more costly procedures are required for stiff problems, the topic of Section 16.5, than for multistep methods.

Specific exercises for this section: Exercises 6–11.

16.4 Multistep methods

Note: It is possible to read almost all of Sections 16.5 and 16.6 without first going through Section 16.4 in detail.

Linear multistep methods are the traditional rival family to RK methods. While the latter are based on repeated evaluations of the function $f(t, y)$ within one step, the methods considered here are simpler in form but use past information as well.

The basic idea behind them is very simple. Observe that at the start of step i we usually have knowledge not only of the approximate solution y_i at t_i but also of previous solution values: y_{i-1} at t_{i-1}, y_{i-2} at t_{i-2}, and so on. So, use polynomial interpolation of these values, or of their corresponding values of $f(t, y)$, in order to obtain cheap yet accurate approximations for the next unknown, y_{i+1} at t_{i+1}. See Figure 16.7.

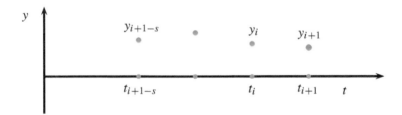

Figure 16.7. *Multistep method: known solution values of y or $f(t, y)$ at the current location t_i and previous locations $t_{i-1}, \ldots, t_{i+1-s}$ are used to form an approximation for the next unknown y_{i+1} at t_{i+1}. (Reprinted from Ascher [3].)*

General form of a linear multistep method

Assuming for simplicity a uniform step size h, so that
$$t_i = a + ih, \, i = 0, 1, \ldots,$$
an s-step method reads
$$\sum_{j=0}^{s} \alpha_j y_{i+1-j} = h \sum_{j=0}^{s} \beta_j f_{i+1-j}.$$

16.4. Multistep methods

Here, $f_{i+1-j} = f(t_{i+1-j}, y_{i+1-j})$ and α_j, β_j are coefficients which specify the actual multistep method; let us set $\alpha_0 = 1$ for definiteness, because the entire formula can obviously be rescaled.

The need to evaluate f at the unknown point y_{i+1} depends on β_0: the method is **explicit** if $\beta_0 = 0$ and **implicit** otherwise. So, for the explicit variant we have the solution at the next step defined by known quantities as

$$y_{i+1} = -\sum_{j=1}^{s} \alpha_j y_{i+1-j} + h \sum_{j=1}^{s} \beta_j f_{i+1-j},$$

whereas the implicit formula can be written as

$$y_{i+1} - h\beta_0 f(t_{i+1}, y_{i+1}) = -\sum_{j=1}^{s} \alpha_j y_{i+1-j} + h \sum_{j=1}^{s} \beta_j f_{i+1-j},$$

with all the unknown terms gathered at the left-hand side.

Such methods are called *linear* because, unlike general RK, the expression in the multistep formula is linear in f. Note that f itself may still be nonlinear in y.

Example 16.9. The forward Euler method is a particular instance of a linear multistep method, with $s = 1, \alpha_1 = -1, \beta_0 = 0$, and $\beta_1 = 1$.

The backward Euler method is also a particular instance of a linear multistep method, with $s = 1, \alpha_1 = -1, \beta_0 = 1$, and $\beta_1 = 0$. Note that $\beta_0 = 0$ for the explicit method and $\beta_0 \neq 0$ for the implicit one.

The other explicit RK methods that we have seen in Section 16.3 are not linear multistep methods. ∎

Let us define the **local truncation error** for the linear multistep method to be

$$d_i = h^{-1} \sum_{j=0}^{s} \alpha_j y(t_{i+1-j}) - \sum_{j=0}^{s} \beta_j y'(t_{i+1-j}).$$

As before, this is the amount by which the exact solution fails to satisfy the difference equation, divided by h. Thus, the method has **accuracy order** q if

$$d_i = \mathcal{O}(h^q)$$

for all problems with sufficiently smooth exact solutions $y(t)$.

The most popular families of linear multistep methods are the *Adams family* and the *backward differentiation formula* (BDF) *family*.

The Adams family

The Adams methods are derived by considering integration of the ODE over the most recent subinterval, yielding

$$y(t_{i+1}) = y(t_i) + \int_{t_i}^{t_{i+1}} f(t, y(t)) dt$$

and then approximating the integrand $f(t, y)$ by an interpolating polynomial through previously computed values of $f(t_l, y_l)$. In the general form of multistep methods we therefore set, for all Adams methods, the α_j values to be

$$\alpha_0 = 1, \alpha_1 = -1, \quad \text{and} \quad \alpha_j = 0, j > 1.$$

The s-step *explicit Adams* method, also called the **Adams–Bashforth** method, is obtained by interpolating f through the previous points $t = t_i, t_{i-1}, \ldots, t_{i+1-s}$, thus giving an explicit method. Since there are s interpolation points which are $\mathcal{O}(h)$ apart we expect (and obtain) order of accuracy $q = s$.

Example 16.10. With $s = 2$ we interpolate f at the points t_i and t_{i-1}. This gives the straight line

$$p(t) = f_i + \frac{f_i - f_{i-1}}{h}(t - t_i),$$

which is then integrated to yield

$$\int_{t_i}^{t_{i+1}} \left(f_i + \frac{f_i - f_{i-1}}{h}(t - t_i) \right) dt$$

$$= \left[f_i t + \frac{f_i - f_{i-1}}{2h}(t - t_i)^2 \right]_{t_i}^{t_{i+1}}$$

$$= h \left(\frac{3 f_i}{2} - \frac{f_{i-1}}{2} \right).$$

Therefore, the obtained formula is

$$y_{i+1} = y_i + \frac{h}{2}(3 f_i - f_{i-1}).$$

This is the two-step Adams–Bashforth method.

The four-step and five-step Adams–Bashforth methods can be similarly derived using polynomial interpolation of degrees 3 and 4, respectively. Any of the representations described in Sections 10.2–10.4 would do. Integration from t_i to t_{i+1} follows. The obtained methods (of orders 4 and 5) are

$$y_{i+1} = y_i + \frac{h}{24}(55 f_i - 59 f_{i-1} + 37 f_{i-2} - 9 f_{i-3}) \quad \text{and}$$

$$y_{i+1} = y_i + \frac{h}{720}(1901 f_i - 2774 f_{i-1} + 2616 f_{i-2} - 1274 f_{i-3} + 251 f_{i-4}).$$

For $s = 1$ we obtain the forward Euler method. ∎

The s-step *implicit Adams* method, also called **Adams–Moulton** method, is obtained by interpolating f through the previous points plus the next one, $t = t_{i+1}, t_i, t_{i-1}, \ldots, t_{i+1-s}$, giving an implicit method. Since there are $s + 1$ interpolation points which are $\mathcal{O}(h)$ apart we expect (and obtain) order of accuracy $q = s + 1$.

Example 16.11. Interpolating (t_{i+1}, f_{i+1}) by a constant and integrating, we clearly obtain the backward Euler method ($s = q = 1$).

Passing a straight line through (t_i, f_i) and (t_{i+1}, f_{i+1}) and integrating, we get, just like in Section 15.1, the implicit trapezoidal method ($s = 1, q = 2$). Thus, there are two one-step methods in the Adams–Moulton family. Life gets more interesting only with $s > 1$.

Exercise 12 asks you to derive the two-step formula by passing a quadratic through the points (t_{i+1}, f_{i+1}), (t_i, f_i), and (t_{i-1}, f_{i-1}). The three-step and four-step Adams–Moulton methods can be similarly derived using polynomial interpolation of degrees 3 and 4, respectively. Integration from

16.4. Multistep methods

t_i to t_{i+1} follows, and the obtained methods (of orders 4 and 5) are

$$y_{i+1} = y_i + \frac{h}{24}(9f_{i+1} + 19f_i - 5f_{i-1} + f_{i-2}) \quad \text{and}$$

$$y_{i+1} = y_i + \frac{h}{720}(251f_{i+1} + 646f_i - 264f_{i-1} + 106f_{i-2} - 19f_{i-3}).$$

Compare these formulas to their explicit counterparts in Example 16.10. ∎

Example 16.12. For the scalar problem

$$y' = -y^2, \qquad y(1) = 1,$$

the exact solution is $y(t) = \frac{1}{t}$. Tables 16.3 and 16.4 list absolute errors at $t = 10$. These should be compared against the RK methods demonstrated in Table 16.2 of Example 16.7. For the additional initial values necessary we use the exact solution, excusing this form of cheating by claiming that we are interested here only in the global error behavior.

Table 16.3. *Example* 16.12: *errors and calculated rates for Adams–Bashforth methods;* (s,q) *denotes the s-step method of order q.*

Step h	$(1,1)$ error	Rate	$(2,2)$ error	Rate	$(4,4)$ error	Rate
0.2	4.7e-3		9.3e-4		1.6e-4	
0.1	2.3e-3	1.01	2.3e-4	2.02	1.2e-5	3.76
0.05	1.2e-3	1.01	5.7e-5	2.01	7.9e-7	3.87
0.02	4.6e-4	1.00	9.0e-6	2.01	2.1e-8	3.94
0.01	2.3e-4	1.00	2.3e-6	2.00	1.4e-9	3.97
0.005	1.2e-4	1.00	5.6e-7	2.00	8.6e-11	3.99
0.002	4.6e-5	1.00	9.0e-8	2.00	2.2e-12	3.99

Table 16.4. *Example* 16.12: *errors and calculated rates for Adams–Moulton methods;* (s,q) *denotes the s-step method of order q.*

Step h	$(1,1)$ error	Rate	$(1,2)$ error	Rate	$(3,4)$ error	Rate
0.2	6.0e-3		1.8e-4		1.1e-5	
0.1	2.4e-3	1.35	4.5e-5	2.00	8.4e-7	3.73
0.05	1.2e-3	1.00	1.1e-5	2.00	5.9e-8	3.85
0.02	4.6e-4	1.00	1.8e-6	2.00	1.6e-9	3.92
0.01	2.3e-4	1.00	4.5e-7	2.00	1.0e-10	3.97
0.005	1.2e-4	1.00	1.1e-7	2.00	6.5e-12	3.98
0.002	4.6e-5	1.00	1.8e-8	2.00	1.7e-13	3.99

> **Theorem: Multistep Method Order.**
> Let
> $$C_0 = \sum_{j=0}^{s} \alpha_j,$$
> $$C_i = (-1)^i \left[\frac{1}{i!} \sum_{j=1}^{s} j^i \alpha_j + \frac{1}{(i-1)!} \sum_{j=0}^{s} j^{i-1} \beta_j \right], \quad i = 1, 2, \ldots.$$
>
> Then the linear multistep method has order p if and only if
> $$C_0 = C_1 = \cdots = C_p = 0, \ C_{p+1} \neq 0.$$
>
> Furthermore, the local truncation error is then given by
> $$d_i = C_{p+1} h^p y^{(p+1)}(t_i) + \mathcal{O}(h^{p+1}).$$

We can see that the observed order of the methods is as advertised. The error constant (i.e., γ in $e(h) \approx \gamma h^q$) is smaller for the Adams–Moulton method of order $q > 1$ than the corresponding Adams–Bashforth method of the same order. This is not surprising: the s interpolation points are more centered with respect to the interval $[t_i, t_{i+1}]$, where the ensuing integration takes place.

The error constant in RK2 is comparable to that of the implicit trapezoidal method in Table 16.4. The error constant in RK4 is smaller than that in the corresponding fourth order methods here. (Can you intuitively explain why?) ∎

A pleasant theoretical advantage of linear multistep methods is that deriving such a method to an arbitrarily high, proven order is straightforward; this is unlike the situation for RK methods. In fact, the rather general theorem given on the current page can be proved in a straightforward way using Taylor expansions of $y(t - jh)$ and $y'(t - jh)$ about t. The theorem gives not only conditions for the method's order but also the leading term of the truncation error.

Example 16.13. For the two-step Adams–Bashforth formula we have $C_1 = -(-1 + 3/2 - 1/2) = 0$, $C_2 = 1/2(-1) + 3/2 - 1 = 0$, $C_3 = -[1/6(-1) + 1/2(3/2 - 2)] = \frac{5}{12}$.
Hence the local truncation error is
$$d_i = \frac{5}{12} h^2 y'''(t_i).$$

See also Exercise 15. ∎

A far less pleasant practical requirement for an s-step method is that of supplying *startup values*: in principle all s initial values $y_0, y_1, \ldots, y_{s-1}$ must be $\mathcal{O}(h^q)$-accurate for a method of order q, and this is an issue when $s > 1$. There are various ways for obtaining the additional initial values $y_1, \ldots, y_{s-1}$ (whereas $y_0 = c$ is specified as usual). This typically involves using another method which requires fewer initial values. Moreover, abandoning the notion of accuracy order in favor of maintaining an a posteriori error estimate under control, sufficiently accurate initial values may be obtained using a lower order method with a smaller step size.

Backward differentiation formulas (BDF)

Here is another family of linear multistep methods. The real power of this family will only be realized in the next section.

16.4. Multistep methods

The s-step BDF method is obtained by evaluating f only at the right end of the current step, (t_{i+1}, y_{i+1}), driving an interpolating polynomial of y (rather than f) through the points $t = t_{i+1}, t_i, t_{i-1}, \ldots, t_{i+1-s}$, and differentiating it. This gives an implicit method of accuracy order $q = s$. Table 16.5 gives the coefficients of the BDF methods for s up to 6.[61] For $s = 1$ we obtain again the backward Euler method.

Table 16.5. *Coefficients of BDF methods up to order 6.*

s	β_0	α_0	α_1	α_2	α_3	α_4	α_5	α_6	q
1	1	1	-1						1
2	$\frac{2}{3}$	1	$-\frac{4}{3}$	$\frac{1}{3}$					2
3	$\frac{6}{11}$	1	$-\frac{18}{11}$	$\frac{9}{11}$	$-\frac{2}{11}$				3
4	$\frac{12}{25}$	1	$-\frac{48}{25}$	$\frac{36}{25}$	$-\frac{16}{25}$	$\frac{3}{25}$			4
5	$\frac{60}{137}$	1	$-\frac{300}{137}$	$\frac{300}{137}$	$-\frac{200}{137}$	$\frac{75}{137}$	$-\frac{12}{137}$		5
6	$\frac{60}{147}$	1	$-\frac{360}{147}$	$\frac{450}{147}$	$-\frac{400}{147}$	$\frac{225}{147}$	$-\frac{72}{147}$	$\frac{10}{147}$	6

Example 16.14. Continuing with Example 16.12, we now compute errors for the same ODE problem using three BDF methods.

The results in Table 16.6 are clearly comparable to those in Tables 16.3 and 16.4, although the error constants for methods of similar order are worse here. ∎

Table 16.6. *Example 16.14: errors and calculated rates for BDF methods; (s, q) denotes the s-step method of order q.*

Step h	$(1,1)$ error	Rate	$(2,2)$ error	Rate	$(4,4)$ error	Rate
0.2	6.0e-3		7.3e-4		7.6e-5	
0.1	2.4e-3	1.35	1.8e-4	2.01	6.1e-6	3.65
0.05	1.2e-3	1.00	4.5e-5	2.00	4.3e-7	3.81
0.02	4.6e-4	1.00	7.2e-6	2.00	1.2e-8	3.91
0.01	2.3e-4	1.00	1.8e-6	2.00	7.8e-10	3.96
0.005	1.2e-4	1.00	4.5e-7	2.00	4.9e-11	3.98
0.002	4.6e-5	1.00	7.2e-8	2.00	1.3e-12	3.99

Predictor-corrector methods

The great practical advantage of higher order multistep methods is their cost per step in terms of function evaluations. For instance, only one function evaluation is required to advance the explicit Adams–Bashforth formula by one step. However, for the higher order Adams–Bashforth methods there are some serious limitations in terms of absolute stability, so much so that they are usually not employed as stand-alone discretizations with $s > 2$.

---

[61] We do not consider BDF with $s > 6$ because then these methods become unstable in a basic way that simply cannot happen for RK methods and will not be discussed further in this text.

For the implicit Adams–Moulton methods, which are naturally more accurate and more stable than the corresponding explicit methods of the same order, we need to solve nonlinear algebraic equations for y_{i+1}. Worse, for ODE systems we generally have a nonlinear system of algebraic equations at each step. So, are we ahead yet?

Fortunately, everything implicit and nonlinear is multiplied by h in the implicit ODE method, because only y and not y' may appear nonlinearly in the ODE. Thus, for h sufficiently small a simple **fixed point iteration** of the form considered in Sections 3.3 and 7.3 converges under fairly mild conditions. This works in practice unless the ODE system is *stiff*; see Section 16.5 for the latter case.

To start the fixed point iteration for a given s-step Adams–Moulton formula we need a starting iterate, and since all those previous values of f are stored anyway we may well use them to apply the corresponding Adams–Bashforth formula. This explicit formula yields a *predicted* value for y_{i+1}, which is then *corrected* by the fixed-point iteration based on Adams–Moulton.

But next, note that the fixed point iteration need not be carried to convergence: all it yields in the end is y_{i+1}, which is just an approximation for $y(t_{i+1})$ anyway. Applying only a fixed number of such iterations creates a new, explicit, *predictor-corrector* method. The most popular predictor-corrector variant, denoted PECE, carries out only one fixed point iteration. The algorithm is given below.

Algorithm: Predictor-Corrector Step (PECE).
At step i, given $y_i, f_i, f_{i-1}, \ldots, f_{i+1-s}$:

1. Use an s-step Adams–Bashforth method to calculate y_{i+1}^0, calling the result the Predicted value.

2. Evaluate $f_{i+1}^0 = f(t_{i+1}, y_{i+1}^0)$.

3. Apply an s-step Adams–Moulton method using f_{i+1}^0 for the unknown, calling the result the Corrected value y_{i+1}.

4. Evaluate $f_{i+1} = f(t_{i+1}, y_{i+1})$.

The last evaluation in the PECE algorithm is carried out in preparation for the next time step and to maintain an acceptable measure of absolute stability, a concept further discussed in Section 16.5.

Example 16.15. Combining the two-step Adams–Bashforth formula with the second order one-step Adams–Moulton formula (i.e., the implicit trapezoidal method), we obtain the following method for advancing one time step.

Given y_i, f_i, f_{i-1}, set

1. $y_{i+1}^0 = y_i + \frac{h}{2}(3f_i - f_{i-1})$,

2. $f_{i+1}^0 = f(t_{i+1}, y_{i+1}^0)$,

3. $y_{i+1} = y_i + \frac{h}{2}(f_i + f_{i+1}^0)$,

4. $f_{i+1} = f(t_{i+1}, y_{i+1})$.

This is an explicit, second order method which has the local truncation error

$$d_i = -\frac{1}{12}h^2 y'''(t_{i+1}) + \mathcal{O}(h^3).$$

16.5. Absolute stability and stiffness

This method looks much like an explicit trapezoidal variant of an RK method. The predictor-corrector approach really shines when more steps are involved, because the same simplicity—and cost of two function evaluations per step!—holds for higher order methods, whereas higher order RK methods get significantly more complex and expensive per step. ∎

In the PECE algorithm given on the facing page the order of accuracy is $s+1$, one higher than that of the predictor. This allows for local error estimation. Error control in the spirit of Section 16.6 is then facilitated. Another common situation is where the orders of the predictor formula and of the corrector formula are the same. In the latter case the principal term of the local truncation error for the PECE method is the same as that of the corrector. It is then possible to estimate the local error in a very simple manner.

In fact, it is also possible to adaptively vary the method order, $q = s$, of the PECE pair. However, varying the step size is more complicated, though certainly possible, than in the case of the one-step RK methods.

> **Note:** Comparing multistep methods to RK methods, the comments made at the end of Section 16.3 are relevant here. Briefly, the important advantages of the multistep family are the cheap high order PECE pairs with the local error estimation that comes for free.
>
> The important disadvantages are the need for additional startup procedure and the relatively cumbersome adjustment to local changes such as lower continuity, event location, and drastically adapting the step size.
>
> In the 1980s, linear multistep methods were the methods of choice for most general-purpose ODE codes. With less emphasis on cost and more on flexibility, however, RK methods have taken the popularity lead since. This is true except for stiff problems, for which BDF methods are still popular.

Specific exercises for this section: Exercises 12–15.

16.5 Absolute stability and stiffness

The methods described in Sections 16.3 and 16.4 are routinely used in many applications, and they often give satisfactory results in practice. An exception is the case of stiff problems, where explicit RK methods are forced to use a very small step size and thus become unreasonably expensive. Adams–Bashforth and Adams predictor-corrector multistep methods meet a similar fate. But before turning to stiffness we reintroduce absolute stability, in a more general context than in Section 16.2.

Complex-coefficient test equation and absolute stability region

In this section we consider the *test equation*

$$y' = \lambda y,$$

where λ is a complex scalar. The exact solution is still $y(t) = e^{\lambda t} y(0)$ (compare to page 491). The solution magnitude depends only on the real part of λ: we have

$$|y(t)| = e^{\Re(\lambda) t} |y(0)|.$$

So $|y(t_{i+1})| \leq |y(t_i)|$, i.e., the solution decays in magnitude as t grows, if $\Re(\lambda) \leq 0$.

For the forward Euler method the absolute stability condition still reads

$$|1 + h\lambda| \leq 1,$$

but the meaning is different from that in Section 16.2 because $z = h\lambda$ now roams in the *complex plane*, and the absolute stability condition is particularly important in the left half of that plane, namely, $\Re(z) \leq 0$. The condition $|1+z| \leq 1$ in fact describes a disk of radius 1 in the z-plane, centered at $z = (-1,0)$ and depicted in red in Figure 16.8. Using a constant step size h, we must choose it so that z stays inside that red disk.

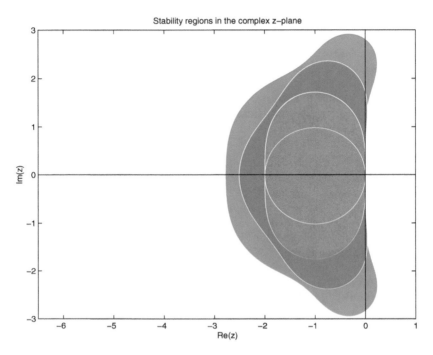

Figure 16.8. *Stability regions for q-stage explicit RK methods of order q, $q = 1,2,3,4$. The inner circle corresponds to forward Euler, $q = 1$. The larger q is, the larger the stability region. Note the "ear lobes" of the fourth-order method protruding into the right half plane.*

Is the test equation too simple to be of practical use?

In general we must consider the nonlinear ODE system (16.1a) given on page 482. Here, absolute stability properties are determined by the Jacobian matrix, defined by

$$J = \frac{\partial \mathbf{f}}{\partial \mathbf{y}} = \begin{pmatrix} \frac{\partial f_1}{\partial y_1} & \cdots & \cdots & \frac{\partial f_1}{\partial y_m} \\ \vdots & \ddots & & \vdots \\ \vdots & & \ddots & \vdots \\ \frac{\partial f_m}{\partial y_1} & \cdots & \cdots & \frac{\partial f_m}{\partial y_m} \end{pmatrix}.$$

Why is that so? Because stability is all about the propagation in time t of perturbations to the solution. So, if our solution $\mathbf{y}(t)$ is a perturbation of a nearby function $\hat{\mathbf{y}}(t)$ satisfying the same differential equation, then we should consider the progress of their difference

$$\mathbf{w}(t) = \mathbf{y}(t) - \hat{\mathbf{y}}(t).$$

Expanding in (what else!) Taylor series in m variables (see page 253), we can write

$$\mathbf{f}(\mathbf{y}) = \mathbf{f}(\hat{\mathbf{y}}) + J(\hat{\mathbf{y}})\mathbf{w} + \mathcal{O}\|\mathbf{w}\|^2,$$

16.5. Absolute stability and stiffness

so the perturbation difference obeys the ODE

$$\mathbf{w}' = J(\hat{\mathbf{y}})\mathbf{w} + \mathcal{O}\|\mathbf{w}\|^2 \approx J(\hat{\mathbf{y}})\mathbf{w}.$$

Hence we obtain, for small perturbations, an ODE system that looks like the test equation with the Jacobian matrix J in place of λ.

A surprisingly good practical estimate of the stability properties of a numerical method for the general nonlinear ODE system is obtained by considering the test equation for each of the eigenvalues of the Jacobian matrix. This is also the reason, though, why we must consider a complex scalar λ: eigenvalues are in general complex scalars. Thus, instead of an absolute stability bound on $z = h\lambda$ as in Section 16.2, we are really looking at an **absolute stability region** in the complex plane.

Let us consider next the application of the explicit trapezoidal method to the test equation. We have $f(y) = \lambda y$, and our task is to find $R(z)$ such that $y_{i+1} = R(z)y_i$. For the first stage we get again

$$Y = (1+z)y_i,$$

and then

$$y_{i+1} = y_i + \frac{z}{2}(y_i + Y) = \left[1 + z + \frac{z^2}{2}\right] y_i.$$

Thus, $R(z) = 1 + z + \frac{z^2}{2}$. A similar expression is obtained for the explicit midpoint method. Exercise 17 reveals the larger picture.

Stability regions for explicit q-stage RK methods of orders q up to 4 are depicted in Figure 16.8.

Example 16.16. Let us return to the mildly nonlinear problem of Example 16.4. The 8×8 Jacobian matrix is

$$J(\mathbf{y}) = \begin{pmatrix} -1.71 & .43 & 8.32 & 0 & 0 & 0 & 0 & 0 \\ 1.71 & -8.75 & 0 & 0 & 0 & 0 & 0 & 0 \\ 0 & 0 & -10.03 & .43 & .035 & 0 & 0 & 0 \\ 0 & 8.32 & 1.71 & -1.12 & 0 & 0 & 0 & 0 \\ 0 & 0 & 0 & 0 & -1.745 & .43 & .43 & 0 \\ 0 & 0 & 0 & .69 & 1.71 & -280y_8 - .43 & .69 & -280y_6 \\ 0 & 0 & 0 & 0 & 0 & 280y_8 & -1.81 & 280y_6 \\ 0 & 0 & 0 & 0 & 0 & -280y_8 & 1.81 & -280y_6 \end{pmatrix}.$$

The eigenvalues of J, like J itself, depend on the solution $\mathbf{y}(t)$. The MATLAB command `eig` reveals that at the initial state $t = 0$, where $\mathbf{y} = \mathbf{y}_0$, the eigenvalues of J up to the first few leading digits are

$$0, \; -10.4841, \; -8.2780, \; -0.2595, \; -0.5058, \; -2.6745 \pm 0.1499\imath, \; -2.3147.$$

There is a conjugate pair of eigenvalues, while the rest are real. To get them all into the disk of absolute stability for the forward Euler method, the most demanding condition is

$$-10.4841 h > -2,$$

implying that $h = .1$ would be a safe choice.

However, integration of this problem with a constant step size $h = .1$ yields a huge error: the integration process becomes unstable. Good results as in Example 16.4 are obtained by carrying out the integration process with the much smaller $h = .005$.

Indeed it turns out that at $t \approx 10.7$, where things are at about their worst, the eigenvalues equal

$$-211.7697, -10.4841, -8.2780, -2.3923, -2.1400, -0.4907, -3 \times 10^{-5}, -3 \times 10^{-12}.$$

The stability bound that the most negative eigenvalue yields is

$$-211.7697 h > -2,$$

implying that $h = .005$ would be a safe choice, but even $h > .01$ may not be. ∎

Stiffness

An intuitive definition of the concept of stiffness is given on the current page. A more precise definition turns out to be surprisingly elusive.

> **Stiffness.**
> The initial-value ODE problem is **stiff** if the step size needed to maintain absolute stability of the forward Euler method is much smaller than the step size needed to represent the solution accurately.

The problem of Examples 16.4 and 16.16 is stiff. The simple problem of Example 16.5 is even stiffer, intuitively speaking.

From Figure 16.8 we see that increasing the order of an explicit RK method does not do much good when it comes to solving stiff problems. In fact this turns out to be true for any of the explicit methods that we have seen.

For stiff problems, we therefore seek other methods. Implicit methods such as backward Euler, implicit trapezoidal, or implicit midpoint become more attractive then, because their region of absolute stability contains the entire left half z-plane. (See Exercise 16.) A method whose domain of absolute stability contains the entire left half plane is called **A-stable**.

> **Note:** The case study below considers a PDE and relies directly on Example 7.1. If you are not familiar with the contents of Chapter 7, especially Example 7.1, then your best bet at this point may well be to just skip Example 16.17. Otherwise please don't, because it demonstrates several important issues.

Example 16.17 (heat equation). The *heat equation* in its simplest form is written as the PDE

$$\frac{\partial u}{\partial t} = \left(\frac{\partial^2 u}{\partial x_1^2} + \frac{\partial^2 u}{\partial x_2^2} \right).$$

Here x_1 and x_2 are spatial variables in the unit square, $0 < x_1, x_2 < 1$, and t is time, $t \geq 0$.

Imagine a square electric blanket that has been heated up to a uniform temperature of $25°$. The blanket is located in a room temperature of $0°C$ (it's a room in the high Andes), and the plug is pulled at the time $t = 0$ when our story begins. The temperature at future times $t > 0$ is governed by the heat equation subject to the *boundary conditions*

$$u(t, x_1, 0) = u(t, x_1, 1) = u(t, 0, x_2) = u(t, 1, x_2) = 0 \quad \forall t \geq 0$$

16.5. Absolute stability and stiffness

and the *initial conditions* specifying that $u(0, x_1, x_2) = 25$ for any $0 < x_1, x_2 < 1$. Thus, it is an initial-boundary value problem.

To estimate the heat distribution over the blanket at positive times we first discretize the spatial derivatives as in Example 7.1. Using a uniform step size $\Delta x = 1/M$ in both spatial directions, we are looking for a two-dimensional *grid function*, $\{u_{i,j}(t), i = 0, \ldots, M, j = 0, \ldots, M\}$, such that $u_{i,j}(t) \approx u(t, i\Delta x, j\Delta x)$. Next, we discretize the second spatial derivatives using a centered three-point formula as in Section 14.1. Thus, holding $x_2 = j\Delta x$ fixed we approximate $\frac{\partial^2 u}{\partial x_1^2}$ at $x_1 = i\Delta x$ by $\frac{1}{\Delta x^2}(u_{i+1,j} - 2u_{i,j} + u_{i-1,j})$, and likewise holding $x_1 = i\Delta x$ fixed we approximate $\frac{\partial^2 u}{\partial x_2^2}$ at $x_2 = j\Delta x$ by $\frac{1}{\Delta x^2}(u_{i,j+1} - 2u_{i,j} + u_{i,j-1})$. The heat equation plus boundary conditions is correspondingly approximated by the ODE system

$$\frac{du_{i,j}}{dt} = \frac{1}{\Delta x^2}(u_{i+1,j} + u_{i,j+1} - 4u_{i,j} + u_{i-1,j} + u_{i,j-1}), \quad 1 \le i, j \le M-1,$$
$$u_{i,j} = 0 \quad \text{otherwise}.$$

This is a linear system of $m = (M-1)^2$ ODEs, subject to the initial conditions

$$u_{i,j}(0) = 25, \quad 1 \le i, j \le M-1.$$

Furthermore, we can reshape the array of $\{u_{i,j}\}$ into a vector $\mathbf{y}(t)$ and write the ODE system as

$$\mathbf{y}' = -A\mathbf{y},$$

the sparse matrix A being the one derived in Example 7.1 with an obvious slight notational modification. Note that for $\Delta x = .01$, say, we obtain $m = 99^2 = 9801$ ODEs. It is not difficult to think of situations where even larger ODE systems are encountered (think three space variables, for one thing).

The next question is how to solve this ODE system. Certainly there is a good incentive here to use an explicit method if possible, thus avoiding the need to solve linear systems. The forward Euler method reads

$$\mathbf{y}_{i+1} = \mathbf{y}_i - hA\mathbf{y}_i, \quad i = 0, 1, \ldots,$$

where h is the step size in time, so at each time step only one matrix-vector multiplication is required.

The eigenvalues of the matrix A, which is scaled by Δx^2 as compared to Example 7.1, are given by

$$\lambda_{l,m} = \frac{1}{\Delta x^2}[4 - 2(\cos(l\pi \Delta x) + \cos(j\pi \Delta x))], \quad 1 \le l, j \le M-1.$$

Thus, $\max \lambda_{l,j} = \lambda_{M-1,M-1} \approx \frac{24}{\Delta x^2}$. (What's important here is the relation to Δx^2, less so the constant 24.) The absolute stability requirement for the forward Euler method then dictates setting

$$h \le \frac{\Delta x^2}{12}.$$

This can hurt if Δx is small. For $\Delta x = .01$ it translates to needing 12,000 time steps just to reach $t = .1$. Note that a similar restriction in essence holds for higher order explicit methods, which works to increase frustration because the separation between what is required for accuracy and what is demanded by stability increases even further.

Instead, consider the implicit trapezoidal method, which in the current context is called the **Crank–Nicolson** scheme. Now there is no absolute stability requirement on h, and accuracy considerations suggest setting it in general to be of the same order as Δx. (In fact, in this particular application it makes sense to set h smaller at first and letting it grow because the solution gets smoother and more uniform with time; but let's not get into such further refinement.)

For a rough comparison, let us set $h = \Delta x$. Please verify that the trapezoidal, or Crank–Nicolson, method reads

$$\left(I + \frac{h}{2}A\right)\mathbf{y}_{i+1} = \left(I - \frac{h}{2}A\right)\mathbf{y}_i.$$

So, in comparison with the forward Euler method the Crank–Nicolson method will be more efficient provided that the solution of the linear system it requires at each step can be achieved in less than $12/\Delta x$ matrix-vector multiplications. Employing an appropriately preconditioned CG method for this purpose (see Section 7.4) can easily achieve this and much more for a fine resolution. Thus, the implicit method wins!

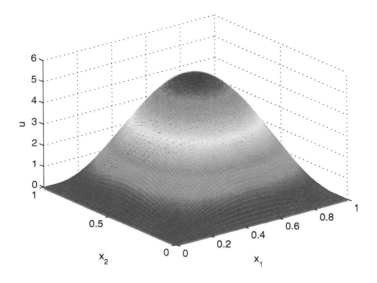

Figure 16.9. *Solution of the heat problem of Example* 16.17 *at* $t = 0.1$.

Figure 16.9 depicts the solution at time $t = .1$. The maximum value is $u(.1,.5,.5) \approx 5.63$. The shape of the temperature distribution is similar to that seen at other times t, the main difference being the maximum value. For an earlier time, $u(.05,.5,.5) \approx 15$, while $u(.2,.5,.5) < 1$. The temperature is rapidly decreasing everywhere towards a uniform zero level. ∎

Backward Euler and L-stability

The backward Euler method, briefly introduced in Section 16.2, is of fundamental importance for stiff problems. It is written here for an ODE system (16.1a) as

$$\mathbf{y}_{i+1} = \mathbf{y}_i + h\mathbf{f}(t_{i+1},\mathbf{y}_{i+1}), \quad i = 1,2,\ldots.$$

For the test equation we obtain $y_{i+1} = y_i + h\lambda y_{i+1}$, and hence

$$y_{i+1} = \frac{1}{1 - h\lambda} y_i.$$

Therefore, the region of absolute stability is defined by the inequality

$$|1 - z| = |1 - h\lambda| \geq 1.$$

16.5. Absolute stability and stiffness

This includes in particular the entire left half z-plane, so the method is A-stable. Even more importantly in cases where the real part of z is large and negative, we have $\frac{1}{1-h\lambda} \to 0$ as $|z| \to \infty$. Therefore $y_{i+1} \to 0$, as does the exact solution $y(h)$ under these circumstances. Such a method is called *L-stable*.

BDF methods are L-stable, whereas the implicit trapezoidal and midpoint methods are not.

Carrying out an integration step

The inherent difficulty in implicit methods is that the unknown solution at the next step, y_{i+1}, is defined implicitly. Using the backward Euler method, for instance, we have to solve an often nonlinear equation at each step. This difficulty gets significantly worse when an ODE system is considered. Now there is a nonlinear system of algebraic equations to solve; see Section 9.1. Even if the ODE is linear there is a system of linear equations to solve at each integration step. This can get costly if the size of the system, m, is large; recall Example 16.17.

For instance, applying the backward Euler method to our ODE system we are facing at the ith step a system of algebraic equations given by

$$\mathbf{g}(\mathbf{y}_{i+1}) \equiv \mathbf{y}_{i+1} - h\mathbf{f}(\mathbf{y}_{i+1}) - \mathbf{y}_i = \mathbf{0}.$$

Unfortunately, the obvious fixed point iteration utilized in Section 16.4 for a predictor-corrector method will not work here unless h is unreasonably small. Fortunately, however, Newton's method as described in Section 9.1 does work also in stiff circumstances. Indeed, with our current notation J for the Jacobian matrix of $\mathbf{f}(\mathbf{y})$, we have the Jacobian matrix $I - hJ$ for an iterative method for $\mathbf{y}_{i+1}$. For a good starting iterate we can take the current approximate solution, $\mathbf{y}_{i+1}^{(0)} = \mathbf{y}_i$. The Newton iteration algorithm is given on this page.

Algorithm: Newton's Method for Stiff ODEs.

Set $\mathbf{y}_{i+1}^{(0)} = \mathbf{y}_i$.
for $k = 0, 1, \ldots$, until convergence

$$\text{solve linear system } \left(I - hJ(\mathbf{y}_{i+1}^{(k)})\right) \mathbf{p}_k = -\mathbf{g}\left(\mathbf{y}_{i+1}^{(k)}\right) \text{ for } \mathbf{p}_k;$$
$$\text{set } \mathbf{y}_{i+1}^{(k+1)} = \mathbf{y}_{i+1}^{(k)} + \mathbf{p}_k.$$

end

All this is to be done at each and every integration step! The saving grace is that the initial guess $\mathbf{y}_{i+1}^{(0)}$ is only $\mathcal{O}(h)$ away from the solution $\mathbf{y}_{i+1}$ of this nonlinear system, so we may expect an $\mathcal{O}(h^2)$ closeness after only one iteration due to the quadratic convergence of Newton's method. Therefore, one such iteration is usually sufficient, at least for the backward Euler method whose local error (see page 516) at each step is $\mathcal{O}(h^2)$ anyway.

Applying one Newton iteration per time step yields the **semi-implicit backward Euler** method

$$(I - hJ(\mathbf{y}_i))\mathbf{y}_{i+1} = (I - hJ(\mathbf{y}_i))\mathbf{y}_i + h\mathbf{f}(\mathbf{y}_i) \quad \text{or}$$
$$\mathbf{y}_{i+1} = \mathbf{y}_i + h(I - hJ(\mathbf{y}_i))^{-1}\mathbf{f}(\mathbf{y}_i).$$

This semi-implicit method still involves the solution of a linear system of equations at each step. Note also that it is no longer guaranteed to be stable for any z in the left half plane. (Can you see why?!)

Example 16.18. The problem of Examples 16.4 and 16.16 is stiff, to recall. With a uniform step size the forward Euler method requires $322/.005 = 64,400$ integration steps in Example 16.4.

A solution that is qualitatively similar to the one displayed in Figure 16.3 is obtained by the semi-implicit backward Euler method using $h = 0.1$, resulting in only $3,220$ steps. The script is

```
h = 0.1; t = 0:h:322;
y = y0 * ones(1,length(t));
for i = 2:length(t)
  A = eye(8) - h*hiresj(y(:,i-1));
  y(:,i) = y(:,i-1) + h* A \ hires(y(:,i-1)));
end
plot(t,y)
```

The function `hiresj` returns the Jacobian matrix given in Example 16.16.

Thus, if an accuracy level of around .01 is satisfactory, then for this example the implicit method is much more efficient than the explicit one despite the need to solve a linear system of equations at each integration step. ∎

For higher order methods, required for the efficient computation of high accuracy solutions for stiff problems, BDF are often the methods of choice, although there are also some popular implicit RK methods that can be manipulated to yield competitive performance. The implementation of BDF methods is not much more complicated than backward Euler, except for the need for additional initial values and the cumbersome step size modification typical of all multistep methods. Their behavior, especially the two- and three-step methods, is generally similar to that of backward Euler as well. In sophisticated codes, even if utilizing more accurate methods than backward Euler, usually less than one Jacobian evaluation per step is applied on average.

There is a MATLAB code called `ode23s` for stiff problems. As the name suggests it uses a pair of methods of order 2 and 3. See also Section 16.6 if you wonder why a pair of methods is employed.

Problems with purely imaginary eigenvalues

Let us shift our attention somewhat. Here we are concerned not with what is usually regarded as a stiff problem, but rather with problems whose Jacobian matrices have purely imaginary eigenvalues. These eigenvalues are such that it is reasonable to select a step size h which yields $|z| = \mathcal{O}(1)$, with z imaginary.

Example 16.19. The test equation $y' = \lambda y = \iota \alpha y$, for a real scalar $\alpha \neq 0$, may look artificial. But it arises already when considering the simple ODE

$$u'' + \alpha^2 u = 0.$$

To see this, note that defining $y_1 = u$ and $y_2 = u'/\alpha$, we have the equivalent first order system

$$\mathbf{y}' = A\mathbf{y}, \quad A = \begin{pmatrix} 0 & \alpha \\ -\alpha & 0 \end{pmatrix}.$$

The skew-symmetric matrix A has $\pm \iota \alpha$ for eigenvalues. ∎

In Figure 16.8 we are now looking at behavior along the imaginary (i.e., the vertical) axis. We see that trouble may arise if the forward Euler or explicit trapezoidal or midpoint method is used for such a problem, because the imaginary axis save the origin is outside the absolute stability regions of these methods. Thus, roundoff error gets amplified by a factor larger than 1 at each time step. The fourth order method RK4 is now much preferred over the simpler explicit methods, not just because of its higher order but because of its superior stability properties!

But, one may ask, are there interesting practical problems with such imaginary eigenvalues?

Yes, there are, and many. The simplest instance is in Example 16.19 and Exercise 8. A far more complex problem of a similar type is discussed in Example 16.21. Indeed, attempting to reproduce Figure 16.12 using the explicit Euler or explicit midpoint method can prove to be a major frustration builder. Finally, a large class of hyperbolic PDEs with smooth solutions yield such ODE systems upon discretizing spatial derivatives using centered differences; see Section 16.8. The classical RK4 is very popular for the numerical solution of such problems.

Specific exercises for this section: Exercises 16–18.

16.6 Error control and estimation

Thus far we have assumed that the step size h is constant, independently of the location t_i of the ith step. But an examination of the form of a typical truncation error (see especially the Multistep Method Order Theorem on page 504) suggests that to keep the local truncation error roughly the same at each step, thus obtaining a quality approximation efficiently, it makes sense to take h small where the solution varies rapidly, and relatively large where the solution variation is slow. Any modern ODE package varies the step size in order to achieve some error control, and we next describe some of the issues involved.

As with quadrature and the function quads described in Section 15.6, it is desirable here to write a mathematical software package where a user would specify only the function $f(t, y)$ and the initial value information a and c, as well as a tolerance tol and a set of output points $\hat{t}_1, \ldots, \hat{t}_{\hat{N}}$ where the solution is desired. The general-purpose code would subsequently calculate $\{\hat{y}_i\}_{i=1}^{\hat{N}}$, a set of output values such that $\hat{y}_i \approx y(\hat{t}_i)$ accurate to within tol for each i. Note that the actual mesh $\{t_0, t_1, \ldots, t_N\}$ used by such a code is expected to contain the output points $\hat{t}_i$ as a subset, but the precise values of all the t_j's is not the user's concern. What the user wants are results, efficiently obtained, and the latter desire dictates a quest for a small N.

Throughout this section we consider a scalar ODE during the presentation to simplify notation. However, the numerical examples all involve ODE systems. The extension should be clear.

Types of error control

Despite superficial similarities with adaptive quadrature, however, here the situation is significantly more complex:

- The global error $e_i = y(t_i) - y_i$ does not provide a direct, simple indication on how to choose the next step size h_i such that $t_{i+1} = t_i + h_i$.

- The global error at t_i is not simply a sum of the local errors made at each previous step j for $j = 0, \ldots, i-1$. Indeed, as the error bound that we have derived for the forward Euler method (page 489) indicates, the global error may grow exponentially in time, which means that the contribution of each local error may also grow in time.

The latter concern can occasionally be traumatic; see, for instance, Example 16.21. There are methods to estimate the global error by recalculating the entire solution on the interval $[a,b]$. But more often in practice the global error does not accumulate nastily. Moreover, the user typically does not have a firm idea of a global error tolerance anyway. Then it is the first concern above that matters, because we still want for efficiency reasons a local error control, i.e., we want to adapt the step size locally! Relating this to the global error is hard, though, so we adjust expectations and are typically content to control the **local error**

$$l_{i+1} = \bar{y}(t_{i+1}) - y_{i+1},$$

where $\bar{y}(t)$ is the solution of the ODE $y' = f(t, y)$ which satisfies $\bar{y}(t_i) = y_i$, but $\bar{y}(0) \neq c$ in general. See Figures 16.10 and 16.2.

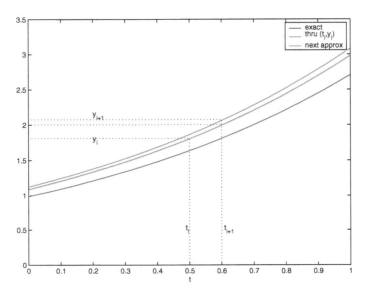

Figure 16.10. *The exact solution $y(t_i)$, which lies on the lowest of the three curves, is approximated by y_i, which lies on the middle curve. If we integrate the next step exactly, starting from (t_i, y_i), then we obtain $(t_{i+1}, \bar{y}(t_{i+1}))$, which also lies on the middle curve. But we don't: rather, we integrate the next step approximately as well, obtaining (t_{i+1}, y_{i+1}), which lies on the top curve. The difference between the two curve values at the argument t_{i+1} is the local error.*

Adaptive step size selection for local error control

Thus, we consider the ith subinterval as if there is no error accumulated at its left end t_i and wonder what might result at the right end t_{i+1}. To control the local error we keep the step size small enough, so in particular, $h = h_i$ is no longer the same for all steps i.

Suppose now that we use a pair of RK methods, one of order q and the other of order $q + 1$, to calculate at t_{i+1} two approximations, both starting from y_i at t_i. Let us denote the obtained values by y_{i+1} and $\hat{y}_{i+1}$, respectively. Then we can estimate

$$|l_{i+1}| \approx |\hat{y}_{i+1} - y_{i+1}|.$$

So, at the ith step we calculate these two approximations and compare against a given error tolerance. If

$$|\hat{y}_{i+1} - y_{i+1}| \leq h \, \text{tol},$$

16.6. Error control and estimation

then the step is accepted: set $y_{i+1} \leftarrow \hat{y}_{i+1}$, i.e., the more accurate of the two values calculated,[62] and $i \leftarrow i+1$.

If the step is not accepted, then we decrease h to $\tilde{h}$ and repeat the procedure starting from the same (t_i, y_i). This is done as follows. Since the local error in the less accurate method is $l_{i+1} \approx \gamma h^{q+1}$, upon decreasing h to a satisfactory $\tilde{h}$ the local error will become $\gamma \tilde{h}^{q+1} \approx \mu \tilde{h}$ tol, where the factor $\mu < 1$ is for safety, say $\mu = 0.9$. Dividing, we get

$$\frac{\tilde{h}^{q+1}}{h^{q+1}} \approx \frac{\mu \tilde{h} \,\texttt{tol}}{|\hat{y}_{i+1} - y_{i+1}|},$$

which then yields the expression

$$\tilde{h} = h \left(\frac{\mu h \,\texttt{tol}}{|\hat{y}_{i+1} - y_{i+1}|} \right)^{\frac{1}{q}}.$$

We caution again that the meaning of `tol` here is different from that in the adaptive quadrature case of Section 15.4. Here we attempt to control the local, not the global, error.

How is the value of h selected upon starting the ith time step? A reasonable choice is the final step size of the previous time step. But then the sequence of step sizes only decreases and never grows! So, some mechanism must be introduced that allows occasional increase of the starting current step size. For instance, if in the past two time steps no decrease was needed, then we can hazard doubling the initial step size for the current step.

The only question left is how to choose the pair of formulas of orders q and $q+1$ wisely. In the multistep PECE context the answer is obvious. For RK methods this is achieved by searching for a pair of formulas which *share the internal stages* Y_j as much as possible. A good RK pair of orders 4 and 5 would use only 6 (rather than 9 or 10) function evaluations per step. Such pairs of formulas are implemented in MATLAB's routine `ode45`, which carries out a local error control in the spirit described above.

Two case studies

The rest of this section is devoted to two longish examples, both involving use of the adaptive routine `ode45` in applications. The first of these demonstrates a great advantage to be had, while the second flashes out some warning signals.

Example 16.20 (astronomical example). The following classical example from astronomy gives a strong motivation to integrate initial value ODEs with local step size control.

Consider two bodies of masses $\mu = 0.012277471$ and $\hat{\mu} = 1 - \mu$ (earth and sun) in a planar motion, and a third body of negligible mass (moon) moving in the same plane. The motion is governed by the equations

$$u_1'' = u_1 + 2u_2' - \hat{\mu} \frac{u_1 + \mu}{D_1} - \mu \frac{u_1 - \hat{\mu}}{D_2},$$

$$u_2'' = u_2 - 2u_1' - \hat{\mu} \frac{u_2}{D_1} - \mu \frac{u_2}{D_2},$$

$$D_1 = ((u_1 + \mu)^2 + u_2^2)^{3/2},$$

$$D_2 = ((u_1 - \hat{\mu})^2 + u_2^2)^{3/2}.$$

---

[62] We are "cheating" here twice. Once by quietly hoping that `tol` relates to the global error, because often in practice $|l_{i+1}| \simeq h|e_{i+1}|$. And then we are cheating by using the more accurate method for the next y_{i+1}, even though the error estimate relates to the less accurate method. But, as the politician said, it's all for the greater good.

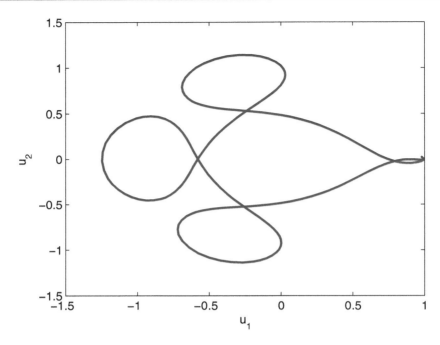

Figure 16.11. *Astronomical orbit using* ode45; *see Example* 16.20. *(Reprinted from Ascher and Petzold [5].)*

Starting with the initial conditions

$$u_1(0) = 0.994, u_2(0) = 0, u_1'(0) = 0,$$
$$u_2'(0) = -2.00158510637908252240537862224,$$

the solution is periodic with period < 17.1. Note that $D_1 = 0$ at $(-\mu, 0)$ and $D_2 = 0$ at $(\hat{\mu}, 0)$, so we need to be careful when the orbit passes near these singularity points.

To apply our codes we write the problem as a first order system, $\mathbf{y}' = \mathbf{f}(\mathbf{y})$, where $\mathbf{y} = (u_1, u_1', u_2, u_2')^T$. The conversion to first order form is similar to that demonstrated in Example 16.2 and need not be repeated here.

The orbit is depicted in Figure 16.11. It was obtained using ode45 with the default absolute error tolerance of 1.e-6 and relative error tolerance of 1.e-3 to integrate the problem on the interval $[0, 17.1]$. This necessitated 309 time steps with max $h_i = .1892$ and min $h_i = 1.5921$e-7. Additional runs with stricter tolerances suggest that the plot in Figure 16.11 is qualitatively correct.

We have also integrated the same problem employing our simple routine rk4 (see page 495) with a uniform step size. Using 1,000 steps yields nonsensical results. Even 5,000 uniform steps yield qualitatively incorrect results. Only 10,000 uniform steps yield a qualitatively correct figure, similar as far as the naked eye is concerned to Figure 16.11. ∎

The success of the adaptive black box code in Example 16.20 is gratifying after all the effort that we have invested in studying numerical methods. But there are other examples, or problem instances, too. Often in practice, a uniform step size is not a bad choice. Perhaps more important, all those shortcuts described above in devising an efficient adaptive step size selection procedure may combine unexpectedly to surprise the unaware. Here is an example.

16.6. Error control and estimation

Example 16.21 (FPU). Consider a chain of $\hat{m}$ mass points connected with springs that have alternating characteristics: the odd ones are soft and nonlinear, whereas the even ones are stiff and linear. (This is often referred to as the *Fermi–Pasta–Ulam* (FPU) problem.) We next describe the details of the corresponding ODE system, which may look a bit complex to the point of being prohibitive. But fear not: all we really want to do here eventually is concentrate on the performance of `ode45`.

There are variables $q_1, \ldots, q_{2\hat{m}}$ and $p_1, \ldots, p_{2\hat{m}}$ in which the ODE system of size $m = 4\hat{m}$ is written as

$$q'_j = \frac{\partial H}{\partial p_j}, \quad p'_j = -\frac{\partial H}{\partial q_j}, \quad j = 1, \ldots, 2\hat{m},$$

where H is a scalar function of the p_j and q_j variables. This function is called the associated Hamiltonian and is given by

$$H(\mathbf{q}, \mathbf{p}) = \frac{1}{4}\left[2\sum_{j=1}^{2\hat{m}} p_j^2 + 2\omega^2 \sum_{j=1}^{\hat{m}} q_{\hat{m}+j}^2 + (q_1 - q_{\hat{m}+1})^4 + (q_{\hat{m}} + q_{2\hat{m}})^4 \right.$$
$$\left. + \sum_{j=1}^{\hat{m}-1} (q_{j+1} - q_{\hat{m}+1+j} - q_i - q_{\hat{m}+j})^4 \right].$$

The parameter $\omega = 100$ relates to the stiff spring constant.

For our purpose here, the details of H are unimportant: what matters is that we are able to construct the corresponding ODE system (16.1a) for the m unknowns

$$\mathbf{y} = \begin{pmatrix} \mathbf{q} \\ \mathbf{p} \end{pmatrix}$$

from the above prescription. Further, let us define

$$I_i = \frac{1}{2}(p_{\hat{m}+i}^2 + \omega^2 q_{\hat{m}+i}^2), \quad I = \sum_{i=1}^{\hat{m}} I_i.$$

Then it turns out that

$$I(\mathbf{q}(t), \mathbf{p}(t)) = I(\mathbf{q}(0), \mathbf{p}(0)) + \mathcal{O}(\omega^{-1})$$

for very long times t. (Such a property makes I an *adiabatic invariant*, but this again is not our concern here.) Thus, we can run an ODE code over a relatively long time interval, plot the total energy I, and observe whether or not its trend to remain almost constant in time is honored in the computation. This provides a qualitative measure for the numerical simulation.

So, let us set $\hat{m} = 3$, yielding an ODE system of size $m = 12$, and select the initial conditions $\mathbf{q}(0) = (1, 0, 0, \omega^{-1}, 0, 0)^T$, $\mathbf{p}(0) = (1, 0, 0, 1, 0, 0)^T$. Integrating this ODE system from $a = 0$ to $b = 500$ using `rk4` with a constant step size $k = .00025$, i.e., using 2,000,000 time steps, yields the qualitatively correct Figure 16.12. The curves depicted in the figure are exact as far as the eye can tell (trust us). The "noise" is not a numerical artifact. Rather, small "bubbles" of rapid oscillations occasionally flare up and quickly die away.

Next, we integrate this ODE system using `ode45` with default tolerances, as in Example 16.20. This requires 112,085 time steps. The result, depicted in Figure 16.13, is a disaster, especially because it does not even look conspicuously wrong: it just *is* wrong.

Running `ode45` with tighter tolerances, specifically setting the relative error tolerance to 1.e-6, which is the same value as the absolute tolerance, does produce good results in 402,045 steps. But the lesson remains: do not trust a complex mathematical software package blindly! ∎

Specific exercises for this section: Exercises 19–20.

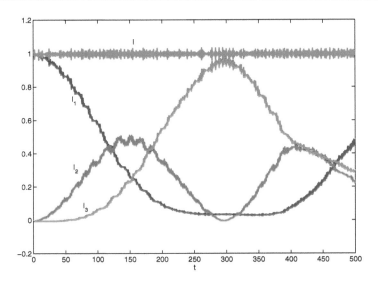

Figure 16.12. *Oscillatory energies for the FPU problem; see Example* 16.21.

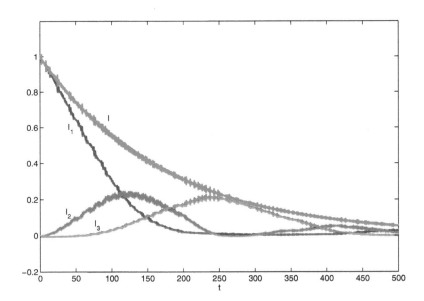

Figure 16.13. *Oscillatory energies for the FPU problem obtained using MATLAB's* `ode45` *with default tolerances. The deviation from Figure* 16.12 *depicts a significant, nonrandom error.*

16.7 *Boundary value ODEs

Generally, an ODE system with m components, given as in (16.1a) by

$$\mathbf{y}' = \mathbf{f}(t, \mathbf{y}), \quad a < t < b,$$

yields a family of solutions that can be specified in principle using m parameters, as in Example 16.1. To obtain a locally unique solution, meaning that an arbitrarily small perturbation does not yield an-

16.7. *Boundary value ODEs

other solution for the same ODE problem, requires the specification of m side conditions, or specific solution values. In the easiest, and fortunately most important, case of an initial value problem all m solution values are specified at the same initial point $t = a$. Thus, at $t = a$ all information about $\mathbf{y}(t)$ is available, and we can subsequently march along in $t > a$, both theoretically and numerically, with the entire solution information at hand. For instance, a step of the forward Euler method yields, starting from knowing an approximation for all components of $\mathbf{y}(t_i)$, an approximation for all components of $\mathbf{y}(t_i + h)$. Theoretically, this allows for conclusions about uniqueness and existence of a solution for the initial value ODE problem under general, mild assumptions. Practically, this allows for devising *explicit* numerical discretizations, an evolutionary solution process, and a local error control and step size determination.

The celebration ends when considering a *boundary value problem* (BVP),[63] where in its simplest incarnation there are l components of $\mathbf{y}$ specified at $t = a$ and $m - l$ components of $\mathbf{y}$ specified at $t = b$. The case $l = m$ brings us back to an initial value problem, but here we concentrate on the case $0 < l < m$. Thus, necessarily $m > 1$, accounting for the boldface notation. Numerically, envisioning a discretization mesh

$$a = t_0 < t_1 < \cdots < t_N = b, \quad h_i = t_i - t_{i-1},$$

on which the solution is sought, we expect all solution values $\{\mathbf{y}_i\}_{i=0}^{N}$ to become known *simultaneously* rather than gradually.

Example 16.22. Returning to Example 9.3, we can write the ODE

$$v'' + e^v = 0, \quad 0 < t < 1,$$

in first order form as in Example 16.2 and (16.1a). This yields

$$\mathbf{y} \equiv \begin{pmatrix} y_1 \\ y_2 \end{pmatrix} = \begin{pmatrix} v \\ v' \end{pmatrix}, \quad \mathbf{f}(t, \mathbf{y}) = \begin{pmatrix} y_2 \\ -e^{y_1} \end{pmatrix}.$$

The boundary conditions are

$$y_1(0) = 0, \quad y_1(1) = 0.$$

If we were given not the value of y_1 at $t = 1$ but instead the value of c_2 such that $y_2(0) = c_2$, then the resulting initial value problem would have had a unique solution over the interval $[0, 1]$ and beyond for any finite value c_2. In contrast, for the BVP there are two distinct solutions; see Figure 9.3.

The simple discretization described in Examples 9.3 (page 257) and 4.17 (page 87) attempts to find all solution values of y_1 on a given mesh, in contrast to the marching strategy used for initial value ODE problems. This means that the size of the algebraic system to be solved for the BVP grows like N, which in turn is governed by accuracy considerations. In general, this algebraic system size is about mN. In contrast, upon applying an implicit method for the initial value problem there are about N algebraic systems to be solved, each of a size that grows like m. Think, say, of 500 mesh points for a system of 5 ODEs to picture the difference. ∎

Finite difference methods

The case of a scalar ODE of second order as in Example 16.22 is special. Having only v and not v' appear explicitly is even more special. The reason that all our previous BVP examples look like this

---

[63] Unfortunately, just about any PDE in practice is subject to boundary conditions, and not just initial conditions, as, for instance, in Examples 16.17 and 7.1. Correspondingly, there are no general-purpose codes, of the quality and generality available for initial value ODEs for any other class of differential problems.

is that a simple, straightforward discretization suggests itself, and this allowed us to concentrate on other solution aspects in previous chapters. Let us now turn to the more general BVP case where the ODE system is in the general first order form, and consider designing finite difference methods.

In principle, there are no explicit *finite difference* discretizations for BVPs. Since all methods are now implicit, and for reasons of maintaining sparsity in the resulting large linear algebraic systems to be solved, **implicit Runge–Kutta** discretizations are typically employed. For instance, consider the implicit midpoint method, for which we have

$$\frac{\mathbf{y}_i - \mathbf{y}_{i-1}}{h_i} = \mathbf{f}\left(t_{i-1/2}, \frac{\mathbf{y}_i + \mathbf{y}_{i-1}}{2}\right), \quad i = 1, \ldots, N.$$

Let us write the boundary conditions as

$$B_a \mathbf{y}_0 = \mathbf{c}_a, \quad B_b \mathbf{y}_N = \mathbf{c}_b,$$

where B_a is an $l \times m$ given matrix and B_b is likewise $(m-l) \times m$. The given boundary values $\mathbf{c}_a$ and $\mathbf{c}_b$ are of size l and $m-l$, respectively. These difference equations plus the boundary conditions then yield a system of $m(N+1)$ algebraic equations in $m(N+1)$ unknowns.

Newton's method comes to mind in the case that $\mathbf{f}$ is nonlinear. However, the resulting algebraic problem may be much harder to solve than the nonlinear algebraic systems encountered earlier in this chapter around page 513 for stiff equations, because here there is no good initial guess in general (such a guess would have to be for the entire solution profile, as in Example 9.3), and because the system is rather large. Moreover, cutting down the step size makes the resulting algebraic system both larger and not necessarily easier to solve, in contrast to certain situations in the stiff initial value ODE context. Methods to handle the nonlinear BVP situation in general fall beyond the scope of the present quick presentation.

Let us concentrate instead on the case where $\mathbf{f}$ is linear, writing it as

$$\mathbf{f}(t, \mathbf{y}) = A(t)\mathbf{y} + \mathbf{q}(t),$$

with A a known $m \times m$ matrix and $\mathbf{q}$ a known vector function of t, both potentially obtained from applying one step of a modified Newton method. Then the difference equations can be rearranged to read

$$R_i \mathbf{y}_i - S_i \mathbf{y}_{i-1} = \mathbf{b}_i, \quad i = 1, \ldots, N,$$
$$R_i = I - \frac{h_i}{2} A(t_{i-1/2}), \quad S_i = I + \frac{h_i}{2} A(t_{i-1/2}), \quad \mathbf{b}_i = h_i \mathbf{q}(t_{i-1/2}).$$

Putting all these linear equations together yields a sparse, block-diagonal linear system of equations given by

$$\begin{pmatrix} B_a & & & & & \\ -S_1 & R_1 & & & & \\ & -S_2 & R_2 & & & \\ & & \ddots & \ddots & & \\ & & & -S_N & R_N & \\ & & & & B_b \end{pmatrix} \begin{pmatrix} \mathbf{y}_0 \\ \mathbf{y}_1 \\ \vdots \\ \vdots \\ \mathbf{y}_{N-1} \\ \mathbf{y}_N \end{pmatrix} = \begin{pmatrix} \mathbf{c}_a \\ \mathbf{b}_1 \\ \vdots \\ \vdots \\ \mathbf{b}_N \\ \mathbf{c}_2 \end{pmatrix}.$$

This matrix can be viewed as banded, and it is amenable to direct linear algebra solution techniques, for which we refer the reader to Section 5.6. The solution cost is only $\mathcal{O}(N)$, so this is usually *not* the Achilles heel of BVP methods.

16.7. *Boundary value ODEs

For a linearized version of Example 16.22, Exercise 24 shows that the obtained linear algebraic system using the midpoint method is pentadiagonal, rather than tridiagonal, as in Examples 4.17 and 9.3. Such is the price paid for generality.

Shooting methods

The above description should make it clear that treating boundary value ODE problems using finite difference techniques is possible, but it is significantly more cumbersome than solving corresponding initial value ODE problems in the form (16.1), i.e., for the same $\mathbf{f}(t,\mathbf{y})$ with $\mathbf{y}(a)$ known. This brings about the following intuitive idea for solving the BVP: find the missing initial values by repeatedly solving the corresponding initial value ODEs for different initial values, matching the boundary values given at the other end b. This yields a generally nonlinear algebraic system of equations of size only $m - l$, the number of missing initial values. The resulting method is called *shooting*. When it works it has an unbeatable charm of apparent simplicity and the direct reliance on past knowledge and available software.

Example 16.23. Let us solve the problem of Examples 16.22 and 9.3 using the shooting method.

For a given value c, denote by $\mathbf{y}(t;c)$ the solution of the initial value problem with $y_1(0) = 0$ and $y_2(0) = c$. Then we are looking for that value of c such that

$$g(c) = y_1(1;c) = 0$$

to match the given boundary condition at $t = 1$. The following MATLAB script uses `fzero` and `ode45` to carry out the shooting method in this very simple case:

```
c0 = input(' guess missing initial value for y_2 : ');

% solve IVP for plotting purposes
tspan = [0 1];
y0 = [0 ; c0];
[tot0,yot0] = ode45(@func, tspan, y0);

% solve BVP using fzero to solve g(c) = y_1(1;c) = 0.
c = fzero(@func14,c0);

% plot obtained BVP solution
y0 = [0 ; c];
[tot,yot] = ode45(@func, tspan, y0);
plot(tot0,yot0(:,1)','m--',tot,yot(:,1)','b')
xlabel('t')
ylabel('y_1')
legend('initial trajectory','BVP trajectory')

function f = func(t,y)
f(1) = y(2);
f(2) = -exp(y(1));
f = f(:);

function g = func14(c)
[tot,yot] = ode45(@func, [0 1], [0 c]);
g = yot(end,1);
```

The script invokes `fzero` to find a root for the function defined in `func14`, and this in turn invokes `ode45` to solve the initial value ODE defined in `func` with $y_2(0) = c$.

The results are plotted in Figure 16.14. The initial guess $c_0 = 1$ yields convergence to $c = .5494$ and Figure 16.14(a), whereas the guess $c_0 = 10$ yields convergence to $c = 10.8472$ and Figure 16.14(b). ∎

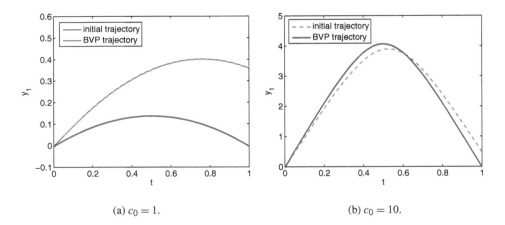

(a) $c_0 = 1$.   (b) $c_0 = 10$.

Figure 16.14. *The two solutions of Examples 16.22 and 9.3 obtained using simple shooting starting from two different initial guesses c_0 for $v'(0)$. Plotted are the trajectories for the initial guesses (dashed magenta), as well as the final BVP solutions (solid blue). The latter are qualitatively the same as in Figure 9.3. Note the different vertical scales in the two subfigures.*

Example 16.23 is misleadingly simple. In the first place, the nonlinear function $g(c)$ is usually not scalar, so normally the relevant solution techniques are not the simple ones of Chapter 3 but rather the much more complex ones of Section 9.1. Moreover, the function fzero which has been successfully employed here does not use the derivative of g and as such is suitable only for very simple problems. More generally the Jacobian matrix of a system of algebraic equations is required, and this leads to the solution of a larger auxiliary initial value ODE system, called the *variational equations*.

Still, the shooting method has attractive generality. However, it also has serious limitations. The first is that it relies on stability of the initial value ODE systems encountered, whereas all that we are legitimately allowed to assume is that the given boundary value ODE problem is stable. The other limitation is that the nonlinear algebraic problem can get exceedingly tough to solve as the interval of integration $[a, b]$ gets longer. Both of these limitations are mitigated by an approach called **multiple shooting**, which can be viewed as a method located somewhere between shooting and finite differences, where the interval $[a, b]$ is subdivided by a mesh as before, and the simple shooting method is applied on each subinterval $[t_{i-1}, t_i]$. The number of shooting points N no longer relates directly to accuracy requirements, and can often (though not always!) be taken significantly smaller than in a finite difference approach. The plot thickens here, however, so we must end our brief exposition lest we get carried away.

Specific exercises for this section: Exercises 24–25.

16.8 *Partial differential equations

The previous sections of this chapter are all concerned with differential equations that depend on just one independent variable. They provide a treatment of several approaches and methods, and also give some insight and useful practical knowledge for solving subproblems associated with PDEs.

16.8. *Partial differential equations

PDEs and their associated numerical methods are in general much more complex than ODEs, and we proceed here to provide but a very quick description of what is today still a vast research area.

Elliptic, parabolic, and hyperbolic PDEs

Some examples of PDEs and their numerical treatment have already made their way into previous chapters and sections of this text. The problem presented in Example 7.1 is the simplest example of an **elliptic** PDE. Endowing this PDE with *boundary conditions* (*not* initial conditions) yields a well-posed PDE problem, amenable to numerical treatment by methods that can be derived from basic principles. Such problems arise when modeling physical processes at steady state, i.e., when there is no dependence on time.

The simple finite difference discretization (7.1) (which may be derived by a direct application of the methods of Section 14.1, yielding second order accuracy) gives rise to a large, sparse system of linear algebraic equations. The latter fact is the reason why it stars in Chapter 7: see Examples 7.3, 7.5, 7.7, 7.8, 7.10, 7.11, and 7.17. More complicated elliptic PDEs are treated in Examples 7.13–7.15 and Exercises 7.18, 9.4, and 9.8.

In Example 16.17 we have discussed some numerical methods for the heat equation. This is a simple example of a **parabolic** PDE, which is typically endowed with both initial and boundary conditions. It is useful to consider a parabolic PDE as having both time and space variables, as in Example 16.17. If there is a steady state for the model problem, then we have a situation where $\frac{\partial u}{\partial t} = 0$, or $\frac{\partial u}{\partial t} \to 0$ as $t \to \infty$, and the parabolic problem (in more than one space variable) becomes an elliptic one.[64]

For a simple boundary value ODE such as that of Example 4.17 it is obvious that the solution $v(t)$ has two more derivatives than the right-hand-side function. Thus, in the notation of that example, if

$$g(t) = \begin{cases} 1, & 0 \leq t < .5, \\ 2, & .5 \leq t \leq 1, \end{cases}$$

then g is (bounded but) not differentiable, and yet v and its derivative are differentiable. This **smoothing property** of the inverse of the differential equation operator is inherited by elliptic PDEs, especially when defined on domains without corners, and it is crucial to the success of preconditioners for iterative methods as described in Sections 7.4–7.6. Likewise, if the boundary conditions are not homogeneous, say, $u(x, y) = \hat{g}(x, y)$ on the domain boundary, and $\hat{g}$ has a jump discontinuity, say, then the solution inside the domain is smoother, even in locations very close to the boundary jump.

Parabolic problems also have a smoothing property. Often in practice the given initial and boundary value functions do not agree where they meet (i.e., at the boundary for $t = 0$; see Example 16.17), but this discontinuity is immediately smoothed inside the domain where the PDE is defined. This property is important for the design of numerical methods for such problems.

There is a third class of scalar, linear PDEs that do not possess a smoothing property. A **hyperbolic** PDE is typically well-posed with initial or initial-boundary conditions, but not with boundary conditions around its entire domain of definition.

Example 16.24. The first order PDE

$$\frac{\partial u}{\partial t} = \frac{\partial u}{\partial x}$$

is an instance of the **advection equation**. Let us consider it on the half plane $t \geq 0$, $-\infty < x < \infty$, with initial conditions $u(t = 0, x) = u_0(x)$.

---

[64] However, a parabolic problem does not have to be in more than one space variable: for instance, the equation $\frac{\partial u}{\partial t} = \frac{\partial^2 u}{\partial x^2}$ is a parabolic PDE.

It is easy to see that the solution of the PDE can be written in general as $u(t,x) = q(t+x)$. Furthermore, setting $t = 0$ we get $q(x) = u_0(x)$, so the solution for this initial value problem is

$$u(t,x) = u_0(t+x).$$

Thus, the initial profile $u_0(x)$ is propagated in time leftward (e.g., setting $t = 1$ the value of u at x is the initial value at $x + 1$) unchanged in shape, like a ripple on the ocean surface or a water wave.

In particular, if the initial value function contains a discontinuity, then this discontinuity propagates unsmoothed into the domain in (t,x). Also, thinking of a finite domain we cannot arbitrarily set boundary values for u at the top (i.e., for some finite $T > 0$) or the left wall (i.e., for some finite x_L such that $x_L \leq x$), because the values there are dictated by u_0. See Figure 16.15(a), which displays the solution for

$$u_0(x) = \begin{cases} (1 - \exp(-x - \pi))/(1 - \exp(-\pi)), & x < 0, \\ 0, & x \geq 0. \end{cases}$$

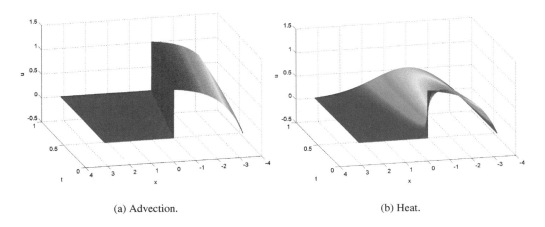

(a) Advection.   (b) Heat.

Figure 16.15. *Hyperbolic and parabolic solution profiles starting from a discontinuous initial value function. For the advection equation the exact solution is displayed. For the heat equation we have integrated numerically, using homogeneous BC at $x = \pm\pi$ and a rather fine discretization mesh.*

The solution profile for the advection equation should be compared with the solution starting from the same initial value function $u_0(x)$ of the parabolic PDE

$$\frac{\partial u}{\partial t} = \frac{\partial^2 u}{\partial x^2}.$$

For the latter we set the boundary conditions $u = 0$ at $x = \pm\pi$; see Figure 16.15(b). Note the immediate smoothing of the initial value profile by the heat equation operator for $t > 0$. ∎

The finite element method

If the domain on which an elliptic PDE is given is simply a rectangle (or a box in three dimensions), then often a discretization based on finite differences can be conveniently whipped up. The challenge then shifts to solving the resulting system of algebraic equations. But what if the domain

16.8. *Partial differential equations

is not geometrically simple, as in Figure 11.13, for instance? Constructing reasonable difference approximations becomes complicated and cumbersome in such a situation.

The finite element method (FEM) is in fact an approach yielding a family of methods that provide an alternative to the finite difference family. It is particularly attractive for problems defined on domains of general shapes whose boundaries do not align with coordinate axes, as it remains general and elegant for such problems, too. Furthermore, the theory behind the FEM family is relatively solid and deep. In fact, in addition to the attention given to computational aspects, there is a large community of scientists who are interested in the mathematical theory behind FEM approximations.

For the prototype Poisson equation, reintroduced towards the end of Section 16.1, the domain Ω can be triangulated into *elements* as in Figure 11.13. An approximate solution is sought in the form of a piecewise polynomial function over the elements. On each triangle this solution reduces to a polynomial of a certain degree, and these polynomials are pieced together along interelement boundaries. Such an approximate solution belongs to a space of functions V_h, where h is a typical element side length.

The functions in V_h have bounded first derivatives, yet second derivatives may be generally unbounded, unlike those of the exact solution u. However, the space in which u is sought can be correspondingly enlarged by considering the **variational form**. If V is the space of all functions satisfying the homogeneous boundary conditions (see Example 7.1 and imagine a nonsquare domain) that are square integrable together with their first derivatives, then for each **test function** $w \in V$ we can integrate by parts to obtain

$$\iint_\Omega -\left(\frac{\partial^2 u}{\partial x^2} + \frac{\partial^2 u}{\partial y^2}\right) w \, dx dy = \iint_\Omega \left(\frac{\partial u}{\partial x}\frac{\partial w}{\partial x} + \frac{\partial u}{\partial y}\frac{\partial w}{\partial y}\right) dx dy.$$

Writing the right-hand side as $b(u, w)$ and defining for the source term g the corresponding expression

$$(g, w) = \iint_\Omega g w \, dx dy,$$

we obtain the variational formulation of the differential problem:
Find $u \in V$ such that for all $w \in V$

$$b(u, w) = (g, w).$$

Now, the **Galerkin** FEM discretizes the variational form above instead of the original PDE! Thus we seek $u_h \in V_h$ such that for any $w_h \in V_h$

$$b(u_h, w_h) = (g, w_h).$$

This FEM is called **conforming** since $V_h \subset V$.

How is the computation mandated by the Galerkin formulation carried out in practice? We describe the piecewise polynomial functions using a **basis**. Let the functions $\phi_i(x, y) \in V_h, i = 1, \ldots, J$, be such that any $w_h \in V_h$ can be uniquely written as

$$w_h(x, y) = \sum_{i=1}^{J} w_i \phi_i(x, y)$$

for a set of coefficients w_i. Then the coefficients u_i that correspondingly define u_h are determined by requiring

$$\sum_{i=1}^{J} b(\phi_i, \phi_j) u_i = b\left(\sum_{i=1}^{J} u_i \phi_i, \phi_j\right) = (g, \phi_j), \quad j = 1, \ldots, J.$$

This yields a set of J linear equations for the J coefficients u_i.

Assembling this linear system of equations, the resulting matrix A is called the **stiffness matrix** for historical reasons. We have $a_{i,j} = b(\phi_i, \phi_j)$, $1 \leq i, j \leq J$. Of course, we seek a basis such that assembling A and solving the resulting linear algebraic equations can be done as efficiently as possible. In one dimension we have seen such bases in Section 11.4. For a piecewise linear approximate solution on a triangulation as in Figure 11.13, this is achieved by associating with each vertex, or *node* (x_i, y_i), a **roof function** (which also goes by the more tribal name *tent function*) $\phi_i \in V_h$ that vanishes at all nodes other than the ith one and satisfies $\phi_i(x_i, y_i) = 1$. This gives three interpolation conditions at the vertices of each triangular element, determining a linear polynomial in two dimensions, and the global solution pieced together from these triangle contributions is thus once differentiable. Then, unless vertex j is an immediate neighbor of vertex i, we have $b(\phi_i, \phi_j) = 0$ because the regions on which these two functions do not vanish do not intersect. Moreover, the construction of nonzero entries of A becomes relatively simple, although it still involves quadrature approximations for the integrals that arise.

Note that
$$u_i = u_h(x_i, y_i).$$

This is called a **nodal** method. The resulting scheme has many characteristics of a finite difference method, including sparsity of A, with the advantages that the mesh has a more flexible shape. Note also that A is symmetric positive definite in this specific case.

The assembly of the FEM equations, especially in more complex situations than for our model Poisson problem, can be significantly more costly than simple finite differences. Moreover, for more complicated PDEs different (and often more involved) elements may be required. Nevertheless, for problems with complex geometries the FEM is the approach of choice, and most general packages for solving elliptic boundary value problems are based on it.

Methods for hyperbolic problems

The numerical solution of hyperbolic PDEs involves considerable challenges, even in one space dimension where complex domain geometry does not arise. Thus, next we consider finite difference methods for such PDEs in time and one space variable. The essential source for numerical difficulties is the lack of smoothing or natural dissipation. Even if the solution is known to be smooth, there is no quick decay to steady state as in Example 16.17, so integration over a long time must be contemplated. Moreover, assuming an appropriate discretization in space, the eigenvalues of the resulting ODE system in time tend to be purely imaginary, which leads to only marginal stability in terms of the absolute stability concept discussed in Section 16.5. Furthermore, discontinuities in initial or boundary value functions propagate into the domain in (t, x) where the PDE is defined, as we have seen in Example 16.24. In fact, it turns out that for nonlinear problems of this type solution discontinuities may form inside the spacetime domain even if the initial and boundary value data are smooth! Who said life can't be exciting?

Consider again the simple advection equation of Example 16.24. We have seen that the exact solution propagates from $x = 0$ at the initial time line with **wave speed** $\frac{dx}{dt} = -1$ along the **characteristic curve**
$$x + t = 0.$$

Another simple instance of a hyperbolic equation is the **classical wave equation**

$$\frac{\partial^2 u}{\partial t^2} - c^2 \frac{\partial^2 u}{\partial x^2} = 0, \quad t \geq 0, \ x_0 < x < x_{J+1},$$

where $c > 0$ is the speed of sound. This equation is subject to two initial conditions

$$u(0, x) = u_0(x), \quad \frac{\partial u}{\partial t}(0, x) = u_1(x),$$

16.8. *Partial differential equations

and to boundary conditions, which we will take to be

$$u(t, x_0) = u(t, x_{J+1}) = 0 \quad \forall t.$$

Here in fact there are two characteristic curves given by

$$\frac{dx}{dt} = \pm c,$$

so there are waves propagating both leftward and rightward at the speed of sound. In fact, it can be verified that the exact solution before boundary conditions get in the way is

$$u(t,x) = \frac{1}{2}[u_0(x-ct) + u_0(x+ct)] + \frac{1}{2c}\int_{x-ct}^{x+ct} u_1(s)ds,$$

which can be used as a sanity check on proposed numerical methods.

The classical wave equation resembles the homogeneous Poisson equation in appearance (if we could only take $c = \imath$), even though its essential properties are rather different (because we can't: the speed of sound is real). Applying a similar centered discretization as in (7.1) we obtain here the **leapfrog** scheme

$$\frac{u_j^{n+1} - 2u_j^n + u_j^{n-1}}{\Delta t^2} = c^2 \frac{u_{j+1}^n - 2u_j^n + u_{j-1}^n}{\Delta x^2}, \quad j = 1, 2, \ldots, J.$$

This is an explicit, two-step scheme in time: we use the initial conditions to determine $u_j^0 = u_0(x_j)$ and $u_j^1 \approx u(\Delta t, x_j)$ for all $j = 1, \ldots, J$, and then march forward in time by calculating u_j^{n+1}, for all j, for $n = 1, 2, \ldots$, using the boundary conditions at each such step to close the system.

Is there a restriction on the time step size Δt that we may take? A look at the exact solution formula indicates that at some point $(n\Delta t, j\Delta x)$, the solution depends on the initial data at $t = 0$ only in the dependence interval $[j\Delta x - cn\Delta t, j\Delta x + cn\Delta t]$. For the numerical solution by the leapfrog scheme the corresponding dependence interval is $[j\Delta x - n\Delta x, j\Delta x + n\Delta x]$. Now, if $\Delta x < c\Delta t$, then the latter interval is strictly contained in the former, so an arbitrary change of $u_0(x)$ in the "skin area" $j\Delta x + n\Delta x < x \leq j\Delta x + cn\Delta t$ will affect the exact solution without having any effect on the numerical one! This is unacceptable, of course, and we therefore must obey the time step restriction

$$c\Delta t \leq \Delta x.$$

This is the famous CFL condition derived by Courant, Friedrichs, and Lewy in 1928 without the use of computing power. As it turns out it agrees with the numerical stability restriction corresponding to Section 16.5 for this method.

Example 16.25. Let us use the leapfrog scheme to calculate solutions to the classical wave equation with $c = 1$, subject to initial conditions $u_0(x) = \exp(-\alpha x^2)$, $u_1(x) = 0$, on the interval $-10 \leq x \leq 10$. We set $\Delta x = .04$ and $\Delta t = .02$, a combination that satisfies the CFL condition.

A "waterfall" plot depicting the progress of the initial pulse in time appears in Figure 16.16. The original pulse splits into two pulses which travel at speeds ± 1, change sign at the boundaries, and return, change sign again, and reunite to form the original shape at $t = 40$. This pattern repeats periodically for $t > 40$. For $\alpha = 1$ the numerical reconstruction is very good.

However, for $\alpha = 20$ the pulse profile is too sharp for this mesh resolution. Additional ripples form in the numerical solution as a result, which is unfortunate since the error looks "physical," so it is possible in more complex situations that a researcher may not be able to tell from the numerical solution shape that it is wrong. ∎

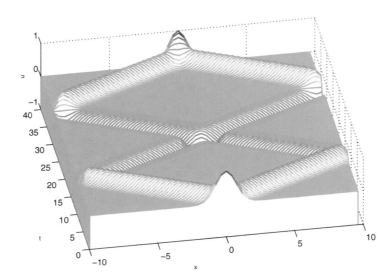

Figure 16.16. *"Waterfall" solutions of the classical wave equation using the leapfrog scheme; see Example* 16.25. *For $\alpha = 1$ the solution profile is resolved well by the discretization.*

The leapfrog scheme is simple and second order accurate in both time and space, provided that the approximate solution is smooth. This follows from the derivation of difference approximations in Section 14.1 combined with a classical convergence-stability theorem that is beyond the scope of our quick PDE-world tour. But if the solution is not smooth, then no such result holds for a simple centered scheme. Indeed, the derivations in Chapter 14 all use one form or another of local polynomial approximation. Thus, numerical differentiation across a mesh element in spacetime where the solution is discontinuous could prove to be a very bad idea.

Consider discretizing the advection equation in Example 16.24. A numerical solution is sought at the point (t_{n+1}, x_j) based on knowledge of current numerical solution values at (t_n, x_i), for all i, where $t_n = n \Delta t$ and $x_i = i \Delta x$. But the exact solution follows the characteristic curve, and hence $u(t_{n+1}, x_j) = u(t_n, x_j + \Delta t)$. The CFL condition here indicates that we must select $\Delta t \leq \Delta x$, so $x_j \leq x_j + \Delta t \leq x_{j+1}$. We can therefore interpolate the two mesh values u_j^n and u_{j+1}^n for the value at the foot of the characteristic that leads to the point of interest at the next time level. This gives the explicit one-sided method

$$u_j^{j+1} = u_j^n + \frac{\Delta t}{\Delta x}(u_{j+1}^n - u_j^n).$$

More generally, for the advection equation

$$\frac{\partial u}{\partial t} + a \frac{\partial u}{\partial x} = 0,$$

we construct the **upwind scheme** by

$$u_j^{j+1} = u_j^n - \frac{\Delta t}{\Delta x} \begin{cases} a(u_{j+1}^n - u_j^n), & a < 0, \\ a(u_j^n - u_{j-1}^n), & a > 0. \end{cases}$$

This scheme goes with the flow, and not just in a manner of speech.

The correct version of the upwind scheme for the nonlinear **Burgers equation**

$$\frac{\partial u}{\partial t} + .5 \frac{\partial u^2}{\partial x} = 0$$

reads
$$u_j^{j+1} = u_j^n - \frac{\Delta t}{2\Delta x} \begin{cases} (u_{j+1}^n)^2 - (u_j^n)^2, & u_j^n < 0, \\ (u_j^n)^2 - (u_{j-1}^n)^2, & u_j^n \geq 0. \end{cases}$$

(Note that the PDE can be written as $\frac{\partial u}{\partial t} + u \frac{\partial u}{\partial x} = 0$, so u plays the role of a in the advection equation.) This yields reasonable first order approximations for a problem whose solution may develop discontinuities even when starting from smooth initial data. But justifying all this is well beyond the scope of our text, and here is a good spot to stop.

16.9 Exercises

0. **Review questions**

    (a) For a given ODE system, what are an initial value problem, a terminal value problem, and a boundary value problem?

    (b) What are the basic differences between the numerical solution of ODEs and numerical integration?

    (c) Distinguish between the forward Euler and backward Euler methods.

    (d) Define local truncation error and order of accuracy and show that both the forward Euler and the backward Euler methods are first order accurate.

    (e) How does the global error relate to the local truncation error?

    (f) What is an explicit RK method? Write down its general form.

    (g) Define convergence rate, or observed order.

    (h) Name three advantages that RK methods have over multistep methods and three advantages that multistep methods have over RK methods.

    (i) Why is it difficult to apply error control and step size selection using the global error?

    (j) In what sense are linear multistep methods linear?

    (k) What are the two Adams families of methods? What is the main distinguishing point between them?

    (l) Write down the PECE method for the two-step formula pair consisting of the two-step Adams–Bashforth method and the one-step, second order Adams–Moulton method.

    (m) Define region of absolute stability and explain its importance.

    (n) What is a stiff ODE problem? Why is this concept important in the numerical solution of ODEs?

    (o) Define A-stability and L-stability and explain the difference between these concepts.

1. The ODE that leads to Figure 16.2 is $y' = 10y$. The exact solution satisfies $y(0) = 1$, and Euler's method is applied with step size $h = 0.1$.

    What are the initial values $y(0)$ for the other two trajectories, depicted in Figure 16.2 in dashed lines?

2. (a) Use the Taylor expansion
$$y(t_{i+1}) = y(t_i) + hy'(t_i) + \frac{h^2}{2}y''(t_i) + \frac{h^3}{6}y'''(t_i)$$
$$+ \frac{h^4}{24}y^{(iv)}(t_i) + \frac{h^5}{120}y^{(v)}(t_i) + \mathcal{O}(h^6)$$

to derive a corresponding series expansion for the local truncation error of the forward Euler method.

(b) Manipulating the forward Euler method written for the step sizes h and $h/2$, apply extrapolation (Section 14.2) to obtain a second order one-step method.

(c) Manipulating the forward Euler method written for the step sizes h, $h/2$, and $h/3$, apply extrapolation to obtain a third order one-step method.

3. Show that the backward Euler method obeys similar convergence theorem as the one given on page 489 for the forward Euler method.

4. Derive convergence results (i.e., a bound on the global error in terms of a bound on an appropriate derivative of the exact solution $y(t)$ and h^q) for the extrapolation methods of orders $q = 2$ and $q = 3$ derived in Exercise 2.

5. Consider the ODE
$$\frac{dy}{dt} = f(t, y), \qquad 0 \le t \le b,$$
where $b \gg 1$.

   (a) Apply the *stretching* transformation $t = \tau b$ to obtain the equivalent ODE
$$\frac{dy}{d\tau} = b\, f(\tau b, y), \qquad 0 \le \tau \le 1.$$

   (Strictly speaking, y in these two ODEs is not quite the same function. Rather, it stands in each case for the unknown function.)

   (b) Show that applying the forward Euler method[65] to the ODE in t with step size $h = \Delta t$ is equivalent to applying the same method to the ODE in τ with step size $\Delta \tau$ satisfying $\Delta t = b \Delta \tau$. In other words, the same stretching transformation can be equivalently applied to the discretized problem.

6. (a) Show that the explicit trapezoidal method for $y' = f(y)$ is second order accurate.

   (b) Show that the explicit midpoint method for $y' = f(y)$ is second order accurate.

   [Hint: You need to compare derivatives of y with their expressions in terms of f and show that terms that are not $\mathcal{O}(h^2)$ cancel in the expression for d_i. Use the identities $y' = f$ and $y'' = \frac{\partial f}{\partial y} y' = f_y f$.]

7. Show that the methods obtained by extrapolation in Exercise 2 are special cases of explicit RK methods, and argue that in general there is nothing particular to recommend them, except for the ease of ascertaining their order of accuracy.

8. To draw a circle of radius r on a graphics screen, one may proceed to evaluate pairs of values $x = r\cos(\theta)$, $y = r\sin(\theta)$ for a succession of values θ. But this is computationally expensive. A cheaper method may be obtained by considering the ODE
$$\dot{x} = -y, \qquad x(0) = r,$$
$$\dot{y} = x, \qquad y(0) = 0,$$
where $\dot{x} = \frac{dx}{d\theta}$, and approximating this using a simple discretization method. However, care must be taken to ensure that the obtained approximate solution looks right, i.e., that the approximate curve closes rather than spirals.

---
[65] The same holds for any of the discretization methods in this chapter.

Carry out this integration using a uniform step size $h = .02$ for $0 \le \theta \le 120$, applying forward Euler, backward Euler, and the implicit trapezoidal method. Determine if the solution spirals in, spirals out, or forms an approximate circle as desired. Explain the observed results.

[Hint: This has to do with a certain invariant function of x and y, rather than with the accuracy order of the methods.]

9. The observed order (rate) defined before Example 16.7 is based on the assumption that the calculation is carried out once with $h_1 = h$ and once with $h_2 = 2h$. Show that, more generally,

$$\text{Rate}(h) = \log_2\left(\frac{e(h_2)}{e(h_1)}\right) / \log_2\left(\frac{h_2}{h_1}\right).$$

10. Write the explicit and implicit trapezoidal methods, as well as the classical RK method of order 4, in tableau notation.

11. The three-stage RK method given by the tableau

0	0	0	0
$\frac{1}{2}$	$\frac{1}{2}$	0	0
1	0	1	0
	$\frac{1}{6}$	$\frac{2}{3}$	$\frac{1}{6}$

is based, like the classical RK method, on the Simpson quadrature rule.

Implement this method and add a corresponding column to Table 16.2. What is the observed order of the method?

12. Derive the two-step Adams–Moulton formula

$$y_{i+1} = y_i + \frac{h}{12}(5f_{i+1} + 8f_i - f_{i-1}).$$

13. Propose a way to initialize (i.e., specifying $y_1, \ldots, y_{s-1}$) the Adams–Bashforth methods of orders 1, 2, 4, and 5.

14. Show that the local truncation error of the four-step Adams–Bashforth method is $d_i = \frac{251}{720} h^4 y^{(v)}(t_i)$, that of the five-step Adams–Bashforth method is $d_i = \frac{95}{288} h^5 y^{(vi)}(t_i)$, and that of the four-step Adams–Moulton method is $d_i = \frac{-3}{160} h^5 y^{(vi)}(t_i)$.

15. The **leapfrog** method for the ODE $y' = f(t, y)$ is derived like the forward and backward Euler methods, except that centered differencing is used. This yields the formula

$$\frac{y_{i+1} - y_{i-1}}{2h} = f(t_i, y_i).$$

Show that this is an explicit linear two-step method that is second order accurate and does not belong to the Adams or BDF families.

16. Verify that the implicit midpoint and trapezoidal methods are both A-stable.

[Some of that suppressed knowledge you have about complex arithmetic may come in handy here!]

17. (a) Show that the application of an explicit s-stage RK method to the test equation can be written as $y_{i+1} = R(z)y_i$, where $z = \lambda h$ and $R(z)$ is a polynomial of degree at most s in z.

    (b) Further, if the order of the method is $q = s$, then any such method for a fixed s has the same amplification factor $R(z)$, given by

    $$R(z) = \sum_{j=0}^{s} \frac{z^j}{j!}.$$

    [Hint: Consider the Taylor expansion of the exact solution of the test equation.]

18. (a) Show that no explicit RK method can be A-stable or L-stable.

    (b) Show that the implicit midpoint and trapezoidal methods are not L-stable.

19. Suppose we have at mesh points $a = t_0 < t_1 < \cdots < t_{N-1} < t_N = b$ the result of a calculation of a solution for a given smooth initial value ODE using a one-step method of order q.

    (a) By repeating the calculation using the same method at the mesh points $t_0, \frac{t_0+t_1}{2}, t_1, \frac{t_1+t_2}{2}, t_2, \ldots t_{N-1}, \frac{t_{N-1}+t_N}{2}, t_N$, i.e., with each of the original step sizes $h_i = t_{i+1} - t_i$ halved, show how an estimate for the *global* error may be obtained.

    (b) Estimate global errors for the results of Example 16.7 without using the known exact solution. Compare your estimates with the exact errors given in Table 16.2.

20. The ODE $u'' = \frac{1}{t}u' - 4t^2 u$ has the solution $u(t) = \sin(t^2) + \cos(t^2)$.

    (a) Plot the exact solution over the interval $[1, 20]$.

    (b) Integrate this problem on the interval $[1, 20]$ using ode45 with default tolerances (specify initial values $u(1)$ and $u'(1)$ based on the given exact solution), and record the absolute error in u at $t = 20$ and the number of steps required to get there.

    Then use our rk4 with a uniform step size $h = .01$ for the same purpose. Compare errors and step counts, and comment.

    (c) Now use MATLAB's odeset to change the relative error tolerance in ode45 to 1.e-6. Repeat the run as above. Repeat also the rk4 run, this time with a uniform step size $h = .002$. Compare errors and step counts, and comment further. Do not become cynical.

21. The ODE system given by

    $$y_1' = \alpha - y_1 - \frac{4y_1 y_2}{1 + y_1^2},$$

    $$y_2' = \beta y_1 \left(1 - \frac{y_2}{1 + y_1^2}\right),$$

    where α and β are parameters, represents a simplified approximation to a chemical reaction. There is a parameter value $\beta_c = \frac{3\alpha}{5} - \frac{25}{\alpha}$ such that for $\beta > \beta_c$ solution trajectories decay in amplitude and spiral in phase space into a stable fixed point, whereas for $\beta < \beta_c$ trajectories oscillate without damping and are attracted to a stable limit cycle. (This is called a *Hopf bifurcation*.)

(a) Set $\alpha = 10$ and use any of the discretization methods introduced in this chapter with a fixed step size $h = 0.01$ to approximate the solution starting at $y_1(0) = 0$, $y_2(0) = 2$, for $0 \leq t \leq 20$. Do this for the parameter values $\beta = 2$ and $\beta = 4$. For each case plot y_1 vs. t and y_2 vs. y_1. Describe your observations.

(b) Investigate the situation closer to the critical value $\beta_c = 3.5$. (You may have to increase the length of the integration interval b to get a better look.)

22. In molecular dynamics simulations using classical mechanics modeling, one is often faced with a large nonlinear ODE system of the form

$$M\mathbf{q}'' = \mathbf{f}(\mathbf{q}), \quad \text{where } \mathbf{f}(\mathbf{q}) = -\nabla U(\mathbf{q}).$$

Here $\mathbf{q}$ are generalized positions of atoms, M is a constant, diagonal, positive mass matrix, and $U(\mathbf{q})$ is a scalar potential function. Also, $\nabla U(\mathbf{q}) = (\frac{\partial U}{\partial q_1}, \ldots, \frac{\partial U}{\partial q_m})^T$. A small (and somewhat nasty) instance of this is given by the Morse potential where $\mathbf{q} = q(t)$ is scalar, $U(q) = D(1 - e^{-S(q-q_0)})^2$, and we use the constants $D = 90.5 \cdot 0.4814\text{e-}3$, $S = 1.814$, $q_0 = 1.41$, and $M = 0.9953$.

(a) Defining the velocities $\mathbf{v} = \mathbf{q}'$ and momenta $\mathbf{p} = M\mathbf{v}$, the corresponding first-order ODE system for $\mathbf{q}$ and $\mathbf{v}$ is given by

$$\mathbf{q}' = \mathbf{v},$$
$$M\mathbf{v}' = \mathbf{f}(\mathbf{q}).$$

Show that the Hamiltonian function

$$H(\mathbf{q}, \mathbf{p}) = \mathbf{p}^T M^{-1} \mathbf{p}/2 + U(\mathbf{q})$$

is constant for all $t > 0$.

(b) Use a library nonstiff RK code based on a 4(5) embedded pair such as MATLAB's `ode45` to integrate this problem for the Morse potential on the interval $0 \leq t \leq 2000$, starting from $q(0) = 1.4155$, $p(0) = \frac{1.545}{48.888} M$. Using a tolerance $\text{tol} = 1.\text{e-}4$, the code should require a little more than 1000 times steps. Plot the obtained values for $H(q(t), p(t)) - H(q(0), p(0))$. Describe your observations.

23. The first order ODE system introduced in the previous exercise for $\mathbf{q}$ and $\mathbf{v}$ is in *partitioned form*. It is also a Hamiltonian system with a separable Hamiltonian; i.e., the ODE for $\mathbf{q}$ depends only on $\mathbf{v}$ and the ODE for $\mathbf{v}$ depends only on $\mathbf{q}$. This can be used to design special discretizations. Consider a constant step size h.

(a) The *symplectic Euler* method applies backward Euler to the ODE $\mathbf{q}' = \mathbf{v}$ and forward Euler to the other ODE. Show that the resulting method is explicit and first order accurate.

(b) The *leapfrog*, or *Verlet*, method can be viewed as a staggered midpoint discretization and is given by

$$\mathbf{q}_{i+1/2} - \mathbf{q}_{i-1/2} = h\,\mathbf{v}_i,$$
$$M(\mathbf{q}_{i+1/2})(\mathbf{v}_{i+1} - \mathbf{v}_i) = h\,\mathbf{f}(\mathbf{q}_{i+1/2}).$$

Thus, the mesh on which the $\mathbf{q}$-approximations "live" is staggered by half a step compared to the $\mathbf{v}$-mesh. The method can be kick-started by

$$\mathbf{q}_{1/2} = \mathbf{q}_0 + h/2\mathbf{v}_0.$$

To evaluate $\mathbf{q}_i$ at any mesh point, the expression

$$\mathbf{q}_i = \frac{1}{2}(\mathbf{q}_{i-1/2} + \mathbf{q}_{i+1/2})$$

can be used.

Show that this method is explicit and second order accurate.

24. Consider a linearized version of Example 16.22, given by

$$v'' + a(t)v = q(t), \quad v(0) = v(1) = 0.$$

(a) Converting the linear ODE to first order form as in Section 16.7, show that

$$A(t) = \begin{pmatrix} 0 & 1 \\ -a(t) & 0 \end{pmatrix}, \quad \mathbf{q}(t) = \begin{pmatrix} 0 \\ q(t) \end{pmatrix},$$

$$B_a = B_b = \begin{pmatrix} 1 & 0 \end{pmatrix}, \quad c_a = c_b = 0.$$

(b) Write down explicitly the linear system of algebraic equations resulting from the application of the midpoint method. Show that the obtained matrix is banded with five diagonals.

25. Consider the problem

$$v'' + \tau e^v = 0, \quad v(0) = v(1) = 0,$$

where $\tau > 0$ is a parameter. For $\tau = 1$ we saw in Examples 16.22–16.23 that there are two distinct solutions. These can be distinguished by their function norm defined by

$$\|v\|^2 = \int_0^1 v^2(s)\,ds.$$

It turns out that as τ is increased the two solutions approach one another: for a critical value τ^* they coincide, and for $\tau > \tau^*$ there is no solution for this problem anymore.

Solving this problem numerically for values of τ near τ^* can be challenging. Instead, we embed it first in the larger problem

$$v'' + \tau e^v = 0, \quad v(0) = v(1) = 0,$$
$$\tau' = 0,$$
$$w' = v^2, \quad w(0) = 0, \; w(1) = \mu.$$

This nonlinear system has a unique solution for each $\mu = \|v\|^2$.

Solve this enlarged boundary value ODE by a method of your choice for a sequence of values of $\mu \in [.1, 10]$, using the obtained solution for the previous μ as an initial guess for the next boundary value problem in the sequence. This is called a **simple continuation**. Plot μ vs. τ. You should clearly see the turning point where the number of solutions for a given τ changes.

16.10 Additional notes

There are many books devoted to the numerical solution of initial value ODEs. The material in Sections 16.2–16.7 is covered more thoroughly in Ascher and Petzold [5]. Deeper, more encyclopedic books are Hairer, Norsett, and Wanner [36] and Hairer and Wanner [37].

Many iterative methods in optimization and linear algebra, including most of those described in Chapters 3, 7, and 9, can be written as

$$\mathbf{y}_{i+1} = \mathbf{y}_i + h_i \mathbf{f}(\mathbf{y}_i), \quad i = 0, 1, \ldots,$$

where h_i is a scalar step size. This reminds one of Euler's method for the ODE system

$$\frac{d\mathbf{y}}{dt} = \mathbf{f}(\mathbf{y}).$$

The independent variable t is an "artificial time" variable. Much has been made of such connections recently, and this simple observation does prove important in some instances. But caution should be exercised here: always ask yourself if the "discovery" of the artificial ODE actually adds something in your quest for better algorithms for your given problem.

Much research on methods for stiff problems was carried out in the 1970s and 1980s. Despite the simplicity of the test equation there is significant general complication in stiff problems, essentially because some fast scales that are present in the given ODE are not approximated well in cases where these scales don't show up as a fast variation in the sought solution. This is fundamentally different from the usual nonstiff scenario, where the discretization typically approximates all scales well. Pioneering work was done by Gear [28]. The most exhaustive reference known to us remains [37].

The problem described in Example 16.4 is one in a set of initial value ODE applications used for testing research codes and maintained by F. Mazzia and F. Iavernaro in http://pitagora.dm.uniba.it/~testset/.

A lot of attention has been devoted to numerical methods for *dynamical systems*; see Stuart and Humphries [66] and Strogatz [65].

Significant recent research work has been carried out in the context of *geometric integration*, and we refer the reader to Hairer, Lubich, and Wanner [35] and Leimkuhler and Reich [49] for comprehensive accounts on this topic. There is a lighter version in Ascher [3].

A relatively readable coverage of numerical methods for boundary value ODEs is [5]. An earlier, pioneering work is Keller [46]. A deeper, more encyclopedic treatment can be found in Ascher, Mattheij, and Russell [4].

There is vast literature on numerical methods for PDEs. Two recent textbooks are LeVeque [50] and [3]. We mention further only Trefethen [69] for spectral methods, Elman, Silvester, and Wathen [23] for an emphasis on linear iterative solvers, and Trottenberg, Oosterlee, and Schuller [71] for multigrid methods. Let us repeat that our present treatment of this topic in Section 16.8 is meant to give just a taste, and other texts (and more advanced courses) are required to cover it properly.

Bibliography

[1] R. ADAMS AND J. FOURNIER, *Sobolev Spaces*, 2nd ed., Academic Press, New York, 2003.

[2] D. A. ARULIAH, U. M. ASCHER, E. HABER, AND D. OLDENBURG, *A method for the forward modelling of 3-D electromagnetic quasi-static problems*, Math. Models Methods Appl. Sci., 11 (2001), pp. 1–21.

[3] U. M. ASCHER, *Numerical Methods for Evolutionary Differential Equations*, SIAM, Philadelphia, 2008.

[4] U. M. ASCHER, R. M. M. MATTHEIJ, AND R. D. RUSSELL, *Numerical Solution of Boundary Value Problems for Ordinary Differential Equations*, SIAM, Philadelphia, 1995.

[5] U. M. ASCHER AND L. R. PETZOLD, *Computer Methods for Ordinary Differential Equations and Differential-Algebraic Equations*, SIAM, Philadelphia, 1998.

[6] M. BENZI, G. H. GOLUB, AND J. LIESEN, *Numerical solution of saddle point problems*, Acta Numerica, 14 (2005), pp. 1–137.

[7] M. W. BERRY AND M. BROWNE, *Understanding Search Engines: Mathematical Modeling and Text Retrieval*, 2nd ed., SIAM, Philadelphia, 2005.

[8] L. BIEGLER, O. GHATTAS, M. HEINKENSCHLOSS, AND B. VAN BLOEMEN WAANDERS, eds., *Large-Scale PDE-Constrained optimization*, vol. 30 of Lecture Notes in Computational Science and Engineering, Springer-Verlag, Berlin, 2003.

[9] L. BLUM, F. CUCKER, M. SHUB, AND S. SMALE, *Complexity and Real Computation*, Springer-Verlag, Berlin, 2001.

[10] D. R. BRILLINGER, *Time Series: Data Analysis and Theory*, SIAM, Philadelphia, 2001.

[11] R. L. BURDEN AND J. D. FAIRES, *Numerical Analysis*, 8th ed., Brooks/Cole, Pacific Grove, CA, 2004.

[12] W. CHENEY AND D. KINCAID, *Numerical Mathematics and Computing*, 6th ed., Brooks/Cole, Pacific Grove, CA, 2008.

[13] S. D. CONTE AND C. DE BOOR, *Elementary Numerical Analysis, an Algorithmic Approach*, McGraw–Hill, New York, 1972.

[14] T. CORMEN, C. LEISERSON, AND R. RIVEST, *Introduction to Algorithms*, MIT Press, Cambridge, MA, 1990.

[15] G. DAHLQUIST AND A. BJÖRCK, *Numerical Methods*, Prentice–Hall, Englewood Cliffs, NJ, 1974.

[16] I. DAUBECHIES, *Ten Lectures on Wavelets*, SIAM, Philadelphia, 1992.

[17] P. DAVIS AND P. RABINOWITZ, *Methods of Numerical Integration*, 2nd ed., Academic Press, New York, 1985.

[18] P. J. DAVIS, *Circulant Matrices*, John Wiley & Sons, New York, 1979.

[19] T. A. DAVIS, *Direct Methods for Sparse Linear Systems*, SIAM, Philadelphia, 2006.

[20] C. DE BOOR, *A Practical Guide to Splines*, Springer-Verlag, New York, 1978.

[21] J. W. DEMMEL, *Applied Numerical Linear Algebra*, SIAM, Philadelphia, 1997.

[22] J. E. DENNIS, JR., AND R. B. SCHNABEL, *Numerical Methods for Unconstrained Optimization and Nonlinear Equations*, SIAM, Philadelphia, 1996.

[23] H. C. ELMAN, D. J. SILVESTER, AND A. J. WATHEN, *Finite Elements and Fast Iterative Solvers: With Applications in Incompressible Fluid Dynamics*, Oxford University Press, New York, 2005.

[24] H. W. ENGL, M. HANKE, AND A. NEUBAUER, *Regularization of Inverse Problems*, Kluwer, Dordrecht, Netherlands, 1996.

[25] R. FLETCHER, *Practical Methods of Optimization*, 2nd ed., John Wiley & Sons, New York, 1987.

[26] B. FORNBERG, *A Practical Guide to Pseudospectral Methods*, Cambridge University Press, London, 1998.

[27] W. GANDER AND W. GAUTSCHI, *Adaptive quadrature revisited*, BIT, 40 (2000), pp. 84–101.

[28] C. W. GEAR, *Numerical Initial Value Problems in Ordinary Differential Equations*, Prentice–Hall, Englewood Cliffs, NJ, 1973.

[29] P. E. GILL, W. MURRAY, AND M. H. WRIGHT, *Practical Optimization*, Academic Press, New York, 1981.

[30] G. H. GOLUB AND C. F. VAN LOAN, *Matrix Computations*, 3rd ed., Johns Hopkins University Press, Baltimore, MD, 1996.

[31] R. GRAHAM, D. KNUTH, AND O. PATASHNIK, *Concrete Mathematics*, 2nd ed., Addison–Wesley, Reading, MA, 1994.

[32] A. GREENBAUM, *Iterative Methods for Solving Linear Systems*, SIAM, Philadelphia, 1997.

[33] A. GRIEWANK, *Evaluating Derivatives: Principles and Techniques of Algorithmic Differentiation*, SIAM, Philadelphia, 2000.

[34] I. GRIVA, S. G. NASH, AND A. SOFER, *Linear and Nonlinear Optimization*, 2nd ed., SIAM, Philadelphia, 2009.

[35] E. HAIRER, C. LUBICH, AND G. WANNER, *Geometric Numerical Integration*, Springer-Verlag, New York, 2002.

[36] E. HAIRER, S. P. NORSETT, AND G. WANNER, *Solving Ordinary Differential Equations* I: *Nonstiff Problems*, Springer-Verlag, Berlin, 1993.

[37] E. HAIRER AND G. WANNER, *Solving Ordinary Differential Equations II: Stiff and Differential-Algebraic Problems*, 2nd ed., Springer-Verlag, Berlin, 1996.

[38] T. HASTIE, R. TIBSHIRANI, AND J. FRIEDMAN, *The Elements of Statistical Learning*, Springer-Verlag, New York, 2001.

[39] M. T. HEATH, *Scientific Computing: An Introductory Survey*, 2nd ed., McGraw–Hill, New York, 2002.

[40] N. J. HIGHAM, *Accuracy and Stability of Numerical Algorithms*, 2nd ed., SIAM, Philadelphia, 2002.

[41] F. S. HILL, *Computer Graphics Using Open GL*, 2nd ed., Prentice–Hall, Englewood Cliffs, NJ, 2001.

[42] B. HORN, *Robot Vision*, MIT Press, Cambridge, MA, 1986.

[43] H. HUANG, *Efficient Reconstruction of 2D Images and 3D Surfaces*, Ph.D. thesis, University of British Columbia, Vancouver, Canada, 2008.

[44] H. HUANG AND U. ASCHER, *Fast denoising of surface meshes with intrinsic texture*, Inverse Problems, 24 (2008), article 034003.

[45] H. HUANG, D. LI, R. ZHANG, U. ASCHER, AND D. COHEN-OR, *Consolidation of unorganized point clouds for surface reconstruction*, ACM Trans. Graphics (SIGGRAPH Asia), 28(5) (2009).

[46] H. B. KELLER, *Numerical Solution of Two Point Boundary Value Problems*, SIAM, Philadelphia, 1976.

[47] A. N. LANGVILLE AND C. D. MEYER, *Google's PageRank and Beyond: The Science of Search Engine Rankings*, Princeton University Press, Princeton, NJ, 2006.

[48] R. B. LEHOUCQ, D. C. SORENSEN, AND C. YANG, *ARPACK Users' Guide*, SIAM, Philadelphia, 1998.

[49] B. LEIMKUHLER AND S. REICH, *Simulating Hamiltonian Dynamics*, Cambridge University Press, London, 2004.

[50] R. J. LEVEQUE, *Finite Difference Methods for Ordinary and Partial Differential Equations*, SIAM, Philadelphia, 2007.

[51] D. LUENBERGER, *Introduction to Linear and Nonlinear Programming*, Addison–Wesley, New York, 1973.

[52] S. MALLAT, *A Wavelet Tour of Signal Processing: The Sparse Way*, 3rd ed., Academic Press, New York, 2009.

[53] S. MEHROTRA, *On the implementation of a primal-dual interior point method*, SIAM J. Optim., 2 (1992), pp. 575–601.

[54] R. E. MOORE, R. B. KEARFOTT, AND M. J. CLOUD, *Introduction to Interval Analysis*, SIAM, Philadelphia, 2009.

[55] N. M. NACHTIGAL, S. C. REDDY, AND L. N. TREFETHEN, *How fast are nonsymmetric matrix iterations?*, SIAM J. Matrix Anal. Appl., 13 (1992), pp. 778–795.

[56] F. NATTERER, *The Mathematics of Computerized Tomography*, SIAM, Philadelphia, 2001.

[57] J. NOCEDAL AND S. WRIGHT, *Numerical Optimization*, 2nd ed., Springer-Verlag, New York, 2006.

[58] M. L. OVERTON, *Numerical Computing with IEEE Floating Point Arithmetic*, SIAM, Philadelphia, 2001.

[59] W. PRESS, S. TEUKOLSKY, W. VETTERLING, AND B. FLANNERY, *Numerical Recipes*, 3rd ed., Cambridge University Press, London, 2007.

[60] T. RIVLIN, *An Introduction to the Approximation of Functions*, Dover, New York, 1981.

[61] W. RUDIN, *Functional Analysis*, McGraw–Hill, New York, 1991.

[62] Y. SAAD, *Iterative Methods for Sparse Linear Systems*, 2nd ed., SIAM, Philadelphia, 2003.

[63] G. W. STEWART, *Matrix Algorithms. Vol. II: Eigensystems*, SIAM, Philadelphia, 2001.

[64] G. STRANG, *Introduction to Linear Algebra*, 4th ed., SIAM, Philadelphia, 2009.

[65] S. H. STROGATZ, *Nonlinear Dynamics and Chaos*, Addison–Wesley, Reading, MA, 1994.

[66] A. M. STUART AND A. R. HUMPHRIES, *Dynamical Systems and Numerical Analysis*, Cambridge University Press, London, 1996.

[67] A. N. TIKHONOV AND V. Y. ARSENIN, *Methods for Solving Ill-Posed Problems*, John Wiley & Sons, New York, 1977.

[68] L. N. TREFETHEN, *Lax-stability vs. eigenvalue stability of spectral methods*, in Numerical Methods for Fluid Dynamics III, K. Morton and M. Bains, eds., Clarendon Press, Oxford, UK, 1988, pp. 237–253.

[69] L. N. TREFETHEN, *Spectral Methods in MATLAB*, SIAM, Philadelphia, 2000.

[70] L. N. TREFETHEN AND D. BAU, III, *Numerical Linear Algebra*, SIAM, Philadelphia, 1997.

[71] U. TROTTENBERG, C. OOSTERLEE, AND A. SCHULLER, *Multigrid*, Academic Press, New York, 2001.

[72] H. VAN DER VORST, *Iterative Krylov Methods for Large Linear Systems*, Cambridge University Press, London, 2003.

[73] C. R. VOGEL, *Computational Methods for Inverse Problem*, SIAM, Philadelphia, 2002.

[74] D. WATKINS, *Fundamentals of Matrix Computations*, 2nd ed., John Wiley & Sons, New York, 2002.

[75] J. H. WILKINSON, *The Algebraic Eigenvalue Problem*, Oxford University Press, New York, 1988.

Index

(Page numbers set in **bold** type indicate where the statement of the corresponding theorem or the detailed account of the corresponding algorithm are provided.)

A-stability, 510
absolute stability, 484, 491
absolute stability region, 509
active set method, 275
Adams method, 501
Adams–Bashforth method, 502
Adams–Moulton method, 502
adaptive algorithm, 307, 463
adaptive quadrature, 462
adiabatic invariant, 519
advection equation, 525
algorithm
    Arnoldi, 194
    backward substitution, 95
    BFGS, 268
    Cholesky decomposition, 116
    conjugate gradient, 184
    continuous Gram–Schmidt, 374
    continuous least squares, 370
    cubic spline interpolation, 342
    discrete cosine transform, 403
    discrete Fourier transform, 389
    fixed point iteration, 46
    forward substitution, 96
    Gauss–Newton least squares, 269
    Gaussian elimination, 99
    inverse iteration, 228
    Lagrange polynomial interpolation, 305
    least squares via normal equations, 145
    least squares via QR, 156
    least squares via SVD, 235
    LU decomposition, 103
        for banded matrices, 120
    modified Gram–Schmidt, 159
    multigrid, 206
    Newton's method, 51
        for stiff ODEs, 513
        for systems, 254
        for unconstrained minimization, 261
    polynomial interpolation Newton form, 321
    power method, 222
    preconditioned conjugate gradient (PCG), 188
    predictor-corrector, 506
    QR with shifts, 241
    Rayleigh quotient iteration, 228
    Romberg integration, 471
    secant method, 53
algorithm property, 9–14
    adaptivity, 307
    efficiency, 42
AMD, *see* approximate minimum degree
approximate inverse, 187
approximate minimum degree (AMD), 125
approximation, 295
    continuous least squares, 366
    general, 365
    local vs. global, 316
    polynomial, 298, 363
    spline, 363
Arnoldi algorithm, 192, **194**
artificial time, 537
augmented Lagrangian method, 275
augmented matrix, 97
automatic differentiation, 424, 439

B-spline, 347, 349, 363
backtracking, 264
backward difference operator, 327, 410, 416
backward differentiation formulas (BDF), 501
backward error analysis, 130
backward Euler, 485, 492

semi-implicit, 513
backward substitution, 94
    algorithm, **95**
bandwidth, 119
    upper and lower, 119
barrier method, 275
barycentric
    interpolation, 305
    weight, 305
basis
    monomial, 297
    orthogonal, 192, 371
    piecewise polynomial, 344
basis function, 297, 366, 527
Bayesian method, 479
BDF method, *see* backward differentiation formulas
best approximation, 365
best lower rank approximation theorem, **231**
Bézier polynomial, 352, 362
BFGS algorithm, **268**
BiCGSTAB, 202
bidiagonalization, 244
bisection algorithm, 43
bit, 18
BLAS, 113
boundary value problem, 256, 481, 484
    finite difference, 522
    implicit Runge–Kutta, 522
    multiple shooting, 524
    shooting, 523
Burgers' equation, 530
Butcher tableau, 499

CAGD, *see* computer aided geometric design
CG, *see* conjugate gradient
CGLS, *see* conjugate gradient for least squares
characteristic curve, 528
Chebyshev, 316
    collocation, 429
    differentiation matrix, 428
    extremum points, 328, 428
    interpolation, 316, 394
    points, 316, 382, 394
    polynomial, 316, 330, 376–378
Chebyshev min-max property theorem, **379**
checkerboard ordering, 176
Cholesky decomposition, 115
    algorithm, **116**

chopping, 22
compact difference, 420
complementarity slackness, 277
complex number
    conjugate, 71
    imaginary part, 71
    magnitude, 71
    real part, 71
complexity theory, 37
composite
    Gaussian quadrature, 477
    midpoint quadrature, 449, 453, 476
    Simpson quadrature, 448, 449, 463
    trapezoidal quadrature, 448, 449
computational geometry, 357
computer aided geometric design (CAGD), 331, 347, 351, 357, 363
computer graphics, 351, 357, 363
computer vision, 406
condition number, 93, 128, 153
conditioning, 11, 129, 301, 304
    ill-conditioned problem, 11, 12, 69, 301
    well-conditioned problem, 12, 55, 69
conjugate gradient (CG), 167, 182, 184–190, 330
    algorithm, **184**
    convergence theorem, **187**
    for least squares (CGLS), 191
    preconditioned, *see* preconditioned conjugate gradient
constrained minimization conditions theorem, **274**
constrained optimization, 271–276
constraint qualification, 273
continuation, 536
continuous Gram–Schmidt algorithm, **374**
continuous Gram–Schmidt orthogonalization theorem, **375**
continuous least squares, **370**
contraction, 47
convection-diffusion equation, 199
convergence, 46
    linear, 52
    order, 52
    quadratic, 52, 54
    rate, 10, 49, 52
    superlinear, 52, 54
convex function, 58
convolution, 387, 401
corrected trapezoidal rule, 476

CPLEX, 293
Crank–Nicolson scheme, 511
critical point, 56, 260
cubic interpolation algorithm, **342**
curve, 353

data compression, 404
data fitting, 83, 85, 141, 146, 295, 365
DCT, *see* discrete cosine transform
deblurring, 401
Delaunay triangulation, 357
descent direction, 262
DFT, *see* discrete Fourier transform
differential equation, 9, 85, 256, 481
differentiation matrix, 409
    Chebyshev, 428
    Fourier, 429
    polynomial, 428
digital signal processing, 384
directional derivative, 259
discrete cosine transform (DCT), 403
    algorithm, **403**
    fast, 404
    inverse transform, 403
discrete Fourier transform (DFT), 388, 389
    algorithm, **389**
discrete orthogonality, 388
divide-and-conquer, 399, 463
divided difference, 308, 312, 313, 320
    formula, 308
    table, 308, 323
divided difference and derivative theorem, **312**
double precision, 19, 422
drop tolerance, 189
duality, 276
duality gap, 278
dynamical systems, 537

eigenfunction, 242
eigenvalue, 69, 219
    algebraic vs. geometric multiplicity, 70
    inverse iteration, 219, 227
    power method, 219
    QR method, 237
eigenvector, 69
energy norm, 90
error, 3
    a posteriori estimate, 463
    a priori estimate, 463
    absolute, 3, 6
    approximation, 5
    cancellation, 20, 26
    convergence, 5
    discretization, 5, 6
    estimate, 415, 463, 472
    Gaussian quadrature, 457
    global, 488, 517
    in input data, 5
    in model, 4
    in problem, 4
    in quadrature, 443
    local, 517
    local truncation, 488, 501
    order, 335
    relative, 3, 19
    roundoff, 5, 17, 55, 492
    Simpson rule, 445
    trapezoidal rule, 444
    truncation, 410, 416
error tolerance, 462
    absolute, 44
    relative, 44
Euler's method
    symplectic, 535
extrapolation, 296, 409, 413, 414, 470

fast Fourier transform (FFT), 396, 400, 406
    algorithm, 400
fast matrix-vector product, 400, 428
feasible solution, 271
FEM, *see* finite element method
FFT, *see* fast Fourier transform
finite difference, 256
    method, 168
finite element method (FEM), 168, 346, 382, 527
finite volume method, 356
fixed point, 45
    theorem, **47**, 49, 59
fixed point iteration, 45, 52, 172, 506
    algorithm, **46**
floating point
    double precision, 30
    guard digit, 19
    IEEE standard, 29, 36
    long word, 30
    mantissa, 33
    normalization, 22
    representation, 18
    representation error theorem, **24**

single precision, 30
flop, 10
flux function, 438
FOM, *see* full orthogonalization method
forward difference operator, 326, 410, 416, 485
forward Euler, 485
    convergence theorem, **489**
forward substitution, 96, 307
    algorithm, **96**
Fourier transform, 383
    continuous, 383
    cutoff frequency, 384
    discrete, 383, 388, 389, 450
    discrete inverse, 389
    fast, *see* fast Fourier transform
    inverse, 387, 389, 396
    inverse fast, *see* IFFT
    series, 384, 387
frequency, 384
full orthogonalization method (FOM), 197
function minimization, 79
    one variable, 40
function norm, 365
function space, 366

Gauss points, 376, 382, 455
Gauss–Jordan elimination, 134
Gauss–Newton method, 269
    algorithm, **269**
Gauss–Seidel method, 175, 205
Gaussian elimination, 93, 94, 96, 301, 343
    algorithm, **99**
    partial pivoting (GEPP), 106
Gaussian noise, 424
Gaussian quadrature, 441, 454–456, 477
generalized minimum residual (GMRES), 197
    algorithm, 198
    restarted, 199
geometric integration, 537
GEPP, *see* Gaussian elimination, partial pivoting
Gibbs phenomenon, 395
Givens rotation, 159, 198
global minimization, 58
GMRES, *see* generalized minimum residual
golden section search algorithm, 64
Google, 220
gradient, 79, 259

gradient descent, 167, 182, 214, 263
Gram–Schmidt algorithm, 157, 158
Gram–Schmidt orthogonalization, 374, 375
greedy algorithm, 265
grid, 168, 463
guidepoint, 352

Hamiltonian system, 535
hat basis function, 345, 346
heat equation, 510
Hermite polynomial, 376
Hessian matrix, 259
high frequency component, 204
Hilbert matrix, 368
Hopf bifurcation, 534
Horner's rule, 10, 59
Householder reflection, 154, 157, 159

IC, *see* incomplete Cholesky
IEEE floating point standard, 19, 29, 36
IFFT, 401
ill-posed problem, 439
ILU, *see* incomplete LU
image deblurring, 401
image processing, 407
incomplete Cholesky (IC), 188
    drop tolerance, 189
incomplete factorization, 188
incomplete LU (ILU), 203
inexact Newton method, 267
initial value ODE, 481
initial value problem, 481, 482
inner product, 110
integral
    definite, 441
    indefinite, 441
integration by parts, 368
interior point method, 275, 278
intermediate value theorem, **10**, 46
interpolant
    construction, 298
    evaluation, 298
interpolation, 83, 295
    bicubic, 354
    bilinear, 206, 354
    break point, 332, 333
    broken line, 333, 334
    error, 335
    Hermite, 320
    Hermite piecewise cubic, 336, 344, 362
    Lagrange form, 302

Index 547

linear form, 297
parametric, 349
parametric Bézier polynomial, 362
parametric Hermite, 351, 362
patch, 354
piecewise constant, 335
piecewise cubic, 335
piecewise linear, 333, 335
piecewise polynomial, 298, 332
polynomial, 297, 299, 324, 415
rational, 363
trigonometric, 298, 388, 406
interval arithmetic, 37
inverse iteration, 219, 227
inverse iteration algorithm, **228**
iterated integral, 473
iterated least squares, 292
iteration matrix, 179
iterative method, 5, 41, 172

Jacobi's method, 174, 205
damped, 178, 206, 213
under-relaxed, 178
Jacobian matrix, 508

Karush–Kuhn–Tucker (KKT) conditions, 273
Kronecker product, 200
Krylov
orthogonal basis, 192
subspace, 186, 192
subspace method, 167, 191

L-BFGS, 267
L-stability, 513
lagged steepest descent method, 266
Lagrange multiplier, 273
Lagrange polynomial, 302, 304
Lagrange polynomial interpolation algorithm, **305**
Lagrangian, 273
Laguerre polynomial, 376
Lanczos algorithm, 195
latent semantic analysis, 231
latent semantic indexing, 231
law of large numbers, 475
leapfrog method, 529, 535
least squares, 191, 366
nonlinear, 268
normal equations, 143
pseudo-inverse, 144
QR, 153

theorem, **144**
via normal equations, **145**
via QR, **156**
via SVD, **235**
Legendre polynomial, 371
level set, 272
lexicographic ordering, 176
limit cycle, 498, 534
limited memory method, 267
line search, 263
linear independence, 67
linear programming (LP), 151, 166, 276
center path, 278
dual form, 276
duality gap, 278
interior point method, 278
predictor-corrector algorithm, 279
primal form, 276
primal-dual method, 278
simplex, 277
linear system
direct method, 93
iterative method, 93
linearly independent functions, 297, 366
linearly independent vectors, 67
Lipschitz continuity, 489
Lipschitz continuous function, 60
Lobatto points, 461
local maximizer, 56
local minimizer, 56
local refinement, 463
local truncation error, 494, 501
long difference, 412
low pass filter, 385
LP, *see* linear programming
LU
decomposition, 93, 103, 168
algorithm, **103**
for banded matrices algorithm, **120**
incomplete, 203

machine epsilon, 19, 23
machine precision, 19, 23, 31, 421, 422
mantissa, 18
Maple, 438, 479
mathematical model, 1
mathematical software, 11, 463
MATLAB, 2
adaptive quadrature, 464
backslash, 84, 111, 150

bisec, 45
bisect, 44, 48, 59
circshift, 389
cond, 129
dct, 403
delaunaytri, 357
eig, 73, 237, 509
fft, 400, 430
fft2, 401
fzero, 39, 523
ginput, 362
ifft, 401, 430
ldl, 136
myfft, 400
norm, 75
ode23s, 514
ode45, 517, 518, 523
plot, 362
polyfit, 150, 164
qr, 154
quad, 381, 462
quad2d, 478
quadl, 461
quads, 515
rk4, 495, 497
spline, 164, 339, 363
svd, 81
symamd, 126
symrcm, 126
triplot, 357
triscatteredinterp, 357
matrix
    approximate inverse, 173
    banded, 119, 522
    Cholesky decomposition, 114, 115
    column stochastic, 224
    condition number, 93, 128, 153
    defective, 226
    dense, 96
    determinant, 67, 104
    diagonal, 94
    diagonally dominant, 109, 175, 343
    eigenpair, 69
    eigenvalue, 69, 77, 219
    eigenvector, 69
    fill-in, 121, 168
    graph, 124
    Hermitian, 137
    Hessenberg, 139
    Hilbert, 368
    inverse, 103
    lower triangular, 96, 307
    Markov, 246
    norm, 73
    nullspace, 68
    orthogonal, 65, 80, 108, 131
    permutation, 107
    principal minor, 79, 114
    projection, 245
    pseudo-inverse, 235
    range space, 68
    similar, 72
    singular, 67
    singular value, 77, 78, 81, 219, 229
    singular value decomposition, 80, 229
    skew-symmetric, 90, 216, 248
    sparse, 86, 93, 117
    spectral radius, 77
    spectrum, 69
    stochastic, 246
    symmetric, 78, 114
    symmetric positive definite, 65, 78, 114, 343
    tridiagonal, 119, 343
    unitary, 165
    upper Hessenberg, 139
    upper triangular, 94
    Vandermonde, 84, 148
mean value theorem, **10**
    for integrals, **475**
memory hierarchy, 113
mesh, 463
method
    Chebyshev collocation, 429
    finite element, 527
    nodal, 528
    pseudospectral, 429
    simultaneous relaxation, 174
    spectral, 430
midpoint rule, 443, 455, 476
min-max, 151, 316, 379, 382
minimum residual, 197
minimum residual (MINRES), 197, 198
modified Gram–Schmidt algorithm, **159**
molecular dynamics, 535
monomial basis function, 297
Monte Carlo method, 474, 479
multigrid method, 167, 204
    algorithm, **206**
multiple root, 54, 55

Index   549

multiresolution, 173, 407
multistep method, 500
multistep method order theorem, **504**

NaN, 19, 31
nested form, 10, 298
Newton and secant convergence theorem, **54**
Newton backward difference, 327, 330
Newton forward difference, 327, 330
Newton's method, 50, 522
    algorithm, **51**
    for stiff ODEs, **513**
    for systems, **254**
    for unconstrained minimization, **261**
    geometric interpretation, 50
Newton–Cotes, 446
    closed vs. open, 443
nodal method, 346
noise, 424, 425
    filter, 424
nonlinear equation, 39
    contraction, 47
    root (zero), 39
nonlinear least squares, 166, 268
nonsingular matrix, 67
norm, 73
    L_1, L_2 and L_∞, 366
    ℓ_1 and ℓ_∞, 151
    ℓ_2, 141
    function, 366
    matrix, 75
    sup vs. max, 366
    vector, 73
normal direction, 88
normal equations, 143, 191, 367, 374
nullspace, 68
numerical algorithm, 1
numerical analysis, 1
numerical differentiation, 6, 337, 409, 418
    backward, 410
    centered, 411, 412
    compact, 420
    forward, 410
    general formula, 417
    long, 412
    one-sided, 410
numerical integration, 441
    multidimensional, 472, 479

objective function, 55
ODE, *see* ordinary differential equation

operation count
    flop, 95
operator splitting, 432
optimization, 55, 258
    active set, 272
    active set method, 275
    augmented Lagrangian method, 275
    barrier method, 275
    complementarity slackness, 277
    constrained, 271, 293
    continuous, 293
    convexity, 58
    discrete, 293
    feasible solution, 271
    global, 58, 293
    interior point method, 275, 278
    KKT, 273
    Lagrange multiplier, 273
    Lagrangian, 273
    linear programming, 276
    local, 56
    objective function, 55
    penalty method, 275
    primal-dual form, 277
    primal-dual method, 278
    quadratic programming, 274
    saddle point, 260, 273, 289
    simplex, 277
    SQP, 276
    unconstrained, 258, 293
order, 501
order notation, 7
order of accuracy, 449
ordinary differential equation (ODE), 481
    autonomous, 483
    boundary value problem, 484, 521
    FPU, 519
    initial condition, 482
    initial value problem, 482, 521
    nonlinear, 483
    stiff, 492, 506, 507, 510
    terminal value problem, 484
    test equation, 491
    trajectory, 482
ordinary differential equation (ODE) method
    A-stable, 510, 513
    absolute stability, 491, 505
    absolute stability region, 509
    Adams method, 501
    Adams–Bashforth, 502, 505

Adams–Moulton, 502, 505
backward differentiation formulas (BDF), 505
backward Euler, 502, 510, 512
classical Runge-Kutta, 495
explicit, 487, 493, 501
explicit midpoint, 494
explicit trapezoidal, 494
extrapolation, 531, 532
forward Euler, 484, 485, 492
global error, 488
implicit, 487, 493, 501, 510
implicit midpoint, 494
implicit trapezoidal, 493, 502
L-stable, 513
leapfrog, 535
local truncation error, 488, 494
midpoint, 510
multistep, 500
order, 501
order of accuracy, 488
predictor-corrector, 506
Runge–Kutta, 493
semi-implicit backward Euler, 513
trapezoidal, 510
verlet, 535
orthogonal
basis, 371
function, 366, 371
iteration, 239
polynomial, 366, 370, 455
projection, 368
vectors, 144
orthogonal decomposition, 153
orthonormal basis, 373
osculating interpolation, 319
outer product, 110
overflow, 20, 26, 49

PageRank, 220
parametrization, 350
partial differential equation, 481
partial differential equation (PDE), 168, 484
elliptic, 525
hyperbolic, 525
parabolic, 525
partial differential equation (PDF), 199
PCA, *see* principal component analysis
PCG, *see* preconditioned conjugate gradient
PDE, *see* partial differential equation

penalty method, 275
pendulum, 482
periodic function, 383
piecewise polynomial
basis, 344
interpolation error theorem, **336**
pivoting, 93
complete, 109
partial, 106
scaled partial, 108
point cloud, 88
Poisson equation, 168
polynomial
Bézier, 352
Chebyshev, 376, 378
Hermite, 376, 479
Laguerre, 376, 479
Legendre, 371, 375, 455
monic, 379
nested form evaluation, 10, 59
orthogonal, 365, 374
trigonometric, 373, 383
polynomial interpolation, 83, 299, 324, 409, 442
equispaced, 326, 327, 330
error, 313, 416
error theorem, **314**
existence and uniqueness theorem, **301**
Hermite, 320
Hermite cubic, 320
Lagrange form, 302, 305, 416, 442
Newton form, 306, 308
Newton form algorithm, **321**
osculating, 319, 321
parametric, 351
trigonometric, 389
uniqueness, 301
power method, 219
algorithm, **222**
precision, 18, 446, 449, 454
double, 19, 421
single, 19, 421
preconditioned conjugate gradient (PCG), 187–190
algorithm, **188**
preconditioner, 167
predator-prey model, 497
predictor-corrector algorithm, **506**
primal-dual method, 278
principal component analysis (PCA), 87

principal error term, 462
problem solving environment, 2
product notation, 301
projector, 245
pseudo-inverse, 144, 235
pseudo-spectral method, 429

QMR, 202
QR
    eigenvalue algorithm, 240
    iteration, 241
QR decomposition, 153
quadratic form, 259
quadratic programming, 274
quadrature, 441
    abscissa, 441
    adaptive, 462
    basic rule, 441
    composite rule, 446
    degree of accuracy, 446
    error, 443
    Gaussian, 454, 477
    Lobatto, 461
    midpoint rule, 443
    Newton–Cotes, 443
    panel, 448
    precision, 446
    Radau, 461
    Romberg integration, 469
    Simpson, 533
    Simpson rule, 443, 463
    trapezoidal rule, 443
    weights, 441, 442
quadrature errors theorem, **453**
quasi-Newton method, 53, 267
    algorithm, 268

Radau points, 461
radial basis function (RBF), 358
range space, 68
rate of convergence, 10, 49, 180
Rayleigh quotient, 222
Rayleigh quotient iteration, **228**
RCM, see reverse Cuthill McKee
recursion, 44, 399
red-black ordering, 176
reflection, 160
regularization, 230, 402, 424
relaxation method, 174, 205
reverse Cuthill McKee (RCM), 125
Richardson extrapolation, 413, 470, 478

Ritz value, 196
Rolle theorem, **10**
Romberg integration, 441, 469, **471**
root of unity, 398
rotation, 159
roughing, 453
rounding, 22
    exact, 19
rounding unit, 19, 23, 31, 33, 55, 69, 421, 424, 460
roundoff error, 17, 492
Runge example, 316
Runge–Kutta method, 493
Runge–Kutta tableau notation, 499

saddle point, 274
SAXPY, 113
scientific computing, 1
search algorithm, 220
search direction, 182
secant method, 53, 267
secant method algorithm, **53**
sensitivity, 11
sequential quadratic programming (SQP) method, 276
shift: explicit and implicit, 243
significant digits, 23, 31
similarity transformation, 72
simplex, 277
Simpson rule, 443
simultaneous iteration, 239
sinc function, 62
singular value decomposition (SVD), 80, 153, 219, 229
    economy size, 236
    truncated, 231, 234
smoothing, 453
smoothing factor, 217
smoothness, 335
Sobolev space, 382
soliton, 433
SOR method, see successive over-relaxation method
sparse solution, 285
spectral
    accuracy, 319, 328, 393, 394
    collocation, 330
    method, 407, 430, 439
spline interpolation, 164
    boundary conditions, 342

clamped, 339, 343, 361
complete, 339, 362
cubic, 337, 339
free boundary, 339
natural, 339, 342, 344
not-a-knot, 339, 343, 360
splitting, 173
splitting method, 432
SQP method, *see* sequential quadratic programming method
SSOR, *see* symmetric SOR
stability, 11, 12, 108, 507
staggered mesh, 426
stationary method convergence theorem, **180**
steepest descent method, 167, 183, 214, 266
step size, 57, 182, 500
stiff ODE, 461, 492
stiff problem, 484, 507, 510
Stirling centered difference, 330
stretching transformation, 532
subnormal number, 25
subspace, 67
subspace iteration, 239
successive over-relaxation (SOR) method, 177
sufficiently smooth, 410
support
    local, 346
    of a function, 345
surface, 353
surface mesh, 353
SVD, *see* singular value decomposition
symbolic computing, 438, 479
symmetric SOR (SSOR), 179, 188

tangent plane, 88
Taylor polynomial, 320
Taylor series, 5, 6, 50, 302, 320
    for vector functions theorem, **253**
    in several variables theorem, **259**
    theorem, **5**
tensor product, 354
theorem
    best lower rank approximation, 231
    Chebyshev min-max property, 379
    conjugate gradient convergence, 187
    constrained minimization conditions, 274
    continuous Gram–Schmidt orthogonalization, 375
    divided difference and derivative, 312
    fixed point, 47

floating point representation error, 24
forward Euler convergence, 489
intermediate value, 10
least squares, 144
mean value, 10
mean value for integrals, 475
multistep method order, 504
Newton and secant convergence, 54
piecewise polynomial interpolation error, 336
polynomial interpolant existence and uniqueness, 301
polynomial interpolation error, 314
quadrature errors, 453
Rolle, 10
stationary method convergence, 180
Taylor series, 5
Taylor series for vector functions, 253
Taylor series in several variables, 259
unconstrained minimization conditions, 260
thin plate spline, 425
Thomas algorithm, 121
time series data analysis, 406
tomography, 406
trajectory, 482
trapezoidal rule, 443, 461
trigonometric basis function, 383
trigonometric polynomial, 373, 383
truncated SVD, 231, 234
trust region, 263

unconstrained minimization conditions theorem, **260**
underflow, 20, 26
unit vector, 104
upstream method, 530
upwind method, 530

variational form, 527
vector
    orthogonal, 79, 144
    orthonormal, 79
Verlet method, 535
vertex degree, 125

wave number, 387
wave speed, 528
wavelet, 407
    transform, 407
weak line search, 264
weighted least squares, 373